ANTIDIABETIC POTENTIAL OF PLANTS IN THE ERA OF OMICS

ANTIDIABETIC POTENTIAL OF PLANTS IN THE ERA OF OMICS

Edited by
Deepu Pandita
Anu Pandita
Chander Bhanu

AAP | APPLE ACADEMIC PRESS

First edition published 2023

Apple Academic Press Inc.
1265 Goldenrod Circle, NE,
Palm Bay, FL 32905 USA

760 Laurentian Drive, Unit 19,
Burlington, ON L7N 0A4, CANADA

CRC Press
6000 Broken Sound Parkway NW,
Suite 300, Boca Raton, FL 33487-2742 USA

4 Park Square, Milton Park,
Abingdon, Oxon, OX14 4RN UK

Apple Academic Press exclusively co-publishes with CRC Press, an imprint of Taylor & Francis Group, LLC

Library and Archives Canada Cataloguing in Publication

Title: Antidiabetic potential of plants in the era of omics / edited by Deepu Pandita, Anu Pandita, Chander Bhanu.
Names: Pandita, Deepu, 1977- editor. | Pandita, Anu, 1977- editor. | Bhanu, Chander, 1973- editor.
Description: First edition. | Includes bibliographical references and index.
Identifiers: Canadiana (print) 20220284695 | Canadiana (ebook) 20220284741 | ISBN 9781774910085 (hardcover) | ISBN 9781774910092 (softcover) | ISBN 9781003282860 (ebook)
Subjects: LCSH: Diabetes—Alternative treatment. | LCSH: Diabetes. | LCSH: Medicinal plants. | LCSH: Genomics. | LCSH: Hypoglycemic agents.
Classification: LCC SB293 .A583 2023 | DDC 633.8/8—dc23

Library of Congress Cataloging-in-Publication Data

CIP data on file with US Library of Congress

ISBN: 978-1-77491-008-5 (hbk)
ISBN: 978-1-77491-009-2 (pbk)
ISBN: 978-1-00328-286-0 (ebk)

About the Editors

Deepu Pandita, MPhil
Senior Lecturer, Government Department of School Education, Jammu, Union Territory of Jammu and Kashmir, India

Deepu Pandita is working as a Senior Lecturer in the Government Department of School Education, Jammu, Union Territory of Jammu and Kashmir, India. Deepu Pandita has more than 18 years of teaching experience and earned her Master of Botany (MSc) at the University of Kashmir, Jammu and Kashmir, India, and Master of Philosophy (MPhil) in Biotechnology from the University of Jammu, Jammu and Kashmir, India. Deepu Pandita has a number of international and national courses to her credit. Deepu Pandita has qualified fellowships including JRF NET and SRF from the Council of Scientific & Industrial Research (CSIR), New Delhi, India, Biotechnology Fellowship, Government Department of Science and Technology, Jammu & Kashmir, India; and IASC-INSA-NASI Summer Research Teacher Fellowship, India. Deepu Pandita has presented her research at work at both the national and international conferences and was awarded a Best Oral Presentation Award in the International Conference on Biotechnology for Better Tomorrow, a Women Researcher Award, and Research Excellence Award by two professional associations in India. Deepu Pandita is a life member of various scientific societies. Deepu Pandita acts as a reviewer (18 journals) and associate editor and editor of a number of international journals. Deepu Pandita has published a number of editorials, 25 book chapters (Springer, Elsevier, CRC, etc.), reviews, and research articles in various journals of national and international repute like Cells (MDPI) (IF=7.666), *Frontiers in Plant Sciences* (IF=5.753), and *Frontiers in Physiology* (IF=4.566) and has currently six books in the process of production.

Anu Pandita, MSc
Dietician, Vatsalya Clinic, New Delhi, India

Anu Pandita is working as a Senior Dietician at Vatsalya Clinic, E-5/11, Krishna Nagar, New Delhi, India. Previously she worked as a Lecturer at the Bee Enn College of Nursing, Talab Tillo, Jammu, India, and Dietician at the Ahinsa Dham Bhagwan Mahavir Charitable Health Centre, New Delhi, India. Anu Pandita has done her MSc internship and a course in the Dietetics Department of PGI, Chandigarh, India. Anu Pandita has taken a case study at the Pediatric Gastroenterology Ward at "Nehru Hospital" PGI, Chandigarh, India, on a patient suffering from chronic liver disease. She earned a certificate for a course in food and nutrition as well. She has a number of training sessions, refresher courses, and workshops to her credit. Anu Pandita is a life-time member of the Indian Dietetic Association and Indian Science Congress Association, Kolkata, India. Anu Pandita has presented her research at work at both the national and international conferences. Anu Pandita has published various book chapters in Springer, CRC, etc., publications and has a number of review and research articles in various journals of national and international repute, including *Cells* (MDPI) (IF=7.666) and *Frontiers in Physiology* (IF=4.566). She currently has six books in the process of production.

Chander Bhanu, MBBS
Senior Consultant Physician, Vatsalya Clinic; Dr. Kailash Chander Jain Memorial Clinic; Ahinsa Dham Bhagwan Mahavir Medical Center, New Delhi, India

Dr. Chander Bhanu is a Senior Consultant Physician at the Vatsalya Clinic at the Dr. Kailash Chander Jain Memorial Clinic and at the Ahinsa Dham Bhagwan Mahavir Medical Center, all in New Delhi, India. Previously he was a neuro associate at the Institute of Brain and Spine; resident medical officer in at the Vidya Sagar Institute of Mental Health and Neurosciences Hospital, New Delhi; Chief Medical Officer at Primus Hospital, where he handled international cases from various embassies; and a medical officer in the Neurology and Neurosurgery Departments at Fortis Hospital. He has expertise and experience in various clinical procedures such as pleural tapping, ascitic tapping, lumbar puncture, suturing, and other procedures both in pediatrics and adult groups and deals with stroke and accident cases as well. Dr. Bhanu earned his Bachelor of Medicine and Bachelor of Surgery (MBBS) from Govt. Medical College, Calicut, Kerala, and a Post Graduate Diploma in Diabetes and Clinical Endocrinology (P.G.D.D.C.E) from the Royal College of Physicians, United Kingdom. Dr. Chander Bhanu has certification in diabetes mellitus accredited by Royal Liverpool Academy, United Kingdom.

Contents

Contributors

Payal Agarwal
Amity institute of Biotechnology, Amity University, Mumbai, Maharashtra, India

Ahmad Ali
Department of Life Sciences, University of Mumbai, Vidyanagari, Santacruz (East), Mumbai – 400098, Maharashtra, India, E-mail: ahmadali@mu.ac.in

Shaza Hussiny Aly
Assistant Professor, Department of Pharmacognosy, Badr University, Cairo, Egypt

Lekha Bhagtaney
Sunandan Divatia School of Science, NMIMS, Vile Parle (West), Mumbai – 400056, Maharashtra, India, E-mail: lekhabhagtaney@gmail.com

Lepakshi Md. Bhakshu
Assistant Professor, Department of Botany, PVKN Government College (Autonomous), Chittoor – 517002, Andhra Pradesh, India

K. Arun Bhanu
Senior Research Fellow, University of Trans-Disciplinary Health Science and Technology, Yelahanka, Bengaluru, Karnataka, India, E-mail: dr.arunbhanu@gmail.com

Suman Ganger
Department of Microbiology, Vivekanand Education Society's College of Arts, Science, and Commerce, Mumbai, Maharashtra, India, E-mail: sumanganger@gmail.com

Abdul Hadi
Department of Botany, University of Kashmir, Hazratbal, Srinagar – 190006, Jammu and Kashmir, India

Younis Ahmad Hajam
Department of Life Sciences and Allied Health Sciences, Sant Baba Bagh Singh University, Khalia Padhiana, Jalandhar, Punjab–144030, India, Mobile: +91-6006101812, E-mail: younismajeed64@gmail.com

Geetanjali Harale
Department of Biotechnology, Vivekanand Education Society's College of Arts, Science, and Commerce, Mumbai, Maharashtra, India, E-mail: haralegeetanjali@gmail.com

Harikrishnan
Department of Chemistry, School of Arts and Sciences, Vinayaka Mission Research Foundation-Aarupadai Veedu (VMRF-AV) Campus, Paiyanoor, Chennai – 603104, Tamil Nadu, India, E-mail: harich4@gmail.com

Pamela Jha
Department of Biological Sciences, Sunandan Divatia School of Science, NMIMS Deemed to be University, Vile Parle (West), Mumbai 400 056., India, E-mail: pamelajha@gmail.com

Renitta Jobby
Amity institute of Biotechnology, Amity University, Mumbai, Maharashtra, India

Neelesh Kapoor
Department of Fingerprinting, College of Biotechnology, Sardar Vallabhbhai Patel University of Agriculture and Technology, Meerut – 250110, Uttar Pradesh, India,
E-mail: drneeleshbiotech@gmail.com

Guru Lakshmi Kola
Molecular Genetics and Functional Genomics Laboratory, Department of Biotechnology, School of Life Sciences, Yogi Vemana University, Kadapa – 516005, Andhra Pradesh, India

Anand Kumar
Molecular Genetics and Functional Genomics Laboratory, Department of Biotechnology, School of Life Sciences, Yogi Vemana University, Kadapa – 516005, Andhra Pradesh, India

Rajesh Kumar
Department of Biosciences, Himachal Pradesh University, Shimla, Himachal Pradesh – 175001, India,
E-mail: drkumar83@rediffmail.com

Rohit Kumar
Advanced Centre for Treatment, Research, and Education in Cancer, Kharghar, Navi Mumbai – 410210, Maharashtra, India, E-mail: rohitbiomed@hotmail.com

Parama Majumdar
Department of Botany, Hislop College, Nagpur, Maharashtra, India, E-mail: parama.majumdar@gmail.com

Javid Ahmad Malik
Department of Zoology Guru Ghasidas Vishwavidyalaya, Bilaspur, Chhattisgarh – 495001, India, Phone: 7006970581; 7051625284, E-mail: javidmalikr@gmail.com

Gunti Mallikarjuna
Molecular Genetics and Functional Genomics Laboratory, Department of Biotechnology, School of Life Sciences, Yogi Vemana University, Kadapa – 516005, Andhra Pradesh, India

Lavanya Mohan
Intedent Healthcare, Bengaluru, Karnataka, India

C. Nagendra
Yogi Vemana University, Kadapa – 516005, Andhra Pradesh, India

Anu Pandita
Vatsalya Clinic, Krishna Nagar, New Delhi, India, E-mail: dt.anunischal46@gmail.com

Deepu Pandita
Lecturer, Government Department of School Education, Jammu, Jammu and Kashmir, India,
E-mail: deepupandita@gmail.com

Sheetal Pardeshi
Department of Microbiology, PES Modern College of Arts, Science, and Commerce (Autonomous), Shivajinagar, Pune, Maharashtra, India, E-mail: sheetalppardeshi@gmail.com

Pravallika
Department of Biotechnology, School of Applied Sciences, REVA University,
Rukmini Knowledge Park, Kattigenahalli, Yelahanka Post, Bengaluru – 5600064, Karnataka, India

Sameer Quazi
Department of Genetics, Indian Academy Degree College, Bangalore, Karnataka, India

R. R. Venkata Raju
Department of Botany, Sri Krishnadeveraya University, Ananthapuramu – 515003, Andhra Pradesh, India

Raksha Rani
Department of Biosciences, Division Zoology, Career Point University, Hamirpur – 17604, Himachal Pradesh, India, E-mail: rakshasharma429@gmail.com

K. Venkata Ratnam
Assistant Professor, Department of Botany, Rayalaseema University, Kurnool – 518007, Andhra Pradesh, India, E-mails: drvenkatapkd@gmail.com; kvratnam@rayalaseemauniversity.ac.in

A. Madhusudhana Reddy
Yogi Vemana University, Kadapa – 516005, Andhra Pradesh, India, E-mail: grassced@gmail.com

P. Chandra Obul Reddy
Plant Molecular Biology Laboratory, Department of Botany, School of Life Sciences, Yogi Vemana University, Kadapa – 516005, Andhra Pradesh, India

A. Chandra Sekhar
Molecular Genetics and Functional Genomics Laboratory, Department of Biotechnology, School of Life Sciences, Yogi Vemana University, Kadapa – 516005, Andhra Pradesh, India, E-mails: chandrasekhar9@yahoo.com; acsekhar@yogivemanauniversity.ac.in

J. Selvakumar
Department of Chemistry, Wright State University, 3640 Colonel Glenn HWY, Dayton, OH45435, USA

Sameena Shaik
Molecular Genetics and Functional Genomics Laboratory, Department of Biotechnology, School of Life Sciences, Yogi Vemana University, Kadapa – 516005, Andhra Pradesh, India

Nikhat Shaikh
Research Officer Unani, RRIUM, Mumbai Under CCRUM, Ministry of AYUSH, Government of India, New Delhi, India, E-mail: drnikhat.unani@gmail.com

Lata Sharma
Molecular Diagnosis Laboratory, Post-Graduate Institute of Veterinary Education and Research, Jaipur – 302031, Rajasthan, India

Riya Sharma
Department of Biosciences, Division Zoology, Career Point University, Hamirpur – 17604, Himachal Pradesh, India, E-mail: riyasantoshsharma@gmail.com

Charanjit Singh
Consultant Medicine, Associated Hospital, Government Medical College, Baramulla – 193101, Jammu and Kashmir, India

Seema Singh
Department of Botany, University of Kashmir, Hazratbal, Srinagar – 190006, Jammu and Kashmir, India, E-mail: seemakuniv@gmail.com

Priya Sundarrajan
Associate Professor, Department of Life Science and Biochemistry and Director, Caius Research Laboratory, St. Xavier's College (Autonomous), Mumbai – 400001, Maharashtra, India, E-mail: priya.s@xaviers.edu

B. S. Supreetha
Department of Biotechnology, School of Applied Sciences, REVA University, Rukmini Knowledge Park, Kattigenahalli, Yelahanka Post, Bengaluru – 5600064, Karnataka, India

V. Veena
Department of Biotechnology, School of Applied Sciences, REVA University, Rukmini Knowledge Park, Kattigenahalli, Yelahanka Post, Bengaluru – 5600064, Karnataka, India, Tel.: 080-66226622, Fax: 66226622, E-mails: veena.v@reva.edu.in; btveenavijaykumar@gmail.com

Abbreviations

4CL	4-coumaroyl CoA ligase
AAL	α-amylase
AAV	adeno-associated virus
ACAE	acarbose equivalents
ACCORD	action to control risk in diabetes
AD	Alzheimer's disease
ADA	American Diabetes Association
AFLP	amplified fragment length polymorphism
AGL	α-glucosidase
AME	amentoflavone
AMPK	adenosine monophosphate-activated protein kinase
AMV	alfalfa mosaic virus
ANOVA	one-way analysis of variance
ANR	anthocyanidin reductase
APCI	chemical ionization
APD	allyl propyl disulfide
ATP	adenosine triphosphate
AZ	artificial zinc-finger protein
BA	benzyladenine
BAN	banylus
BBB	blood-brain-barrier
BCAA	branched-chain amino acid
BGL	blood glucose levels
BMI	body mass index
BUSCO	benchmarking universal single-copy orthologs
CAGE	cap analysis of gene expression
CaMV	cauliflower mosaic virus
CC	*Coptis chinensis*
CE	capillary electrophoresis
CHS	chalcone synthase
CKD	chronic renal disease
CLEt	*Cassia auriculata* leaf extracts
CoDA	context-dependent assembly
CPMV	cowpea mosaic virus

CPP	cell-penetrating peptides
CVD	cardiovascular disease
DC	diabetic control
DM	diabetes mellitus
DMAPP	dimethylallyl diphosphate
DPP 4	dipeptidyl peptidase 4
DPP	daily practice problems
DPP-IV	dipeptidyl peptidase IV
DR	diabetic retinopathy
DSB	double-stranded breaks
EBE	effector-binding element
ECM	extracellular matrix
ESI	electrospray ionization
ESRD	end-stage kidney disease
EST	expressed sequence tag
FAIMS	field asymmetric waveform ion mobility spectrometry
FBG	fasting blood glucose
FBS	fasting blood sugar
FFA	free fatty acid
FINS	fasting insulin
FPG	fasting plasma glucose
FT-ICR	Fourier transform ion cyclotron resonance
FTIR	Fourier-transform infrared
G6Pase	glucose 6 phosphatase
GC	gas chromatography
GCGR	G-protein coupled receptor
GDPP	geranylgeranyl diphosphate
GEOs	genome-edited
GFP	green fluorescent protein
GI	gastrointestinal
GIPs	gastric-inhibitory polypeptides
GIT	gastrointestinal tract
GLP	glucagon-like peptide
GLUT	glucose transporters
GMOs	genetically modified
GP	glycogen phosphorylase
GRE	germinated rice extract
GSH	glutathione
GSV	GLUT-4 storage vesicles

GUS	β-glucuronidase
HCA	hierarchical clustering analysis
HDL-C	high-density lipoprotein cholesterol
HDR	homology-directed repair
HFD	high-fat diet
HLA	human leukocyte complex
HMP	hexose monophosphate
HNF4α	hepatic nuclear factor 4 alpha
HPLC	high-performance liquid chromatography
HR	homologous recombination
HSCs	hepatic stellate cells
ICA	islet cell antibodies
IDDM	insulin-dependent diabetes mellitus
IFN-γ	interferon-γ
IGF	impaired fasting glucose
IGT	impaired glucose tolerance
IL-1β	interleukin-1β
IL-6	interleukin 6
INGAP	islet neogenesis associated protein
IPNI	international plant name index
IPP	isopentenyl diphosphate
IR	insulin receptors
ISSRs	inter simple sequence repeats
KIN	kinetin
LBP	*Lycium barbarum* polysaccharide
LC	liquid chromatography
LDL	low–density lipoproteins
LDL-C	low-density lipoprotein cholesterol
LLB	*Lycium barbarum* L.
LMIC	low-middle income countries
LSM	lifestyle management
Luc	luciferase
MAS	marker-assisted selection
MDA	malondialdehyde
MDE	*Melastoma dodecandrum*
MEP	methylerythritol phosphate
MI	myocardial infarction
MMW	medium molecular weight
MODY	maturity-onset diabetes of the young

MPCSC	medicinal plant cell suspension cultures
MS	mass spectrometry
MVA	mevalonate
NADH	nicotinamide adenine dinucleotide
NAG	N-acetyl glycoproteins
NarCh	naringenin chalcone
NDPP	national diabetes prevention program
NE	North East
NGS	next-generation sequencing
NGT	normal glucose tolerance
NHEJ	non-homologous end-joining
NLS	nuclear localization signal sequence
NMR	nuclear magnetic resonance
NOS	nopaline synthase
OGTT	oral glucose tolerance test
OPEN	oligomerized pool engineering
OPLS-DA	orthogonal partial least squares discriminant analysis
OS	orthosiphon stamineus
OX-LDL	oxidized low-density lipoprotein
PAL	phenylalanine ammonia-lyase
PCA	principal components analysis
PCR	polymerase chain reaction
PDK-1	3-phosphatidylinositol-dependent kinase-1
PEG	polyethylene glycol
PEPCK	phospho-enol pyruvate carboxykinase
PGZ	pioglitazone
PI3-K	phosphatidylinositol 3 kinase
PIP-3	phosphatidylinositol-3,4,5-triphosphate
PLS	partial least squares
POWO	plants of the world online
PPAR	peroxisome proliferator-activated receptor
PTB	phospho-tyrosine binding
PTP1B	protein tyrosine phosphatase 1B
PVX	potato virus X
QTL	quantitative trait loci
RAPD	random amplification of polymorphic DNA
RdRp	RNA polymerase
RFLPs	restriction fragment length polymorphism
RG	*Rehmannia glutinosa*

RI AGL	rat intestinal α-glucosidase
RNAi	RNA interference
RNA-Seq	RNA sequencing
ROS	reactive oxygen species
RR	relative risks
RVDs	repeat variable di-residues
SACS	S-allyl cysteine sulfoxide
SAGE	serial analysis of gene expression
SB	steviolbioside
SCAR	sequence characterized amplified regions
SEF	six essential factors
SER	smooth endoplasmic reticulum
SG	steviol glycosides
SGAs	steroidal glycoalkaloids
SGLT	sodium glucose cotransporter
SGLT 2	sodium glucose cotransporter 2
SGU	splanchnic glucose uptake
SH 2	Src homology 2
SM	secondary metabolites
SNPs	single nucleotide polymorphisms
SR	scutellariae radix
SSH	sequence subtractive hybridization
SSLP	simple sequence length polymorphism
SSR	simple sequence repeats
STR	short tandem repeat
STZ	streptozotocin
SVG	steviol glucuronide
T1DM	type 1 diabetes mellitus
T2DM	type 2 diabetes mellitus
TAL	transcription activator-like
TAL	tyrosine ammonia-lyase
TALENs	transcription activator-like effector nucleases
TC	total cholesterol
TCM	traditional Chinese medicine
TDZ	thiadiazuron
TG	triglyceride
TIA	terpene indole alkaloid
TMAO	trimethylamine N-oxide
TMV	tobacco mosaic virus

TP	teucrium polium
UCP-1	uncoupling protein-1
USDA-APHIS	US Department of Agriculture Animal and Plant Health Inspection Service
VADT	veterans affairs diabetes trial
VDCC	voltage-gated calcium channels
VIP	variable importance projections
VNTR	variable number tandem repeat
WHO	World Health Organization
ZDF	Zucker diabetic fatty
ZFN	zinc finger nucleases
ZFP	zinc finger proteins

Preface

The symptoms of diabetes mellitus (DM) are an increase in thirst, frequent urination, tiredness, and other associated health complications. From times immemorial, medicinal plants were used for treating these combined symptoms. Currently, the traditional medicine system is practiced in most of the Middle East, Asia, as well as in developed countries. Over 800 plant species are reported as antidiabetic. More than 400 plants and 700 recipes and compounds have been scientifically assessed for the treatment of type 2 diabetes to date. Metformin, which is the most accepted antidiabetic drug, was developed from the biguanide compound present in the French lilac plant. Medicinal plants contain a range of bioactive compounds and exhibit manifold actions on insulin production and insulin action mechanisms of sensitization, mimicking, and inhibition of digestion and absorption of carbohydrate in intestines. This book begins with a detailed introduction to diabetes mellitus, and also discusses current plant-based treatments for this disease and available methods for studying the antidiabetes mellitus activities of scientifically developed plant products, mechanisms of action, their therapeutic superiority, and current genome editing research perspectives and biotechnological approaches.

Part I

Introduction of Metabolic Disorder Diabetes

CHAPTER 1

Diabetes Mellitus: History, Diagnosis, Classification, Pathophysiology, and Risk Factors

YOUNIS AHMAD HAJAM,[1] JAVID AHMAD MALIK,[2] DEEPU PANDITA,[3] and RAKSHA RANI[4]

[1]*Department of Life Sciences and Allied Health Sciences, Sant Baba Bagh Singh University, Khalia Padhiana, Jalandhar, Punjab–144030, India, Mobile: +91-6006101812, E-mail: younismajeed64@gmail.com (Y. A. Hajam)*

[2]*Department of Zoology, Guru Ghasidas Vishwavidayalaya, Bilaspur – 495001, Chhattisgarh, India*

[3]*Government Department of School Education, Jammu, Jammu and Kashmir, India*

[4]*Department of Biosciences, Division Zoology, Career Point University, Hamirpur – 17604, Himachal Pradesh, India*

ABSTRACT

Diabetes or diabetes mellitus (DM) is derived from the Greek word which means "diabino" and which signifies "I pass through the diabetes," it is a situation through which running of fluid occurs. Following the footsteps of Areateus, it is possible to esteem the medical presentation and analysis of DM. It is a magnificent love, and it is uncommon in men. However, track is familiar one such as the kidneys and the bladder, in those patients who does not stop making water, but the flow of water is persistent and the disease

Antidiabetic Potential of Plants in the Era of Omics. Deepu Pandita, PhD, Anu Pandita, MSc, & Chander Bhanu, MBBS (Eds.)

becomes chronic and disease takes prolonged time to form; however, life of patients becomes short-term, once the symptoms of the disease become absolutely developed; for the melting is fast and immediate death. The diagnosis of DM remains unclear for many years which results in chronic effects due to the persistent hyperglycemia. It is a polygenic disorder that produces because of the complex interaction among various genes and environmental aspects. DM is associated with "aging and lifestyle conditions," such as no physical activity, sedentary habits, smoking of cigarette and continuous intake of alcohol significantly contributes in the progression of DM severity. This chapter summarizes the various historical perspectives, diagnostic features, risk factors, and classification criteria.

1.1 HISTORY

Diabetes mellitus (DM) is commonly called diabetes. In *diabetes mellitus*, the word *diabetes* means to pass through in Greek and the word *mellitus* means sweet in Latin. If there is an elevation in the sugar level in blood and urine leads to diabetes. During the era of the 17th-century people used to call it the *pissing evil.* The term "diabetes" was coined around 250 BC by Apollonius of Memphis. It was documented in English firstly, in the form of *diabetes;* in medical terminology, it was coined around 1,425. The word "*mellitus*" was added to diabetes in 1,675 by Thomas Willis. When there is the occurrence of diabetes in the human body, the taste of urine gets sweet. That is why the name Mellitus was given because ancient Greeks observed a sweet taste in urine. Chinese, Egyptians, Indian, and Persians are well documented in their literature.

1.2 HIGHLIGHTS IN THE HISTORY OF DIABETES

1.2.1 THE FIRST ACCURATE DESCRIPTION OF DIABETES BY ARETAEUS OF CAPPADOCIA (2ND CENTURY AD)

After Hippocrates, Aretaeus Cappadocian was the highest physician of the Greco-Roman antiquity and at the slightest equal of Galen. Aretaeus was born in Cappadocian, a small province in Eastern Asia Minor, and he studied medicine in Alexandria in the 2nd century AD; he started practicing in Rome. Areataeus was practicing on the rules of the Pneumatic School of believing not only in the fundamental role of pneuma (air) but also studied the theory of the four humors such as heat, cold, moisture, and dryness was also accepted. His two treatises, "De causis et signis morboruma cutorum et diuturnorum" (in this, he explained about the causes and symptoms of diseases that are chronic or acute) and "De curatione morborum acutorum et diuturnorum" (in this, he explained about cure of chronic diseases), these both treatises are documented in ionic dialect. In other places, he explained in a perfect approach during his time, about all these diseases, which include leprosy, asthma, pneumonia, cancer, tetanus, hysteria, epilepsy (Peumery, 1987; Laios et al., 2012). Before Aretaeus, some other ancient Greek medical authors like Rufus of Ephesus (C. 1st century AD) and (130-c, 201) were reported that diabetes is characterized by "increased thirst, polyuria, emaciation (thinning)" of the body of human beings and may cause death.

Considering the symptoms of polyuria, another name was given by Golen to the disease, i.e., diarrhea of the urine because it occurs twice. Afterwards, Aretaeus proposed a medical nomenclature and introduced the term diabetes. Diabetes is derived from the Greek word which means "diabino" and which signifies I pass through the DM, the situation through which running of fluid occurs. Following the footsteps of Areateus, it is possible to esteem the medical presentation and analysis of diabetes: Diabetes is a magnificent love and it is uncommon in men. However, track is familiar one such as the kidneys and the bladder, in those patients who does not stop making water, but the flow of water is persistent and the disease becomes chronic and takes prolonged time to form; however, life of patients becomes short-term, once the symptoms of the disease becomes absolutely developed; for the melting is fast and immediate death. In addition to this, life becomes abhorrent and panic; thirst, surplus drinking, which results in frequent and large production of urine. However, if the patient stops to drink more and more water his/her mouth and body becomes dry; and may result in "nausea, restlessness, and burning thirst" and may lead to death. If the thirst conditions remain prolonged time, it can lead to various complications such as gastrointestinal (GIT), shriveling in the abdomen, bulging of veins, and thinning of the body.

1.2.2 "MELLITUS" [THOMAS WILLIS (1621–1675)]

Thomas Willis, an English anatomist and physician, was a well-recognized and renowned physician in the 17th century. He was a resident of England in a period of political and religious turmoil. Due to this, he faced time to time interruptions in his studies. Wilis was a student of classics and then studied medicine at Oxford University, and in the same in university he was recruited as "Professor in Natural Philosophy" to the greatly esteemed "Sedleian chair." Various books and articles are credited to his career and he also worked on the brain anatomy and nervous system. Wilis explained the detailed anatomy of the nervous system, the spinal cord, the brain vasculature (circle of Wilis) and the various accessory nerves (Wilis nerve) (Furdell, 2009).

T. Wilis, a physician, at the Latrochemical School of medicine who supposed that chemistry is the fundamental unit of functions of the human body. Regarding the DM, Wilis wrote a chapter on the title "Pissing evil" in his *Pharmaceutice rationalis*. Here is explained about sweetening in the urine samples of patients who are diabetic and hence coined the term "mellitus." However, it was rediscovered in the 7th century, the "Indian Physician Sushruta" declared that the "sweet urine of the disease." Unfortunately, Wilis was not aware about this work. Hence, he was the first European medical writer who declared that during DM the taste of urine becomes sweet. He assigned it to the eating style and psychological state "an ill method of living and predominantly persevering and excessive drinking of cider, beer, and sharp wines; on some occasion's sadness, long sorrow." He also documented diabetic nephropathy in the patients (time to time "contractions or convulsions," "twinging or tightening" of the tendons and muscles and other imbalances) (Furdell, 2009; Williamson, 1923). In the time span of 24 hours, he emptied a gallon and a half of limpid, clear, and magnificent sweet water in taste like that of honey. Wilis considered honey as a therapeutic option for the disease, due to its viscous nature and fairly cooling diet and cordials" and he explained that green vegetables, rice, and starch may improve the health status of diabetic patients. Implantation is the therapeutic advices proposed by Wilis; here, the condition of patients become improved within a month but instantaneously to his past habits (Furdell, 2009). However, Willis cannot describe "why the urine is wonderfully sweet like sugar or honey?" The English physician Matthew Dobson (1732–1784) gave the reason for this about 100 years back, and he performed the experiment. He boiled the urine up to dryness and observed a crystalline material, which tasted like brown sugar and hence confirmed the presence of sugar in urine (Williamson, 1923).

1.2.3 DISCOVERY ABOUT THE GLYCOGENIC ACTION OF THE LIVER BY CLAUDE BERNARD (1813–1878)

Claude Bernard was born in low middle-class family in "Beaujolais province," located in southern France. At the age of 19, Claude Berbard was trained to become an apothecary. His dedication towards the theater made him interesting in writing two plays, which were "La Rose du Rhone" and "Arthur de Bretagne" however, after his onset in Paris, he was disheartened by the literary opponent and politician Saint-Marc Girardin (1801–1373) as Saint-Marc Giradin convinced him to join in medicine. During his studies in "Medical School of Paris," he was not recognized as an intelligent student and showed no interests in practicing medicine, Bernard was recruited as "Assistant Professor in Physiology" and "pioneer of experimental physiology" Francosi Magendie (1783–1855). Later on, Bernard gained huge success in research field. Bernard was appointed as member of the Science Academy in 1854. Among Bernard's various discoveries are cited as: "the vasomotor innervations," "the principle of physiological determinism," "the concept of internal secretion," "the concept of milieu interieur," and "internal environment."

Unfortunately, the animal vivisection is the only way to understand and discover the phenomenon and have better knowledge of physiology. The leading contribution of Bernard in exploring the metabolism and diabetes observed during the post-mortem observations of the disease atrophic or stone-filled pancreases. During the 19th century, scientists predicted the contributions of the pancreas in the physiopathology of diabetes on the basis of observation made during the post-mortem observation of diseased, atrophic, or stone filed pancreases. Whereas, they thought it to be an exocrine organ, and they analyzed these post-mortem results as a probability phenomenon. Claude Bernard (French experimental physiologist)was determined to reexamine this hypothesis (Kahn and Weir, 2005; Grmek, 1997).

Initially, Bernard wrongly supposed that "diabetes was a nervous affection of the lungs." Later on, he conducted an experiment in which he administered glucose injection into the vena jugularis of a dog, and simultaneously, he collected a blood sample from the 'carotid artery.' After examination of the collected blood sample, he observed that in the higher blood sugar level in the sample and then he came to know that damage of glucose was not occurred in lungs, the reason is that it is necessary that blood must be passed through these organs, allowing its movement from the vena jugularis to the 'carotid artery.' Moreover, then Bernard fed dogs with food which was rich in carbohydrates. Moreover, he extended his study, Bernard confirmed that in the blood of animals, sugar is present, although if food is not provided to the animal. Surprisingly, Bernard observed an increased quantity of sugar in 'hepatic veins.' A similar experiment has been repeated in the other control group, in which animals were fed on meat. After the dissection, he harvested the liver tissues and analyzed that a large amount of glucose was present, however glucose was absent in other organs. Bernard observed that liver is a storage house of water-insoluble starchy substance known as glycogen, it gets converted into sugar or glucose, which is then released into blood. He predicted that excess secretion of glucose causes diabetes. Bernard used needle and stimulated floor of the 'fourth brain ventricle' and generated short-lived "artificial diabetes," which lasted for less than a day. Bernard termed this process "Piqure diabetique," and first time he associated "glucose homeostasis" and the pathogenesis in the brain.

1.2.4 OSKAR MINKOWSKI (1858–1931) AND JOSEPH VON MERING (1849–1908)

The findings of "Pancreatic Diabetes" 1889 are considered as crossroads in the history of DM. In 1886, Von Mering revealed that "phlorizin" a glucoside has ability to cause temporary "glucuresis." "Professor Bernard Naunyn (1839–1925)," visited the institute to search for several chemical-related books in the library. Unintentionally he met Von Mering and discussed "Lipanin," which is an oil having free fatty acids (FFAs) which were given to the patients experiencing digestive complications. However, Minkowski was not in support of intake of Lipanin oil as their transformation turned on whatever the pancreas contributes in digestion and absorption of fat. The consequences of the meeting were conduction of pancreatectomy in a dog on the same evening in "Naunyn's laboratory." The animals remained alive and were examined almost completely by Minkowski. Immediately prior to the operation, the dog exhibited excessive urination. The examination of urine was done by Minkowski and observed 12% sugar. Progression of diabetes was observed primarily by Minkowski in the dog because of the fact that von Mring has administered phlorizin for prolonged time. Then the pancreas of three more dogs were removed, but before the operation there was absence of sugar in the urine samples of dogs and it was found that there is presence of reducing sugar in the urine of all the dogs (Von, 1889; Von and Minkowski, 1889). In addition to this, Minkowski implanted a little part of pancreas via subcutaneous route, in dogs that does not have pancreas, and found that hyperglycemia was inhibited until the pancreatic implant was removed or had destroyed subcutaneously (Von, 1889). The experiment conducted by Minkowski and von Mering explained that the pancreas was a gland that is essential for the internal secretions for balance the glucose homeostasis. They forced the Banting to conduct good experiment and to get success.

1.2.5 THE DISCOVERY OF INSULIN: JOHN MACLEOD (1876–1935), FREDERICK BANTING (1891–1941), JAMES BERTRAM COLLIP (1892–1965) AND CHARLES BEST (1899–1978)

Insulin was discovered by Frederick Banting and Macleod and they were awarded Nobel Prize in Medicine in 1923. In fact, it was a success story that aggravated a higher conflict among scientists. Frederick Banting was recognized as a young "Canadian surgeon," was working on diabetes in the research laboratory of the great biochemist, as he has an interest in diabetes, "Professor John Macleod," in the Toronto University (Von, 1889). Moses Barron was a physician in Minnesota, and he printed a research article entitled "The Relation of the Islets of Langerhans to Diabetes" in 1920, giving specific preference to patients suffering from pancreatic lithiasis. Barron (1920) declared that the repetition of the experiments of Minkowski and von Mering may lead to the innovation of a product having the potential to control diabetes. Banting become influenced by this article and started on diabetes (Von, 1889). Ernest Starling (1886–1927), a distinguished English Physiologist was stating that we did not know until now we could know how pancreas impacts on the generation of sugar or its use in the same animals. Commonly, it is supposed that the pancreas secretes a hormone in the circulatory blood which could transport into the tissues and capable them to use sugar or transport it into the liver and prevent the generation of sugar in this organ. However, we were found incapable of reproducing the function of the pancreas still in vascular link with the body, though the treatment of the extracts of this organs" (Medvei, 1993). Banting initiated collaboration with Charles Best on 16 May 1921, a young medical student. From the initial experiments conducted on dogs by ligatating their pancreatic ducts, attaining deteriorating the exocrine part of pancreas and within the 10 weeks they removed that disintegrated pancreas. The atrophied pancreatic glands were crushed and homogenized onto a chilled surface or mortar and then froze the mixture in water containing salt. The mass of the gland was allowed to settle down and 100 ml of physiological salt was added to it. Then they administered 5 ml quantity of the extract through intravenous route into a dog having depangreatizing. It was observed that after 2 hours sugar level was dropped significantly. The experiment was repeated many times along with other dogs having diabetes, and correlated outcomes were obtained, and experiments were also conducted on fetal calf pancreas using various approaches of treatment through subcutaneous and rectal routes (Banting et al., 1922). In the end, in 1921, the well-trained and skillful chemist "James

Collip" came into the team and designed a good technique for "extraction and purification." Initially, the obtained product was termed "insletin" and afterwards Macleod named it "insulin" (Von, 1989). Another phase was testing of insulin on humans, on "11 January 1922." Insulin treatment was given to Leonard Thompson (14-year-old boy) to treat diabetes in "Toronto Hospital" (Von, 1989). It has got great significance that prior to the administration of Apollinaire Bouchardat's (1806–1886) revolutionary dietetic supplement to treat diabetes, doctor of medicine recurring in various races of diabetes his motto: "mangez le moins possible" (consume as a smaller amount as possible) (Karamanou et al., 2014; Bouchardat, 1875). Thomson strictly followed the abstaining diet pattern projected by "Frederick Madison Allen (1879–1964)" and he stayed in a thoughtful state. 15 ml insulin was injected in his buttock however swelling occurred at the site of injection and eventually became sicker. Moreover, Collip enhanced the insulin quality and on 23 January, second injection was administered to Thompson. The obtained results were outstanding that blood glucose level ranged between 520 mg/dL fell to "120 mg/dL" in 24 h and urinary ketones vanished. Thompson administered insulin continuously and survived for 13 years. However, died due to pneumonia at the age of 27 years (Von, 1989). She is the daughter of a great politician of United States Charles Evan Hughes, at the age of 11 years, she was diagnosed with diabetes. In the beginning, she was given the treatment by Allen and then, in August-1922 treatment with insulin was started for her. She lived, did her graduation from college, married, gave birth to three children and accidently died due to heart attack at the age of 74 years (Cooper, 2010). The revolutionary effort of "Banting and Best" protected life of millions of diabetic patients and helped them to live a normal life. "Lilly Pharmaceutical Company" works together with two great researchers, and finally, in the year 1923 launch lletin, it is the world's first marketable and available insulin (Von, 1989). Banting got very furious because he supposed that he should share the prize with best inspite of "MacLoeod" and then "Banting" made a decision to share with "Best" his cash award. During his tenure, "MacLeod" shared his award with "Collip" as well. One more bad impression during the discovery of insulin molecule history was also the discovery of other molecule called "pancreatin," it is a bovine pancreas extract discovered by "Romanian Professor of Physiology Nicolae Constantin Paulescu (1869–1931)" in the year 1916, in printed some years late because of the war in 1921 and then it was patented in April 1922. Although, "Paulescu" was the initial one to provide a thorough explanation of the antidiabetic and anti-ketogenic impact of a pancreatic extract, "pancreatin" was not used in humans and passed over noiselessly. An important phase involving the antiquity of diabetes has been accomplished. In the coming years, purification of insulin techniques enhanced and new formulations of insulin were established like as "Protamine-zinic insulin," a long-acting insulin in 1930s, "neutral protamine Hagedorn" in 1940s and Lente series in 1950s.

1.3 CLASSIFICATION

1.3.1 TYPE 1 DIABETES MELLITUS (T1DM)

T1DM is a condition during which pancreatic β-cells does not produce enough insulin to metabolize the surplus glucose and control hyperglycemic condition. T1DM affects 5 to 10% of people and is mostly prevalent in children, therefore also known as "juvenile diabetes" or "insulin-dependent diabetes" (Dabelea et al., 2014; Maahs et al., 2010). Various auto-antibodies are produced during the T1DM such as islet cell antibodies (ICA) is first autoantibody observed in patients suffering from T1DM. Other antibodies against the pancreas included GADA, IA2A, and IAA. ICA are specially an IgG reacts in contrast to the cytoplasm α and β pancreatic cells, in 70% population of type 1 diabetic patients ICA auto-antibodies are found. Before the supplementation of exogenous insulin IAA antibodies were detected in T1DM and produced against the insulin and proinsulin. IA2A is produced in the extracellular portion of β cells, tyrosine phosphatase has been detected in 50–75% population of T1DM patients. GADA has been detected in 70–80% population of T1DM patients prior to the clinical symptoms of the diseases and remains positive for extended time; they mark GAD (protein having mol. Wt. 64,000 Da) which takes part in the production of γ-aminobutyrate in pancreatic cells. GAD65 is an antigenic target for T1DM.

The "human leukocyte complex (HLA)," particularly the HLA-DR and the HLA-DQ genes, contributes significantly to the pathogenesis of TIDM (American Diabetes, 2014). Genes including "HLA-DQA1,"

"HLA-DQB1" and "HLA-DRB1" belongs to the HLA (Human Leukocyte Antigen) family. The HLA (Human Leukocyte Antigen) complex supports the immune system of the body to differentiate between native proteins and proteins manufactured by external pathogens including "viruses" and "bacteria." Various genetic loci are associated with TIDM such as "CeC motif chemokine receptor-5," "cytotoxic T-lymphocyte associated protein," "forkhead box P3," "major histocompatibility complex class II DQ alpha 1," "major histocompatibility complex class II DQ beta 1," "HNF homeobox A," "interleukin 2 receptor subunit alpha," interleukin-6, "insulin," "2′-5′-oligoadenylate synthetase 1," "protein tyrosine phosphatase," "non-receptor type 22" and "small ubiquitin-like modifier 4." Variations in these genes are responsible for modulation in the synthesis of insulin. Various environmental factors, viral infections and stress are well recognized that implicates the etiology of TIDM (Vermeulen et al., 2011) (Figure 1.1).

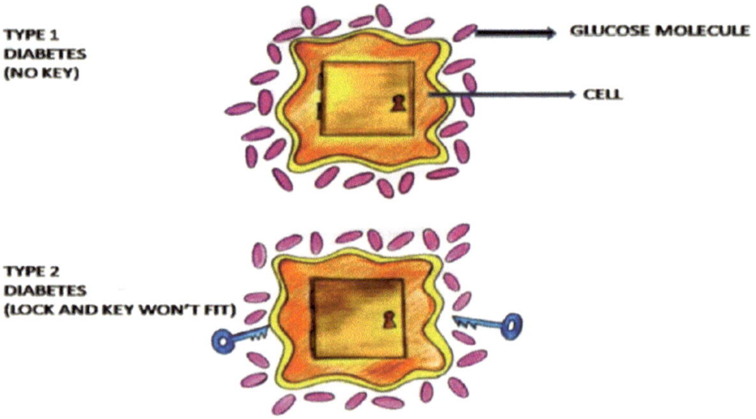

FIGURE 1.1 Type 1 diabetes mellitus and type 2 diabetes mellitus.

1.3.2 TYPE 2 DIABETES MELLITUS (T2DM)

T2DM is commonly known as "adult-onset diabetes," generally occurs during the old age and it comprises approximately 95% of the diabetic population (Bhatti et al., 2017; Fradkin et al., 2012). The features of T2DM including decreasing production of insulin and sometimes pancreatic β-cell failure. It led to the reduction in transport of glucose into the liver, muscles, and adipocytes (Olokoba et al., 2012). The diagnosis of TIDM remain unclear for many years which results in chronic effects due to the persistent hyperglycemia. It is a polygenic disorder that produces because of the complex interaction among various genes and environmental aspects. However, type 2 diabetes is associated with "aging and lifestyle conditions," such as no physical activity, sedentary habits, smoking of cigarette and continuous intake of alcohol significantly contributes in the progression of T2DM severity. T1DM and type 2 diabetes are having genetic background; however, the genetic basis of TIDM is stronger in comparison to type 2 diabetes. Ali et al. (2017) reported that the threatened risk factors for T2DM are not area or organ dependent, but develops due to the multiple gene interaction located in the entire genome. These genes include "TCF7L2, PPARG, CDKAL1, IGF2BP2, SLC30A8, JAZF1, and HHEX. KCNJ11" (potassium inwardly-rectifying channel, subfamily J, member 11), encodes the islet ATP-sensitive potassium channel (transcription factor 7-like 2) that modulates "proglucagon gene expression" and hence leads to the production of "glucagon-like peptide-1" (McCarthy et al., 2010) Figure 1.1.

1.3.3 TYPE 3 DIABETES (T3D)

Type 3 diabetes (T3D) is generally called "neuroendocrine disorder" that signifies the progression of type 2 diabetes to "Alzheimer's disease (AD)" (Maechler et al., 2010). During the type 1 diabetes, various signaling pathways become altered, such as "insulin growth factor signaling," "acetyl-choline esterase activity," "inflammatory responses," "ApoE4A allele on synaptic plasticity," and "vascular dysregulation of brain capillaries." Link between Systematic dysfunctioning like diabetes, impairment in neural system, weight gain, dementia, and other various metabolic diseases was established in recent studies (Stoeckel et al., 2016). Individuals suffering from "AD" reveals downregulation of insulin and "neuronal insulin receptors" in comparison to the age-matched controls. These defected pathways go to gradual irregularity in the whole insulin signaling cascade, which results in the progression of insulin resistance (Supale et al., 2012). Stoeckel and co-authors investigated that there is a link between derangement of "carbohydrates, proteins, lipids, brain dysfunction, and cognitive impairment" (Stoeckel et al., 2016).

1.3.4 GESTATIONAL DIABETES MELLITUS

During pregnancy, due to some irregular metabolic functions, glucose level increases (hyperglycemia) and develops a diabetic like conditions in the mother, which directly affects the developing fetus known as gestational diabetes. A hyperglycemic condition in gravid women elevates the chance of hostile motherly, fetal, and newborn consequences. The characteristics of "gestational diabetes" are intolerance of carbohydrates during the time of pregnancy (Kharroubi et al., 2015). Women suffering from GDM (Gestational diabetes mellitus) are having higher threat for the consequent progressions of frank diabetes. GDM (Gestational diabetes mellitus) could be a risk factor of newborn death, because motherly hyperglycemic condition forces the developing embryo to conceal more and more insulin hormone that leads to the hyper-stimulus of embryo growth and abnormal elevation in natal mass. After birth, this hyperglycemic might be reversed in women (Boles et al., 2017; Lorenzo et al., 2010).

1.3.5 DIABESITY

Recent studies from the USA revealed that basal metabolic index in the age group of 18% (6–11 years) and 21% (12–19) are having "BMI $\geq 95^{th}$ percentile," and are having "pediatric obesity" and are having 45% chances to develop new-onset type 2 diabetes. Term "diabesity" is used to explain the pathophysiological inter-link between type 2 diabetes and excessive weight gain. In the early 1970s, Sim et al. (1973) first coined the term "diabesity" in early 170s to describe the link between diabetes and obesity when both metabolic dysfunction are present in individual. Excessive weight gain is related to fat accumulation in visceral cavities that elevates the incidence of insulin resistance and leads to type 2 diabetes mellitus. However, many drugs are available for successful management of type 2 diabetes mellitus, but management of the obesity at the clinical level is always problematic to medical practitioners. Some surgical procedures are used, such as bariatric surgery, has proven the best approach for the management of obesity, if untreated, may develop diabetes in many patients (Leung et al., 2017). Assessment of body weight is not an accurate for obesity, but "body mass index (BMI)" that measures your weight in relation to your height is general parameters to assess the obesity and its severity. World Health Organization (WHO) has kept some criteria for the assessment of obesity such as person having BMI higher than 30 is considered as "class 1 obesity," and BMI above 35 is considered as Class II (indicates serious obese condition) and BMI more than 40 is considered as class III (indicates that obesity is very serious). Patients suffering from type 2 diabetes and having BMI above 32.5 should go for minimally invasive bariatric weight-loss and diabetes surgery or if they suffer from any other obesity-associated with co-morbidity; or if your BMI is more than 37.5 without any co-morbidity.

1.3.6 DIAGNOSIS OF "DIABETES MELLITUS (DM)"

The recognition of patients suffering from "diabetes or pre-diabetes" by showing allows for previous interference, with possible decline in perspective complications rate, even though randomized trials are missing to definitely reveal profit. The patients explained in the vignette have risk aspects like excessive weight gain, high blood pressure, and diabetes family history and ought to be monitor (Cryer, 2012; Forbes and Cooper, 2013; Harris et al., 1992). It has been reported that 25 patients out of hundreds are suffering from T2DM previously have small vessel impediments at the time of examination (American Diabetes Association, 2011; Cox and Elelman, 2009). There are different methods to diagnose diabetes in the different persons. In 1997 "American Diabetes Association (ADA)" proposed some recommendations to diagnose DM considering the "fasting plasma glucose (FPG)," however WHO considering the "oral tolerance test (OGTT)" (Gillett, 2009).

1.4 RANDOM PLASMA TEST FOR THE DIAGNOSIS OF T1DM AND T2DM

Before taking the test, it does not need any fasting because it is a simple test. If the level of glucose in the blood is 200 or >200 mg/dl then, it would be "diabetes." However, it needs final confirmation.

1.4.1 "FASTING PLASMA GLUCOSE (FPG) TEST"

Before taking the test, it needs 8 hours fasting because it is not simple test. Blood glucose above 126 mg/dl on two or more tests carry out on various days proves the diagnosis of diabetes (Gillett, 2009). "Fasting glucose level" is estimated, and then gives 75 gm of glucose, 100 gm for "pregnant women." Blood testing is done each and every 30 minutes to 1 hour for 2–3 hours. This test is normal if blood glucose level at 2 hours is >140 mg/dl. When fasting glucose level of >126 mg/dl and two-hour glucose level is >200 mg/dl proves the detection of diabetes (Gillett, 2009).

1. **Glycated Proteins:** The derivatives of glucose and fructose are generated by impulsive reaction of protein in circulatory blood with glucose. The amount of glycated proteins is regulated through the glucose concentration in blood and by the number of reactive

amino groups present in the protein that are available for glucose reaction. Proteins having reactive sites are susceptible for glycation and the quantity of "glycated proteins" can be measured in blood is an indicator for the detection of changes in blood glucose levels (BGL) during some time period. Considering the medical diagnostic statement "glycated proteins" are having long lifetime in blood are of attention, because they show the exposure of these proteins to glucose for prolonged time.

2. **Glycated Hemoglobin:** The hemoglobin is having duration of 90–120 days under in vivo conditions. At the time of formation of glycated hemoglobin-A forms, hemoglobin-A reacts with glucose and there is a generation of ketoamine compound. Various subfractions of "glycated hemoglobin" have been separated. Among these "glycated hemoglobin-A fraction HbA1c" is of having great attention and serving as a demonstrative detector of the standard glucose level. "HbA1c" is suggested as an important indicator for the assessment of blood sugar level. "HbA1c $\geq$ 6.5%" in blood is thought to be a diagnostic feature of diabetes (Selvin et al., 2010).

1.4.2 ORAL GLUCOSE TOLERANCE TEST (OGT)

If random plasma glucose ranges between "160–200 mg/dl" and the FPG level ranges from "110–125 mg/dl" then this test is conducted (Guidance for Industry, 2008). OGT is evaluated to check the response of body towards the glucose and it requires fasting conditions of minimum 8–16 hours.

1.4.3 FRUCTOSAMINE TEST

One of the main components of "plasma proteins" is albumin. In albumin free amino groups and glucose reacts non-enzymatically in blood plasma. Thus, glycated albumin acts as an indicator to assess the glucose level in blood. Generally, glycated albumin is considered to give an explainable approach to measure average concentration of glucose level in blood for a period of 1–3 weeks. Reference interval: 205–285 micro mol/L.

Diagnosis of "gestational diabetes mellitus" occurs after the 6 weeks after the end of pregnancy; the women shall assess for "oral glucose tolerance test" and can be considered diabetic. Women with previous positive family history, record of gestation diabetes, noticeable obesity, and greater cultural threat must be examined very soon. If negative result comes out after initial screening, they shall be re-examining after 24–48 weeks. GDM is diagnosis is carried out plasma glucose level met or increased. It has not been confirmed that whether the threat of complications of diabetes varies as per the diagnostic approach/test that whether the basis of diagnosis is only "FPG testing" or only "glycated hemoglobin testing." Initial data from a huge, society-based aspect cohort reported that the "glycated hemoglobin level," which combines fasting and non-fasting glucose levels for prolonged time, maybe a good mark of some impediments, particularly cardiovascular diseases (CVDs) (Shafizadeh et al., 2011). Moreover, it is not clear whether the treat of diabetes varies among identified patients suffering from pre-diabetes through the quantification of "glycated hemoglobin" evaluation and those recognized through "FPG testing." These threatening factors possibly differs depends on the test finally used for the diagnosis. The current research is evaluating the data of risk scores that includes not only glycemic measures, but other indicators, and threatening factors to evaluate the risk of diabetes (Salomaa et al., 2010; American Diabetes Association, 2012).

"Impaired fasting glucose (IGF)" is a condition when fasting glucose level ranges between 100–125 mg/dl (5.6–6.9 mmol/liter). "Elevated glycated hemoglobin (IGH)" is a condition arises when glycated hemoglobin level varies between 5.7–6.4%. Confirmation of diabetes depends on the repetition of tests on different days or via alternative tests (i.e., glycated hemoglobin inspite of FPG or vice versa) on the similar day or on a different day. When the consequences of the repeated tests comes under the range of prediabetic condition, then the patient shall be advised or shall be treated during the prediabetic condition. When the consequences of repeated testing are whole normal, it should be rescreened after 6 months.

1.4.3.1 MANAGEMENT OF GLYCEMIC

One of the major confronts that health practitioners are facing today is dealing with consistent requires and demands of the patients suffering from prolonged sickness such as diabetes (UKPDS, 1998). The necessity of continuous follow-up of diabetic patients in association with healthcare practitioners is of much importance in preventing any prolonged complications. It has been studied that stringent control on metabolic functions can restrict or prevent the development of other complications induced by diabetes (Ohkubo et al., 1995).

In the large randomized trials results, there is involvement of patients with T1DM or newly recognized or developed T2DM and it reveals that modulation of glycemic condition can restrict the occurrence and development of small vessel complications, such as "nephropathy," "retinopathy," and "neuropathy" (Ismail-Beigi et al., 2010; Shobana et al., 2005). In diabetic patients in addition to controlling the glycemic condition but also to prevent its complications; disability limitation and treatment. Various reports from India showed that highly poor adherence to administered regimens because of the poor approach towards the ailment and deprived health literacy in the common people (Chew, 2004; Coban et al., 2004). Various factors shall be taken into consideration for evaluating the glycemic goals, such as psychosocial drawback (Whitmer et al., 2009). Glycemic goal in patients having "hypoglycemic unawareness" shall be comfortable for chronic periods that pertains to the possible reversal of the state (De Galan et al., 2009; Stumvoll et al., 2005). Patients suffering from serious contemporaneous situation that might interfere with the implementation of the management approach, the targets are to prevent the clinically considerable renal-glycosuria, loss of water and electrolytes, infection, and the progression of hyperosmolar non-ketotic coma. For the T1DM and T2DM patients having insulinopenia, insulin is used as an indicator within which "hyperglycemia" does not respond to diet therapy either alone or combined with "oral hypoglycemic drugs" (Hermansen et al., 2008). Regulation of glucose by the administration of insulin is only because of its glucose-lowering potential (Kroon et al., 2010). Several patients suffering from T2D sometimes requires insulin therapy. However, T2D is linked with "insulin resistance," requirements of insulin can increase by 1 unit/kg/day (Riddlem, 2005).

1.5 DIABETES MELLITUS (DM) PATHOPHYSIOLOGY

Hyperglycemia and physiological and behavioral responses are directly linked with each other. At the time of hyperglycemia, some specific biochemicals are released by our brain through the nerve's impulses to the pancreas and various body organs to decrease its impact (Patidar, 2011).

1.5.1 TYPE 1 DIABETES MELLITUS (T1DM)

T1D is portrayed by autoimmune damage of insulin cells by T helper cells and cytotoxic T lymphocytes cells in addition to macrophage cells invading the pancreatic islets (Al-Homsi and Lukic, 1992). A few features of T1DM which portray it as the disorder of immune system (Hussain and Vincent, 2007) such as:

- Presence of immune-competent and accessory cells islet of pancreas;
- Relation of susceptibility to illness with the class II genes of the "major histocompatibility complex;"
- Adjustments of T cell-mediated immune regulation;
- The contribution of monokines and TH1 cells in the production of interleukins;
- Immunotherapy response; and
- Continuous incidence of auto-immune diseases in affected people or in their relatives.

Circulating antibodies are present in around 85% of patients, anti-insulin antibodies are also present in most of the individuals prior to accepting therapy of insulin. Most islet antibodies are harmonized against "glutamic acid decarboxylase" inside beta cells of pancreas (Raju and Raju, 2010). The auto-immune damage of beta cells of the pancreas cause lack of insulin release bringing about the metabolic derangements related to T1D. Additionally in the insulin secretion deficiency, the role of alpha cells of the pancreas is additionally anomalous and secretes exorbitant glucagon in T1D patients. Typically, "hyperglycemia" prompts declined secretion of glucagon, however in T1DM individuals; the release of glucagon is not muted hyperglycemia (Holt, 2004). The resultant raised glucagon levels aggravate the metabolic irregularity due to the insulin deficiency. Insufficiency in insulin secretion is the important irregularity in T1D, however there is also an imperfection in the administration of insulin.

Deficiency in insulin causes "uncontrolled lipolysis," and further, there is an increase in the level of free unsaturated fatty acid in the plasma, which chokes the glucose metabolism in peripheral tissues; skeletal muscle is an example of it (Holt, 2004). This stops utilization of glucose and insulin insufficiency, in addition reduces the expression of different genes vital for target tissues to respond to insulin on a regular basis. Holt (2004) clarified that the most metabolic derangements resulting from insulin insufficiency in T1D and decrease in lipid glucose and protein digestion.

1.5.2 TYPE 2 DIABETES MELLITUS (T2DM)

In T2D, the two fundamental obsessive deformities are weakened insulin secretion due to the malfunctioning of β-cells, and impeded insulin activity by "insulin resistance" (American Diabetes Association, 2010). In conditions where resistance to insulin occurs, the beta cells go through a change, elevating the insulin supply. Totally, the insulin level in plasma for most of the time is elevated, the plasma level of insulin is deficient to keep up typical homeostasis regarding glucose. "Insulin resistance" and "hyperinsulinemia" at last causes "impaired glucose tolerance" (Mahler and Adler, 1999). Except of "maturity-onset diabetes of the young (MODY)," the mode of inheritance for T2DM is indistinct. Maturity onset diabetes of the young is an autosomal dominant attribute, may be a consequence of mutations in glucokinase quality on chromosome 7p. Maturity onset diabetes of the young can be characterized as hyperglycemia accessed before 25 years of age and treatable for greater than 5 years without insulin in the cases where ICA are negative (Sekikawa et al., 1993).

1.6 INSULIN RESISTANCE

The essential occasions are accepted to be an underlying deficit in insulin emission and in a number of patients comparative insulin inadequacy in relationship with "peripheral insulin resistance" (Botero and Wolfsdorf, 2005). Insulin resistance will bring about impaired insulin-mediated uptake of glucose in the periphery, deficient suppression of liver glucose yield and debilitated triglyceride take-up by fat. Endogenous glucose formation is accelerated in patients with T2D or hindered fasting glucose. Since this increment happens within the sight of hyper-insulinemia, in any event in the early and transitional illness phases, liver insulin resistance is the prominent force of T2D hyperglycemia.

1.6.1 RISK FACTORS

1.6.1.1 OBESITY (OVERALL AND ABDOMINAL)

Asians are having less incidence of overweight and also fatness compared to their Western partners, based on some conventional terms. Regardless of reduction in BMI, various Asian nations have comparable or much greater pervasiveness of diabetes in comparison to the Western nations (Yoon et al., 2006). In Asia countries like India, Pakistan, the predominance of overweight increases with BMI (Shera et al., 2007). The expanding pattern of adolescence obesity in Asia places numerous youthful people at high stake for T2D in early adulthood. As per US Centers for Disease Control and Prevention, 22.9% of young men and 10.4% of young ladies going to metropolitan schools in China were overweight (Zhou et al., 2006). Likewise, 17.8% of young men and 15.8% of young ladies were overweight in metropolitan South India (Ramachandran, 2005). Similar observations were revealed in Malaysia, (Disease Control Division, Ministry of Health Malaysia, 2009) Korea, (Kim et al., 2005, 2006) and Thailand (Likitmaskul et al., 2003)

Thus, midsection periphery reflecting focal stoutness is a helpful proportion of weight related danger of type 2 diabetes, particularly in people with ordinary BMI standards (Aekplakorn et al., 2006; Balkau et al., 2007). This information reveals that the higher risk of type 2 diabetes in Asians might be credited to alarming stomach and instinctive adiposity for a given BMI. Indian babies have greater subcutaneous fat, leptin as well as insulin levels (Yajnik et al., 2002) which have been related with expanded danger of insulin opposition and diabetes (Ruderman et al., 1994). Moreover, expanded waist has been related with considerably expanded danger of 410 diabetes (Aekplakorn et al., 2006; Cheung et al., 2007), just as expanded danger of cardiovascular and all reason mortality, autonomous of BMI (Zhang et al., 2007; Wang et al., 2008).

1.6.1.2 NOURISHMENT TRANSITION AND CHANGES IN DIET AND LIFESTYLE

Rapid financial advancement has prompted a simultaneous move in framework, innovation, and food supply that advances over sustenance and stationary ways of life. Conventional dietary examples are vanishing as people adjust to progressively modern and metropolitan conditions coming about of globalization. Quick nutrition transition has left numerous nations dealing with existing together issues of over nourishment and undernutrition (Misra and Khurana, 2008). In India the animal food and heavy fatty food consumption was more seen among metropolitan inhabitants, who devoured 32% of the energy from fat in contrast to 17% from rural areas (Shetty, 2002). Vegetable ghee, like *Dalda*—and explained spread ordinarily utilized in catering in India along with in other southeastern Asian nations—contains trans unsaturated fat levels up to 50%. Higher admission of trans-fatty acids is related with weight acquires expanded cardio-metabolic danger, and insulin resistance (Haag et al., 2005). Polished rice and refined wheat pose with high glycemic record and glycemic load values (Ludwig, 2002). Intake of "sugary beverages," a significant patron of dietetic glycemic burden as well as abundance calories, has expanded exponentially around the world, especially in Asia (Malik et al., 2006). In developing nations, a quick take-up of advancements has been joined by expanding shifts from farming and higher services in the field of industrialized as well as services (Popkin et al., 2001). In Asia, autos are quickly supplanting bikes as the essential method of transportation. In the previous decade, the yearly pace of expansion in vehicle possession in India maybe around 11% (85). "Psychosocial pressure," sorrow, and lesser relaxing hours and got progressively basic in non-industrial nations going through quick financial turns of events, have been related with more risk of the metabolism-related disorder and diabetes (Ko et al., 2007; Takeuchi et al., 2009). Meta-investigation and sadness was related with 60% of type 2 diabetic patients, while the latter was related with a 15% to depressed persons (Mezuk et al., 2008). The concurrence of both "diabetes and depression" was correlated having 50 to 100% risk of all-cause death (Katon et al., 2005).

1.6.1.3 CIGARETTE SMOKING

Recent meta-analysis stated that current smoking has been found to be related with 44% to the threat of developing diabetes (Willi et al., 2007; Chen et al., 2008). Smoking is well recognized to stimulate insulin resistance and lacking balances the secretion of insulin. People having typical BMI, smokers were found to pose a greater "abdominal obesity" in compared to nonsmoker ones (Willi et al., 2007). Most Indians utilize smokeless tobacco items, like betel quid, and 40% use bidis—little, regularly seasoned, nontaxable cigarettes—the creation of which gives work to numerous urban poor (Kabir et al., 2007).

1.6.1.4 PANCREATIC BETA CELL FUNCTION

In the 1980s, scientists originally unwound that decreased early insulin reaction was a free indicator for diabetes (Kadowaki et al., 1984). Fukushima et al. (2004) discovered, every phase of "glucose intolerance," Japanese people had decreased early and later periods of insulin reactions. In Japanese adult males with typical glucose resilience, even a little expansion in BMI delivered a decline in beta cell work lopsided to the one in insulin sensitivity (Kuroe et al., 2003). Among a sample of type 2 diabetic patients, 50% were having normal weight, with lower BMI linked with lesser degrees of fasting plasma C-peptide which acts as an indicator of low insulin discharge), higher glycated hemoglobin contents (Chan et al., 2004). Taken together, in few populaces, deficient beta-cell reaction to expanding insulin resistance brings about losing glycemic control and expanded danger of diabetes with still a slight weight gain.

1.6.1.5 DEVELOPMENTAL ORIGINS OF DIABETES

Numerous adults witnessed extraordinary difficulty in war or common agitation in earlier periods of life is presently encountering stamped changes in way of life. Likewise, "lower birth weight" as well as openness to under nourishment in utero are regular in several populaces, particularly in India, where about 30% of babies are below average weight (Yajnik, 2009).

Stresses in the intrauterine period can prompt lasting alterations in structure, digestion, and physiology by genome modifications without any changes with respect to the DNA codes, an interaction known as epigenetic (Gluckman et al., 2008). Such early life occasions may later induce the vulnerability to diabetes, the metabolic disorder, and cardio-renal illnesses. Imminent examinations have shown the effect of fetal undernourishment (low birth weight) just as over-nutrition (i.e., baby of a diabetic mother) on prospective danger of diabetes (Yajnik, 2009). In India, slenderness in earliest stages as well as higher weight at 12 years age are related with expanded danger of creating "IGT or diabetes in young adulthood" (Bhargava et al., 2004).

1.6.1.6 HEREDITARY SUSCEPTIBILITY

Among lean, sound people with similar "age, BMI, waist expansion, birth weight, and present eating regimen, Asians had more elevated 'postprandial glycemia' levels and lower insulin affectability" (Dickinson et al., 2002). These discoveries increase the likelihood that "Asians" having hereditarily defenseless to IR as well as diabetes compared to whites. Approximately 10% of the patients conveyed hereditary variations coding for pancreatic beta-cell pathways, also including "transcription factors and amylin, or mitochondrial polymorphisms." These discoveries give additional proof that beta-cell dysfunction assumes a basic part in the improvement of diabetes.

1.6.1.7 OTHER RISK FACTORS

Evidences propose the openness toward the ecological aggravations, like persistent natural toxins, is related with expanded "insulin resistance," the metabolic diseases and DM (Lee et al., 2008). People with thalassemia qualities were accounted for to have a few crease expanded danger of gestational diabetes (Lao et al., 2001) "insulin resistance," (Tong et al., 2002) and glucose intolerance (Chern et al., 2001). "Chronic hepatitis B" transporters influenced by adult-onset disease likewise have a previous time of finding and four times high danger of terminal stage renal sickness (ESRD) than on carriers (Cheng et al., 2006). A similar relationship with diabetes and diabetes related kidney illness has been accounted for in ongoing hepatitis-C carriers (Mehta et al., 2003). Other contaminations endemic in Asia, for example, tuberculosis (Leung et al., 2008) have additionally been related with expanded danger of 'diabetes' and serious clinical course of the infection.

KEYWORDS

- **classification**
- **diabetes**
- **diagnosis**
- **historical perspectives**
- **pathogenesis**

REFERENCES

Aekplakorn, W., Bunnag, P., Woodward, M., Sritara, P., Cheepudomwit, S., Yamwong, S., & Rajatanavin, R., (2006). A risk score for predicting incident diabetes in the Thai population. *Diabetes Care, 29*, 1872–1877.

Al Homsi, M. F., & Lukic, M. L., (1992). *An Update on the Pathogenesis of Diabetes Mellitus*. Department of pathology and medical microbiology (immunology unit) faculty of medicine and health sciences, UAE University, Al Ain, United Arab Emirates.

American Diabetes Association, (2010). Diagnosis and classification of diabetes mellitus. *Diabetes Care, 33*, 62–69.

American Diabetes Association, (2011). Standards of medical care in diabetes-2011. *Diabetes Care, 34*, 11–61.

American Diabetes Association, (2012). Standards of medical care in diabetes-2012. *Diabetes Care, 35*, 11–63.

Balkau, B., Deanfield, J. E., & Despres, J. P., (2007). International Day for the evaluation of abdominal obesity (IDEA): A study of waist circumference, cardiovascular disease, and diabetes mellitus in 168,000 primary care patients in 63 countries. *Circulation, 116*, 1942–1951.

Banting, F. G., Best, C. H., & Macleod, J. J. R., (1922). The internal secretion of the pancreas. *American Journal of Physiology, 59*, 479.

Banting, F. G., Best, C. H., Collip, J. B., Cambell, W. R., Fletcher, A. A., Macleod, J. J. R., & Noble, E. C., (1922). The effect produced on diabetes by extractions of pancreas. *Transact. Association of American Physicians, 37*, 337.

Barron, M., (1920). The relation of the islets of Langerhans to diabetes. *Surgery, Gynecology and Obstetrics, 31*, 437–448.

Baynes, H. W., (2015). Classification, pathophysiology, diagnosis and management of diabetes mellitus. *Journal of Diabetes and Metabolism, 6*, 1–9.

Bhargava, S. K., Sachdev, H. S., Fall, C. H., Osmond, C., Lakshmy, R., Barker, D. J., & Reddy, K. S., (2004). Relation of serial changes in childhood body-mass index to impaired glucose tolerance in young adulthood. *N. Engl. J. Med., 350*, 865–875.

Botero, D., & Wolfsdorf, J. I., (2005). Diabetes mellitus in children and adolescents. *Arch. Med. Res., 36*, 281–290.

Bouchardat, A., (1875). De la glycosurieoudiabètesucré. Son traitement hygiénique. Paris: Baillière.

Chan, W. B., Tong, P. C. Y., Chow, C. C., So, W. Y., Ng, M. C. Y., Ma, R. C. W., Osaki, R., et al., (2004). The associations of body mass index, C peptide and metabolic status in Chinese type 2 diabetic patients. *Diabet. Med., 21*, 349–353.

Chen, C. C., Li, T. C., Chang, P. C., Liu, C. S., Lin, W. Y., Wu, M. T., Li, C. I., et al., (2008). Association among cigarette smoking, metabolic syndrome, and its individual components: The metabolic syndrome study in Taiwan. *Metabolism, 57*, 544–548.

Cheng, A. Y., Kong, A. P., Wong, V. W., So, W. Y., Chan, H. L., Ho, C. S., Lam, C. W., et al., (2006). Chronic hepatitis B viral infection independently predicts renal outcome in type 2 diabetic patients. *Diabetologia, 49*, 1777–1784.

Chern, J. P., Lin, K. H., Lu, M. Y., Lin, D. T., Lin, K. S., Chen, J. D., & Fu, C. C., (2001). Abnormal glucose tolerance in transfusion-dependent betathalassemic patients. *Diabetes Care, 24*, 850–854.

Cheung, B. M., Wat, N. M., Man, Y. B., Tam, S., Thomas, G. N., Leung, G. M., Cheng, C. H., et al., (2007). Development of diabetes in Chinese with the metabolic syndrome: A 6-year prospective study. *Diabetes Care, 30*, 1430–1436.

Chew, L. D., (2004). The impact of low health literacy on diabetes outcomes. *Diabetes Voice, 49*, 30–32.

Coban, E., Ozdogan, M., & Timuragaoglu, A., (2004). Effect of iron deficiency anemia on the levels of hemoglobin A1c in nondiabetic patients. *Acta Haematol., 112*, 126–128.

Cooper, T., & Ainsberg, A., (2010). *Breakthrough: Elizabeth Hughes, the Discovery of Insulin, and the Making of a Medical Miracle*. New York: St Martin's Press.

Cox, E. M., & Elelman, D., (2009). Test for screening and diagnosis of type 2 diabetes. *Clinical Diabetes, 4*, 132–138.

Cryer, P. E., (2012). Minireview: Glucagon in the pathogenesis of hypoglycemia and hyperglycemia in diabetes. *Endocrinology, 153*, 1039–1048.

De Galan, B. E., Zoungas, S., Chalmers, J., Anderson, C., Dufouil, C., Pillai, A., Cooper, M., et al., (2009). Cognitive function and risks of cardiovascular disease and hypoglycaemia in patients with type 2 diabetes: The action in diabetes and vascular disease: Preterax and diamicron modified-release controlled evaluation (ADVANCE) trial. *Diabetologia, 52*, 2328–2336.

Dickinson, S., Colagiuri, S., Faramus, E., Petocz, P., & Brand-Miller, J. C., (2002). Postprandial hyperglycemia and insulin sensitivity differ among lean young adults of different ethnicities. *J. Nutr., 132*, 2574–2579.

Disease Control Division, Ministry of Health Malaysia. NCD risk factors in Malaysia, (2009). World Health Organization Web site. http://www.who.int/chp/steps/MalaysiaSTEPSReport.pdf (accessed on 18 December 2021).

Forbes, J. M., & Cooper, M. E., (2013). Mechanisms of diabetic complications. *Physiol. Rev., 93*, 137–188.

Fukushima, M., Usami, M., Ikeda, M., Nakai, Y., Taniguchi, A., Matsuura, T., Suzuki, H., et al., (2004). Insulin secretion and insulin sensitivity at different stages of glucose tolerance: A cross-sectional study of Japanese type 2 diabetes. *Metabolism, 53*, 831–835.

Furdell, E. L., (2009). *Fatal Thirst: Diabetes in Britain Until Insulin* (pp. 81–92). Leiden: Brill.

Gillett, M. J., (2009). International expert committee report on the role of the A1c assay in the diagnosis of diabetes: *Diabetes Care, 32*, 1327–1334.

Gluckman, P. D., Hanson, M. A., Cooper, C., & Thornburg, K. L., (2008). Effect of in utero and early-life conditions on adult health and disease. *N. Engl. J. Med., 359*, 61–73.

Grmek, M. D., (1997). *Le Legs de Claude Bernard.* Paris: Fayard.

Guidance for Industry, (2008). *Diabetes Mellitus: Developing Drugs and Therapeutic Biologics for Treatment and Prevention* (p. 3). U.S. Department of Health and Human Services Food and Drug Administration Center for Drug Evaluation and Research (CDER).

Haag, M., & Dippenaar, N. G., (2005). Dietary fats, fatty acids and insulin resistance: Short review of a multifaceted connection. *Med. Sci. Monit., 11,* RA359–RA367.

Harris, M. I., Klein, R., Welborn, T. A., & Knuiman, M. W., (1992). Onset of NIDDM occurs at least 4–7 yr before clinical diagnosis. *Diabetes Care, 15*, 815–819.

Hermansen, K., Mortensen, L. S., & Hermansen, M. L., (2008). Combining insulin with oral antidiabetic agents: Effect on hyperglycemic control, markers of cardiovascular risk and disease. *Vascu. Health Risk Manag., 4*, 561–574.

Holt, R. I., (2004). Diagnosis, epidemiology and pathogenesis of diabetes mellitus: An update for psychiatrists. *Br. J. Psychiatry Suppl., 47*, 55–63.

Hussain, A. N., & Vincent, M. T., (2007). *Type 1 Diabetes Mellitus.*

Ismail-Beigi, F., Craven, T., Banerji, M. A., Basile, J., Calles, J., Cohen, R. M., Cuddihy, R., et al., (2010). Effect of intensive treatment of hyperglycaemia on microvascular outcomes in type 2 diabetes: An analysis of the ACCORD randomized trial. *The Lancet, 376*, 419–430.

Kabir, Z., Clancy, L., & Connolly, G. N., (2007). Tobacco control efforts: Where is India now? *Lancet, 370,* 134.

Kadowaki, T., Miyake, Y., Hagura, R., Akanuma, Y., Kajinuma, H., Kuzuya, N., Takaku, F., & Kosaka, K., (1984). Risk factors for worsening to diabetes in subjects with impaired glucose tolerance. *Diabetologia, 26*, 44–49.

Kahn, C. R., & Weir, G. C., (2005). *Joslin's Diabetes Mellitus* (14th edn.). Philadelphia: Lippincott.

Karamanou, M., Koutsilieris, M., Laios, K., Marineli, F., & Androutsos, G., (2014). Apollinaire Bouchardat (1806–1886): Founder of modern diabetology. *Hormones (Athens), 13*, 296–300 [PMID: 24776631].

Katon, W. J., Rutter, C., Simon, G., Lin, E. H., Ludman, E., Ciechanowski, P., Kinder, L., et al., (2005). The association of comorbid depression with mortality in patients with type 2 diabetes. *Diabetes Care, 28*, 2668–2672.

Kim, D. M., Ahn, C. W., & Nam, S. Y., (2005). Prevalence of obesity in Korea. *Obes. Rev., 6*, 117–121.

Kim, H. M., Park, J., Kim, H. S., Kim, D. H., & Park, S. H., (2006). Obesity and cardiovascular risk factors in Korean children and adolescents aged 10–18 years from the Korean national health and nutrition examination survey, 1998 and 2001. *Am. J. Epidemiol., 164*, 787–793.

Ko, G. T., Chan, J. C., Chan, A. W., Wong, P. T., Hui, S. S., Tong, S. D., Ng, S. M., et al., (2007). Association between sleeping hours, working hours and obesity in Hong Kong Chinese: The "better health for better Hong Kong" health promotion campaign. *Int. J. Obes. (Lond), 31*, 254–260.

Kroon, L., Goldfine, I. D., & Tanyolac, S., (2010). *Insulin-Pharmacology, Types of Regimens, and Adjustments.* San Francisco: University of California.

Kuroe, A., Fukushima, M., Usami, M., Ikeda, M., Nakai, Y., Taniguchi, A., Matsuura, T., et al., (2003). Impaired beta-cell function and insulin sensitivity in Japanese subjects with normal glucose tolerance. *Diabetes Res. Clin. Pract., 59*, 71–77.

Laios, K., Karamanou, M., Saridaki, Z., & Androutsos, G., (2012). Aretaeus of Cappadocia and the first description of diabetes. *Hormones, 11*, 109–113.

Lao, T., & Ho, L. F., (2001). Alpha-thalassaemia trait and gestational diabetes mellitus in Hong Kong. *Diabetologia, 44*, 966–971.

Lee, D. H., Steffes, M. W., & Jacobs, D. R., (2008). Can persistent organic pollutants explain the association between serum gamma-glutamyltransferase and type 2 diabetes? *Diabetologia, 51*, 402–407.

Leung, C. C., Lam, T. H., Chan, W. M., Yew, W. W., Ho, K. S., Leung, G. M., Law, W. S., et al., (2008). Diabetic control and risk of tuberculosis: A cohort study. *Am. J. Epidemiol., 167*, 1486–1494.

Likitmaskul, S., Kiattisathavee, P., Chaichanwatanakul, K., Punnakanta, L., Angsusingha, K., & Tuchinda, C., (2003). Increasing prevalence of type 2 diabetes mellitus in Thai children and adolescents associated with increasing prevalence of obesity. *J. Pediatr. Endocrinol. Metab., 16*, 71–77.

Ludwig, D. S., (2002). The glycemic index: Physiological mechanisms relating to obesity, diabetes, and cardiovascular disease. *JAMA., 287*, 2414–2423.

Mahler, R. J., & Adler, M. L., (1999). Clinical review 102: Type 2 diabetes mellitus: Update on diagnosis, pathophysiology, and treatment. *J. Clin. Endocrinol. Metab., 84*, 1165–1171.

Malik, V. S., Schulze, M. B., & Hu, F. B., (2006). Intake of sugar-sweetened beverages and weight gain: A systematic review. *Am. J. Clin. Nutr., 84*, 274–288.

Mayfield, J., (1998). Diagnosis and classification of diabetes mellitus: New criteria. *Am. Fam. Physician, 58*, 1369–1370.

Medvei, V. C., (1993). *The History of Clinical Endocrinology: A Comprehensive Account of Endocrinology from Earliest Times to the Present Day* (p. 253). New York: Parthenon.

Mehta, S. H., Brancati, F., Strathdee, S., Pankow, J. S., Netski, D., Coresh, J., Szklo, M., & Thomas, D. L., (2003). Hepatitis C virus infection and incident type 2 diabetes. *Hepatology, 38*, 50–56.

Mezuk, B., Eaton, W. W., Albrecht, S., & Golden, S. H., (2008). Depression and type 2 diabetes over the lifespan: A meta-analysis. *Diabetes Care, 31*, 2383–2390.

Misra, A., & Khurana, L., (2008). Obesity and the metabolic syndrome in developing countries. *J. Clin. Endocrinol. Metab., 93,* S9–S30.
Ohkubo, Y., Kishikawa, H., Araki, E., Miyata, T., Isami, S., Motoyoshi, S., Kojima, Y., et al., (1995). Intensive insulin therapy prevents the progression of diabetic microvascular complications in Japanese patients with non-insulin-dependent diabetes mellitus: A randomized prospective 6-year study. *Diabetes Res. Clin. Pract., 28*, 103–117.
Patidar, D., (2011). *Pharmacology III* (2nd edn., pp. 113, 114). Meerut: Shree Sai Prakashan.
Peumery, J. J., (1987). *Histoireillustrée du Diabète: De l'Antiquité à Nosjours*. Paris: Les Éditions Roger Dacosta.
Raju, S. M., & Raju, B., (2010). *Illustrated Medical Biochemistry* (2nd edn.). Jaypee Brothers Medical Publishers Ltd., New Delhi, India.
Ramachandran, A., (2005). Epidemiology of diabetes in India, three decades of research. *J. Assoc. Physicians India, 53*, 34–38.
Riddlem, M. C., (2005). Glycemic management of type 2 diabetes: An emerging strategy with oral agents, insulins, and combinations. *Endocrinol. Metabol. Clin. North America, 34*, 77–98.
Ruderman, N., Chisholm, D., Pi-Sunyer, X., & Schneider, S., (1998). The metabolically obese, normal-weight individual revisited. *Diabetes, 47*, 699–713.
Salomaa, V., Havulinna, A., Saarela, O., Zeller, T., Jousilahti, P., Jula, A., Muenzel, T., et al., (2010). Thirty-one novel biomarkers as predictors for clinically incident diabetes. *PloS One, 5,* 10100.
Sekikawa, A., Tominaga, M., Takahashi, Eguchi, H., Igarashi, M., Ohnuma, H., Sugiyama, K., et al., (1993). Prevalence of diabetes and impaired glucose tolerance in Funagata area, Japan. *Diabetes Care, 16*, 570–574.
Selvin, E., Steffes, M. W., Zhu, H., Matsushita, K., Wagenknecht, L., Pankow, J., Coresh, J., & Brancati, F. L., (2010). Glycated hemoglobin, diabetes, and cardiovascular risk in nondiabetic adults. *New England J. Med., 362*, 800–811.
Shafizadeh, T. B., Moler, E. J., Kolberg, J. A., Nguyen, U. T., Hansen, T., Jorgensen, T., Pedersen, O., & Borch-Johnsen, K., (2011). Comparison of accuracy of diabetes risk score and components of the metabolic syndrome in assessing risk of incident type 2 diabetes in Inter99 cohort. *PloS One, 6*, 22863.
Shera, A. S., Jawad, F., & Maqsood, A., (2007). Prevalence of diabetes in Pakistan. *Diabetes Res. Clin. Pract., 76*, 219–222.
Shobana, R., Augustine, C., Ramachandran, A., & Vijay, V., (2005). Improving psychosocial care: The Indian experience. *Diabetes Voice, 50*, 19–21.
Stumvoll, M., Goldstein, B. J., & Van, H. T. W., (2005). Type 2 diabetes: Principles of pathogenesis and therapy. *Lancet, 365*, 1333–1346.
Takeuchi, T., Nakao, M., Nomura, K., & Yano, E., (2009). Association of metabolic syndrome with depression and anxiety in Japanese men. *Diabetes Metab., 35*, 32–36.
Tong, P. C., Ng, M. C., Ho, C. S., So, W. Y., Li, J. K., Lam, C. W., Cockram, C. S., Chan, J. C., 2002). C reactive protein and insulin resistance in subjects with thalassemia minor and a family history of diabetes. *Diabetes Care, 25*, 1480–1481.
UK Prospective Diabetes Study (UKPDS), (1998). Group Intensive blood-glucose control with sulphonylureas or insulin compared with conventional treatment and risk of complications in patients with type 2 diabetes (UKPDS 33). *Lancet, 352*, 837–853.
Von, E. D., (1989). *Diabetes: Its Medical and Cultural History* (pp. 306–319, 350–358, 411–426). Berlin: Springer-Verlag.

Von, M. J., & Minkowski, O., (1889). Diabetes mellitus nach pankreas extirpation. *Archives of Experimental Pathology and Pharmacology, 26,* 371. doi: 10.1007/BF01831214.

Wang, C., Hou, X., Bao, Y., Pan, J., Zuo, Y., Zhong, W., Jia, W., & Xiang, K., (2008). The metabolic syndrome increased risk of cardiovascular events in Chinese—A community based study. *Int. J. Cardiol.*

Whitmer, R. A., Karter, A. J., Yaffe, K., Quesenberry, C. P. Jr., & Selby, J. V., (2009). Hypoglycemic episodes and risk of dementia in older patients with type 2 diabetes mellitus. *J. Amer. Med. Assoc., 301*, 1565–1572.

Willi, C., Bodenmann, P., Ghali, W. A., Faris, P. D., & Cornuz, J., (2007). Active smoking and the risk of type 2 diabetes: A systematic review and meta-analysis. *JAMA, 298*, 2654–2664.

Williamson, R. T., (1923). *English Physicians of the Past* (pp. 47–52). Newcastle: Andrew Reid and Company.

Willis, T., (1694). *Opera Omnia* (p. 460). Coloniae: Sumptibus Gasparis Storti.

Yajnik, C. S., (2009). Nutrient-mediated teratogenesis and fuel-mediated teratogenesis: Two pathways of intrauterine programming of diabetes. *Int. J. Gynaecol. Obstet., 104,* S27–S31.

Yajnik, C. S., Lubree, H. G., Rege, S. S., Naik, S. S., Deshpande, J. A., Deshpande, S. S., Joglekar, C. V., & Yudkin, J. S., (2002). Adiposity and hyperinsulinemia in Indians are present at birth. *J. Clin. Endocrinol. Metab., 87*, 5575–5580.

Yoon, K. H., Lee, J. H., Kim, J. W., Cho, J. H., Choi, Y. H., Ko, S. H., Zimmet, P., & Son, H. Y., (2006). Epidemic obesity and type 2 diabetes in Asia. *Lancet, 368*, 1681–1688.

Zhang, X., Shu, X. O., Yang, G., Li, H., Cai, H., Gao, Y. T., & Zheng, W., (2007). Abdominal adiposity and mortality in Chinese women. *Arch Intern. Med., 167*, 886–892.

Zhou, H., Yamauchi, T., Natsuhara, K., Yan, Z., Lin, H., Ichimaru, N., Kim, S. W., et al., (2006). Overweight in urban schoolchildren assessed by body mass index and body fat mass in Dalian, China. *J. Physiol. Anthropol., 25*, 41–48.

CHAPTER 2

Diabetes Mellitus: Signs and Symptoms, Epidemiology, Current Prevention, Management Therapies, and Treatments

YOUNIS AHMAD HAJAM,[1] RAKSHA RANI,[2] JAVID AHMAD MALIK,[3] ANU PANDITA,[4] RIYA SHARMA,[2] and RAJESH KUMAR[5]

[1]*Department of Life Sciences and Allied Health Sciences, Sant Baba Bagh Singh University, Khalia, Padhiana, Jalandhar, Punjab–144030, India, Mobile: +91-6006101812, E-mail: younismajeed64@gmail.com (Y. A. Hajam)*

[2]*Department of Biosciences, Division Zoology, Career Point University, Hamirpur – 17604, Himachal Pradesh, India*

[3]*Department of Zoology, Guru Ghasidas Vishwavidayalaya, Bilaspur – 495001, Chhattisgarh, India*

[4]*Vatsalya Clinic, Krishna Nagar, New Delhi, India*

[5]*Department of Biosciences, Himachal Pradesh University, Shimla, Himachal Pradesh – 175001, India*

ABSTRACT

"Diabetes mellitus (DM)" is a major endocrine and metabolic syndrome, due to the irregular metabolic processes. Hyperglycemic condition due to the deficiency of insulin leads to irregularities in the metabolism of carbohydrates, fats, and proteins. This hyperglycemia state arises because of lower insulin secretion or cells develops resistance against the binding of insulin on their receptors present on the membrane, or it might be the

Antidiabetic Potential of Plants in the Era of Omics. Deepu Pandita, PhD, Anu Pandita, MSc, & Chander Bhanu, MBBS (Eds.)

amalgamation of both. It was shown in the epidemiologic studies that in the world 150 million persons are suffering from diabetes and it might reach up to 300 million or more than 300 million up to 2025. Epidemiologic parameters include sedentary lifestyle, intake of high energy foods, urbanization, less physical work. It has threatened the entire world especially Asian and European countries. It was reported in the previous study that almost 10% of the population affected with diabetes reported that diabetes is affecting nearly 10% of the population in the world. Diabetes has now become the first leading cause of mortality between chronic degenerative diseases and estimated 16.7% death. In addition to hyperglycemia, it also causes hyperlipidemia, oxidative stress, excessive urination (polyuria), elevation in ketone bodies (ketosis), dysfunctioning of nerves (neuropathy), damages to the nephron (nephropathy), and various heart diseases. In low-middle income countries (LMIC) about 200 million people are suffering from diabetes. There are a number of allopathic drugs available in the market to cure diabetes. But, continued utilization of these allopathic medicines cause several adverse effects, which leads to secondary complications like low blood sugar, feeling of vomiting, low blood sodium, diarrhea, headache, increase in body weight, lactic acidosis, lightheadedness, arthritis, dyspepsia. In addition to allopathic, there are various ayurvedic and plants extract available for the cure and impediment of diabetes. Therefore, the present chapter summarizes the epidemiological status, prevents measures, and various treatments available for the cure of diabetes.

2.1 INTRODUCTION

"Diabetes mellitus" (DM) is the term which is given by the Greek physician "Aertaeus." In Greek, the word diabetes means "to pass through" and mellitus is the Latin word which means "honey or sweetness." DM is a common metabolic disorder. It is diagnosed only on the basis of hyperglycemia and beta cells unable to secrete insulin or inappropriate action of insulin. Hyperglycemic condition due to the deficiency of insulin leads to irregularities in the metabolism of carbohydrates, fats, and proteins. DM is a major endocrine and metabolic disorder, due to the irregular metabolic processes (carbohydrates and lipids). The main cause of diabetes is improper insulin secretion or action or due to the factors that oppose the function of insulin (Watkins, 2003), results in the elevation in blood glucose level (hyperglycemia). In the list of endocrine disorders, DM is on the top cause numerous deaths in humans. Diabetes is clinically and genetically associated with assembly of diseases distinguished by significant amount of glucose in blood (Johar et al., 2018). This hyperglycemia state arises because of lower insulin secretion or cells develops resistance against the binding of insulin on their membrane receptors, or it might be the amalgamation of both (Johar et al., 2018). Diabetes negatively affects each and every physiological and biochemical functions in the body of an individual by changing the normal use and glucose metabolism.

The high amount of blood glucose is a major threatened aspect for *arteriosclerosis*, or diseases that causes inflammation of the arterial walls, because of the buildup of fats, cholesterol in artery walls. It causes stroke or cardiac arrest. Angiogenesis and healing of wounds negatively affected by diabetes (Soybir et al., 2003), supply of ascorbic acid to the retina and brain is also inhibited by diabetes which leads to the blood-brain-barrier (BBB) permeability (Prasad et al., 2014). "World Health Organization" evidenced that DM is the 8th leading cause of deaths and morbidness (Marth, 2008). The unmetabolized glucose gets self-oxidized and leads in the production of free radicals and in turn damages cellular membranes. The self-oxidized glucose causes different complications such as hepatorenal (Rai et al., 2016; Rai, 2019), neuronal damages (Hajam et al., 2017), and also reproductive complications (Hajam et al., 2017).

2.2 RISK FACTORS FOR DIABETES

It has been shown in various studies that diabetes is associated with many impediments, and it is very low in those peoples who particularly live in the rural areas (Ramachandran, 2005). Therefore, there is an immediate necessity to create consciousness in the populations about the diabetes and other associated long term disorders. Epidemiologic studies in India revealed the presence of various hazards which can be simply diagnosed by noninvasive threats (Ramachandran, 2005; Mohan and Anbalagan, 2013). These foremost risk factors include family have positive history of diabetes, age below 35 years, obesity, and overweight, puffy waist and adiposity in the upper body, occurrence of high blood pressure, concurrent increase in body weight, resting habits (Sedentary lifestyle) or diabetes during gestational period.

2.3 CLASSIFICATION OF DM

1. **Insulin-Dependent Diabetes Mellitus or Type-1 Diabetes:** It occurs in all communities irrespective of age, however mostly occurs in childhood before the 25 years of age. Prevalence of T1DM (type 1 diabetes mellitus) is equal in both the sexes and estimates about 10–15% of total diabetes. The most common factor is insufficient secretion of insulin because of the autoimmune degeneration of pancreatic β-cells. Previously T1DM was called "insulin-dependent diabetes mellitus" or "Juvenile diabetes."
2. **Non-Insulin Dependent Diabetes Mellitus or Type-2 Diabetes:** It occurs because of the receptor dysfunctions, insulin confrontation (insulin receptors on the cells become insensitive to insulin or response poorly towards the insulin (*see* Figure 1.1). Therefore, it is commonly called "non-insulin-dependent diabetes mellitus" or "adult-onset diabetes." The principal cause is too much increase in the body weight (*see* Figure 1.1).
3. **Gestational Diabetes:** The third type of diabetes is the "gestational diabetes" it takes place during pregnancy irrespective of earlier history of diabetes developed due to high blood sugar levels.

Clinically diabetes is diagnosed by polyuria, albuminuria, renal enlargement, elevated glomerular filtration rate and an augmentation in serum creatinine level (Grover et al., 2003; Handelsman et al., 2015). Additionally, diabetes also causes neuronal disorders, damage in the visceral, somatic, and cranial nerves and spinal nerve (Anjaneyulu and Chopra, 2004; Hajam et al., 2017). Ravi et al. (2005) estimated that deficiency of insulin results in the progression of different metabolic complications such as irregular carbohydrate, lipid metabolism, protein glycation, membrane disintegration and dyslipidemia are major common impediments in DM in 40% of diabetic patients. Dyslipidemia includes elevation in the cholesterol levels, triglyceride levels, low-density lipoprotein, very low-density lipoprotein (Soltani et al., 2007), but decrease in HDL (good cholesterol) (Rai et al., 2016). Unusual elevation in lipids in blood serum signifies the possibility and threats for cardiovascular disease (CVD) (Rajasekaran et al., 2006).

2.4 SIGNS AND SYMPTOMS OF DIABETES

The symptoms include repetitive urination, surge in thirst and appetite which results in aggregation of various dysfunctions.

As we know diabetes is a disease of prolonged development of symptoms, hence the sign and symptoms are ignored many times. Consequences of hyperglycemia are not expressed instantaneously so that people do not deliberate diabetes as a serious problem. People are unaware of the damage that can start several years before symptoms turn out to be observable. Therefore, to avoid several complications it is necessary to recognize the initial symptoms of this disease to keep it under control (Figure 2.1).

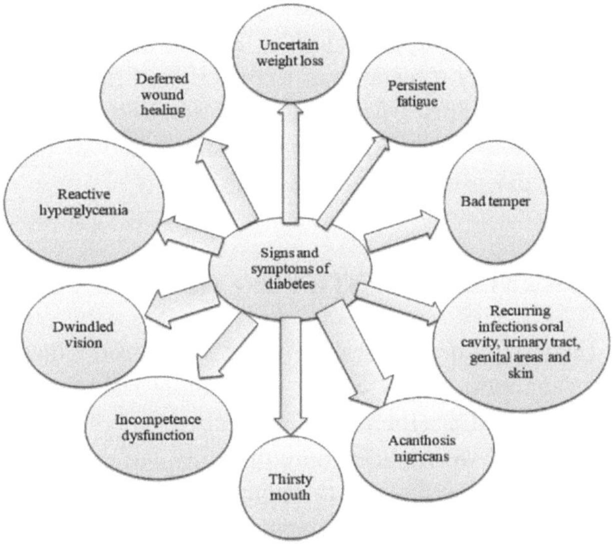

FIGURE 2.1 Signs and symptoms of diabetes mellitus.

Some symptoms include polyphagia, polyuria, and polydipsia are the characteristic symptoms of type 1 diabetes, having quick progress in severe hyperglycemia, while type 2 diabetes has rising points of hyperglycemia. Excessive weight loss is the primary symptom of T1DM and T2DM. Uncertain loss in the body weight, tiredness, and body ache are also considered as symbols of unobserved DM. Sometimes, mild, or slow progress of various symptoms which go on unobserved can cause diabetes.

2.5 SCREENING TEST FOR DIABETES

A person suffering from above mentioned risk issues have to undergo a screening test for DM. One of the frequently used screening tests for is an "Oral Glucose Tolerance Test or OGTT." This test is mostly used to screen

T2DM. Detection of "Impaired fasting glucose or IFG" can be done through fasting tests and 2 h post glucose tests. Fasting glucose should be greater than 110–140–126 mg/dl and 2 hours glucose should be greater than 200 mg/dl. Additional validation by OGTT is necessary in conditions when blood glucose value is >150 mg/dl. In recent times, an HbA1 test (Glycosylated hemoglobin) may be in use to diagnose diabetes. Blood glucose values between 5.7–6.4% taken by HbA1 test directed the presence of prediabetes (American Diabetes Association, 2011). Standard diagnostic criteria are used to screen undiagnosed T2DM in women with various risk factors at their early prenatal stays. Period of 24–28 weeks is considered as the period of detection for "gestational diabetes mellitus or GDM" in women not having earlier history of diabetes, due to the asymptomatic nature of gestational diabetes (11). But, history of gestational diabetes brings high risk for evolving the disease.

2.6 SIGNIFICANCE OF IDENTIFYING PREDIABETES

"Impaired fasting glycemia or IFG" and "impaired glucose tolerance or IGT" are pre-diabetic stages which can initiate risk for vascular difficulties mainly CVDs. IDF (International Diabetes Federation) estimated the current assessments; worldwide, more than 6.9% of people (316 million) are suffered from "impaired glucose tolerance (IGT)," in which more than 70% of the community live in those countries which have low or middle income.

At this time, it is noticeable that one third of the total population having IGT (impaired glucose tolerance) is live in the prolific age of 20–39 years. Hence, they are at elevated risk of evolving diabetes and its other complications (IDF Diabetes Atlas, 2014). Reactive hypoglycemia is a phase in person with prediabetes occurs after 2–3 hrs. of a meal. It is an indication of weakened insulin metabolism which may cause diabetes in the near future. Thus, time to time medical checkup in people with several risk factors of diabetes would helpful to ease the threats of undiagnosed diabetes. It would result in rising awareness among people about their metabolic abnormalities related to diabetes and ultimately improves the health status of many.

2.7 EFFECT OF DIABETES ON AGING OF LIVER AND BONES

Stable changes of hepatic structure and functions comprising changes in sinusoidal endothelial cells interrelated with occurrence of aging (Martinez et al., 2008). Onset of aging rises the probabilities of a number of liver ailments and act as an adverse projecting factor which initiates an increase in the death rate (Floreani, 2007; Amarapurkar et al., 2007; Sheedfar et al., 2013). In liver, aging influence the blood flow as it slowly declines. Studies reported that, there is the occurrence of 20–40% reduction in volume of liver with respect to aging (Schmucker et al., 2005; Kim et al., 2015). Aging promotes the building up of "cholesterol," "γ-glutamyl transferase," "LDL," "alkaline phosphate" and "serum" (Kim et al., 2015).

Expressions of TGF-β1 raises in animal models as diabetes cause damage to the liver (Sahai et al., 2004). An increase in expressions of TGF-β1 isoform has been used as a sign of damaged liver cells due to inflammation (Weng et al., 2007). In liver, TGF-β1 comes out as a significant profibrogenic cytokine. It helps in the triggering of "hepatic stellate cells (HSCs)," promotes death of liver cells. Studies reported a significant rise in expression of hepatic CYP24 linked with enduring diabetes in rat model. In case of any damage to the liver, there is stimulation of HSCs through their alteration from inactive hepatic stellate cells (HSCs) into proliferative, fibrogenic, and contractile myofibroblast (Vuicaa et al., 2016). Variation in extracellular matrix (ECM) configuration can directly stimulate fibrogenesis. Likewise, diabetes stimulates injury of hepatocytes which consequently causes liver fibrosis and inflammation. Different studies have revealed that aged rats have severe liver injury than the young rats in response to various toxics. Hence, expression of TGF-β1 increases with respect to "aging," "inflammatory processes" and "diabetes mellitus" (Vuicaa et al., 2016). Cytoplasm of different hepatic cells also recognize with increased expressions. In hepatocyte, effective cytoplasmic TGF-β1 expression was mostly confined with strong nuclear expression of CYP24. Weakening of bones and diabetes linked with aging may be occur due to the irregular increased expression of CYP24 as it is capable of calcitriol catabolism in large amounts (Vuicaa et al., 2016).

2.8 DIABETES AND METABOLIC FUNCTIONS

Diabetes disturbs the normal levels of glucose and metabolism in the body. Abnormality in glucose level leads to diabetes. Reduction of insulin is the key source of all kinds of DM. When the pancreas is unable to secrete insulin or secrete little insulin, then it is called insulin deficiency while poor response of cells towards insulin known as insulin resistance. These irregularities then result in metabolic deviations of "lipids," "ketones," "carbohydrates" and "amino acids."

Insulin deficient persons and those with resistance towards insulin having hyperglucagonemia, there is a production of glucose inside the liver, decline in outer glucose uptake and reduction in the formation of glycogen from glucose occurs. Protein and lipid metabolism has a major impact of the deviations in secretion of insulin and glucagon. This stabilizes the lipolytic action of catecholamine, cortisol, growth hormone and a substrate for gluconeogenesis (glycerol) concentration and also decreases frees fatty acids (a substrate for formation of ketone bodies that increased ketogenesis) and a necessary fuel for gluconeogenesis. Oxidation of free fatty acids (FFAs) to acetyl Coenzyme A in the hepatocytes produces ketone bodies and then acetyl Coenzyme A is then converted to acetoacctate and β-hydroxybutyrate and in the absence of proper treatment, might be led to ketonemia and acidosis.

Insulin hydrolyzes triglycerides present in "VLDL or very low density lipoproteins" and chylomicrons, which results in the discharge of "IDL or intermediate-density lipoproteins" particles. The intermediate-density lipoproteins particles are transformed by the liver to the LDL (low-density lipoproteins) which is cholesterol-rich. Therefore, hyperglycemia, and hypercholesterolemia frequently happens in the diabetic patients those remains untreated or under treated. Additionally, insulin deficiency might be associated with increased generation of very low density lipoproteins. Extensive lipolysis in fat tissues resulted in the increase in non-esterified fatty acid availability (Ebbert and Jensen, 2013). Ageing and long-term DM leads to the considerable variations in the metabolism and content of lipid in the liver (Berlanga et al., 2014). Insulin encourages amino acids uptake and protein synthesis during protein metabolism, and in tissues and muscles it terminates protein degradation also. Hence, insulin promotes less absorption of various amino acids in blood. "Alanine" and "glutamine" are the key amino acids mainly associated with gluconeogenesis. Euglycemia is a condition in which insulin reduces the alanine concentration while increment in alanine consumption leads to the reduction in alanine level of marginal tissues. In uninhibited hyperglycemia, there is an increment in the conversion of alanine to glucose which is causing an increase in the rate of gluconeogenesis.

Stiffening of the vessel's basement membrane and many other vascular variations are the uniqueness of pathogenesis of DM. The accumulative effect is the progressive contraction of the vessel lumina which results insufficient perfusion in some serious areas of some body organs. In many blood vessels matrix gets expanded in the retinal basement membrane and in the renal glomerulus mesangial cells (Kruegel and Miosge, 2010).

2.9 EFFECT OF DIABETES ON IMMUNE SYSTEM

Determination of a strong and systematized routine for the nourishment and disease free life directly linked with a system named as immune system (Arora et al., 2013). There are number of cells involved in this process and interconnect with each other to promote the working ability of system (Xie and Davis, 2013). During all type of diabetes in humans, certain changes in functioning of T-lymphocytes leads to infections of urinary tract, respiratory tract, skin, and mucous membrane (Sakowicz-Burkiewicz and Pawelczyk, 2011).

2.10 EFFECT OF DIABETES ON T-LYMPHOCYTES

During diabetes, change in leading complex of immune cells occurs due to establish functions of lymphocytes as they directly promote the hyperglycemia or hypoinsulinemia. "Insulin-dependent diabetes mellitus" inhibits the proliferative reaction of CD4+ T-cells against crucial antigens (Schloot et al., 1997; Eibl et al., 2002). Various studies on diabetic patients revealed that peripheral blood mononuclear cells shows a reduction in the basal formation of cytokines (Pickup et al., 2000). In diabetic patients, increased rate of lymphocyte apoptosis has been observed (Otton et al., 2010).

2.11 FUNCTIONS OF THE IMMUNITY IN THE PATHOGENESIS OF DIABETIC NEPHROPATHY

In adults, "diabetic nephropathy" is the chief communal reason behind the end stage kidney disease (ESRD) and around 50% of chronic renal disease (CKD) (Collins et al., 2012). Around 15–40% of diabetic patients is suffered from kidney related complications (Ritz and Orth, 1999; de Boer et al., 2011). Hyperfiltration has been affected in diabetic nephropathy, as deviations such as apoptosis, accumulation of Mg and ECM (extracellular matrix) in basement membrane take place in glomerulus (Fioretto et al., 1995; Pagtalunan et al., 1997). Changes in glomerulus then cause microalbuminuria, reduction in glomerular filtration and lastly ESRD (Kanwar et al., 2005; Pambianco et al., 2006).

2.12 EFFECT OF T-LYMPHOCYTES IN DIABETIC NEPHROPATHY

Presence of stimulated T-cells in the kidney of diabetic patient might be helpful in increment of proteinuria (protein in urine). In T1DM patients, highly suffered from proteinuria, accumulation of activated T-cells in the renal parenchyma tissue has been observed (Mensah-Brown et al., 2005; Moon et al., 2012). Penetration of T-cells inside the diabetic kidney support the generation of pro-inflammatory cytokines mainly "TNF-α," "interleukin-6" and "interferon-γ," these T cells then cause destruction and inflammation in local tissues and ultimately increment in diameter of glomerular and albuminuria (Eller et al., 2011; Moon et al., 2012).

2.13 COMPLEMENT SYSTEM AND DIABETIC NEPHROPATHY

Increment in glucose levels, i.e., hyperglycemia, prompts the development of intracellular ROS (reactive oxygen species) activates complement (Ostergaard et al., 2005). There is a reduction in mesangial development and proteinuria by complement inhibitor in rats in substantial means. Hence, it was clear that cascade is activated through damaged glomeruli, which results in diabetic glomerulosclerosis impairment.

2.14 DIABETIC NEPHROPATHY

Anti-insulin antibodies are the auto-antibodies which have been recognized in diabetic persons (Silva et al., 2011). Inside the basement membrane of glomeruli, accumulation of auto-antibodies has taken place. These auto-antibodies construction also stimulates the inflammation and damages in the configurations of cells by activating complement and recruitment of macrophages. In primary detection of diabetes, these auto-antibodies play a major role as they deliver valued, non-offensive serum biomarkers (Figure 2.2).

2.15 EPIDEMIOLOGY

Epidemiologic studies evidenced that in the whole world 150 million persons are suffering from diabetes and it might reach up to 300 million or more than 300 million up to 2025 (Bastaki, 2005). Burke et al. (2002) estimated that almost 10% of the population is affected with DM in the world. WHO examined that above 180 million people suffer from diabetes in the world (Petersen, 2008). Diabetes has now become the first leading cause of mortality between chronic degenerative diseases and estimated 16.7% death.

In addition to hyperglycemia, it also causes hyperlipidemia (Pushparaj et al., 2000, 2001), oxidative stress, excessive urination (polyuria), elevation in ketone bodies (ketosis), dysfunctioning of nerves (neuropathy), damages to the nephron (nephropathy) and various heart disease. In low-middle income countries (LMIC) about 200 million people are suffering from diabetes. In 2008, diabetes has been led to about 1.256 million deaths in the world. "World Health Organization" in 1980 estimated that over 108 million adults were suffering from diabetes in the world, and in year 2014 it has been risen over 422 million. The incidence of diabetes has increased from 4.7–8.5% irrespective of regions. However, most cases of diabetes were reported in poor and middle income countries during the era. Prevalence of diabetes in different countries were reported as follows 3.1% in Africa (1980), 7.1% in 2014 (World Health Organization, 2016). The occurrence of diabetes was observed high in China, i.e., 114.4 million, followed by 72.9 million in India, 30.2 million in United States, Brazil, 12.4 million in Brazil, 12.0 million in Mexico, and 10.3 million in Indonesia (Duarte and Golubnitschaja, 2020). Whereas, regarding age-adjusted incidence, the highest load of the disease was reported in various Pacific Islands countries for instance 30% adults in the Marshall Island, Mauritius (22%), Papua-New Guinea (17.7%), followed by other Arab countries (17%–18% of adults) various countries in the Middle East.

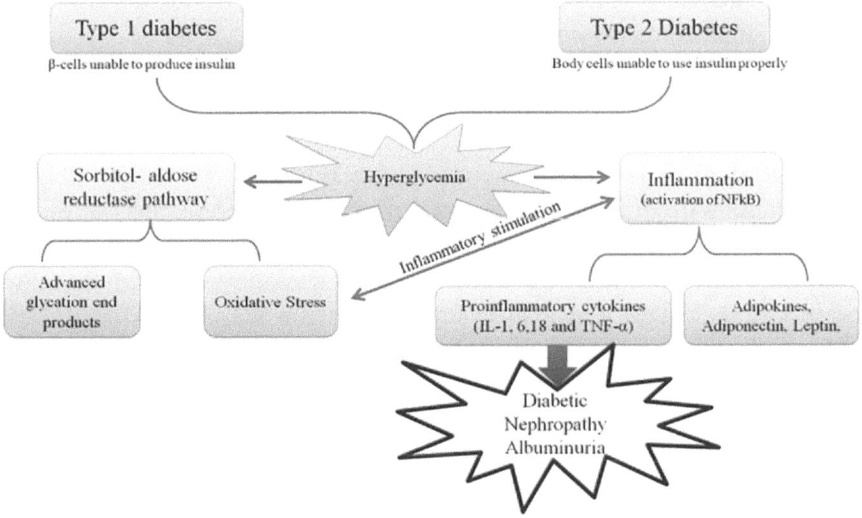

FIGURE 2.2 Diabetic nephropathy caused by T1DM and T2DM.

Accurate extents of diabetes subtypes occurrence are very problematic due to the absence of variation between T1DM and T2DM in maximum overall statistical bases. Assessments revealed a shocking fraction of people with T2DM and about 5% of people have T1DM. In United States different statistical sources assessed the load of diabetes. These assessments depend specially on self-reports like "Behavioral Risk Factor Surveillance System," i.e., phone based system and NHIS (National Health Interview Survey), i.e., in-person interview in which doctor usually make inquiries to person about diabetes and then developed the diagnosed diabetes data sources. The frequency of identified diabetes among grown-ups aged is less than equal to 20 years was 4% in 1980; 6% in 2000, and 8% in 2012, a recurrence in ailment frequency in approx. 30 years (Geiss et al., 2014). In recent times, the above data exhibit noticeable variances in the topographical spreading of diabetes. According to the current assessments in ≥45 year's adults, variance in diabetes is ranging from 11.2% in Colorado to 26.8% in Puerto Rico (Pickens et al., 2018). Likewise, in United States self-testified data have steadily revealed racial alterations in the incidence of diabetes. According to the report of "National Diabetes Statistic" of year 2015, 30 million people expected to have diabetes in which 23% was predictable to be undiagnosed. The occurrence of any ailment is related to the frequency and the period of its existence in living ones. Improvement in screening of disease leads to variations in disease frequency. According to various sources, including potential group studies the occurrence of diabetes mostly T2DM has been growing in the previous 30 years (Geiss et al., 2104; Burke et al., 1999, 2002; Fox et al., 2006).

A survey had taken by "Centers for Disease Control and Prevention's" Division of the 2016 US National Health Interview in which interested participants with diabetes were asked about the different sorts of diabetes they had, i.e., T1DM, T2DM or Gestational diabetes and what medicines were they used. Population were considered to have T1DM those who express both T1DM and insulin use; estimation were made on the basis of these data is that population of United States that has been diagnosed for T1DM is 0.55%, for T2DM it is 8.58%, and for other type of diabetes it is 0.31% (Bullard et al., 2018). In a total population of diabetic people, 6% of them have T1DM while 91% have T2DM. Worldwide prevalence of T1DM has extensive changeability, which is shown in the prevalence ranges from 0.1 per 100,000 every year in China to 36.5 per 100,000 in Finland (Karvonen, 2000). Further, renewed data of China from 2010–2013 directs really low prevalence rate, i.e., 0.93/100,000 (Weng et al., 2018). In the United States, the finest assessments on diabetes in youth come from "SEARCH" for Diabetes in Youth study. The "SEARCH" study is a large observational study examining the prevalence of non-gestational diabetes amongst youths aged 0–19. This study centered the prevalence of type 1 diabetes mellitus in children in 2012 was 21.7% per 100,000. In the middle of 2002 and 2012, there is substantial elevation in the prevalence of both types of diabetes, i.e., T1DM and T2DM among youth (Mayer-Davis et al., 2017). According to the "National Diabetes Statistics Report" of United States, about 1.5 million new cases of diabetes were present there affecting adult and older populations (Centers for Disease Control and Prevention).

2.16 PREVENTION OF DIABETES MELLITUS (DM)

DM can be prevented by adopting four types of categories: (i) primitive prevention, (ii) primary prevention, (iii) secondary prevention, and (iv) tertiary prevention. Primitive prevention involves changes help in blocking of the appearance of the risk factors of the disease. Primordial prevention starts during normal glucose tolerance (NGT) phase, containing healthy nutritional habits and body exercise to maintain normal body weight. Preventions that one should take during the period of pre-diabetes are primary preventions. "Impaired fasting glycemia" and "impaired glucose tolerance" are pre-diabetic stages of DM. Secondary prevention is a preventive approach to diabetes and also prevent it from getting worsen. Secondary preventions help to stop the expansion of enduring impediments of DM. Tertiary preventions are trying to improve the quality of life when complications have already set in. Justification regarding preventive measures of T2DM is related with severe health problems. Cardiovascular risks such as coronary artery disease are also the risk factor of diabetes which further causing blindness, kidney failure and non-traumatic notion. Meanwhile, there is no cure for diabetes, controlling of blood sugar through diet, lifestyle management (LSM) and primary preventions is utmost important. This change would result in a substantial decline in community load and financial expenses. Possibility of prevention of "type 2 diabetes mellitus" has a long symptomless period in history. A number a risk issues are responsible for the development of diabetes. A person showing the presence of one or more risk issues is more susceptible to grow diabetes than the normal one. Simple and tolerant screening tests are available to identify the risk of diabetes. Change in lifestyle will prove effective against the development of DM. “Alcohol consumption,” “mental stress,” “reduced physical activity,” “tobacco smoking,” and “obesity” are the issue factors promoting prevalence of type 2 diabetes mellitus. According to Nurses’ Health study report, obesity, and overweight were major issues promoting type 2 diabetes mellitus. The Relative Risks (RR) for BMI less than 35 kg/m^2 was 38.9 and for body mass index (BMI) of 30.0 to 34.9 kg/m^2 was 20.1, as compared with a BMI < 0.001) when compared with placebo (Karvonen et al., 2000). It is examined that effectiveness of LSM (lifestyle management) to cure diabetes is much more than the drug as it can cause several side effects.

Lifestyle management (LSM), nutritious diet and daily exercise were useful to cure diabetes and also decreases the risk of many other long term ailments diseases like “colon cancer” and “coronary artery disease” (Hu et al., 2001). Survey of LSM (lifestyle management) mediation studies focused on IGT (impaired glucose tolerance) between 3–20 years (Joseph et al., 2016). While the western studies (Burke et al., 1999) directed reduction in weight in obese patients by lifestyle management, the Asian studies (Geiss et al., 2014; Burke et al., 2014) had comparatively thinner patients and the reduction in the weight was not a main contributor to better consequence of lifestyle management. Healthy changes in dietary schedules have positive effects on diabetes. Higher eating of fruits, vegetables, poultry, fish, grains linked with uncertain reduction in threats of T2DM while eating of meat, red meat, fatty dairy products, sweets linked with great threats of T2DM (Willi et al., 2007; Knowler et al., 2002). A survey has conducted by Nurse’s Health Study and found remarkable impact on the prevalence of diabetes via the eating of cereal fiber, trans fats and polyunsaturated fatty acids (Hu et al., 2001). Extensive gap between systematic knowledge taken out from research and in its implementation is the only reason behind diabetes prevalence. An expansive community health outlook is required to regulate the diabetes widespread. The CDC-led “National Diabetes Prevention Program (NDPP)” is an indication-based program that can alter the lifestyle for the prevention of T2DM (Tuomilehto et al., 2001), based on the results of the DPP (daily practice problems). This program has duration of one year and it helps the patients to make changes in the lifestyle including healthy diet, opting physical activity on their everyday routine and problem-solving and management skills. It has been shown that by opting for lifestyle management, 58% of the prevalence of T2DM was reduced and by taking metformin for 2.8 years and 31% metformin follow up for up to 2.8 years. According to 10-year accepted papers, reduction of DM was found in LSM (lifestyle management) group by 34% (Grundy et al., 2004). People with age >60 were found more successful in decrement of risk of diabetes by 49% (Grundy et al., 2004). Even though results were revolutionary, charges of serious patient therapy model in lifestyle management group was higher, i.e., ($1,400/participant) (Gujral et al., 2018). Yet, these results could show a lower cost in a public-based group setting in the YMCA Diabetes Education and Prevention with a Lifestyle Intervention Offered at the YMCA study (Y-DPP), thus decreasing expenditures to $275–325/participant (Bertoni et al., 2010). In comparison with western population, Asian people have high incidence of IGT (impaired glucose tolerance) into diabetes, i.e., 18.3% per year with an increasing incidence 55.0% in 3 years, as compared to Europe (Finnish – 6%; US–11 persons per year) and China (11.3%) (Burke et al., 2002). Weight loss does not consider as a sign of success. In India diabetes treatment costs $ 140–150/ person/year (Joseph et al., 2016). As we know, LSM (lifestyle management) is effective in the prevention of diabetes; it would result in a gain in healthcare funds (Christine et al., 2015). Development of realistic, cost-effective approaches is mandatory to provide benefits to common people. Most recent studies were tested information technology like mobile-phone based short message services for this purpose (Franks and Merino, 2018). Recently, a two year program was conducted in India based on the efficacy of short message service for guiding and inspiring men with impaired glucose tolerance to follow lifestyle management and it was found to be greatly operational with decline in the risks for up to 36% as compared to those patients who had given only standard care (Franks and Merino, 2018).

Improvement in management of non-infectious diseases, development of cost effective, innovative, accessible techniques to screen various diseases to improve public health problems are considered to be fulfilled in the future (Joseph et al., 2016).

2.17 TREATMENT

Treatment starts with the control of sugar levels in both T1DM and T2DM. Treatment is given to maintain sugar level normal by daily monitoring and insulin therapy to the patients having T1DM, while in T2DM, changes in lifestyle, oral medication, reduction in body weight and healthy diet may primarily necessary to decrease the sugar level of blood. "UKPDC (United Kingdom Prospective Diabetes Study)" and "DCCT (Diabetes Control and complication trial)" have revealed the results that dropping glycemic marks show a reduction in number of microvascular difficulties. Though, in early 21st century three clinical trials was done and results of these trial showed a serious glycemic control when examination was done. The results of "The action to control risk in diabetes (ACCORD)," "Action in Diabetes" and "Vascular Disease," "Preterax and Diamicron MR Controlled Evaluation (ADVANCE)" and "Veterans Affairs Diabetes Trial (VADT)" demonstrated that low glycemia cannot significantly decrease the CVDs risks, hence possible risk may balance benefits in high-risk persons (Fowkes et al., 2013). At last, a personalized approach in the duration of disease demographics including age and risk of hypoglycemia, disorders would help the person in suitable blood sugar controlling (American Diabetes Association, 2018). Moreover, proper observation of complications and control in blood pressure and lipid formation is highly suggested. There are chances of T1DM and T2DM can be cured in some conditions.

In T1DM, there are chances of pancreas and islet cell transplantation has been used for patients suffering with disease. Hence, this approaches having substantial checks in attaining successful engraftment and limited availability of donors with immune-suppression complications (Bruniet al., 2014). In T2DM, a drop in blood sugar level was attained in a number of sugar patients, connected with a proper lifestyle changes (Gregg et al., 2017), and is usually achieved with bariatric surgery (Cummings and Rubino, 2018).

2.18 ALLOPATHIC FORMULATION

There are a number of allopathic drugs available in the market to cure diabetes. But, continued use of these medications cause several adverse effects which leads to secondary complications like low blood sugar, feeling of vomiting, low blood sodium, diarrhea, headache, increase in body weight, lactic acidosis, lightheadedness, arthritis, dyspepsia, etc. (Table 2.1).

TABLE 2.1 Different Allopathic Drugs, Their Cellular and Physiological Actions

Class	Agents (Route of Administration)	Cellular Mechanism	Physiologic Key Action
Biguanides	Metformin (Glucophage glucophage XR, Fortamet, Glumetza	AMP-kinase pathway activated by these drugs	↓Hepatic glucose production ↓Insulin resistance
Sulfonylureas	Tolbutamide (Orinase) Glimepiride (Amaryl) Glipizide (Glucotrol) Glucotrol XL Glyburide (Micronase, DiaBeta Glyburide, Micronized GlynasePresTab	Closes KATP channels on beta-cell plasma membranes Stimulates Pancreatic β-cells to release Insulin Control blood glucose level	↑Insulin secretion
Thiazolidine-diones (TZDs)	Pioglitazone (Actos) Rosiglitazone (Avandia)	Nuclear transcription factor PPAR activated by these drugs Improve sensitivity to insulin	↑Insulin secretion
GLP1-Analogs	Exenatide (Byetta) Liraglutide (Victoza) Albiglutide (Tenzeum) Dulaglutide (Trelicity)	Activates GLP-1 receptors	↑Insulin secretion (glucose-dependent) ↓Glucagon secretion (glucose-dependent) Slows gastric emptying ↑Satiety
DPP-4 Inhibitors	Alogliptin (Nesina) Sitagliptin (Onglyza) Linagliptin (Tredjenta)	DPP-4 activity is inhibited, postprandial incretin (GLP-1, GIP) concentrations are also elevated by these drugs	↑Insulin secretion (glucose-dependent) ↓Glucagon secretion (glucose-dependent)
α-Glucosidase Inhibitors	Acarbose (Precose)	Inhibits α-glucosidase activity	Absorption of carbohydrates ↓Digestion

2.19 THE ROLE OF GLUCOSE-LOWERING AGENTS IN THE MANAGEMENT OF DIABETES MELLITUS (DM)

Pharmacological treatments in DM is directed only when fasting glucose level go beyond 1,600 mg/L. Oral pills are used for the management of T2DM as they help in reducing the blood glucose levels (BGL). Exercise and dietary changes are not helpful in every patient with T2DM. Hence, oral treatments are recommended to control type 2 diabetes mellitus. Even though preliminary results of oral pills may be effective, but in a number of diabetes patients, these oral hypoglycemic mediators lost their efficiency. *Sulphonylureas* is one of the oral hypoglycemic mediators. It is useful in decreasing blood glucose level by endorsing insulin secretion through beta cells of the pancreas in patients having existing beta-cell function. All cells are finely captivated and their half-life and period of act vary by mediator. Oral treatment comprises first-generation drugs mainly "chlorpropamide (Diabinese)," "tolbutamide (Orinase)," "acetohexamide (Dymelor)" and "tolazamide (Tolinase)," second-generation drugs including "glyburide (Diabeta, Glynase, Micronase, and Pres Tab)" and "glipizide (Glucotrol or Glucotrol XL)" and third-generation drug include "glimepiride (Amaryl)." These agents cause mild hypoglycemia but serious hypoglycemia is less common (Blumenschein and Johannesson, 1996). Sulphonylureas severally instigate pancreatic β cells to secrete insulin. Frequent use of these kind of oral mediators enhance the β cell sensitivity, uptake of glucose by cells and then decrement in gluconeogenesis (Kaur and Valecha, 2014). Patients suffering from type 2 diabetes mellitus, having age less than 40 with less than 3,000 mg/L fasting blood glucose (FBG) level respond best to the agent Sulphonylureas. Only two-third part of total patients react positively towards sulphonylureas therapy.

On the other hand, 20% of the patients also need supplementary treatments. Switching from one sulphonylurea mediator to other, can be effective in some patients with uninhibited DM. The utilization of agent chlorpropamide, with a longer half-life in the elderly patients also have impairment in kidneys is dejected because the hypoglycemia risk is amplified (Srinivas et al., 2014). Furthermore, sulphonylureas also have some side effects including "hypothyroidism," "cardiovascular disorders," "hematological disorders," "cholestatic jaundice," "hyponatremia," "skin infections" and "water retention." Sulphonylureas therapy also linked with weight gain as it stimulates hyperinsulinemia and these complications then associated with secondary drug failure. Biguanides mediators proliferate the insulin sensitivity by reducing the gluconeogenesis of liver, combined skeletal muscle glucose uptake, decreasing triglycerides level in plasma and LDL-Cholesterol levels, and aggregating peripheral insulin sensitivity. Biguanides mediators don't affect the level of insulin. Hence, no weight gain is there. They do not cause hypoglycemia if used alone (Blumenschein and Johannesson, 1996). Metformin is an example of Biguanides and derivative of medicinal plant "*Galega officinalis.*" For managing type 2 diabetes mellitus, one can use it as solitary or in combination with sulphonylureas. According to "Asian Journal of Biochemical and Pharmaceutical Research Issue 2" (Vol. 2), metformin when used as monotherapy is called anti-hyperglycemic because of its hypoglycemic functions. The use of metformin is contradicted in patients who have kidney problems like serum creatinine level in men is 15 mg/L and in women it is 14 mg/L. Use of oral Tagamet helps to reject the clearance of metformin from kidneys and also stimulates its outcome. Patients those are giving anticoagulant drug orally and metformin, need high dose of warfarin to gain antithrombotic effect (Power). Though, metformin have certain adverse effects like loss of muscle strength, lethargy, lightheadedness, dyspnea, vomiting, lactic acidosis and renal toxicity. Acarbose and Miglitol agents are well known to inhibit α-glucosidase. These agents are useful in decreasing the glucose level after meal by delaying the breakdown of polysaccharides and stooping in absorption of simple carbohydrates from the gastrointestinal tract (GIT) (Tom Walley). This enzyme is helpful in digesting carbohydrates in the intestines, therefore, stopping, and making blood glucose peaks less effective. The undigested carbohydrates are now transported to the colon, where these undigested carbohydrates get converted into short-chain fatty acids, CH_4, CO_2, and H_2 (Blumenschein and Johannesson, 1996). Less than 2% of drugs is used or observed as active drugs. These drugs can be used as single or in combination with another mediator such as "sulfonylurea," thiazolidinedione," "metformin" or "insulin." They are advised to take with food to reduce negative impacts (Rohilla).

Dose-dependent liver toxicity is related to this class of drugs so, the examination of liver function tests should be done very carefully in those patients who are getting a high dose of these drugs (for example, ≥50 mg three times daily). Drugs like cholestyramine and Questran are prone to bind with other drugs are taken 2–4 hours despite α-glucosidase inhibitors to avoid interactions of drugs (Srinivas et al., 2014). Though, the major adverse effects are their major side effects are gassiness and diarrhea. Thiazolidinedione are a different type of class of drugs, i.e., "insulin sensitizers" that endorse the uptake of glucose through skeletal muscle

(Sule et al., 2002). They increase the insulin sensitivity in muscles and in the liver, declining triglyceride levels in plasma, due to these reductions there is weight gain and increase in LDL-cholesterol level was observed. These drugs are very expensive. The binding site for these drugs is in the peroxisome there is a receptor named "peroxisome proliferator-activator receptor-γ (PPAR-γ)" is present, which is mainly located on adipocytes (Walley and Davey, 1995). In patient's having T2DM, thiazolidinedione therapy is related to a decrease in circulation. On the other hand, no former study has observed whether the reduction in fat content of liver and FFA concentration in plasma is correlated to upgrade splanchnic glucose uptake (SGU) following treatment with thiazolidinedione in patients with T2DM." Mainly troglitazone, rosiglitazone, and pioglitazone are the drugs that are included. Patients those who are receiving large amount of insulin on daily basis (more than 30 units per day) their DM is still not under control. In March 2000, US Food and Drug Administration was directed that troglitazone can be utilized with diabetic remedies only. Efficacy of pills is lowered in addition to troglitazone administration. Along with the troglitazone, about 150 cases of liver damage are analyzed. During treatment, the functioning of the liver is observed in 8 months and also further analyzed for 4 month subsequently. Rosiglitazone is considered as an insulin disinfector having significant effect to excite assimilation of glucose in adipose tissue or skeletal muscle. Rosiglitazone is decreasing the amount of plasma insulin. Pioglitazone enhances the target cell reaction via insulin unless multiply the emission of insulin from the pancreas or lower the hepatic glucose output or may enlarge the insulin-dependent glucose utility in skeletal muscle and liver. *V. Meglitinides* are the agents which are short-acting insulin medicines. In pancreatic beta-cells meglitinides affects ATP dependent K^+ channels, admit the opening of Ca^+ channels or raise the insulin liberation. During 1998, the first non-sulfonylurea meglitinideis found. A drug repaglinide derivative of benzoic acid also reported in early 1998s. Action mechanism of repaglinide and the corollary profile are same as sulfonylureas. It can be used along with food on a daily basis at 2–4 times. It is an important alternative for sulfa allergy patients. The remedy is utilized as monotherapy along with metformin. And can be measured carefully in elderly patients or the patients having renal or hepatic dysfunction. It has many side effects like weight gain, gastrointestinal disturbances, or hypoglycemia. Insulin therapy: insulin is added up in oral agent in case of glycemic controls inadequate at high doses of oral medications. There are the adverse effects of insulin treatment such as increased body weight or hypoglycemia. Excess insulin treatment can promote the risk of thermogenesis. Type-1 diabetic patients need continuous medication assist glucose utilization. Some insulin arrangements are utilized, such as short, intermediate, or long-acting. From long times, beef pork insulin were used but currently it is replaced by use of recombinant human insulin. A mixture of 70–30 or 50–50 insulin is used, which is commercially produced. The patients having type-1 diabetes need a mixed insulin regimen. Through a multiple injection regimen, patient will increase or decreases their doses in response to immediate blood glucose level prior to food. The reconciliation of doses are significant in patient having renal, or hepatic dysfunction (Table 2.2).

TABLE 2.2 List of Medicines Having Hypoglycemic Effect

SL. No.	Medications Stimulate Hypoglycemic Effects of Insulin	Medications Prevent Hypoglycemic Effects of Insulin
1.	Tetracyclines	Acetazolamide
2.	Alcohol	AIDS antivirals
3.	ACE inhibitors	Asparaginase
4.	Calcium	Phenytoin
5.	β-blockers	Nicotine
6.	Lithium carbonate	Isoniazid
7.	Anabolic steroids	Diltiazem
8.	Pyridoxine	Diuretics
9.	Alicylates	Corticosteroids
10.	Monoamine oxidase inhibitors (MAOIs)	Thiazide diuretics
11.	Mebendazole	Estrogens
12.	Sulfonamides	Calcitonin
13.	Phenylbutazone	Oral contraceptives
14.	Chloroquine	Diazonide
15.	Clofibrate	Phenothiazines
16.	Fenfluramine	Cyclophosphamide
17.	Guanethidine	Dextrothyroxine
18.	Octreotide	Epinephrine
19.	Pentamidine	Morphine sulfate
20.	Sulfinpyrazone	Niacin

2.20 AYURVEDIC MEDICINES

Ayurvedic Medicines are used worldwide about 70% obtained from medicinal plants. Ayurvedic elaboration is utilized to cure DM and also it has good market value in India there are four Ayurvedic formulation is utilized by most of the Indian people until the primordial period. Recently, people incline to Ayurvedic formulation because of tendency with regards to natural sources or healthy life style. "Allopathic medicines" are very costly, having side effects and complexity, so people go along with the alternative therapies or Ayurvedic medicines because it is affordable or have no side effects. Many diseases are cured by Ayurvedic formulations, DM also which called madhumeha in Ayurveda. It is a metabolic disorder described by "carbohydrate," "lipid," or "protein metabolism." About 1.3% people are enduring world widely by this disease. Mostly the hypoglycemic agent or hypolipidemics are used in allopathic practice for the DM treatment and in case of "hyperlipidemia" many side effects forms when used in excess for long period of time. Hence, effective research is required to overcome these problems.

2.21 ACCORDING TO AYURVEDA

Madhumeha is the ancient ayurvedic remedies among all the therapies, also in prameha. Pramehas is a urinary disorder. The main cause of prameha is less exercise or unbalanced food, categorized in ushna, snigdha, or guru, examples are fish and curd. Meal which enhance "Kapha," "Medhas," or "Moothra" are etiological factors for "Prameha." Pra means excess and Meha means excretion of urine. The prameha is the flow of extreme urine and it is turbid in color. Madhumeha known as DM is considered as worldwide problem in modern science. "Diabetes mellitus" is a major disease which affects almost all body parts and cells of human physiology. Sweetness of urine is one of the major symptoms, also having the connection along with 5 sheaths of body such as Anamaya Kosha, Pranamaya Kosha, Manomaya Kosha, Vijnanamaya Kosha, or Anandamaya Kosha reported by Indian physician. Various classical study reports prameha. It is obtained from root mihsechane means watering (dilution of everything in the body except urine) (Srinivas et al., 2014) (Table 2.3).

TABLE 2.3 Showing the Classification of Prameha

Sahaja (Hereditary)	Apathyanimittaja (unpleasant things like food and exercise, etc.)
According to the physical management	
lean diabetic patients is described under Apatharpanauthajaprameha.	Obese diabetic patients are described under Santharpanauthajaprameha.
According to the Doshic causes	
Vatajaprameha**s**: There are total 4 vatajapramehas.	
Pittajapramehas: There are total 6 pittajapramehas.	
Kaphajapramehas: There are total 10 kaphajapramehas,	
(DM is termed as "Madhumeha")	
Samprapthighatakas: (good things for disease)	
Dosha (humur): Vata, Pitta, Kapha	
Dushya: Meda, Mamsa, Kleda, Rakta, Vasa, Majja, Lasika, Rasa, and Ojas	
Srotas (channel): Mootravaha	
Srotodusti: Atipravrutti	
Agni: Dhatvagni	
Udhbhavasthana: Kostha	
Vyaktasthana: Mootravahasrotas (urinary tract)	

2.22 ROLE OF PLANTS FOR THE TREATMENT OF DIABETES

There are various medicinal plants and their parts which are used to treat diabetes. These plants have various medicinal properties as discussed in Figure 2.3.

2.22.1 FICUS RELIGIOSA

Ficus religiosa or (holy fig) is a species of fig comes under the family Moraceae. It is commonly called the peepal tree in India. The tree is useful to cure DM in traditional medicine. On the basis of various studies, *Ficus religiosa* has revealed large scale in vivo and in vitro pharmaceutical properties including hypolipidemic, antidiabetic, anti-inflammatory, pain-relieving, anti-convulsant, anti-viral, anti-microbial, anti-tumor, antioxidant, anti-helmintic, anti-anxiety, anti-ulcer, immunomodulatory, antiasthmatic, apoptosis inducer, cognitive enhancer, antihypertensive, endothelin receptor antagonist and estrogenic (Singh et al., 2011).

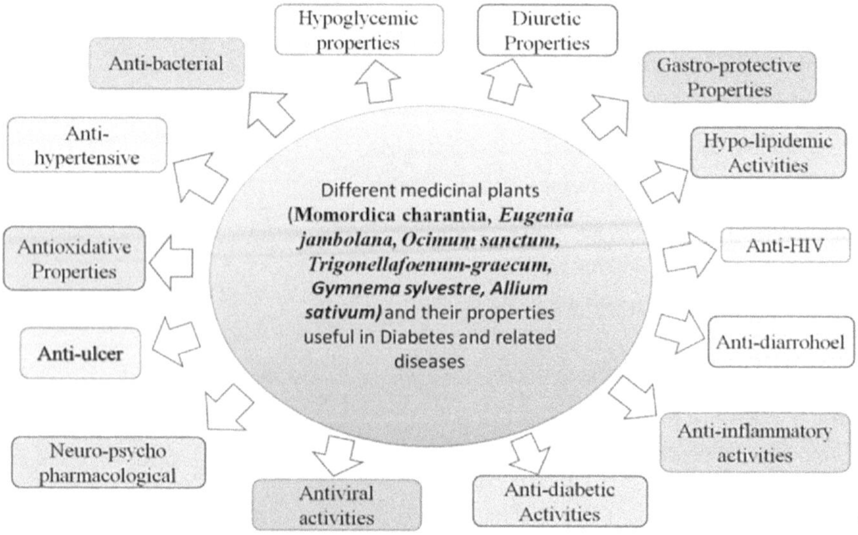

FIGURE 2.3 Various medicinal properties of plants.

Tree bark has been used to prepare decoction to manage diabetes. Several Phytoconstituents like polyphenols, saponins, tannins, sterols, and flavonoids are found in the plant. Plant bark includes sitosterol-d-glucoside which is thought to stimulate hypoglycemic action in rabbits (Ambike and Rao, 1967). Most common bio-active ingredients of *Ficus religiosa* are "Leucocyandin 3-O-beta-d-galactosyl cellobioside" and "leucopelargonidin 3-O-alpha-L rhamnoside" (Bnouham et al., 2006; Ayodhya et al., 2010). The phytoconstituents existent in Peeple tree can report a substantial antidiabetic effect. Tannins, phytosterols, flavonoids, and furanocoumarin products, viz., bergapten, and bergaptol have been found in *Ficus* when tested (Swami and Bisht, 1996). The leaves of *Ficus religiosa* have also been used for anti-hyperglycemic activity (Deshmukh et al., 2007). 21 Days oral assimilation of aqueous extract of different plant parts of *Ficus* instigated a major let down in blood glucose/sugar levels, also amplified insulin levels has been identified. *Ficus religiosa* plant extract helps to maintain muscle and liver glycogen levels in the body as DM is responsible for their decrement (Bouche et al., 2004; Grover et al., 2002). Certain other complications of DM including "hypercholesteremia" and "hypertriglyceridemia" were brought into being decrement through considerably reduction in the triglycerides of serum and total cholesterol concentrations in streptozocin-diabetic rats (Ananthan et al., 2003). Studies has revealed that intake of aqueous plant bark extract at the dose of 500 mg/kg is capable to recover blood glucose concentrations, lipid, and liver enzymes concentrations in streptozocin-induced diabetic rats (Gayathri and Kannabiran, 2008). One of the major complications of T2DM is generation of oxidative stress. *Ficus religiosa* has testified to restrain the enzymes of antioxidant defense system for conflicting oxidative stress. Antioxidant properties of *Ficus religiosa* include rebuilding of glutathione (GSH) and preventing the amount of malondialdehyde (Kirana et al., 2009).

2.22.2 EUGENIA JAMBOLANA

Eugenia jambolana comes under the family Myrtaceae. Bark, fruits, leaves, and seeds are the different plant parts that are used mostly. Tree has smooth, shiny turpentine-smell of leaves. Plant bear forked trunk with scaly gray bark. Plants have aromatic white flowers along with purplish-black edible fruits/berries. Each berry has only single seed inside it. From the ancient times, Eugenia jambolana was grown in India and other parts of South Asia. It has been reported that *Eugenia jambolana* was used in many equivalent and substitute medicine systems of India. It was used as a main antidiabetic drug not only in India but in Europe also. Jamun seeds are boiled in water and used in many drinks in various traditional systems of medicine (Baliga et al., 2013). Various types of bioactive components is present in this plant which includes "anthocyanins," "kaempferol," and "hydrolysable tannins (1-0-galloyl castalagin and casuarinin)," etc. Alkaloid jambosine and glycoside jamboline are also present in the seeds of *Eugenia jambolana* (Ayyanar et al., 2012). Betulinic acid, Bsitosterol, ellagic, and gallic acid are present in the bark of plant (Chaudhuri et al., 1996), bergenin (Bhatia and Bajaj, 1975), tannins (Srivastava, 1953), and flavonoids. Plant berries contain glucose, fructose, raffinose (Lewis et al., 1956), malic acid (Ravi et al., 2005), and anthocyanins (Grover et al., 2000); acylated flavonol glycosides are present in huge amount in its leaves (Bhargava et al., 1972), quercetin shows hypoglycemic activity.

It was reported that its effect of lowering the blood glucose level may be due to elevation in the insulin secretion from beta cells or inhibition of insulin secretion (Aybar et al., 2001). It was also reported that it is responsible in lowering the effect of lipid confirmed by a decrease in the level of cholesterol in blood, decrease in triglyceride level (Sagrawat et al., 2006). It is because of the presence of various flavonoids, saponins, and glycosides in its extract and it was reported that the activity of *3*-hydroxy-*3*-methylglutaryl coenzyme A reductase in liver reduced (Ravi et al., 2004). In the seed extract of *Eugenia jambolana* there is a presence of ellagic acid and decrease the blood pressure. It was reported that alloxan-induced diabetic rats when treated with ethanolic seed extract of *Eugenia jambolana* showed a reduction in the level of blood sugar and progress in pancreatic islet histopathology (Singh and Gupta, 2007). It has also been reported that glycosuria and blood urea level was also reduced. Various studies were also done on dogs and rabbits were showed the same kind of results (Shrotri, 1963; Kedar and Chakrabarti, 1983). Fruit juice of *Eugenia jambolana* was reported to have diuretic properties, and it has also been providing gentle effect to the human digestive system (Chaturvedi et al., 2007). Its seeds have also shown the gastroprotective effect. Flavonoids are present in its seed extract and it could also have protective activity against gastric ulcer (Chaturvedi et al., 2007). It has been reported that it has antiviral activity against influenza virus and goat pox (Sood et al., 2012; Bhanuprakash et al., 2008). Preclinical and clinical trial has also been done on the efficacy of *Eugenia jambolana* for studying various properties like (Sharma et al., 2003; Kopanski and Schnelle, 1988) hypolipidemic (Chakrabarty et al., 1987), anti-inflammatory, (Chaturvedi et al., 2009), neuropsycho-pharmacological (Bhuiyan et al., 1996), antiulcer (Kusumoto et al., 1995), antibacterial, anti-HIV (Cirqueira and Alves, 2005) antidiarrheal and antihypertensive activities (Bhuiyan et al., 1996).

2.22.3 MOMORDICA CHARANTIA

Common name of "*Momordica charantia*" is Karela. It comes under the family Cucurbitaceae. Whole plant's fruit or mainly seeds of the fruit are commonly used plant parts for medicinal purposes. *Momordica charantia* is widely used for the purpose of treatment of diabetes. It also used to treat CVDs and other complications among inhabitants of East Africa, South America, and Asia. Karela is commonly used as a vegetable in food. Bitter gourd has antidiabetic properties because of the presence of bio-active compounds including vicine, charantin, triterpenoids, and some antioxidants (Krawinkel and Keding, 2006). Antidiabetic and hypoglycemic effects of *Momordica charantia* was shown in various preclinical trials (Leung et al., 2009). It has been reported that *Momordica charantia* showed antidiabetic, anti-cancer, anti-viral, and anti-bacterial properties (Grover and Yadav, 2004; Basch et al., 2003). Different animal studies have been reported through bitter guard hence, promoting metabolic syndrome in which diabetes is the major issue (Gadang et al., 2011; Shih et al., 2009; Wang et al., 2011). A research among Taiwanese grown persons has been taken and resulted in decline in circumference of waist, expansion in diabetes, symptoms of metabolic syndrome (Tsai et al., 2012). It has also been observed that bitter gourd could also acts as hypoglycemic and lipid-lowering. *Momordica charantia* may have the potential to repair beta cells thus stimulate insulin secretion (Saxena and Vikram, 2004; Chaturvedi, 2012) and could also improve the insulin sensitivity (Wang et al., 2011). It limits the glucose absorption by stopping glucosidase and also destroys the disaccharidases activity in the intestine (Chaturvedi, 2012). It was also reported that in normal or streptozotocin-induced diabetic rat's ethanolic extract of *Momordica charantia* showed anti-hyperglycemic effect (Shibib et al., 1993). Triterpenoids are the hypoglycemic components that are present in karela may be responsible for the AMP-activated protein kinase (Cheng et al., 2008). It has also been reported in several animal models that it showed the blood-glucose-lowering activity (Malik et al., 2011). Bitter gourd is also helpful in reducing the body weight or obesity. It also declines the epididymal weight and retroperitoneal white adipose tissues (Basch et al., 2004). For the enhancement of skeletal muscle strength, it was found very effective (Gadang et al., 2011). However, after injecting the dose of *Momordica charantia*, it was found that mice loaded with glycogen showed inhibited effect on elevating postprandial blood glucose level. α-amylase and α-glucosidase may be responsible for stimulating the secretion of insulin (Figure 2.4) (Keller et al., 2011).

2.22.4 OCIMUM SANCTUM

Ocimum sanctum is also called tulsi plant comes under the family Lamiaceae. Each and every plant part has therapeutic uses and is also used to cure various diseases. Tulsi is cultivated throughout the world. Chemical and nutritional configurations of plant increase its capabilities. Medicinal properties of *Ocimum sanctum* is because of the presence of active compound Eugenol (Pattanayak et al., 2010). Tulsi consist of various bio-active complexes including vicenin, orientin, cirsimaritin, isothymonin, isothymusin, apigenin, rosmarinic acid, triterpenoids, tannins, saponins, and flavonoids. Additionally, extracted oil from tulsi leaves contain "eugenol," "ursolic acid," "carvacrol," "linalool," "limatrol," and "caryophyllene" whereas seeds oil constitute bio-active compounds including fatty acids and sitosterolbut seed mucilage comprises more or less sugars. *Ocimum sanctum* leaves possess anthocyanins. Furthermore, phytoconstituents such as vitamins, minerals, chlorophyll, and many other bio-active complexes are present inside the tulsi plant. In Ayurveda tulsi has recognized as a very valuable antidiabetic plant (Khan et al., 2012). In streptozotocin-induced diabetic rats' reduction in blood glucose, glycosylated hemoglobin, elevation in glycogen, hemoglobin, and protein has also been found (Narendhirakannan et al., 2006). *Ocimum sanctum* L. leaf extract has been testified to incite the biological passageways of insulin emission (Hannan et al., 2006). For a period of 30 days, *Ocimum sanctum* extract was given to normal rats and results exhibit the reduction of blood sugar concentrations (Nair et al., 2005). Tulsi plants show antiperoxidative effect which is responsible for the reduction in the serum level of cortisol and glucose in male mice (Gholap and Kar, 2004). In normal rats, it has been recorded that the exogenous insulin action increases (Chattopadhyay, 1993). Dyslipidemia has also been resolved with the use of Tulsi plant extract as reported by Singh et al. (2012).

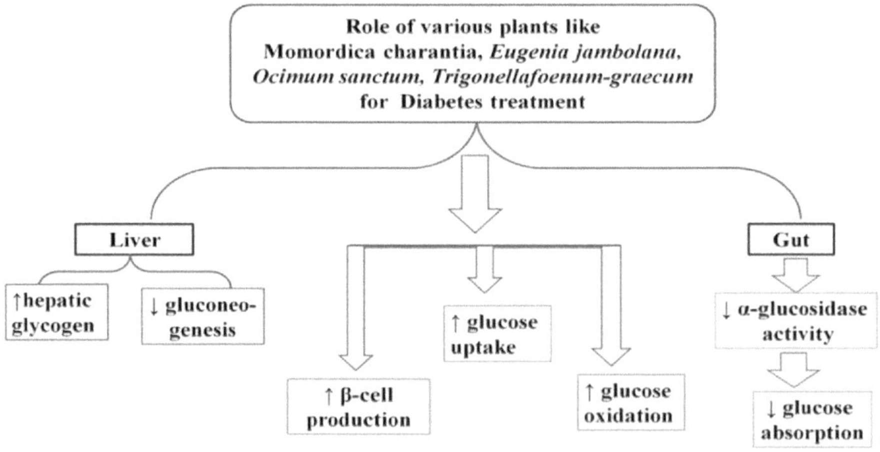

FIGURE 2.4 Potential of different plants in elevating different biochemical attributes with respect to diabetes.

Chloroform extracts of above ground, i.e., areal parts of holy basil have been able to improve the instabilities in metabolism of lipids instigated because of DM in alloxan-influenced diabetic rats. Expressive reduction in the elevated levels of serum glucose, triglycerides, cholesterol, and LDL standards can be done with extract (Patil et al., 2011). The hydroalcoholic extract of *Ocimum sanctum* L. inhibit the prolonged sturdy stress induced rise in plasma cAMP amounts, myocardial superoxide dismutase, and catalase activities (Sood et al., 2006). It has been reported that Ursolic acid isolated from *Ocimum sanctum* L. protect heart cells from lipid peroxidation induced by Adriamycin (Balanehru and Nagarajan, 1992). A noticeable reduction in serum cholesterol, triacylglycerol, and LDL + VLDL cholesterol as linked to untreated cholesterol-fed group was experimental in cholesterol-fed rabbits when accompanied with *Ocimum sanctum* L. seed oil for four weeks (SallesTrevisan et al., 2006). It shows antidiabetic, cardioprotective effects, antifungal, antimicrobial effects also (Singh et al., 2011), analgesic (Khanna and Bhatia, 2003), anti-helmintic (Asha et al., 2001), anti-stress (Modak et al., 2007), anti-fertility (Ahmed et al., 2002), anti-inflammatory (Kelm et al., 2000), antioxidant (Salles et al., 2006; Yanpallewar et al., 2004), gastro-protective (Goel et al., 2005), immunomodulatory (Mukherjee et al., 2005), anti-thyroidic (Panda and Kar, 1998) anticancer (Prakash and Gupta, 2000) and radioprotective effects (Bhartiya et al., 2006; Subramanian et al., 20005). Tulsi is attested to deliver a protection for the nervous system (Subramanian et al., 2005) and against sexually transmitted diseases (Shokeen et al., 2005).

2.22.5 *PTEROCARPUS MARSUPIUM*

Pterocarpus marsupium also called Bijasar. This plant comes under the family Fabaceae. The whole plant is useful but generally used parts including its heartwood, flowers, leaves, bark, and gum. It is found in India, Sri Lanka, and Nepal. By means of Ayurveda, *Pterocarpus marsupium*is one of the multipurpose medicinal plant having an extensive range of natural actions. Different plant parts of kino tree have been recognized for their huge pharmaceutical activities. *Pterocarpus marsupium* is rich in phytochemicals hence; act as a good source of polyphenols. This plant also comprises phenolic and terpenoids complexes including epicatechins, aurone glycosides, iso-flavonoids, lupenol, and β sitosterol (Pterocarpus marsupium, 1983; Rao et al., 2001). Plant also shows antidiabetic activities (Ambujakshi and Ganapaty, 2011). In addition, with antidiabetic properties, this tree has also been reported effective in contrast to various diseases. It has anti-hyperlipidemic, antiobesity (Hougee et al., 2005), anti-helmintic, antinflamatory (Salunkhe et al., 2005; Hougee et al., 2005), anti-tumor, antioxidative, and antiulcerative properties (Nair et al., 2005; Grover et al., 2005). It has been estimated that *Pterocarpus marsupium* protects β cell and also have regenerative properties (Chakravarthy et al., 1981). During various inspected models of diabetes, plant has found effective as it promotes rebuilding of insulin in blood by rejuvenating beta cells of the pancreas (Manickam et al., 1997; Chakravarthy et al., 1982). Augmentation of methanolic extract of *Pterocarpus marsupium* in streptozocin-diabetic rats for a period of 7 to 14 days exhibited stabilization in streotozocin-distressed serum glucose levels (Gupta and Gupta, 2009). Tannate, is the bio-active compound present in the plant extract which is recognized to promote reduction in blood sugar levels. Marsupin, pterosupin, and liquiritigenin present in the plant subsidized to promote antihyperlipidemic activities in living organisms (Farboodniay et al., 1993). Epicatechin has been discovered to have insulinogenic possessions by improving insulin proclamation and transformation of proinsulin into insulin. It has been reported that insulin-like activity was also shown by Epicatechin present in *Pterocarpus marsupium* (Rizvi and Zaid, 2001). Activation of important protein was done by Epicatechin thereby strengthen the insulin signaling and regulating glucose production through adenosine monophosphate-activated protein kinase (Cordero-Herrera et al., 2013).

2.22.6 TRIGONELLAFOENUM-GRAECUM

Trigonellafoenum-graecum also called fenugreek or methi comes under the family Fabaceae. Plant is cultivated worldwide as a semiarid crop. Methi has been used as a vegetable and a spice as well, due to its multipurpose properties. Also, Fenugreek gives very strong aroma and is used as a spice in food (Pandey et al., 2011). Strong antidiabetic properties of fenugreek have been demonstrated during the different animal model experimental trials (Kumar et al., 2012; Roberts et al., 2011). Various experimental studies have been verified the use of methi in the decrease in glucose and lipid levels in blood (Sharma et al., 1990). Numerous observations have been confirmed in humans and other diabetic animals that different plant parts of fenugreek, including seed extract, mucilage of seeds, and leaves can help in cholesterol and glucose level reduction (Gupta et al., 2001; Vats et al., 2002). Presence of saponins (Petit et al., 1995), trigonelline alkaloid (Raghuram et al., 1994), 4-hydroxyleucine (Sauvaire et al., 1998) and higher fiber content (Ali et al., 1995) mainly promotes the pharmaceutical properties of fenugreek. The anti-hyperglycemic effect of the plant related to the reduction in the level of plasma glucagon and somatostatin (Ribes et al., 1986). It has been estimated that methi seed powder has the potential to stabilize the action of skeletal muscles, creatinine kinase of liver and hearts of diabetic rats (Genet et al., 1999). 4-hydroxyisoleucine, showed anti-hyperglycemic effect, it is an amino acid present in fenugreek and it also promote glucose uptake and insulin sensitivity in peripheral tissues (Singh et al., 2010). It has been estimated that steroids present in fenugreek promote reduction in BGL in diabetic rats (Hamden et al., 2010). A significant increase in the area of insulin-immunoreactive beta cells has been observed in studies (Hamden et al., 2010). According to a study on renal and intestinal disaccharides action in streptozocin-induced rats, fenugreek seed proves effective in reduction of maltase action during diabetes (Suresh Kumar et al., 2005). A noticeable decrease in kidney diseases has been observed when fenugreek oil is assimilated in the food of alloxanized rats (Suresh et al., 2005).

2.22.7 GYMNEMA SYLVESTRE

Gymnema sylvesrtre (gurmar) is a woody climbing shrub comes under family Asclepiadaceae. It is a native plant of tropical forests in India, Africa, Australia, and Sri Lanka. This woody plant has huge climbing properties with the presence of roots at nodes. Plant has been used in ayurvedic formulations due to its effective antidiabetic properties. Antidiabetic capabilities of plants have been observed in animal models (Suresh et al., 2005), also collective use of herb with acarbose is reported to decline maltose transport in intestines of rats (Luo et al., 2001). In rats, there is a reduction in the concentration of free oleic acid (Wang et al., 1998). Aqueous plant extract has been attested and it results reversible increment in intracellular secretions of insulin and calcium from the suffered β cells of human and rat with type 2 diabetes (Asare-Anane et al., 2005). Rejuvenation in pancreatic cells might be helpful in raising insulin levels (Persaud et al., 1999). In mice, *Gymnema sylvestre* promotes reduction in BGL by inhibiting stimulation of liver by adrenal hormones, as the liver produces and raises the glucose concentration (Gholap and Kar, 2003). *Gymnema sylvestre* contains triterpene saponins including gymnema saponins and gymnemic acids. These saponins are responsible for the testified pharmaceutical properties of the plant. Oral intake of

Gymnema is observed to be effective in contrast to long term inflammatory problems (Leach, 2007), obesity (Preuss et al., 2004; Cefalu et al., 2008) and also work against various dysfunctions of beta cells of pancreas (Al-Romaiyan et al., 2010). Against alloxan-induced diabetic albino male rats the suspension of *Gymnema sylvestre* showed significant potential toward diabetes (Karthic et al., 2012). Hypoglycemic effect of ethanolic decoction of *Gymnema sylvestre* has been observed as it arouse insulin secretion from the beta cells of the pancreas (Persaud et al., 1999; Liu et al., 2009; Saltiel and Kahn, 2001). Studies observed that plant extract is able to maintain association between normal phospholipid levels and glycemic control (Jain et al., 1998). Studies have observed that oral assimilation of *Gymnema sylvestre* in rats promote consumption of blood glucose by reduction in movement of fat (Karthic et al., 2012). Changes including substantial decrease in body weight, hemoglobin levels and plasma proteins have been observed (Karthic et al., 2012).

2.22.8 ALLIUM SATIVUM

Allium sativum or garlic comes under family Amaryllidaceae, commonly called LaShun in India. Generally used parts of the garlic are its leaves and bulb. According to Ayurveda, garlic is an incredible plant which used in contrast to various complications mainly tumors, headache, insect bites, intestinal worms (Block et al., 1985). From a long time, garlic is used as an established medicine to control diabetes, fungal disease, cardiac diseases, and cancer (Ali et al., 2000). Typically, present bio-active compounds in garlic are ajoene, allixin, allicin, and organosulfur complexes. Presence of organosulfur complexes in garlic promote pharmaceutical characters (Augusti and Mathew, 1974). Presence of bio-active compounds believed to show different biological functions comprising antimicrobial properties, cholesterol, and blood glucose-lowering properties as well as cancer anticipation (Rose et al., 2005). Various studies have demonstrated that, intake of garlic induces dropping down of fasting blood sugar (FBS)/glucose concentrations (Thomson et al., 2012). Diallyl trisulfide has been attested to develop regulation of BGL in streptozocin-induced diabetic rats (Liu et al., 2005). Garlic juice assimilation mainly improve the use of blood glucose in glucose tolerance tests executed in rabbits, though allicin at a measured quantity of 250 mg/kg remained 60% as effective as tolbutamide in alloxan-induced diabetic rabbits (Mathew and Augusti, 1973).

Garlic promotes the emission of insulin hormone from the beta cells of the pancreas. Hence, it might have the antidiabetic means (Jain and Vyas, 1975). Allicin is thought to boost serum insulin by linking with cysteine and sparing it from SH group reactions (Mathew and Augusti, 1973). Garlic constitutes N-acetylcysteine called an organosulfur compound having antioxidant properties which positively affect serum, lipids, and glucose concentrations. It has been studied that N-acetylcysteine decreases the oxidative stress because of improvement in endogenous defenses (Diniz et al., 2006). Garlic promotes hypoglycemic properties due to the presence of Allicin. It is a sulfur containing compound with a pungent flavor. Effects such as increased hepatic metabolism, insulin secretion are believed as a result of the hypoglycemic properties of garlic (Bever and Zahnd, 1979; Al-Awadi and Gumaa, 1987). Regulation of high BGL and lipid peroxidation in rats is due to the S-allyl cysteine sulfoxide the precursor of allicin. Studies have revealed that the use of garlic helps to prevent cardiovascular complications of diabetes (Patumraj et al., 2001). The presence of saponins helps in decrement of serum cholesterol levels (Matsuura, 2000). In case of alloxan-induced diabetic rats, garlic juice represents anti-hyperglycemic and antioxidant effects (El-Demerdash et al., 2005). Anti-oxidative possessions of phytochemicals in garlic can be observed through its scavenging properties on reactive oxygen species (ROS) (Borek et al., 2005) also increment in antioxidant enzymes including CAT or catalase, SOD, or superoxide dismutase and GPx or glutathione peroxidase takes place (Liu et al., 1992). In case of oxidative stress in streptozocin (STZ)-nicotinamide induced diabetic rats, presence of only garlic and with turmeric and ginger exhibited 26–37% increment in the production of insulin and 60–97% decrement in lipid peroxidation (Madkor et al., 2011). Fresh garlic homogenate was observed and it was found to standardize hepatic TBARS (Thiobarbituric acid reactive substances) and GSH concentrations which ultimately progress insulin sensitivity and oxidative stress in fructose-fed rats (Padiya et al., 2011). Several studies investigated the superannuated garlic extract, and blocking in the development of glycation-derived free radicals have been found. Longevous garlic constitute a bio-active compound called S-allyl cysteine is well known for its antioxidant properties to control synthesis of AGEs (Ahmad and Ahmed, 2006). Role of different plants and their bioactive components is enlisted in Table 2.4.

TABLE 2.4 Herbal Treatment for the Diabetes Mellitus

SL. No.	Plant Name	Phytoconstituents (Bioactive Components)	Pharmacological Activity
1.	*Ficus religiosa*	Leucocyandin 3-O-beta-d-galactosyl cellobioside, Leucopelargonidin-3-O-alpha-L rhamnoside	Antidiabetic, hypolipidemic, anti-inflammatory, analgesic, antioxidant, immunomodulatory, estrogenic, endothelin receptor antagonist, apoptosis inducer and cognitive enhancer
2.	*Eugenia jambolana*	Anthocyanins, glucoside, Ellagic acid, Isoquercetin, Kaempferol, Myricetin, and Hydrolysable tannins (1-0-galloyl castalagin and casuarinin)	Hypolipidemic, anti-inflammatory, neuro-psychopharmacological, and antihypertensive
3.	*Momordica charantia*	Vicine, Charantin, and Triterpenoids	Hypoglycemic, lipid-lowering properties, antidiabetic
4.	*Ocimum sanctum*	Eugenol, Ursolic acid, Carvacrol, Linalool, Limatrol, Caryophyllene, Rosmarinic acid, Apigenin, Isothymonin, Orientin, Vicenin	Antidiabetic, cardioprotective, antifertility, anti-inflammatory, antioxidant, immunomodulatory, and antithyroidic
5.	*Pterocarpus marsupium*	*β*-sitosterol, Lupenol, Aurone glycosides, Epicatechins, Iso-flavonoids	Antiobesity, antihyperlipidemic, anti-inflammatory, antioxidative, antitumorigenic, and antidiabetic
6.	*Trigonella foenum-graecum*	4-hydroxyisoleucine, Trigonelline	Antidiabetic, anti-inflammatory, anti-cancer, and antioxidant
7.	*Gymnema sylvestre*	Gymnemic acid, Gymne-marerin, Triterpenoids	Antidiabetic, antifertility, anti-inflammatory, antioxidant, apoptosis inducer and hypolipidemic
8.	*Allium sativum*	Allicin, Allixin, Ajoene	Antidiabetic, anti-inflammatory, and antioxidant

KEYWORDS

- **allopathic**
- **diabetes mellitus**
- **herbal**
- **prevention**
- **treatments**

REFERENCES

Ahmad, M. S., & Ahmed, N., (2006). Antiglycation properties of aged garlic extract: Possible role in prevention of diabetic complications. *J. Nutr., 136,* 796S–799S.

Ahmed, M., Ahamed, R. N., Aladakatti, R. H., & Ghosesawar, M. G., (2002). Reversible anti-fertility effect of benzene extract of *Ocimum sanctum* leaves on sperm parameters and fructose content in rats. *J. Basic Clin. Physiol. Pharmacol., 13,* 51–60.

Al-Awadi, F. M., & Gumaa, K. A., (1987). Studies on the activity of individual plants of an antidiabetic plant mixture. *Acta Diabetol. Lat., 24,* 37–41.

Ali, L., Khan, A. K. A., Hassan, Z., Mosihuzzaman, M., Nahar, N., Nasreen, T., Nur-e-Alam, M., & Rokeya, B., (1995). Characterization of the hypoglycemic effects of *Trigonella foenumgraecum* seed. *Planta Med., 61,* 358–360.

Ali, M., Thomson, M., & Afzal, M., (2000). Garlic and onions: Their effect on eicosanoid metabolism and its clinical relevance. *Prostaglandins Leukot. Essent. Fatty Acids, 62,* 55–73.

Al-Romaiyan, A., Liu, B., Asare-Anane, H., Maity, C. R., Chatterjee, S. K., Koley, N., Biswas, T., et al., (2010). A novel *Gymnema sylvestre* extract stimulates insulin secretion from human islets *in vivo* and *in vitro*. *Phytotherapy Research, 24,* 1370–1376.

Amarapurkar, D. N., Hashimoto, E., Lesmana, L. A., Sollano, J. D., Chen, P. J., & Goh, K. L., (2007). Asia–Pacific working party on NAFLD1. How common is non-alcoholic fatty liver disease in the Asia–Pacific region and are there local differences? *J. Gastroenterol. Hepatol., 22,* 788–793.

Ambike, S. H., & Rao, M., (1967). Studies on a phytosterol in from the bark of *Ficus religiosa*. *Indian J. Pharm., 29,* 91.

Ambujakshi, H. R., & Ganapaty, S., (2011). Anti-obese activity of *Pterocarpus marsupium* barks extract in experimentally induced obese rats. *Inventi. Impact: Nutraceuticals.*

American Diabetes Association, (2018). Glycemic targets: Standards of medical care in diabetes. *Diabetes Care, 41,* 55–S64.

Ananthan, R., Latha, M., Ramkumar, K. M., Pari, L., Baskar, C., & Bai, V. N., (2003). Effect of *Gymnema montanum* leaves on serum and tissue lipids in alloxan diabetic rats. *Exp. Diabetes Res., 4,* 183–189.

Anjaneyulu, M., & Chopra, K., (2004). Fluoxetine attenuates thermal hyperalgesia through 5-HT1/2 receptors in streptozotocin-induced diabetic mice. *Eur. J. Pharmacol., 497,* 285–292.

Arora, M. K., & Singh, U. K., (2013). Molecular mechanisms in the pathogenesis of diabetic nephropathy: An update. *Curr. Vasc. Pharmacol., 58*, 259–271.

Asare-Anane, H., Huang, G., Amiel, S., Jones, P., & Persaud, S., (2005). Stimulation of insulin secretion by an aqueous extract of *Gymnema sylvestre*: Role of intracellular calcium. In: *196th Meeting of the Society for Endocrinology and Society for Endocrinology Joint Endocrinology and Diabetes Day* (Vol. 10). BioScientifica.

Asha, M. K., Prashanth, D., Murali, B., Padmaja, R., & Amit, A., (2001). Anthelmintic activity of essential oil of *Ocimum sanctum* and eugenol. *Fitoterapia, 72*, 669–670.

Augusti, K. T., & Mathew, P. T., (1974). Lipid lowering effect of allicin (diallyldisulfide-oxide) on long term feeding to normal rats. *Experientia, 30*, 468–470.

Augusti, K. T., & Sheela, C. G., (1996). Anti-peroxide effect of S-allyl cysteine sulfoxide, an insulin secretagogue, in diabetic rats. *Experientia, 52*, 115–119.

Aybar, M. J., Riera, A. N. S., Grau, A., & Sánchez, S. S., (2001). Hypoglycemic effect of the water extract of *Smallantus sonchifolius* (yacon) leaves in normal and diabetic rats. *J. Ethnopharmacol., 74*, 125–132.

Ayodhya, S., Kusum, S., & Anjali, S., (2010). Hypoglycaemic activity of different extracts of various herbal plants. *Int. J. Res. Ayurveda Pharm., 1*, 212–224.

Ayyanar, M., & Subash-Babu, P., (2012). *Syzygium cumini* (L.) skeels: A review of its phytochemical constituents and traditional uses. *Asian Pac. J. Trop. Biomed., 2*, 240–246.

Balanehru, S., & Nagarajan, B., (1992). Intervention of Adriamycin induced free radical damage. *Biochem. Int., 28*, 735–744.

Baliga, M. S., Fernandes, S., Thilakchand, K. R., D'souza, P., & Rao, S., (2013). Scientific validation of the antidiabetic effects of *Syzygium jambolanum* DC (black plum), a traditional medicinal plant of India. *J. Altern. Complement. Med., 19*, 191–197.

Basch, E., Gabardi, S., & Ulbricht, C., (2003). Bitter melon (*Momordi cacharantia*): A review of efficacy and safety. *Am. J. Health-Syst. Pharm., 60*, 356–359.

Bastaki, A., (2005). Diabetes mellitus and its treatment. *Int. J. Diabetes. Metab., 13,* 111.

Berlanga, A., Guiu-Jurado, E., Porras, J. A., & Auguet, T., (2014). Molecular pathways in non-alcoholic fatty liver disease. *Clin. Exp. Gastroenterol., 7,* 221.

Bertoni, A. G., Burke, G. L., Owusu, J. A., Carnethon, M. R., Vaidya, D., Barr, R. G., Jenny, N. S., et al., (2010). Inflammation and the incidence of type 2 diabetes: The multi-ethnic study of atherosclerosis (MESA). *Diabetes Care, 33*, 804–810.

Bever, B. O., & Zahnd, G. R., (1979). Plants with oral hypoglycaemic action. *Int. J. Crude. Drug Res., 17*, 139–196.

Bhanuprakash, V., Hosamani, M., Balamurugan, V., Gandhale, P., Naresh, R., Swarup, D., & Singh, R. K., (2008). In vitro antiviral activity of plant extracts on *Goatpox* virus replication. *IJEB 46*(2), 120–127.

Bhargava, K. K., Dayal, R., & Seshadri, T. R., (1974). Chemical components of *Eugenia jambolana* stem bark. *Curr. Sci.*

Bhartiya, U. S., Raut, Y. S., & Joseph, L. J., (2006). Protective effect of *Ocimum sanctum* L after high-dose 131 iodine exposure in mice: an in vivo study. *Indian Journal of Experimental Biology, 44,* 647–652.

Bhatia, I. S., & Bajaj, K. L., (1975). Chemical constituents of the seeds and bark of *Syzygium cumini. Planta Med., 28*, 346–352.

Bhuiyan, M. S. A., Mia, M. Y., & Rashid, M. A., (1996). Antibacterial principles of the seeds of *Eugenia jambolana. Bangladesh J. Bot., 25*, 239–241.

Block, E., (1985). The chemistry of garlic and onions. *Sci. Am., 252*, 114–121.

Blumenschein, K., & Johannesson, M., (1996). Economic evaluation in healthcare. *Pharmaco. Economics, 10*, 114–122.

Bnouham, M., Ziyyat, A., Mekhfi, H., Tahri, A., & Legssyer, A., (2006). Medicinal plants with potential antidiabetic activity: A review of ten years of herbal medicine research (1990–2000). *Int. J. Diabetes Metab., 14*, 1.

Borek, C., (2001). Antioxidant health effects of aged garlic extract. *J. Nutr., 131*, 1010S–1015S.

Bouche, C., Serdy, S., Kahn, C. R., & Goldfine, A. B., (2004). The cellular fate of glucose and its relevance in type 2 diabetes. *Endocr. Rev., 25*, 807–830.

Bruni, A., Gala-Lopez, B., Pepper, A. R., Abualhassan, N. S., & Shapiro, A. J., (2014). Islet cell transplantation for the treatment of type 1 diabetes: Recent advances and future challenges. *Diabet. Metab. Synd. Ob., 7*, 211.

Bullard, K. M., Cowie, C. C., Lessem, S. E., Saydah, S. H., Menke, A., Geiss, L. S., Orchard, T. J., et al., (2018). Prevalence of diagnosed diabetes in adults by diabetes type—United States, 2016. *Mortal. Wkly. Rep., 67*, 359.

Burke, J. P., O'Brien, P., Ransom, J., Palumbo, P. J., Lydick, E., Yawn, B. P., Joseph, M. III. L., & Leibson, C. L., (2002). Impact of case ascertainment on recent trends in diabetes incidence in Rochester, Minnesota. *Am. J. Epidemiol., 155*, 859–865.

Burke, J. P., Williams, K., Gaskill, S. P., Hazuda, W. P., Haffner, S. M., & Stern, M. P., (1999). Rapid rise in the incidence of type 2 diabetes from 1987 tow 1996: Results from the San Antonio heart study. *Int. Arch. Med., 159*, 1450–1456.

Cefalu, W. T., Ye, J., & Wang, Z. Q., (2008). Efficacy of dietary supplementation with botanicals on carbohydrate metabolism in humans. *Endocrinol. Metab., 8*, 78–81.

Chakraborty, D., Mahapatra, P. K., & Chaudhuri, A. N., (1986). A neuropsychopharmacological study of *Syzygium cuminii*. *Planta Medica, 52*, 139–143.

Chakravarthy, B. K., Gupta, S., & Gode, K. D., (1982). Functional beta cell regeneration in the islets of pancreas in alloxan induced diabetic rats by (−)-epicatechin. *Life Sci., 31*, 2693–2697.

Chakravarthy, B. K., Gupta, S., Gambhir, S. S., & Gode, K. D., (1981). The prophylactic action of (−)-epicatechin against alloxan induced diabetes in rats. *Life Sci., 29*, 2043–2047.

Chattopadhyay, R. R., (1993). Hypoglycemic effect of *Ocimum sanctum* leaf extract in normal and streptozotocin diabetic rats. *Indian J. Exp. Biol., 31*, 891–893.

Chaturvedi, A., Bhawani, G., Agarwal, P. K., Goel, S., Singh, A., & Goel, R. K., (2009). Antidiabetic and antiulcer effects of extract of *Eugenia jambolana* seed in mild diabetic rats: Study on gastric mucosal offensive acid-pepsin secretion. *Indian J. Physiol. Pharmacol., 53*, 137–146.

Chaturvedi, A., Kumar, M. M., Bhawani, G., Chaturvedi, H., Kumar, M., & Goel, R. K., (2007). Effect of ethanolic extract of *Eugenia jambolana* seeds on gastric ulceration and secretion in rats. *Indian J. Physiol. Pharmacol., 51*, 131.

Chaturvedi, P., (2012). Antidiabetic potentials of *Momordi cacharantia*: Multiple mechanisms behind the effects. *J. Med. Food, 15*, 101–107.

Chaudhuri, A. N., Pal, S., Gomes, A., & Bhattacharya, S., (1990). Anti-inflammatory and related actions of *Syzygium cuminii*seed extract. *Phytother. Res., 4*, 5–10.

Cheng, H. L., Huang, H. K., Chang, C. I., Tsai, C. P., & Chou, C. H., (2008). A cell-based screening identifies compounds from the stem of *Momordi cacharantia* that overcome insulin resistance and activate AMP-activated protein kinase. *J. Agric. Food Chem., 56*, 6835–6843.

Christine, P. J., Auchincloss, A. H., Bertoni, A. G., Carnethon, M. R., Sánchez, B. N., Moore, K., Adar, S. D., et al., (2015). Longitudinal associations between neighborhood physical and social environments and incident type 2 diabetes mellitus: The Multi-Ethnic Study of Atherosclerosis (MESA). *JAMA Intern. Med., 175*, 1311–1320.

Cirqueira, R. T., & Alves, M. J. Q. F., (2005). Hypotensive and diuretic effects of pitanga (*Eugenia uniflora* L.) and jambos (*Eugenia jambolana* Lam.) aqueous extracts in normotensive anesthetized rats. *Rev. Bras. Plantas Med.*

Collins, A. J., Foley, R. N., Chavers, B., Gilbertson, D., Herzog, C., Johansen, K., Kasiske, B., et al., (2012). US renal data system 2011 annual data report. *Am. J. Kidney Dis., 59*, A7.

Cordero-Herrera, I., Martín, M. A., Bravo, L., Goya, L., & Ramos, S., (2013). Cocoa flavonoids improve insulin signaling and modulate glucose production via AKT and AMPK in Hep G2 cells. *Mol. Nutr. Food. Res., 57*, 974–985.

Cummings, D. E., & Rubino, F., (2018). Metabolic surgery for the treatment of type 2 diabetes in obese individuals. *Diabetologia, 61*, 257–264.

De Boer, I. H., Rue, T. C., Hall, Y. N., Heagerty, P. J., Weiss, N. S., & Himmelfarb, J., (2011). Temporal trends in the prevalence of diabetic kidney disease in the United States. *JAMA, 305*, 2532–2539.

Deshmukh, T. A., Yadav, B. V., Badole, S. L., Bodhankar, S. L., & Dhaneshwar, S. R., (2007). Anti-hyperglycaemic activity of petroleum ether extract of *Ficus racemosa* fruits in alloxan induced diabetic mice. *Pharmacologyonline, 2*, 504–515.

Diniz, Y. S., Rocha, K. K., Souza, G. A., Galhardi, C. M., Ebaid, G. M., Rodrigues, H. G., NovelliFilho, J. L. V., et al., (2006). Effects of N-acetylcysteine on sucrose-rich diet-induced hyperglycaemia, dyslipidemia and oxidative stress in rats. *Eur. J. Pharmacol., 543*, 151–157.

Duarte, A. A., & Golubnitschaja, O., (2020). Diabetes forecasts and statistics for the coming decades. *Diabetes Obes. Metab.*, 165–173.

Ebbert, J. O., & Jensen, M. D., (2013). Fat depots, free fatty acids, and dyslipidemia. *Nutrients, 5*, 498–508.

Eibl, N., Spatz, M., Fischer, G. F., Mayr, W. R., Samstag, A., Wolf, H. M., Schernthaner, G., & Eibl, M. M., (2002). Impaired primary immune response in type-1 diabetes: Results from a controlled vaccination study. *J. Clin. Immunol., 103*, 249–259.

El-Demerdash, F. M., Yousef, M. I., & El-Naga, N. I. A., (2005). Biochemical study on the hypoglycemic effects of onion and garlic in alloxan-induced diabetic rats. *Food Chem. Toxicol., 43*, 57–63.

Eller, K., Kirsch, A., Wolf, A. M., Sopper, S., Tagwerker, A., Stanzl, U., Wolf, D., et al., (2011). Potential role of regulatory T cells in reversing obesity-linked insulin resistance and diabetic nephropathy. *Diabetes, 60*, 2954–2962.

Fioretto, P., Steffes, M. W., Sutherland, D. E., & Mauer, M., (1995). Sequential renal biopsies in insulin-dependent diabetic patients: Structural factors associated with clinical progression. *Kidney Int., 48*, 1929–1935.

Floreani, A., (2007). Liver diseases in the elderly: An update. *Dig. Dis. Sci., 25*, 138–143.

Fowkes, F. G. R., Rudan, D., Rudan, I., Aboyans, V., Denenberg, J. O., McDermott, M. M., Norman, P. E., et al., (2013). Comparison of global estimates of prevalence and risk factors for peripheral artery disease in 2000 and 2010: A systematic review and analysis. *The Lancet, 382*, 1329–1340.

Fox, C. S., Pencina, M. J., Meigs, J. B., Vasan, R. S., Levitzky, Y. S., & D'agostino, R. B., (2006). Trends in the incidence of type 2 diabetes mellitus from the 1970s to the 1990s. *Circulation, 113*, 2914–2918.

Franks, P. W., & Merino, J., (2018). Gene-lifestyle interplay in type 2 diabetes. *Curr. Opin. Genet. Dev., 50*, 35–40.

Gadang, V., Gilbert, W., Hettiararchchy, N., Horax, R., Katwa, L., & Devareddy, L., (2011). Dietary bitter melon seed increases peroxisome proliferator-activated receptor-γ gene expression in adipose tissue, down-regulates the nuclear factor-κB expression, and alleviates the symptoms associated with metabolic syndrome. *J. Med. Food, 14*, 86–93.

Gayathri, M., & Kannabiran, K., (2008). Antidiabetic and ameliorative potential of Ficus bengalensis bark extract in streptozotocin induced diabetic rats. *Indian J. Clin. Biochem., 23*, 394–400.

Geiss, L. S., Wang, J., Cheng, Y. J., Thompson, T. J., Barker, L., Li, Y., Albright, A. L., & Gregg, E. W., (2014). Prevalence and incidence trends for diagnosed diabetes among adults aged 20 to 79 years, United States, 1980–2012. *JAMA, 312*, 1218–1226.

Genet, S., Kale, R. K., & Baquer, N. Z., (1999). Effects of vanadate, insulin and fenugreek (*Trigonella foenumgraecum*) on creatine kinase levels in tissues of diabetic rat. *Indian J Exp Biol. 37*(2), 200–202.

Gholap, S., & Kar, A., (2003). Effects of *Inula racemosa* root and *Gymnema sylvestre* leaf extracts in the regulation of corticosteroid induced diabetes mellitus: Involvement of thyroid hormones. *Int. J. Pharm. Sci. Res., 58*, 413–415.

Gholap, S., & Kar, A., (2004). Hypoglycaemic effects of some plant extracts are possibly mediated through inhibition in corticosteroid concentration. *Die Pharmazie-Int. J. Pharm. Sci. Res., 59*, 876–878.

Goel, R. K., Sairam, K., Dorababu, M., Prabha, T., & Rao, C. V., (2005). Effect of standardized extract of *Ocimum sanctum* Linn. on gastric mucosal offensive and defensive factors. *Indian Journal of Experimental Biology 43*(8), 715–721.

Gregg, E. W., Chen, H., Wagenknecht, L. E., Clark, J. M., Delahanty, L. M., Bantle, J., Pownall, H. J., et al., (2012). Association of an intensive lifestyle intervention with remission of type 2 diabetes. *JAMA, 308*, 2489–2496.

Grover, J. K., & Yadav, S. P., (2004). Pharmacological actions and potential uses of *Momordi cacharantia*: A review. *J. Ethnopharmacol., 93*, 123–132.

Grover, J. K., Vats, V., & Rathi, S. S., (2000). Anti-hyperglycemic effect of *Eugenia jambolana* and *Tinospora cordifolia* in experimental diabetes and their effects on key metabolic enzymes involved in carbohydrate metabolism. *J. Ethnopharmacol., 73*, 461–470.

Grover, J. K., Vats, V., & Yadav, S. S., (2005). *Pterocarpus marsupium* extract (Vijayasar) prevented the alteration in metabolic patterns induced in the normal rat by feeding an adequate diet containing fructose as sole carbohydrate. *Diabetes, Obesity and Metabolism, 7*, 414–420.

Grover, J. K., Vats, V., & Yadav, S., (2002). Effect of feeding aqueous extract of *Pterocarpus marsupium* on glycogen content of tissues and the key enzymes of carbohydrate metabolism. *Mol. Cell. Biochem., 241*, 53–59.

Grover, J. K., Yadav, S. P., & Vats, V., (2003). Effect of feeding *Murraya koeingii* and *Brassica juncea* diet kidney functions and glucose levels in streptozotocin diabetic mice. *J. Ethnopharmacol., 85*, 1–5.

Grundy, S. M., Brewer, Jr. H. B., Cleeman, J. I., Smith, Jr. S. C., & Lenfant, C., (2004). Definition of metabolic syndrome: Report of the national heart, lung, and blood institute/

American heart association conference on scientific issues related to definition. *Circulation, 109*, 433–438.

Gujral, U. P., Weber, M. B., Staimez, L. R., & Narayan, K. V., (2018). Diabetes among non-overweight individuals: An emerging public health challenge. *Curr. Diab. Rep., 18*, 1–10.

Gupta, A., Gupta, R., & Lal, B., (2001). Effect of *Trigonella foenum-graecum* (Fenugreek) seeds on glycaemic control and insulin resistance in type 2 diabetes. *J. Assoc. Physicians India, 49*, 1057–1061.

Gupta, R., & Gupta, R. S., (2009). Effect of *Pterocarpus marsupium* in streptozotocin-induced hyperglycemic state in rats: Comparison with glibenclamide. *Diabetologia Croatica, 38*, 39–45.

Hajam, Y. A., & Rai, S., (2019). Melatonin and insulin modulates the cellular biochemistry, histoarchitecture and receptor expression during hepatic injury in diabetic rats. *Life Sci., 239*, 117046.

Hajam, Y. A., Rai, S., Roy, A., Basheer, M., & Ghosh, H., (2017). Repossession of brain complications in a streptozotocin induced diabetic rat by exogenous melatonin administration. *Int. J. Zool. Res, 13*, 64–73.

Hajam, Y. A., Rai, S., Shree, S., Basheer, M., & Ghosh, H., (2017). Retrieval of reproductive complications by exogenous melatonin treatment in streptozotocin induced diabetic rat model. *Res. Rev: J. Zool. Sci, 5*, 96–104.

Hamden, K., Jaouadi, B., Carreau, S., Aouidet, A., El-Fazaa, S., Gharbi, N., & Elfeki, A., (2010). *Potential Protective Effect on key Steroidogenesis and Metabolic Enzymes and Sperm Abnormalities by Fenugreek Steroids in Testis and Epididymis of Surviving Diabetic Rats, 116*, 146–155.

Hamden, K., Masmoudi, H., Carreau, S., & Elfeki, A., (2010). Immunomodulatory, β-cell, and neuroprotective actions of fenugreek oil from alloxan-induced diabetes. *Immunopharmacol. Immunotoxicol., 32*, 437–445.

Handelsman, Y., Bloomgarden, Z. T., Grunberger, G., Umpierrez, G., Zimmerman, R. S., Bailey, T. S., Blonde, L., et al., (2015). American Association of Clinical Endocrinologists and American College of Endocrinology–clinical practice guidelines for developing a diabetes mellitus comprehensive care plan–2015—executive summary. *Endocr. Pract., 21*, 413–437.

Hannan, J. M. A., Marenah, L., Ali, L., Rokeya, B., Flatt, P. R., & Abdel-Wahab, Y. H. A., (2006). *Ocimum sanctum* leaf extracts stimulate insulin secretion from perfused pancreas, isolated islets and clonal pancreatic β-cells. *J. Endocrinol., 189*, 127–136.

Hougee, S., Faber, J., Sanders, A., De Jong, R. B., Van, D. B. W. B., Garssen, J., Hoijer, M. A., & Smit, H. F., (2005). Selective COX-2 inhibition by a pterocarpus marsupium extract characterized by pterostilbene, and its activity in healthy human volunteers. *Planta Med. 71*(5), 387–392.

Hu, F. B., Manson, J. E., Stampfer, M. J., Colditz, G., Liu, S., Solomon, C. G., & Willett, W. C., (2001). Diet, lifestyle, and the risk of type 2 diabetes mellitus in women. *N. Engl. J. Med., 345*, 790–797.

Jain, R. C., & Vyas, C. R., (1998). Garlic in alloxan-induced diabetic rabbits. *Am. J. Clin. Nutr., 28*, 684, 685.

Johar, D., Maher, A., Aboelmagd, O., Hammad, A., Morsi, M., Warda, H. F., Awad, H. I., et al., (2018). Whole-food phytochemicals antioxidative potential in alloxan-diabetic rats. *Toxicol. Rep., 5*, 240–250.

Joseph, J. J., Echouffo-Tcheugui, J. B., Golden, S. H., Chen, H., Jenny, N. S., Carnethon, M. R., Jacobs, D., et al., (2016). Physical activity, sedentary behaviors and the incidence of type 2 diabetes mellitus: The multi-ethnic study of atherosclerosis (MESA). *BMJ Open Diabetes Res. Care, 4*, 000185.

Kanwar, Y. S., Akagi, S., Sun, L., Nayak, B., Xie, P., Wada, J., Chugh, S. S., & Danesh, F. R., (2005). Cell biology of diabetic kidney disease. *Nephron. Exp. Nephrol., 101*, e100–e110.

Karthic, R., Nagaraj, S., Arulmurugan, P., Seshadri, S., Rengasamy, R., & Kathiravan, K., (2012). *Gymnema sylvestre* R. Br. suspension cell extract show antidiabetic potential in Alloxan induced diabetic albino male rats. *Asian Pac. J. Trop. Biomed., 2*, 930–S933.

Karvonen, M., Viik-Kajander, M., Moltchanova, E., Libman, I., LaPorte, R., & Tuomilehto, J., (2000). Incidence of childhood type 1 diabetes worldwide. Diabetes mondiale (DiaMond) project group. *Diabetes Care, 23*, 1516–1526.

Kaur, V., & Valecha, V., (2014). Diabetes and antidiabetic herbal formulations: An alternative to allopathy. *Eur. J. Med.*, 226–240.

Kedar, P., & Chakrabarti, C. H., (1983). Effects of jambolan seed treatment on blood sugar, lipids and urea in streptozotocin induced diabetes in rabbits. *Indian J. Physiol. Pharmacol., 27*, 135–140.

Keller, A. C., Ma, J., Kavalier, A., He, K., Brillantes, A. M. B., & Kennelly, E. J., (2011). Saponins from the traditional medicinal plant *Momordi cacharantia* stimulate insulin secretion *in vitro*. *Phytomedicine, 19*, 32–37.

Kelm, M. A., Nair, M. G., Strasburg, G. M., & DeWitt, D. L., (2000). Antioxidant and cyclooxygenase inhibitory phenolic compounds from *Ocimum sanctum* Linn. *Phytomedicine, 7*, 7–13.

Khan, V., Najmi, A. K., Akhtar, M., Aqil, M., Mujeeb, M., & Pillai, K. K., (2012). A pharmacological appraisal of medicinal plants with antidiabetic potential. *J. Pharm. Bioallied Sci., 4*, 27.

Khanna, N., & Bhatia, J., (2003). Antinociceptive action of *Ocimum sanctum* (Tulsi) in mice: Possible mechanisms involved. *J. Ethnopharmacol., 88*, 293–296.

Kim, H., Kisseleva, T., & Brenner, D. A., (2015). Aging and liver disease. *Curr. Opin. Gastroenterol., 31*, 184.

Kirana, H., Agrawal, S. S., & Srinivasan, B. P., (2009). Aqueous extract of *Ficus religiosa* Linn. reduces oxidative stress in experimentally induced type 2 diabetic rats. *Indian J Exp Biol. 47*(10), 822–826.

Knowler, W. C., Barrett-Connor, E., Fowler, S. E., Hamman, R. F., Lachin, J. M., Walker, E. A., & Nathan, D. M., (2002). Reduction in the incidence of type 2 diabetes with lifestyle intervention or metformin. *N. Engl. J. Med., 346*, 393–403.

Kopanski, L., & Schnelle, G., (1988). Isolation of bergenin from barks of *Syzygium cumini. Planta Medica, 54*, 572.

Krawinkel, M. B., & Keding, G. B., (2006). Bitter gourd (*Momordi cacharantia*): A dietary approach to hyperglycemia. *Nutr. Rev., 64*, 331–337.

Kruegel, J., & Miosge, N., (2010). Basement membrane components are key players in specialized extracellular matrices. *Cell. Mol. Life Sci., 67*, 2879–2895.

Kumar, G. S., Shetty, A. K., & Salimath, P. V., (2005). Modulatory effect of fenugreek seed mucilage and spent turmeric on intestinal and renal disaccharidases in streptozotocin induced diabetic rats. *Plant Foods Hum. Nutr., 60*, 87–91.

Kumar, P., Kale, R. K., & Baquer, N. Z., (2012). Antihyperglycemic and protective effects of *Trigonella foenumgraecum* seed powder on biochemical alterations in alloxan diabetic rats. *Eur. Rev. Med. Pharmacol. Sci., 16*, 18–27.

Kusumoto, I. T., Nakabayashi, T., Kida, H., Miyashiro, H., Hattori, M., Namba, T., & Shimotohno, K., (1995). Screening of various plant extracts used in ayurvedic medicine for inhibitory effects on human immunodeficiency virus type 1 (HIV-1) protease. *Phytother. Res., 9*, 180–184.

Leach, M. J., (2007). *Gymnema sylvestre* for diabetes mellitus: A systematic review. *J. Altern. Complement. Med., 13*, 977–983.

Leung, L., Birtwhistle, R., Kotecha, J., Hannah, S., & Cuthbertson, S., (2009). Antidiabetic and hypoglycaemic effects of *Momordi cacharantia* (bitter melon): A mini review. *Br. J. Nutr., 102*, 1703–1708.

Lewis, Y. S., Dwarakanath, C. T., & Johar, D. S., (1956). Acids and sugars in *Eugenia jambolana*. *J. Sci. Ind. Res., 15*, 280, 281.

Liu, B., Asare-Anane, H., Al-Romaiyan, A., Huang, G., Amiel, S. A., Jones, P. M., & Persaud, S. J., (2009). Characterization of the insulinotropic activity of an aqueous extract of *Gymnema sylvestre* in mouse β-cells and human islets of Langerhans. *Cell. Physiol. Biochem., 23*, 125–132.

Liu, C. T., Hse, H., Lii, C. K., Chen, P. S., & Sheen, L. Y., (2005). Effects of garlic oil and diallyltrisulfide on glycemic control in diabetic rats. *Eur. J. Pharmacol., 516*, 165–173.

Liu, J. Z., Lin, X. Y., & Milner, J. A., (1992). Dietary garlic powder increases glutathione content and glutathione-s-transferase activity in rat liver and mammary tissue. *FASEB J., 4*, 1493.

Luo, H., Le Feng, W. T. I., & Hiji, Y., (2001). Inhibitory effect and mechanism of acarbose combined with gymnemic acid on maltose absorption in rat intestine. *World J. Gastroenterol., 7*, 9.

Madkor, H. R., Mansour, S. W., & Ramadan, G., (2011). Modulatory effects of garlic, ginger, turmeric and their mixture on hyperglycaemia, dyslipidaemia and oxidative stress in streptozotocin–nicotinamide diabetic rats. *Br. J. Nutr., 105*, 1210–1217.

Malik, Z. A., Singh, M., & Sharma, P. L., (2011). Neuroprotective effect of *Momordi cacharantia* in global cerebral ischemia and reperfusion induced neuronal damage in diabetic mice. *J. Ethnopharmacol., 133*, 729–734.

Manickam, M., Ramanathan, M., FarboodniayJahromi, M. A., Chansouria, J. P. N., & Ray, A. B., (1997). Antihyperglycemic activity of phenolics from *Pterocarpus marsupium*. *J. Nat. Prod., 60*, 609–610.

Martinez, I., Nedredal, G. I., Øie, C. I., Warren, A., Johansen, O., Le Couteur, D. G., & Smedsrød, B., (2008). The influence of oxygen tension on the structure and function of isolated liver sinusoidal endothelial cells. *Comp. Hepatol., 7*, 1–11.

Mathers, C., (2008). *The Global Burden of Disease: 2004 Update*. WHO.

Mathew, P. T., & Augusti, K. T., (1973). Studies on the effect of allicin (diallyldisulfide-oxide) on alloxan diabetes. I. Hypoglycaemic action and enhancement of serum insulin effect and glycogen synthesis. *Indian J. Biochem. Biophys., 10*, 209–212.

Mayer-Davis, E. J., Lawrence, J. M., Dabelea, D., Divers, J., Isom, S., Dolan, L., Imperatore, G., et al., (2017). Incidence trends of type 1 and type 2 diabetes among youths, 2002–2012. *N. Engl. J. Med., 376*, 1419–1429.

Mensah-Brown, E. P. K., Obineche, E. N., Galadari, S., Chandranath, E., Shahin, A., Ahmed, I., Patel, S. M., & Adem, A., (2005). Streptozotocin-induced diabetic nephropathy in rats: The role of inflammatory cytokines. *Cytokine, 31*, 180–190.

Modak, M., Dixit, P., Londhe, J., Ghaskadbi, S., & Devasagayam, T. P. A., (2007). Indian herbs and herbal drugs used for the treatment of diabetes. *J. Clin. Biochem. Nutr., 40*, 163–173.

Mohan, V., & Anbalagan, V. P., (2013). Expanding role of the Madras diabetes research foundation-Indian diabetes risk score in clinical practice. *Indian J. Endocrinol. Metab., 17*, 31.

Moon, J. Y., Jeong, K. H., Lee, T. W., Ihm, C. G., Lim, S. J., & Lee, S. H., (2012). Aberrant recruitment and activation of T cells in diabetic nephropathy. *Am. J. Nephrol., 35*, 164–174.

Mukherjee, R., Dash, P. K., & Ram, G. C., (2005). Immunotherapeutic potential of *Ocimum sanctum* (L) in bovine subclinical mastitis. *Res. Vet. Sci., 79*, 37–43.

NAIR, R., KALARIYA, T., & Chanda, S., (2005). Antibacterial activity of some selected Indian medicinal flora. *Turk. J. Biol., 29*, 41–47.

Narendhirakannan, R. T., Subramanian, S., & Kandaswamy, M. M., (2006). Biochemical evaluation of antidiabetogenic properties of some commonly used Indian plants on streptozotocin-induced diabetes in experimental rats. *Clin. Exp. Pharmacol. Physiol., 33*, 1150–1157.

Østergaard, J., Hansen, T. K., Thiel, S., & Flyvbjerg, A., (2005). Complement activation and diabetic vascular complications. *Clin. Chim. Acta, 361*, 10–19.

Otton, R., Marin, D. P., Bolin, A. P., Dos, S. R. D. C. M., Polotow, T. G., Sampaio, S. C., & De Barros, M. P., (2010). Astaxanthin ameliorates the redox imbalance in lymphocytes of experimental diabetic rats. *Chem. Biol. Interact., 186*, 306–315.

Padiya, R., Khatua, T. N., Bagul, P. K., Kuncha, M., & Banerjee, S. K., (2011). Garlic improves insulin sensitivity and associated metabolic syndromes in fructose fed rats. *Nutr. Metab., 8*, 1–8.

Pagtalunan, M. E., Miller, P. L., Jumping-Eagle, S., Nelson, G., Myers, B. D., Rennke, H. G., Coplon, N. S., et al., (1997). Podocyte loss and progressive glomerular injury in type II diabetes. *J. Clin. Investig., 99*, 342–348.

Pambianco, G., Costacou, T., Ellis, D., Becker, D. J., Klein, R., & Orchard, T. J., (2006). The 30-year natural history of type 1 diabetes complications: The Pittsburgh epidemiology of diabetes complications study experience. *Diabetes, 55*, 1463–1469.

Panda, S., & Kar, A., (1998). *Ocimum sanctum* leaf extract in the regulation of thyroid function in the male mouse. *Pharmacol. Res., 38*, 107–110.

Pandey, A., Tripathi, P., Pandey, R., Srivatava, R., & Goswami, S., (2011). Alternative therapies useful in the management of diabetes: A systematic review. *J. Pharm. Bioallied Sci., 3*, 504.

Patil, R. N., Patil, R. Y., Ahirwar, B., & Ahirwar, D., (2011). Evaluation of antidiabetic and related actions of some Indian medicinal plants in diabetic rats. *Asian Pacific J. Trop. Med., 4*, 20–23.

Pattanayak, P., Behera, P., Das, D., & Panda, S. K., (2010). *Ocimum sanctum* Linn. A reservoir plant for therapeutic applications: An overview. *Pharmacogn. Rev., 4*, 95.

Patumraj, S., Tewit, S., Amatyakul, S., Jariyapongskul, A., Maneesri, S., Kasantikul, V., & Shepro, D., (2000). Comparative effects of garlic and aspirin on diabetic cardiovascular complications. *Drug Deliv., 7*, 91–96.

Persaud, S. J., Al-Majed, H., Raman, A., & Jones, P. M., (1999). *Gymnema sylvestre* stimulates insulin release *in vitro* by increased membrane permeability. *J. Endocrinol., 163*, 207–212.
Petersen, P. E., (2008). World health organization global policy for improvement of oral health-world health assembly 2007. *Int. Dent. J., 58*, 115–121.
Petit, P. R., Sauvaire, Y. D., Hillaire-Buys, D. M., Leconte, O. M., Baissac, Y. G., Ponsin, G. R., & Ribes, G. R., (1995). Steroid saponins from fenugreek seeds: Extraction, purification, and pharmacological investigation on feeding behavior and plasma cholesterol. *Steroids, 60*, 674–680.
Pickens, C. M., Pierannunzi, C., Garvin, W., & Town, M., (2018). Surveillance for certain health behaviors and conditions among states and selected local areas—behavioral risk factor surveillance system, United States, 2015. *MMWR Surveill. Summ., 67*, 1.
Pickup, J. C., Chusney, G. D., Thomas, S. M., & Burt, D., (2000). Plasma interleukin-6, tumor necrosis factor α and blood cytokine production in type 2 diabetes. *Life Sci., 67*, 291–300.
Prakash, J., & Gupta, S. K., (2000). Chemopreventive activity of *Ocimum sanctum* seed oil. *J. Ethnopharmacol., 72*, 29–34.
Prasad, S., Sajja, R. K., Naik, P., & Cucullo, L., (2014). Diabetes mellitus and blood-brain barrier dysfunction: An overview. *J. Pharmacovigil., 2*, 125.
Preuss, H. G., Bagchi, D., Bagchi, M., Rao, C. S., Dey, D. K., & Satyanarayana, S., (2004). Effects of a natural extract of (–)-hydroxycitric acid (HCA-SX) and a combination of HCA-SX plus niacin-bound chromium and *Gymnema sylvestre* extract on weight loss. *Diabetes, Obesity and Metabolism, 6*, 171–180.
Pushparaj, P. N., Tan, B. K. H., & Tan, C. H., (2001). The mechanism of hypoglycemic action of the semi-purified fractions of *Averrhoa bilimbi* in streptozotocin-diabetic rats. *Life Sci., 70*, 535–547.
Pushparaj, P., Tan, C. H., & Tan, B. K. H., (2000). Effects of *Averrhoa bilimbi* leaf extract on blood glucose and lipids in streptozotocin-diabetic rats. *J. Ethnopharmacol., 72*, 69–76.
Raghuram, T. C., Sharma, R. D., Sivakumar, B., & Sahay, B. K., (1994). Effect of fenugreek seeds on intravenous glucose disposition in non-insulin dependent diabetic patients. *Phytother. Res., 8*, 83–86.
Rai, S., Hajam, Y. A., Basheer, M., & Ghosh, H., (2016). Biochemical and histopathological inflections in hepato-renal tissues of streptozotocin (STZ) induced diabetic male rats: Impact of exogenous melatonin administration. *J. Clin. Res. Bioeth, 7*, 10–4172.
Rajasekaran, S., Ravi, K., Sivagnanam, K., & Subramanian, S., (2006). Beneficial effects of Aloe vera leaf gel extract on lipid profile status in rats with streptozotocin diabetes. *Clin. Exp. Pharmacol. P., 33*, 232–237.
Ramachandran, A., Snehalatha, C., Vijay, V., Wareham, N. J., & Colagiuri, S., (2005). Derivation and validation of diabetes risk score for urban Asian Indians. *Diabetes. Res. Clin. Pract., 70*, 63–70.
Rao, B. K., Giri, R., Kesavulu, M. M., & Apparao, C. H., (2001). Effect of oral administration of bark extracts of *Pterocarpus santalinus* L. on blood glucose level in experimental animals. *J. Ethnopharmacol., 74*, 69–74.
Ravi, K., Rajasekaran, S., & Subramanian, S., (2005). Antihyperlipidemic effect of *Eugenia jambolana* seed kernel on streptozotocin-induced diabetes in rats. *Food Chem. Toxicol., 43*, 1433–1439.
Ravi, K., Ramachandran, B., & Subramanian, S., (2004). Effect of *Eugenia jambolana* seed kernel on antioxidant defense system in streptozotocin-induced diabetes in rats. *Life Sci., 75*, 2717–2731.

Ribes, G., Sauvaire, Y., Costa, C. D., Baccou, J. C., & Loubatieres-Mariani, M. M., (1986). Antidiabetic effects of subtractions from fenugreek seeds in diabetic dogs. *Exp. Biol. Med., 182*, 159–166.

Ritz, E., & Orth, S. R., (1999). Nephropathy in patients with type 2 diabetes mellitus. *N. Engl. J. Med., 341*, 1127–1133.

Rizvi, S. I., & Zaid, M. A., (2001). Intracellular reduced glutathione content in normal and type 2 diabetic erythrocytes: Effect of insulin and (-) epicatechin. *Indian J. Physiol. Pharmacol., 52*.

Roberts, K. T., (2011). The potential of fenugreek (*Trigonella foenum-graecum*) as a functional food and nutraceutical and its effects on glycemia and lipidemia. *J. Med. Food, 14*, 1485–1489.

Rose, P., Whiteman, M., Moore, P. K., & Zhu, Y. Z., (2005). Bioactive S-alk (en) yl cysteine sulfoxide metabolites in the genus allium: The chemistry of potential therapeutic agents. *Nat. Prod. Rep., 22*, 351–368.

Sagrawat, H., (2006). Pharmacological potential of *Eugenia jambolana*: A review. *Pharmacogn Mag., 2*, 96–105.

Sahai, A., Malladi, P., Pan, X., Paul, R., Melin-Aldana, H., Green, R. M., & Whitington, P. F., (2004). Obese and diabetic db/db mice develop marked liver fibrosis in a model of nonalcoholic steatohepatitis: Role of short-form leptin receptors and osteopontin. *Am. J. Physiol. Gastrointest., 287*, 1035–G1043.

Sakowicz-Burkiewicz, M., & Pawelczyk, T., (2011). Recent advances in understanding the relationship between adenosine metabolism and the function of T and B lymphocytes in diabetes. *J. Physiol. Pharmacol., 62*, 505.

SallesTrevisan, M. T., Vasconcelos, S. M. G., Pfundstein, B., Spiegelhalder, B., & Owen, R. W., (2006). Characterization of the volatile pattern and antioxidant capacity of essential oils from different species of the genus *Ocimum*. *J. Agric. Food Chem., 54*, 4378–4382.

Saltiel, A. R., & Kahn, C. R., (2001). Insulin signaling and the regulation of glucose and lipid metabolism. *Nature, 414*, 799–806.

Salunkhe, V. R., Yadav, A. V., Shete, A. S., Kane, S. R., & Kulkarni, A. S., (2005). Anti-inflammatory activity of hydrogels of extracts of *Pterocarpus marsupium* and *Coccinia indica*. *Indian Drugs, 42*, 319–321.

Sauvaire, Y., Petit, P., Broca, C., Manteghetti, M., Baissac, Y., Fernandez-Alvarez, J., Gross, R., et al., (1998). 4-Hydroxyisoleucine: A novel amino acid potentiator of insulin secretion. *Diabetes, 47*, 206–210.

Saxena, A., & Vikram, N. K., (2004). Role of selected Indian plants in management of type 2 diabetes: A review. *J. Altern. Complement. Med., 10*, 369–378.

Schloot, N. C., Roep, B. O., Wegmann, D., Yu, L., Chase, H. P., Wang, T., & Eisenbarth, G. S., (1997). Altered immune response to insulin in newly diagnosed compared to insulin-treated diabetic patients and healthy control subjects. *Diabetologia, 40*, 564–572.

Schmucker, D. L., (2005). Age-related changes in liver structure and function: Implications for disease? *Exp. Gerontol., 40*, 650–659.

Sharma, S. B., Nasir, A., Prabhu, K. M., Murthy, P. S., & Dev, G., (2003). Hypoglycaemic and hypolipidemic effect of ethanolic extract of seeds of *Eugenia jambolana* in alloxan-induced diabetic rabbits. *J. Ethnopharmacol., 85*, 201–206.

Sheedfar, F., Biase, S. D., Koonen, D., & Vinciguerra, M., (2013). Liver diseases and aging: Friends or foes? *Aging Cell, 12*, 950–954.

Shibib, B. A., Khan, L. A., & Rahman, R., (1993). Hypoglycaemic activity of *Coccinia indica* and *Momordi cacharantia* in diabetic rats: Depression of the hepatic gluconeogenic enzymes glucose-6-phosphatase and fructose-1, 6-bisphosphatase and elevation of both liver and red-cell shunt enzyme glucose-6-phosphate dehydrogenase. *Biochem. J., 292*, 267–270.

Shih, C. C., Lin, C. H., Lin, W. L., & Wu, J. B., (2009). *Momordi cacharantia* extract on insulin resistance and the skeletal muscle GLUT4 protein in fructose-fed rats. *J. Ethnopharmacol., 123*, 82–90.

Shokeen, P., Ray, K., Bala, M., & Tandon, V., (2005). Preliminary studies on activity of *Ocimum sanctum*, *Drynaria quercifolia*, and Annona squamosa against *Neisseria gonorrhoeae. Sexually Transmitted Diseases, 32*, 106–111.

Shrotri, D. S., Kelkar, M., Deshmukh, V. K., & Aiman, R., (1963). Investigations of the hypoglycemic properties of *Vinca rosea, Cassia auriculata* and *Eugenia jambolana. Indian J. Med. Res., 51*, 464–467.

Silva, D. G., Daley, S. R., Hogan, J., Lee, S. K., Teh, C. E., Hu, D. Y., Lam, K. P., et al., (2011). Anti-islet autoantibodies trigger autoimmune diabetes in the presence of an increased frequency of islet-reactive CD4 T cells. *Diabetes, 60*, 2102–2111.

Singh, A. B., Tamarkar, A. K., Shweta, Narender, T., & Srivastava, A. K., (2010). Antihyperglycaemic effect of an unusual amino acid (4-hydroxyisoleucine) in C57BL/KsJ-db/db mice. *Nat. Prod. Res., 24*, 258–265.

Singh, D., Singh, B., & Goel, R. K., (2011). Traditional uses, phytochemistry and pharmacology of *Ficus religiosa*: A review. *J. Ethnopharmacol., 134*, 565–583.

Singh, N., & Gupta, M., (2007). Effects of ethanolic extract of *Syzygium cumini* (Linn) seed powder on pancreatic islets of alloxan diabetic rats. *IJEB 45*(10), 861–867.

Singh, P. K., Baxi, D., Banerjee, S., & Ramachandran, A. V., (2012). Therapy with methanolic extract of *Pterocarpus marsupium* roxb and *Ocimum sanctum* Linn reverses dyslipidemia and oxidative stress in alloxan induced type I diabetic rat model. *Exp. Toxicol. Pathol., 64*, 441–448.

Singh, S., Malhotra, M., & Majumdar, D. K., (2005). Antibacterial activity of *Ocimum sanctum* L. fixed oil. *Indian J Exp Biol,. 43*, 835.

Soltani, N., Keshavarz, M., & Dehpour, A. R., (2007). Effect of oral magnesium sulfate administration on blood pressure and lipid profile in streptozocin diabetic rat. *Eur. J. Pharmacol., 560*, 201–205.

Sood, R., Swarup, D., Bhatia, S., Kulkarni, D. D., Dey, S., Saini, M., & Dubey, S. C., (2012). Antiviral activity of crude extracts of eugenia jambolana lam. against highly pathogenic avian influenza (H5N1) virus. *Indian J Exp Biol. 50*(3), 179–186.

Sood, S., Narang, D., Thomas, M. K., Gupta, Y. K., & Maulik, S. K., (2006). Effect of *Ocimum sanctum* Linn. on cardiac changes in rats subjected to chronic restraint stress. *J. Ethnopharmacol., 108*, 423–427.

Soybİr, G., Topuzlu, C., OdabaŞ, Ö., Dolay, K., Bİlİr, A., & KÖksoy, F., (2003). The effects of melatonin on angiogenesis and wound healing. *Surg. Today, 33*, 896–901.

Srinivas, P., Devi, K. P., & Shailaja, B., (2014). Diabetes mellitus (madhumeha): an ayurvedic review. *Int. J. Pharm. PharmSci, 6*, 107–110.

Srivastava, H. C., (1953). Paper chromatography of fruit juices. *J. Sci. Res.*

Subramanian, M., Chintalwar, G. J., & Chattopadhyay, S., (2005). Antioxidant and radioprotective properties of an *Ocimum sanctum* polysaccharide. *Redox Rep., 10*, 257–264.

Sule, N. S., Nerurkar, R. P., & Kamath, S. A., (2002). Pharmacoeconomics. *J. Assoc. Phys. India, 50*, 1057–1062.

Swami, K. D., & Bisht, N. P. S., (1996). Constituents of *Ficus religiosa* and *Ficus infectoria* and their biological activity. *J. Indian Chem. Soc., 73.*

Tsai, C. H., Chen, E. C. F., Tsay, H. S., & Huang, C. J., (2012). Wild bitter gourd improves metabolic syndrome: A preliminary dietary supplementation trial. *Nutr. J., 11*, 1–9.

Tuomilehto, J., Lindström, J., Eriksson, J. G., Valle, T. T., Hämäläinen, H., Ilanne-Parikka, P., Keinänen-Kiukaanniemi, S., et al., (2001). Prevention of type 2 diabetes mellitus by changes in lifestyle among subjects with impaired glucose M. C. changes in the blood lipid profile after administration of *Ocimum sanctum* (Tulsi) leaves in the norm tolerance. *N. Engl. J. Med., 344*, 1343–1350.

Vats, V., Grover, J. K., & Rathi, S. S., (2002). Evaluation of anti-hyperglycemic and hypoglycemic effect of *Trigonella foenum-graecum* Linn, *Ocimum sanctum* Linn and *Pterocarpus marsupium* Linn in normal and alloxanized diabetic rats. *J. Ethnopharmacol., 79*, 95–100.

Vuica, A., Vukojević, K., Hamzić, L. F., Jerić, M., Puljak, L., Grković, I., & Filipović, N., (2016). Expression pattern of CYP24 in liver during ageing in long-term diabetes. *Acta Histochem., 118*, 486–495.

Walley, T., & Davey, P., (1995). Pharmacoeconomics: A challenge for clinical pharmacologists. *Br. J. Pharmacol., 40,* 199.

Wang, L. F., Luo, H., Miyoshi, M., Imoto, T., Hiji, Y., & Sasaki, T., (1998). Inhibitory effect of gymnemic acid on intestinal absorption of oleic acid in rats. *Can. J. Physiol. Pharmacol., 76*, 1017–1023.

Wang, Z. Q., Zhang, X. H., Yu, Y., Poulev, A., Ribnicky, D., Floyd, Z. E., & Cefalu, W. T., (2011). Bioactives from bitter melon enhance insulin signaling and modulate acyl carnitine content in skeletal muscle in high-fat diet-fed mice. *J. Nutr. Biochem., 22*, 1064–1073.

Watkins, P. J., (2003). *ABC of Diabetes*. BMJ publishing group.

Weng, H. L., Ciuclan, L., Liu, Y., Hamzavi, J., Godoy, P., Gaitantzi, H., Kanzler, S., et al., (2007). Profibrogenic transforming growth factor-β/activin receptor–like kinase 5 signaling via connective tissue growth factor expression in hepatocytes. *J Hepatol., 46*, 1257–1270.

Weng, J., Zhou, Z., Guo, L., Zhu, D., Ji, L., Luo, X., Mu, Y., & Jia, W., (2018). Incidence of type 1 diabetes in China, 2010–13: Population based study. *BMJ, 360.*

Willi, C., Bodenmann, P., Ghali, W. A., Faris, P. D., & Cornuz, J., (2007). Active smoking and the risk of type 2 diabetes: A systematic review and meta-analysis. *JAMA, 298*, 2654–2664.

World Health Organization, (2016). *World Malaria Report 2015*. World Health Organization,

Xie, J., Tato, C. M., & Davis, M. M., (2013). How the immune system talks to itself: The varied role of synapses. *Immunol. Rev., 251*, 65–79.

Yanpallewar, S. U., Rai, S., Kumar, M., & Acharya, S. B., (2004). Evaluation of antioxidant and neuroprotective effect of *Ocimum sanctum* on transient cerebral ischemia and long-term cerebral hypoperfusion. *Pharmacol. Biochem. Behav., 79*, 155–164.

Part II

Antidiabetic Plants in Systems of Traditional Knowledge

CHAPTER 3

The 'Green Healers' of Diabetes Mellitus: An Ayurvedic Eye

K. ARUN BHANU[1] and LAVANYA MOHAN[2]

[1]University of Trans-Disciplinary Health Science and Technology, Yelahanka, Bengaluru, Karnataka, India,
E-mail: dr.arunbhanu@gmail.com

[2]Intedent Healthcare, Bengaluru, Karnataka, India

ABSTRACT

Diabetes is a chronic metabolic disorder indicated by elevated levels of blood glucose. The complexity and comorbidities of diabetes require multi-dimensional systemic interventions. Ayurveda understands diabetes as a condition manifested as excess and turbid urination (prameha) rising from multiple etiological factors such as genetics, improper food habits and sedentary lifestyles. Therefore, Ayurveda management of prameha aims at identifying and correcting the root causes by holistic approaches through multi-component therapeutics, dietary interventions, and lifestyle modifications. Ayurveda gives a lot of emphasis on food-based management in addition to specific drug formulations. Many of the herbal preparations prescribed for prameha management are having base from the culinary practices. Hence, understanding the importance and ayurvedic rationale of using these botanicals could lead to a novel approach in the management of diabetes. In this chapter, we discuss 10 such herbals, that are ingredients of both Ayurveda formulations and food recipes.

Antidiabetic Potential of Plants in the Era of Omics. Deepu Pandita, PhD, Anu Pandita, MSc, & Chander Bhanu, MBBS (Eds.)

3.1 AYURVEDA: A PERPETUAL EVOLUTION

"From kitchen to pharmacy"—a crude definition for understanding the Evolution of Ayurveda (Narayana and Ala, 2012).

Medicines are nothing but a form of food prepared and intended to be consumed in a different manner. Though mostly believed that "All diseases start from indigestion," it is also accepted that "Food is the best medicine." However, Ayurveda advocates that Food is the cure for what is caused by Food.

Ayurveda – The Science of life, is a compilation of various ancient texts, including references from the Vedas. As the motto speaks, "Prevention is better than Cure," Ayurveda has always emphasized on the Concept of Wellness rather than Curing of Diseases (Narayanaswamy, 1981).

It further suggests that the key to Wellness is to bring harmony between body, mind, and spirit. Just like any other Stream of Science, Ayurveda has also been through a series of transformations and adaptations to match every walks of life from olden centuries to the contemporary ages. Nevertheless, with great efforts and dedication, Ayurveda has and is continuing to maintain its efficacy and Sanctity (Sharma, 2001).

3.2 AYURVEDA IN TERMS OF CONTEMPORARY PRINCIPLES

Ayurveda is one of the ancient legacies in the field of medicine known to mankind. It is an evolved science of the primordial period and thus which is much difficult to correlate with the modern scientific parameters (Narayanaswamy, 1981).

"Swasthasyaswasthyarakshanam, Athurasyavikaaraprashamanam cha" – Ayurveda aims at preservation of health of the healthy, while curing the disease of the unhealthy (Susruta Samhita, 4th cent BCE). In the same lines, the modern world seeks to preserve their health in various ways so that they can escape from being sick.

As per Ayurveda, Disease is not merely a physical occurrence but a conglomeration of physical, mental, and environmental phenomena.

3.2.1 THE BASIC PILLARS OF AYURVEDA

The '*Panchamahabhootha*' theory and '*Tridosa*' theory are considered to be the gateway to the system of Ayurveda (Narayanaswamy, 1981).

The '*Panchamahabhootha*' theory deals with the five basic elements – Earth, Water, Fire, Space, and Air. Though seemingly physical in nature, each of these elements have unique qualities and co-exist as energy patterns in the universe (Narayanaswamy, 1981):

- *Earth* is a stable, hard, and solid element that is usually dense and slow. It accounts for all matter in the universe as well as our body.
- *Water* is an enormous yet soft and cool element that is liquid, moist, and cohesive. It is a detoxifying element.
- *Fire* as the word suggests, it is hot, sharp, and dry. It is an intense element that emits heat and light.
- *Space*, denotes the sky or ether, i.e., the nothingness of the entire cosmos. It is vast, soft, subtle, cold, and non-resistant. It is the mother element where all other elements exist.
- *Air* is the wind element, light, mobile, dry, and cold with vigorous motion. It is responsible for any form of movement.

These elements are naturally present independently in this universe. However, they fundamentally co-exist everywhere at all times, as well as in the body, marking their individual importance. All five elements in collaboration with one another are responsible for maintaining the basic structural and functional well-being of the human body.

The five elements theory is not alone in the journey of natural wellness. '*Tridosa*' is another predominant theory of Ayurveda that mainly deals with the three humors (invisible fluids) in the body: *Vata, Pitta, and Kapha.*

These three dosas are the constitutional principles formed as a resultant of the combination of the five elements in various ways:

- *Vata*: Air + Space
- *Pitta*: Fire + Water
- *Kapha*: Earth + Water

1. **Vata:** Since it is the combination of Air and Space elements, it demonstrates a light, dry, cold, and mobile characteristics. The air element influences the movement of principle through vata. Hence all bodily movements, including breathing, chewing, peristalsis, nerve impulses, urination, thinking, muscle movements, etc., are accounted for by Vata. Imbalance of Vata will impact one or more of these movements (Narayanaswamy, 1981).

2. **Pitta:** The union of Fire and water, Pitta exhibits both intensity of fire as well as the vigor of water. It exhibits hot, sharp, light yet penetrating and mobile characteristics. Similar to the property of fire to be able to transform anything it comes in contact with, pitta rules the principle of transformation. Hence pitta accounts for all the functional processes of the body like digestion, metabolism, sensory perception, comprehension as well as temperature balance. Common feature of imbalance in pitta is inflammation and sharpness (Ranawat, 2017).
3. **Kapha:** Marking the union of Earth and Water elements, Kapha naturally exhibits cold, dull, heavy, and dense features. It usually occurs as a static, hard component which predominantly governs the stability and structure of any substance as well as the human body. It accounts for the skeletal, muscular, and fatty molecules such as lipids that constitute the body. Increase in Kapha can lead to heaviness and excess density of the body (Narayanaswamy, 1981).

3.2.2 WHAT IS HEALTH?

Health is defined in *Ayurveda* as the physical and mental well-being. That is the harmony of all the 3 *dosas*, 7 *dhatus*, 3 *malas,* and the pleasure of mind together is known as a healthy condition. Their imbalances will result in diseases (Sharma, 2001).

To achieve a state of health is the aim of *Ayurveda,* rather than the mere absence of diseases. It's not just the health of humans, but the health of other animals, plants, and the environment which we live in and the equilibrium between them is the main concern in *Ayurveda* (Sharma, 2001).

3.2.2.1 AGNI – THE ELEMENT OF DIGESTION

Is there fire in the stomach? Yes, the activity of enzymes is responsible for digestion. As fire transforms all forms in the universe, similarly these enzymes convert the food particles into energy.

Digestion plays a crucial role in the maintenance of human health. Hence, medicines are targeted towards protecting the digestive power. This power of digestion is referred to as *agni* as per Ayurveda (Ranawat, 2017).

Therefore, Ayurveda emphasizes that the imbalance in this agni is the root cause of all diseases in an individual (Narayanaswamy, 1981).

3.2.2.2 PRAKRUTI – BODY CONSTITUTION

Prakruti is the unique constitution or inherent nature of an individual which is acquired by birth and cannot be changed during the lifetime of the individual (Narayanaswamy, 1981).

Literally meaning "Natural form" in Sanskrit, Prakruthi is more of a biodata to one's physical and mental framework (Ranawat, 2017).

Predominance of dosas (vata, pitta, and kapha) during the union of the Shukra (sperm) and shonita (egg) in the garbhashaya (womb) leads to the prakruthi formation of an individual (Sharma, 2001).

Based on the three *dosas* there are seven types of prakruti formed (Sharma, 2001):

- *Vata prakruti*;
- *Pitta prakruti*;
- *Kapha prakruti*;
- *Vata-Pitta prakruti*;
- *Vata-Kapha prakruti*;
- *Pitta-Kapha prakruti*;
- *Vata-Pitta-Kapha prakruti.*

As discussed above, the dosas predominance decides the prakruti of an individual (Ranawat, 2017). Needless to say, any variation from the normal prakruti, i.e., variation in the original dosas predominance of an individual will lead to Disease. To be precise, disease is nothing but variation in the dosa constitution of an individual. Hence bringing back the balance in the original prakruti of an individual will cure the disease and restore health in them (Sharma, 2001).

To be able to restore the balance in the dosas of an individual, it is first important to be able to assess the nature, i.e., the prakruti of the individual needs to be analyzed. By prakruti assessment we can achieve the following benefits (Ranawat, 2017):

- Prevention of lifestyle diseases by adopting the routine;
- Better diet plans according to the *dosa* prevalence;

- Starting wellness therapies to control the future possible ailments.

The above concept explains how Ayurveda has been able to deliver a customized treatment plan for each individual without compromising the quality of the science (Narayanaswamy, 1981).

Ayurveda is a customized form of treatment method and thus no protocol is applicable for any sort of disease. Ayurveda considers every individual and their bodily dysfunction as a unique entity and the suggestion of therapies will vary according to various factors (Ranawat, 2017).

Thus, two individuals, though presenting with similar symptoms, may receive an entirely customized treatment plan with individualized choice of drugs based on the body constitution, place, season, and the strength of the individual as well as the intensity of the disease.

3.3 AYURVEDIC ANGLE OF DIABETES MELLITUS (DM)

The popularly known diabetes mellitus (DM) has been defined as a chronic disease of the metabolism of an individual featuring an elevated blood glucose level. This has been identified by its classic symptoms of Polyuria, Polyphagia, and Polydipsia. Though Diabetes has been a disease in the limelight for quite some time now, it is rather a contemporary concept to Ayurveda (T. Parivallal).

However, Ayurveda clings on to the concepts that explain the cardinal symptoms of DM such as the earlier mentioned Polyuria, Polyphagia, and Polydipsia. They are distinctly elaborated under the heading of *"Prameha"* – which literally means Polyuria (Sridharan et al., 2011).

Prameha is not exactly a disease, but a syndrome that shows clear resemblance to DM. "*Prakarshenamehatianenaiti Prameha*" – A condition of frequent urination is termed as prameha. There is no special reference to any sort of blood related parameters in Ayurveda (Madhumati et al., 2019).

The assessment of well-being is carried out using the physical parameters such as reduction in the flow of urine, improved energy levels, etc. Ayurveda assessment of prameha is purely symptomatic and is sometimes difficult to advocate on the grounds of contemporary lab reports (Madhumati et al., 2019).

There are 20 types of Prameha described in Ayurveda classified according to the three Dosas combinations (Vata, Pitta, and Kapha) and other factors as listed in Table 3.1 (Sharma, 2001).

TABLE 3.1 Types of Prameha (Parivallal, n.d.)

SL. No.	Prameha Type	Dosha Predominance	Symptoms	Possible Correlation	Body Nature	Prognosis
1.	Udakameha	Kapha	Urine similar to water	Diabetes insipidus	Obese	Curable
2.	Iksumeha		Urine resembling juice of sugarcane	Sucrosuria		
3.	Sandrameha		Urine having density	Phosphaturia		
4.	Sandraprasadameha		Urine having density below and transparency in the upper layer			
5.	Suklameha		White color urine	Phosphaturia		
6.	Sukrameha		Urine mixed with seminal fluid	Spermaturia		
7.	Sitameha		Very cold urine	Glucosuria		
8.	Sikatameha		Urine containing sand-like substances	Crystalluria		
9.	Shanaimeha		Slow output of urine	Oliguria		
10.	Lalameha		Urine containing slimy materials like saliva	Albuminuria		
11.	Ksarameha	Pitta	Urine resembles like alkaline solution	Alkaluria	Moderate	Controllable
12.	Kalameha		Black colored urine	Alkaptonuria		
13.	Neelameha		Indigo colored urine	Indicanuria		
14.	Lohitameha		Urine with bloodstains	Hematuria		
15.	Manjishtameha		Reddish urine similar to the color of manjistha plant	Hemoglobinuria		

TABLE 3.1 *(Continued)*

SL. No.	Prameha Type	Dosha Predominance	Symptoms	Possible Correlation	Body Nature	Prognosis
16.	Haridrameha		Yellow-colored urine like turmeric	Choluria		
17.	Vasameha	Vata	Urine mixed with muscle fat	Lipuria	Lean	Curable with difficulty or not curable
18.	Majjameha		Urine containing parts of bone marrow	Pyuria		
19.	Lasikameha		Urine containing lymph	Diabetes insipidus		
20.	Madhumeha		Thick urine resembling to honey and attracts ants due to sweetness	Diabetes mellitus		

3.3.1 *ASHTAVIDHA PARIKSHA IN PRAMEHA (8 FOLD DIAGNOSTIC TOOLS IN PRAMEHA)*

1. ***Nadi* (Pulse):** It may vary according to Dosha, Prakriti, Age, Season, and the severity of illness.
2. ***Mala* (Excretions)-*Purisha* (Stool):** *Asamyaka Pravritti* (Irregular and unsatisfied).

   Sweda (Sweat) – Swedadhikya, Swedadourgandha (Excessive sweating with offensive odor like in diabetic ketoacidosis).

   There will be accumulation of plaques or waste material kind of consistency on teeth, tongue, and eyes.
3. ***Mutra* (Urine):** *Prabhutaavilamutrata* (increased frequency of micturition), increase in urine sugar level [Urine Glucose: >0.8 mmol/lit].
4. ***Jivha* (Tongue):** *Jivhagatamaladhikya, picchilata, mukhamadhurya* (Coated tongue, sliminess of mouth and sweet sensation on mouth).
5. ***Shabda* (speech):** *Mukhashushkata* (dryness of mouth), *Slurred speech.*
6. ***Sparsha* (Palpation):** *Kaphaja-aardra, dehachikannata* (Kaphaprakriti – wet to touch and sandy consistency) *Pittaja-Shithilanga* (Pitta prakriti – Unsteady skin texture and joints)*Vataja-Rukshakharasparsha* (Vataprakriti – Dry and hard to touch).
7. ***Druka* (Examination of Eyes):** *Drushtimandya* (Diminution of vision), *Akshigatamaladhikya Kaphaja – Shweta* (white conjunctiva) *Pittaja-Pitabha* (Pale conjunctiva) *Vataja-Dhusara* (Dusky conjunctiva).
8. ***Aakruti*:** *Sthula Pramehi* (Obese diabetic) and *Krusha Pramehi* (Lean diabetic) (Madhumati et al., 2019).

3.4 AYURVEDIC OUTLOOK ON FUNCTIONAL PRINCIPLES OF BOTANICALS

From time known to mankind, evidence reveals that plants and plant derivatives have been utilized by humans for its healing abilities (Sharma, 2001).

It is one of the classical resources that marked its definitive benefits in the field of medicine. For generations, these plants were utilized in its pure forms due to lack of technology for extracting the benefits alone from the plant products.

But with the advancement in time there has been a remarkable progress in this direction. Today plant and plant products are one the leading ingredients in all fields of medicine. In India, a larger portion of Indian medicines involves the use of herbal therapeutics by Ayurveda, Homeopathy, Yoga, Siddha, Unani, and naturopathy (Sridharan et al., 2011).

Most of the ancient classical texts such as the Rigveda, Atharvaveda, Charaka Samhita, and Sushruta Samhita show substantial evidence of use of herbal therapeutics as a curative measure for several diseases as well as for upgrading lifestyle (Narayanaswamy, 1981).

Ayurveda believes in investing a good amount of time in observing the characteristics of nature, and every element of nature is analyzed for its pros and cons. Every single herb present in nature shows unique qualities that can be utilized to treat various walks of diseases. In some cases, the entire plant is used, while elsewhere, certain parts are extracted. As simple it may sound, these studies require intense research and involvement to be able to make the most of the available treasure (Ranawat, 2017).

The biggest benefit of deriving medicines from plant and plant products is that it is completely natural thereby leaning no fear of inducing side effects from artificially produced chemicals or compounds. It is truly a holistic approach towards medicine. Nevertheless, these botanicals are non-toxic, self-contained, and highly nutritive in nature. They bring about a balance between the body, mind, and soul.

3.5 BOTANICALS RESPONSIBLE FOR RESTORING HEALTH IN DIABETICS

Some of the many plants which exhibit antidiabetic properties are given in Table 3.2.

TABLE 3.2 List of Botanicals

SL. No.	Plant Name	Sanskrit Name	Botanical Name	Chemicals Responsible	Part Used
1.	Bael/Holy fruit tree	Bilva	*Aegle marmelos*	Aegline, marmesin, marmin, and marmelosin (Goel et al., 2014)	Fruit and leaves

TABLE 3.2 *(Continued)*

SL. No.	Plant Name	Sanskrit Name	Botanical Name	Chemicals Responsible	Part Used
2.	Gooseberry	Amalaki	*Emblica officinalis*	Gallic acid, gallotannin, ellagic acid and corilagin (D'souza et al., 2014)	Fruit rind
3.	Onion	Palandu	*Allium cepa*	Allyl propyl disulfide (APD) (Kelkar et al., 2001)	Bulb
4.	Garlic	Lasuna	*Allium sativum*	Alliins, polysaccharides, saponins, diallyl disulfide (allicin) (WHO Monograph, Vol 2, 2004)	Bulb
5.	Ginger	Sunthi	*Zingiber officinale*	α-Zingiberene, β-Sesquiphellandrene, (E,E)-α-farnesene, geranial, ar-curcumene, and γ-terpinene (Otunola et al., 2020)	Dried rhizome
6.	Curry leaves	Kaidarya	*Murraya koenigii*	Carbazole alkaloids, glycosides, and flavonoids (Al-Ani et al. (2017); Palwankar et al. (2020)	Leaves
7.	Turmeric	Haridra	*Curcuma longa*	Curcuminoids and sesquiterpenoids (Hussain et al., 2002; Jeevangi et al., 2013)	Rhizome powder
8.	Gurmar	Madhunashini	*Gymnema sylvestre*	Triterpenoid Saponins, flavonols, and gurmarin (Kumar et al., 2017; Hl et al., 2018)	Leaves
9.	Aloe vera	Kumari	Aloe vera	Polysaccharides and glycoprotein (Yongchaiyudha et al., 1996)	Gel from leaves

TABLE 3.2 *(Continued)*

SL. No.	Plant Name	Sanskrit Name	Botanical Name	Chemicals Responsible	Part Used
10.	Holy basil	Tulsi	*Ocimum sanctum*	Linalool, methyl chavicol, methyl cinnamate, linolen, rosmarinic acid, citral, eugenol, and geraniol (Sethi et al., 2004; Ezeani et al., 2017)	Leaves
11.	Neem	Nimba	*Azadirachta indica*	Nimbolinin, nimbin, nimbidin, nimbidol, sodium nimbinate, gedunin, salannin, and quercetin (Alzohairy, 2014)	Leaves and Stem bark
12.	Kokum	Vrikshamla	*Garcinia cambogia*	Garcinol, flavonoids, saponin glycosides, alkaloids, and tannins (Mali et al., 2017; Deore et al., 2018)	Fruit rind
13.	Giloy	Guduci	*Tinospora cordifolia*	Alkaloids, diterpenoid lactones, glycosides, steroids, sesquiterpenoid, phenolics, aliphatic compounds and polysaccharides (Upadhyay et al., 2010)	Stem

3.5.1 INDIAN BAEL

Indian bael, popular by the name bilva is used in traditional medicine as well as spiritual practices. The leaf is used for ophthalmia, diabetes, and asthmatic complaints (Sastry, 2005). Several researches have revealed the antiamoebic as well as hypoglycemic properties of both the fruit and the root of this plant, while the aqueous extract of the leaf is evidently hypoglycemic (Goel et al., 2014). The antidiabetic action of all organic solvent of bilva extract has aided in the treatment to reduce blood glucose using crude aqueous extract of leaf along with 3-months-old callus. The methanol extract of leaf and callus proved to be the best among the rest (Sevugan et al., 2008). As per Ayurveda, Bilva is known to have a predominantly bitter taste which aids in reducing the excess fat accumulation and improves digestion by targeting the pancreatic activity. It also ceases the production of excess body fat, which is the main causative of prameha (Sastry, 2005).

3.5.2 ONION

A regular food ingredient of Indian kitchen, known as Palandu in Ayurveda is a potential drug to reduce blood glucose. Sulfoxide amino acids from onion inhibit cholesterol and other lipid synthesis. Allium cepa has been found to decrease the blood glucose in a particular threshold dosage (Kelkar et al., 2001). The hypoglycemic effects of onion were established in vivo. Internal administration of onion juice, alcohol extract and water extract reduced alloxan, glucose, and epinephrine induced increase in blood glucose in rabbits and mice (Kelkar et al., 2001). Ayurveda dictates onion to exhibit "Vyavayi" and "Vikashi" properties which means they possess the ability to penetrate. This makes it an excellent carrier of metabolites throughout the body thereby making it readily available for the better functioning of cells. This in turn helps maintain a balance in the blood glucose levels (BGL) (Sastry, 2005).

3.5.3 GARLIC

Allium sativum, commonly known as garlic, otherwise called Lasuna in Sanskrit, flaunts one particular of its several benefits as a hypoglycemic agent. Hence this herb has marked its dominance as an antidiabetic botanical. It has been observed that the allicin-type compounds are responsible for this distinct feature present in garlic. Carson, in the year 1987, found that the hypoglycemic action of garlic is either directly or indirectly stimulating the insulin secretion. In depth research later suggested the presence of sulfur compounds [di (2-propenyl) disulfide and 2-propenyl propyl disulfide], which react with the endogenous thiol molecules such as cysteine, serum albumins and GSH thereby limiting insulin from -SH inactivation. Garlic acts by blocking the absorption of glucose in the intestine (WHO Monograph, Vol 2, 2004). Similar to onion, Ayurveda suggests that garlic also has been proved to be an exceptional transporter of metabolites throughout the body. These herbs use blood as a channel of transportation. Their action is simply that of a catalyst that enhances the absorption of metabolites from blood into the cells.

3.5.4 CURRY LEAVES

Being an integral part of the Indian kitchen, curry leaves (known as kaidarya in Ayurveda) plays a major role in controlling blood sugar levels. The studies show that it is to be eaten and not to be thrown out from the plate. A simple act of adding tadka (seasoning) to complete a dish in southern India has been proved to have several health benefits including antidiabetic properties. Ayurveda advocates the use of curry leaves in regular food as well as in various medicinal formulations to improve digestion. This might be because of its action on the pancreas which in turn stimulates the endocrine part of the pancreas. Curry leaves have been categorized as kapha medohara which literally means burning of the fat molecules. Being a combination of bitter and astringent flavor, curry leaves enhance the digestive secretions, thereby resulting in improved efficacy of digestion of food consumed (Sastry, 2005).

3.5.5 TURMERIC

The yellow medicine, haridra, is a classic ayurvedic ingredient in several prameha preparations. Curcuminoids and sesquiterpenoids are the basic constituents in turmeric and its extract. Turmeric and curcuminoids are effective in controlling blood glucose levels and enzymes of glucose metabolism in alloxan-induced diabetic rats, and suggested that curcumin had a better hypoglycemic effect than turmeric (Arun et al., 2002). Ayurveda explains haridra/turmeric as "vishahara," i.e., anti-toxic. It is a plant that is pungent, bitter, and astringent in taste which helps in enhancing digestion. Prameha is known to cause depletion of bodily vigor and vitality. Haridra acts as an anti-inflammatory and detoxifying agent, thereby blocking the action of prameha. It is known to cause blood purification and improve liver function as well (Sharma, 2006).

3.5.6 GURMAR

Gymnema sylvestre is regarded as Madhunashini in Ayurveda. Madhunashini, as the name suggests, is intended to kill sweet taste. If we chew the leaves of madhunashini, at least for about 2 hours selectively the sweet taste receptors will become inactive or numb. There are several studies which established the hypoglycemic activity of gurmar. Madhunashini is pungent and bitter in taste as per Ayurveda and has fat reducing properties (Sastry, 2005). Research says that gurmar treated diabetes patients have more advantage over conventional drug therapy (Kumar et al., 2017). Madhunashini is believed to possess anti-inflammatory activities and anti-toxic activities. It is found to reduce the body weight and enhances digestive juice secretions and bowel movements.

3.5.7 ALOE VERA

Popular by the name Kumari, aloe vera possesses several benefits related to the female reproductive system as well as controlling blood sugar levels. Aloe vera is bitter in taste and cold in potency which will help in reducing the food intake and hence prevents the excess fat accumulation in the body and subcutaneous tissues. The part used is the gel of aloe vera leaf which is bitter in taste. Aloe vera gel hinders the absorption of glucose by slowing down digestion. It causes cooling of body fluids and thereby controls any sort of inflammatory conditions. Conditions like excess micturition will be brought under control by regular intake of aloe vera juice. Aloe vera is found to have lipid-lowering properties. This also justifies the obesity control in diabetes related conditions (Sharma, 2006).

3.5.8 HOLY BASIL

The sacred Indian herb popular by the name Tulsi, is an important plant of every Indian household. Basil exhibits potent antidiabetic activity. The aerial parts of Tulsi, when extracted have proven to exhibit antidiabetic properties, mostly by restricting glucose absorption by the inhibition of carbohydrate metabolizing enzymes (Ezeani et al., 2017). The leaves of tulsi were found to exhibit both superoxide and hydroxyl free radical scavenging action. However, as per Ayurveda, it possesses catalytic nature by penetrating through the cells and transporting metabolites all through the body. It is mainly bitter and astringent, which aids in improving taste perception. It also improves circulation by dissolving the lipid molecules and enhancing liver function. It is also considered as a potent antioxidant and detoxifying agent but on prolonged use may cause increased body heat leading to burning sensation (Sastry, 2005).

3.5.9 *KOKUM*

The Indian kitchen substitute for tamarind is called Vriksamla in Ayurveda. It possesses potential weight reducing capacity. Kokum is an excellent anti-inflammatory agent and it is meant to cool down the body. Thus, many of the summer drinks constitute kokum as a major ingredient. Garcinia is sour and sweet in taste and causes dryness of the body. Kokum as per Ayurveda, melts down the subcutaneous fat layers by increasing the internal body heat. It penetrates through smaller channels of the body and also improves the efficacy of the digestive fire (Agni) which is responsible for the tissue conversion and controls undesired fat deposition. It is also known to reduce flatulence and resolve constipation (Sastry, 2005).

3.5.10 *GILOY/GULANCHA*

Also referred to as Amrita or Guduchi in Ayurvedic texts meaning ambrosia. Guduchi possesses a bitter and astringent taste which will reduce the body fat and regulate immune response. Several studies have proved giloy to have antidiabetic property due to its rich nutraceutical resource. Guduchi promotes digestion and removes constipation. It is bitter in taste. The bark of the stem is the commonly utilized extract. The starch extract is known as Guduchi sattva, is a potent antioxidant and a proven prameha controlling formulation. Amrita is considered to be a very effective immunomodulatory and anti-inflammatory agent. Most of the prameha patients suffer from irregular bowel habits and neuropathies. The decoction of guduchi alone is a feasible solution for the above mentioned problems. Guduchi is an inevitable part of major prameha controlling formulations and with the support of contemporary techniques, the scientist population were able to demonstrate the antidiabetic properties of giloy as well (Warrier et al., 1993).

3.6 CONCLUSION

Ayurveda, the ancient science has indicated a briefing on various plants and herbs which are efficaciously potent in reducing prameha and exhibit antidiabetic properties. The concept of prameha elucidated by Ayurvedic texts clearly justifies the correlation to diabetes. Evidence states that herbs that are predominantly bitter in taste, in turn are responsible for controlling prameha. The bolstering outcome of this study is that the remedy for diabetes is very much available within the reach of our kitchen. One only needs to understand and utilize the sciences that were substantiated ages ago.

KEYWORDS

- **agni**
- **ayurveda**
- **diabetes**
- **prameha**
- **rasapanchaka**

REFERENCES

Al-Ani, I., Santosa, R., Yankuzo, M., Saxena, A. K., & Alazzawi, S., (2017). The antidiabetic activity of curry leaves "*Murraya koenigii*" on the glucose levels, kidneys, and islets of Langerhans of rats with streptozotocin induced diabetes. *Makara Journal of Health Research, 21*. 10.7454/msk.v21i2.7393.

Ali Hussain, H. E. M., (2002). Hypoglycemic, hypolipidemic and antioxidant properties of combination of curcumin from *Curcuma longa*, Linn, and partially purified product from *Abroma augusta*, Linn. in streptozotocin-induced diabetes. *Indian J. Clin. Biochem., 17*, 33–43. https://doi.org/10.1007/BF02867969.

Alzohairy, M. A., (2016). Therapeutics role of *Azadirachta indica* (Neem) and their active constituents in diseases prevention and treatment. *Evid Based Complement Alternat Med., 2016*, 7382506. doi: 10.1155/2016/7382506.

Arun, N., & Nalini, N., (2002). Efficacy of turmeric on blood sugar and polyol pathway in diabetic albino rats. *Plant Foods Hum. Nutr., 57*, 41–52.

Bunyapraphatsara, N., Yongchaiyudha, S., Rungpitarangsi, V., & Chokechaijaroenporn, O., (1996). Antidiabetic activity of *Aloe vera* L. juice II. Clinical trial in diabetes mellitus patients in combination with glibenclamide. *Phytomedicine, 3*(3), 245–248. doi: 10.1016/S0944-7113(96)80061-4. PMID: 23195078.

D'souza, J. J., D'souza, P. P., Fazal, F., Kumar, A., Bhat, H. P., & Baliga, M. S., (2014). Antidiabetic effects of the Indian indigenous fruit *Emblica officinalis* gaertn: Active constituents and modes of action. *Food Funct., 5*(4), 635–644. doi: 10.1039/c3fo60366k. PMID: 24577384.

Deore, A., Sapakal, V., & Naikwade, N., (2018). Antidiabetic, antihyperlipidemic and renal protective activities of *Garcinia indica* Linn. *Clusiaceae. International Journal of Pharmaceutical Sciences and Drug Research.* 10. 10.25004/IJPSDR.2018.100107.

Ezeani, C., Ezenyi, I., Okoye, T., & Okoli, C., (2017). *Ocimumbasilicum* extract exhibits antidiabetic effects via inhibition of hepatic glucose mobilization and carbohydrate metabolizing enzymes. *J. Intercult. Ethnopharmacol., 6*(1), 22–28. doi:10.5455/jice.20161229054825.

Goel, R., & Bhatia, D., & Gilani, S., & Katiyar, D., (2012). Medicinal plants as antidiabetics: A review. *International Bulletin of Drug Research, 1*, 100–107.

Jeevangi, S., Manjunath, S., Mariguddi, D., Kalashetty, P., Dass, P., & Manjunath, C., (2013). Antidiabetic effects of turmeric in alloxan-induced diabetic rats. *Journal of Evolution of medical and Dental Sciences, 2*, 1669–1679. 10.14260/jemds/442.

Kelkar, S. M., Kaklij, G. M., & Bapat, V. A., (2001). Determination of antidiabetic activity in Allium cepa (onion) tissue cultures. *Indian Journal of Biochemistry and Biophysics, 38.*

Madhumati, S. C., et al., (2019). A review on rog and rogi pariksha in prameha with special reference to diabetes mellitus. *Int. J. Res. Ayurveda Pharm., 10*(1), 4–6.

Mali, K. K., Dias, R. J., Havaldar, V. D., & Yadav, S. J., (2017). Antidiabetic effect of garcinol on streptozotocin induced diabetic rats. *Indian J. Pharm. Sci., 79*(3), 463–468.

Moksha, B. M., et al., (2017). Dashavidha pariksha- a practical approach to clinical method of diagnosis. *International Ayurvedic Medical Journal [Online].*

Narayana, A., (2012). *Handbook of Anupana Pathya and Apathya (Adjuvant Wholesome & Unwholesome Regimen).* Central Council for Research in Ayurveda and Siddha.

Narayanaswamy, V., (1981). "Origin and development of Ayurveda: (a brief history*)." Ancient Science of Life, 1*(1) 1–7.

Otunola, G., & Afolayan, A., (2020). *A Review of the Antidiabetic Activities of Ginger.* 10.5772/intechopen.88899.

Palwankar, S. M., Kale, P. P., Kadu, P. K., & Prabhavalkar, K., (2020). Assessment of antidiabetic activity of combination of *Murraya koenigii* leaves extract and *Vitis vinifera* seeds extract in alloxan-induced diabetic rats. *J. Rep. Pharm. Sci., 9*, 79–85.

Parivallal, T. (2007). Type 2 diabetes in South Asians: Epidemiology, risk factors, and prevention. Diabetes in Ancient India. Jaypee Brothers Medical Publishers, 1–19.

Posada, E., (2017). The Ayurveda natural medicine system and its environmental implications. *Environ. Sci. Ind. J., 13*(4), 144.

Prem, K., Sudha, R., Arunjyothi, B., Chakrapani, P., & Rojarani, A., (2017). Evaluation of Antidiabetic activity of *Gymnema sylvestre* and *Andrographis paniculata* in streptozotocin-induced diabetic rats. *International Journal of Pharmacognosy and Phytochemical Research, 9*(1), 22–25.

Raghavendra, H., & Kumar, V., (2018). Antidiabetic Activity of Gymnema Sylvestris R. Br. *Recent Progress in Medicinal Plants: Metabolic Disorder: Diabetes, Part II, 46*, Studium Press LLC (USA).

Ranawat, A. K., (2017). A brief review about Ayurveda. *Int. J. Complement Alt. Med., 5*(6), 00170. doi: 10.15406/ijcam.2017.05.00170.

Sastry, J. L. N., (2005b). *Dravyaguna Vijnana*, (Vol. II). Chaukhambha Orientalia, Varanasi.

Sethi, J., Sood, S., Seth, S., et al., (2004). Evaluation of hypoglycemic and antioxidant effect of *Ocimum sanctum. Indian J. Clin. Biochem., 19*, 152–155. https://doi.org/10.1007/BF02894276.

Sevugan, A., Subramanian, K., Balamuthu, K., Abdul, B. A. A., Mohammed, A. A., & Mandali, V. R., (2008). Antidiabetic activity of leaf and callus extracts of *Aegle marmelos* in rabbit. *Science Asia, 34*, 317–321.

Sharma, P. V., (2001). *Susruta Samhita* (Vol. 2). Chaukhambha Orientalia, Varanasi.

Sharma, P. V., (2006b). *Dravyaguna Vijnana* (Vol. II). Chowkhambha Bharati Academy, Varanasi.

Sridharan, K., Mohan, R., Ramaratnam, S., & Panneerselvam, D., (2011). Ayurvedic treatments for diabetes mellitus. *Cochrane Database Syst Rev.,* (12), CD008288. doi: 10.1002/14651858.CD008288.pub2.

Upadhyay, A. K., Kumar, K., Kumar, A., & Mishra, H. S., (2010). *Tinospora cordifolia* (Willd.) hook. f. and thoms. (Guduchi) – validation of the ayurvedic pharmacology through experimental and clinical studies. *Int. J. Ayurveda Res., 1*(2), 112–121. doi: 10.4103/0974-7788.64405.

Warrier, P. K., & Nambiar, V. P. K., (1993). *Indian Medicinal Plants: A Compendium of 500 Species* (Vol. I). Orient Blackswan.

WHO (2004). *Monographs on Selected Medicinal Plants, 2.*

Yongchaiyudha, S., Rungpitarangsi, V., Bunyapraphatsara, N., & Chokechaijaroenporn, O., (1996). Antidiabetic activity of *Aloe vera* L. juice. I. Clinical trial in new cases of diabetes mellitus. *Phytomedicine, 3*(3), 241–243. doi: 10.1016/S0944-7113(96)80060-2. PMID: 23195077.

CHAPTER 4

Concept of Diabetes Mellitus and Antidiabetic Plant in the Unani System of Medicine

AHMAD ALI[1] and NIKHAT SHAIKH[2]

[1]*Department of Life Sciences, University of Mumbai, Vidyanagari, Santacruz (East), Mumbai – 400098, Maharashtra, India*

[2]*Research Officer Unani, RRIUM, Mumbai Under CCRUM, Ministry of AYUSH, Government of India, New Delhi, India, E-mail: drnikhat.unani@gmail.com*

ABSTRACT

According to the World Health Organization (WHO), diabetes mellitus (DM) is rapidly becoming one of the most common non-communicable diseases. It is characterized by hyperglycemia, glycosuria, and negative nitrogen balance. It mainly occurs due to lack of insulin secretion from beta cells of Pancreas and desensitization of insulin receptors. In Unani classical text, diabetes is described by many renowned Unani physicians, like Zakaria Al Razi, Ali Ibn Abbas Majoosi, Ibn Sina, Ismail Jurjani, Ibn Zuhar, and Ibn Hubal Baghdadi, with various names such as Ziabetus, Moattasha, Barkarya Qamamees, etc. According to Hippocrate, *Ziabetus Shakri* (DM) is a disease of *Sue Mizaj Haar Yabis*, and hence the *Barid Ratab Mizaj* herbs which exhibit properties of *Muqawwi-e-Gurda* are beneficial. On the other hand, Jalinoos described that the use of alkalizer and *Tabreed kulliya* and *Qabizatwa Hamizat* foods are the only treatment for *Ziabetus Shakri*. In the present chapter, the recent advances in the development of single drugs and compound formulations

Antidiabetic Potential of Plants in the Era of Omics. Deepu Pandita, PhD, Anu Pandita, MSc, & Chander Bhanu, MBBS (Eds.)

that have been reported to possess hypoglycemic and anti-hyperglycemic properties in Unani Pharmacopeias have been discussed.

4.1 INTRODUCTION

"Diabetes" originated from the Greek verb 'Diabeinein' where 'Dia' is a synonym for across/apart and 'Bainein' means to walk or stand with legs asunder. Hence, its derivative 'Diabetes' literally means 'the one that straddles' for the reason that one of the symptoms of DM is frequent urination. Another meaning of diabetes is 'Diabanmo' which means 'to pass through' or 'to run through' or 'drain off.' 'Mellitus' is a Latin word that stands for honey (Ibn, 2007, 2010; Khan, 1983). Diabetes is also known as '*Zayasakoomus*' and '*Qaramees*.' In Arabic, it is called '*Dawwara*' *or* '*Barkarya*' because the diabetic patients pass urine frequently, and hence they are required to drink water frequently. This cycle of drinking water and eliminating the same in the form of urine resembles spin round. *Arastoo* (Aretus)—one of the famous Unani physician, named the disease 'Diabainein,' which means 'a siphon.' The Greek version is '*Ziabetus*,' whereas in Arabic it is known as '*Dolaab*' which means 'water mill well'—the function of which is to draw water from the well and disperse it. In diabetes, the patient drinks plenty of water and expels it via urine, just like a water mill well wheel (Qarshi, 2011). Hence it is also known as '*Moatasha*' (thirst stimulant) and '*Salasul Boul*' (frequent urination) as these are the common manifestation of diabetes (polydipsia and polyuria). Additionally, it is also termed as '*Istisqa-un-Namas*' which means ascites of urinary bladder (Khan, 1983; Majoosi, 2010; Jurjani, 1996; Ibn Zohar, 1986; Ibn Hubal, 2005). Altogether, these terminologies refer to the persistently full bladder and urinary incontinence of diabetic patients (Khan, 1983).

4.1.1 CLASSIFICATION OF DIABETES

Two major classifications of DM (represented in Figure 4.1) are mentioned in Unani classical texts. The *Ziabetus Sada* or *Ghair Shakri* and *Shakri* are the major classification systems based on the presence or absence of urinary sugar. On the other hand, *Ziabetus Haar* and *Baarid* are classifications based on the intensity of signs and symptoms, and duration of disease (Qarshi, 2011). The detailed description of *Ziabetus Sada* or *Ghair Shakri* closely resembles that of the modern term 'Diabetes insipidus' which literally means passing lots of 'insipid' or 'tasteless urine' or 'urine without sugar.' Likewise, *Ziabetus Shakri* is accurately coined for 'diabetes mellitus (DM)'—a metabolic condition characterized with polydipsia, polyphagia, and polyuria (Khan, 2006).

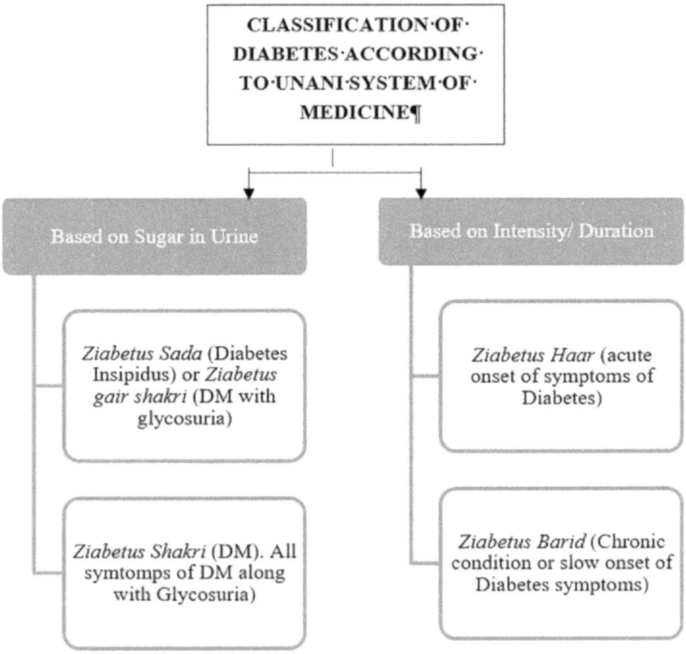

FIGURE 4.1 Classification of diabetes in Unani classical texts.

4.1.2 HISTORICAL BACKGROUND

The historical evidence of DM has been recorded since time immemorial and it has apparently overwhelmed mankind for decades. In the earliest civilizations as well as the Ebers Papyrus, diabetes is described as a disease which represents excessive thirst, loss of weight, excessive urination along with incidences of boils and infections. Aulus Cornelius Celsus introduced the initial clinical presentation of diabetes. Later, Aretaeus of Cappadocia presented the accurate details and coined the word "Diabetes" which means 'to drain off' in Greek. Aretaeus defined diabetes as an alarming condition which leads to progressive melting down of the soft tissues or body fat into urine, consequently causing emaciation in later stages. The affected person is in painful condition with a short lifespan. They experience gluttonous thirst leading to frequent urination and disproportionate nausea, vomiting, and restlessness (Nazamuddin, 2013; Jani et al., 2006; Barthold, 2004; Srinivasan and Ramarao, 2007). The father of Medicine *Buqrat* (Hippocrates) described it as a disease with excessive urination and body emaciation (Savona-Ventura, 2003). 'Diarrhea urinosa' (polyuria) and 'dipsakos' (polydipsia) are the terminologies used to define diabetes by *Jalinoos* (Galen). He reported it as a disease occurring due to *Zoaf Kulliya* (weakness of Kidney) in which the retentive power of kidney is deteriorated (Savona-Ventura, 2003; Sanders, 2002; Singh et al., 2010). *Ibn Sina* (Avicenna), in his famous historical encyclopedia *Al Qanoon Fil Tibb* (Canon of Medicine), labeled the disease as '*Al Dolaab*' (Charkhi or water wheel) and '*Zalqul Kulliya*' (diarrhea of the kidneys). *Ibn Sina* was the first to clearly mention the complications of diabetes like mental troubles, impotence, gangrene, and diabetic coma. He differentiated the clinical features of diabetes from other diseases (Barthold, 2004; Savona-Ventura, 2003; Kirchhof et al., 2008). The Traditional Chinese Medicine (TCM) and Japanese literature also related diabetes with sweet urine which fascinated dogs and insects (Sanders, 2002; Singh et al., 2010). '*Madhumeha*' was a classical term mentioned in Ayurveda classical books like *Sushruta*, *Charaka*, and *Vaghbata*, which represents sweet, sticky urine which engross the insects easily (Eknoyan and Nagy, 2005). The name "Mellitus" is Latin for honey, due to the sweet taste of urine from diabetic patients, and was described by Thomas Willis (Nazamuddin, 2013). Later in 1766, Mathew Dobson evidenced the same fact of excess sugar in urine and blood (Samhita, 2008).

4.2 UNANI PATHOPHYSIOLOGY OF DIABETES MELLITUS (DM)

The causes of DM, mentioned in classical texts, are derangement in the renal functions due to hypertrophy or dilatations of the lumen of renal tubules. Hence, as soon as the glomerular filter enters into the lumen of the kidney, it is excreted in the form of urine (Ibn Sina, 2010). One of the important causes of diabetes is abnormally high heat production in the body with or without the involvement of humors or matter (Ibn Sina, 2010). The excessive heat in the kidneys results in the extraction of water from the liver which, in turn, extracts water from its associated organs. This process produces excessive dehydration in the body leading to thirst (Jurjani, 1996). To quench the thirst, a person drinks more water and passes more urine. Due to this reason, one of the observed symptoms of DM is a decrease in the physiological temperature of organs like liver and kidney, or generalized coldness of the entire body. This change occurs as a result of intake of chilled water or excessive cold which leads to the shrinkage of kidneys. Among the major complications of diabetes mentioned in classical texts is tuberculosis (Rengachary et al., 2008).

4.2 OTHER FACTORS RESPONSIBLE FOR DIABETES MELLITUS (DM) MENTIONED IN UNANI TEXTS

4.2.1 AMRAZ-E-JIGAR (LIVER DISORDERS)

Most of the Unani scholars have the opinion that *Zoaf e Jigar* (weakness of liver) is an important cause of diabetes. The liver is responsible for regulating the blood sugar levels; however, the fat in the liver makes it less responsive to *Haiwani Ajza* and leaves too much glucose in the blood. This leads to type 2 diabetes. Even recent researches have proved that type 2 DM patients are prone to NAFLD (Non-alcoholic fatty liver Diseases) or NASH (Non-alcoholic fatty steatohepatitis).

4.2.2 AMRAZ-E-BANQARAS (PANCREATIC DISORDERS)

A group of scholars had the opinion that the pancreas produces the *Rutubat Batina* (recently known as hormone insulin) which allows sugar from the bloodstream to enter the body's cells where it is used for energy. However, due to the disturbances in the function of *Banqaras* (pancreas), it produces an insufficient amount of *Rutubat Batina* (hormone insulin), or the body cannot use insulin properly, or both. This results in a build-up of glucose in the blood and causes type 2 Diabetes.

4.2.3 AMRAZ-E-NIZAM-E-ASBI (DISORDERS OF CENTRAL NERVOUS SYSTEM)

A group of scholars have the opinion that disturbances in the central nervous system lead to the production of excessive sugar in the liver which gets secreted in blood and urine. Recent researches confirm that the central nervous system plays a vital role in glucose homoeostasis by controlling pancreatic secretion and insulin sensitivity. These mechanisms potentially enhance glucose control.

4.2.4 IV VENTRICLE OF BRAIN

Most of the scholars have the opinion that disturbance in the function of the IV ventricle of the brain results in glycosuria (urine in sugar).

4.2.5 HEMOLYSIS

A group of scholars have the opinion that the breakdown of RBCs is also responsible for type 2 DM; however, the exact mechanism is still unknown.

4.2.6 MUTADI AMRAZ (INFECTIOUS DISEASE)

Few scholars have the opinion that diabetes is an infectious disease caused by a microorganism. The specific microorganism and mechanism is still unknown.

4.2.7 ZOAF-E-AZLAAT (WEAKNESS OF MUSCLES)

Few scholars have the opinion that the lack of *Usara Azliya* (myoglobin) causes type 2 DM. Hence, the intake of raw meat water regularly by patients may be beneficial in reversing the diabetic condition.

4.2.8 AMRAZ-E-GURDA (KIDNEY DISORDERS)

Most of the scholars have the opinion that *Zof e Gurda* (weakness of kidneys), *Ittesae Gurda wa Majrae Bole* (dilatation of renal tubules) and *Sue Mizaj har Gurda* (alterations in temperament of kidneys) are the chief causes of type 2 diabetes.

4.2.9 ZOF-E-MEDAW HAZAM (WEAKNESS OF STOMACH AND INDIGESTION)

The *Zof e Medaw Hazam* is reported as one of the major reasons for the development of DM in Unani classical texts.

4.2.10 *INFEALAT-E-NAFSANIA (EMOTIONAL AND PSYCHOLOGICAL DISTURBANCES)*

The *Infealat-e-Nafsania* which includes conditions like stress, anxiety, strain, and tension are also mentioned as the main causes of type 2 diabetes.

4.2.11 *ASBAB-E-SITTA ZAROORIYAH (SIX ESSENTIAL FACTORS)*

Overeating, sedentary life, lack of exercise, sleeplessness, chronic constipation, and nutritional imbalance are the six essential factors (SEF) that are most imperatively responsible for the development of type 2 diabetes.

4.2.12 *GHIZA (DIET)*

Unhealthy dietary habits including intake of junk food, fast food, soft drinks, and addictions to alcohol and cigarette are accountable factors for type 2 diabetes.

4.3 SIGNS AND SYMPTOMS OF DIABETES MELLITUS (DM) IN CLASSICAL TEXTS

Initially, the excessive thirst, excessive urination and excessive appetite are the most peculiar symptoms of DM mentioned in the classical texts. Other symptoms like weight loss, low sexual urge, infertility, debility, fever, extreme weakness, emaciation, unhealed wound and organopathy are present in later stage of diabetes (Figure 4.2) (Khan et al., 1415).

4.3.1 *FUNDAMENTALS OF TREATMENT*

As per Unani concept, DM is considered to be *Sue Mizaj Haar*. Accordingly, the principal line of treatment of DM is opposite to *Hararat* which means *Tabreed wa Tarteeb* (cold and moist regimens) in all the aspects including diet, regimens, location, and climate (Qureshi and Hikmat, 2011). Another important aspect of line of treatment of Diabetes is *Izala Sabab* means *Islahe Jigar wa Gurda* (correction of dis-temperament of liver and kidney) and *Islahe Quwwa* (retentive power) of kidney (Figure 4.3), which is maintained by dietary modifications, various regimes, single, and compound medicine, etc.

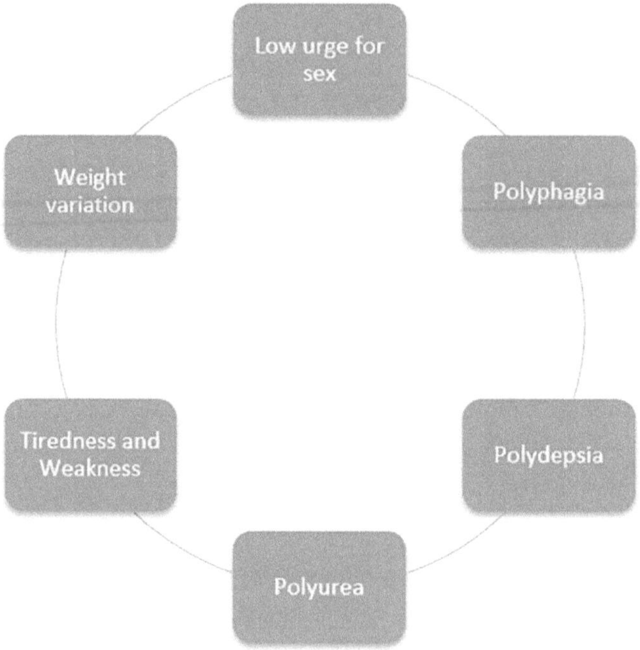

FIGURE 4.2 Common signs and symptoms of diabetes are mentioned in classical Unani texts.

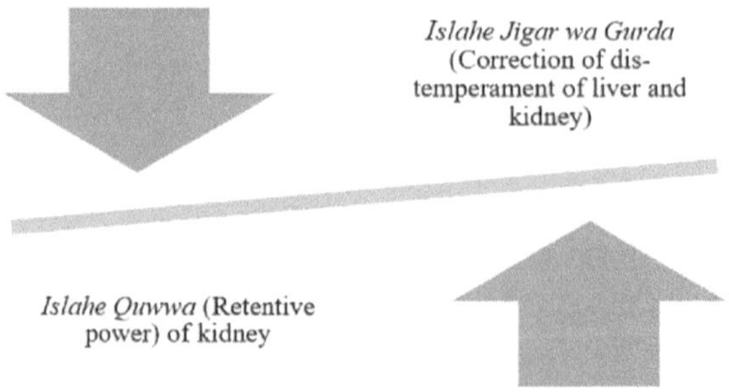

FIGURE 4.3 Possible line of treatment based on causes of diabetes in Unani classical texts.

4.4 THERAPEUTIC APPROACH FOR MANAGEMENT OF DIABETES MELLITUS (DM) IN UNANI SYSTEM OF MEDICINE

Unani system offers several measures to combat diabetes. These include *Ilaj bil Ghiza* (dietary modification), *Ilaj Bil Tadbeer* (exercise, hammam, etc.), *Ilaj Bil Dawa* (pharmacotherapy) and, if necessary, *Ilaj Bil Yad* (surgical procedure). Together, they are known as the four lines of treatment (Figure 4.4) (Rahman et al., 2013; Ahmed and Anwar, 2009).

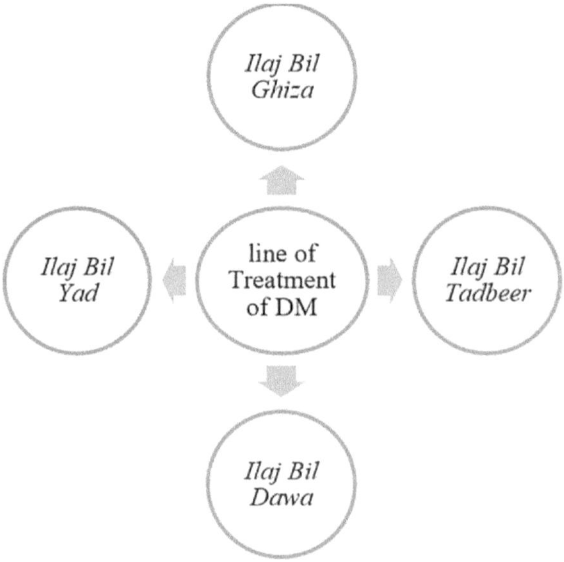

FIGURE 4.4 Four lines of treatment in Unani system of medicine.

4.4.1 *ILAJ BIL GHIZA (DIETOTHERAPY)*

It is basically the modification in dietary habits (daily food and drinks) and lifestyle. These *Tadabeer* are effective in prevention as well as management of diabetes. According to Hippocrates, diabetes is a disease of *Sue Mizaj Haar Yabis*, and hence the use of *Barid Ratab* and *Muqawwi-e-Gurda* food and drinks are beneficial. The best example is *Ma-us-Shayeer* (Barley water). However, according to *Jalinoos*, the use of alkalizer and *Tabreed kulliya* is the sole treatment for diabetes. The use of *Qabiz wa Hamiz* foods and drinks are also beneficial for diabetic patients. These include green leafy vegetables, fruits, and nuts that have cold temperament, e.g., pumpkin, pomegranate fruit or its juice, musk melon, watermelon, pistachio, curd, *Lauki*, *Kheera*, and *Aalobukhara*. At the same time, it is better for diabetic patients to avoid cold drink or water. Sips of *Arqor Joshanda Gulab* are also the most recommended fluid in diabetes. Weight reduction by calorie-restricted diets and increased physical activity are the first-line therapy in the management of diabetes.

4.4.2 ILAJ BIL TADBEER (REGIMENAL THERAPY)

It is also called drugless or non-pharmacological therapy and mainly works by modification in lifestyle (six essential factors) of affected person (Rahman et al., 2013). Almost 30 *Tadabeer* are mentioned in classical text to combat the metabolic diseases like DM which includes *Hijama* (cupping), *Fasd* (venesection), *Irasale Alaq* (leeching), *Isahaal* (purgation), *Qayi* (emesis), *Huqna* (enema), *Riyazat* (exercise), *Taarique* (diaphoresis), etc. (Ahmed and Anwar, 2009). These regimens are used as detoxification methods for diabetes patients.

4.4.3 PRECAUTIONARY MEASURES OF DM IN UNANI SYSTEM OF MEDICINE

The Unani concept of health restoration and disease prevention *Asbab-e-Sitta Zaruriyya* (SEF), represented in Figure 4.5, plays an important role in diabetes prevention and establishment of a coordinated biological regularity mechanism (Jha et al., 2015; Ibn Zohar, 1986; Ibn Hubal, 2005).

4.4.4 ILAJ BILDAWA (PHARMACOTHERAPY) IN MANAGEMENT OF DIABETES MELLITUS (DM)

Unani treasures possess a lot of antidiabetic agents of natural origin. These agents are not only effective in controlling glucose level but also are equally safe, or exhibit negligible side effects, as compared to their modern counterparts. Unani physicians prescribe various *Mufrad* (single) and *Murakkab* (compound) drugs for successful management of DM, particularly type 2 diabetes. It was previously assumed that the causes of diabetes originate due to disorders of the kidneys. Hence all the prescribed drugs in the management of diabetes are based on cooling, and thus the protection, of the kidneys in the old Unani literature. The first principle of the treatment of diabetes, according to old Unani literature, is to quench the thirst. For this purpose, *Arq-e-Gulab* or *Usara-e-Gulab* is the prescribed remedy. Other drugs reported for management of DM are *Ma-us-Shayeer* (Barley water), *Ashreba Mauttiya Mubarrida* (cold syrup), *Qurs-e-Kafoor*, *Qurs-eTabasheer*, *Qurs-e-Ziabetes*, etc. The details of single and compound formulations are listed in subsections.

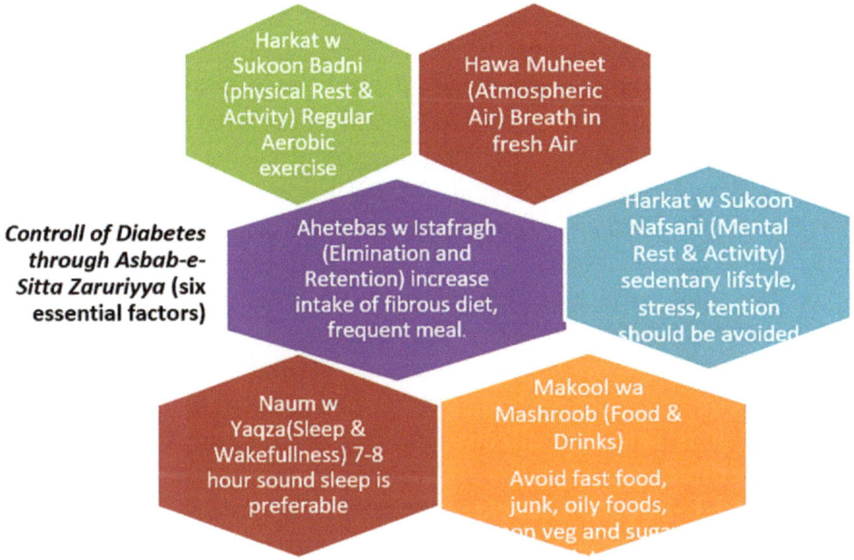

FIGURE 4.5 *Asbab-e-Sitta Zaruriyya* (six essential factors) in Unani system of medicine.

4.4.4.1 SINGLE DRUGS POSSESSING ANTIDIABETIC OR HYPOGLYCEMIC ACTIVITY AND USED IN UNANI SYSTEM OF MEDICINE

1. **Tukhm-E-Hayat (*Withania coagulans dunal*):** *Tukhm-e-Hayat* is commonly known as Indian cheesemaker (Ibn Sina, 2007; Gupta and Keshari, 2013; Chaterjee and Pakrashi, 1995). Dry fruits are commonly used for the treatment of DM in Northern India (Kirtikar and Basu, 1935; Gupta and Keshari, 2013). Many scientific studies have reported that an isolated alkaloid coagulanolide from *W. coagulants* fruits possess anti-hyperglycemic property in experimental

animal models with induced DM (Hemalatha et al., 2004; Akhtar et al., 2016).

2. **Hulba *(Trigonella foenum-gracum)*:** *Hulba*, commonly well-known as Fenugreek (*Methi*), is one of the most promising medicinal herbs used as an antidiabetic agent (Maurya et al., 2008; Raghuram et al., 1994). Ample scientific studies have revealed that the seeds of fenugreek possess hypoglycemic and antidiabetic activities (Modak et al., 2007).
3. **Neem (*Azadirachta indica*):** *Neem* is a member of the Meliaceae family and widely used in Unani, Ayurveda, and TCM, especially in the Indian subcontinent, for various ailments (Neelakantan et al., 2014; Brahmachari, 2004). Various scientific studies have been undertaken to evaluate the antidiabetic activity of Neem root bark extract on diabetes, and statistically significant results has been reported for reduction of blood sugar level (Ketkar and Ketkar, 2004; Patil et al., 2013; Dholi et al., 2011; Bhat et al., 2011).
4. **Asgandh (*Withania somnifera dunal*):** *Asgandh* is a renowned herb mentioned in classical texts for the management of various ailments like joint pain, insomnia, sexual debility, etc. Various scientific studies have been conducted to validate the hypoglycemic, diuretic, and anti-hyperlipidemic activities of *Asgandh* (Akhter et al., 2013; Rasool and Varalakshmi, 2006). The results have demonstrated that it is safe and effective in the management of diabetes and hyperlipidemia.
5. **Gond-E-Babool (*Acacia arabica*):** The studies on antidiabetic effects of *Acacia arabica* in diabetic rats have indicated its hypoglycemic activity (Gopala et al., 2006; Kavishankar et al., 2011; Kundlikrao, 2013).
6. **Bail (*Aegle marmelos L.*):** *Bail* is a well-known medicinal herb used in traditional system of medicine to treat various ailments like gastrointestinal tract related problems, gynecological disorders, infertility, etc. Various scientific studies have revealed that *Bail* comprises of antidiabetic and hypoglycemic activity in diabetic rats (Patel et al., 2012; Arumugam et al., 2008; Kamalakkannan and Prince, 2005; Kesari et al., 2006; Narendhirakannan et al., 2006; Narender et al., 2007).
7. **Gul-E-Ghafis (*Gentiana olivieri grise*):** *Gul e Ghafis*, mentioned in classical texts, is used to treat various ailments including urinary tract, stomach, hepatic, nervous, and reproductive system related disorders. It possesses numerous pharmacological activities such as

antibacterial, antifungal, hepatoprotective, and antidiabetic effect. Various scientific studies have also explored the anti-hyperglycemic, insulin-releasing, and insulin-like activity of *Gul e Ghafis* in diabetic mice (Frode and Medeiros, 2008; Arya et al., 2011).

8. **Lahesun (*Allium sativum*):** *Lahesun* is most commonly used as food and medicine for various metabolic disorders such as atherosclerosis, hyperlipidemia, thrombosis, hypertension, and diabetes (Gray and Flatt, 1997; Banerjee and Maulik, 2002). Numerous scientific studies have also proved the antidiabetic and anti-hyperlipidemic properties of *Allium sativum* (WHO, 1999; Sheela and Augusti, 1992; Ozougwu and Eyo, 2010; Hattori et al., 2005; Liu et al., 2006; El-Demerdash et al., 2005).
9. **Sibr/Elva (*Aloe Vera Extracts*):** In Unani classical text, *Elva* is described in the name of *Sibr* which is spread throughout the domain (Eidi, 2006). Numerous scientific studies have been conducted to evaluate the antidiabetic, anti-hyperlipidemic, and anti-oxidative activities of *Sibr* in the management of diabetes (Racovita et al., 2015; Ramirez et al., 2020; Tanaka et al., 2006). The outcomes have demonstrated that *Sibr* contains a significant amount of chromium, manganese, and zinc which potentiates its antidiabetic, anti-hyperlipidemic, and anti-oxidative activities (Makheswari and Sudarsanam, 2012).
10. **Darchini (*Cinnamomum verum*):** *Darchini* or the bark of cinnamon is traditionally used as a drug as well as spice in India. In Unani classical text, its therapeutic efficacy is well documented (Mohamed, 2011; Sastri, 1950). Additionally, various scientific studies prove the antidiabetic activity of cinnamon in both hyperglycemia and hyperlipidemia (Rahman et al., 2013; Jakhetia et al., 2010).
11. **Tukhm-E-Konch (*Mucuna pruriens*):** *Tukhm e Konch* belongs to the family *Fabaceae*, and is also known as velvet bean. It is an important herb used in the treatment of impotence, DM, bones fractures, cough, dog bite, cancer, and psychosis (Taher et al., 2006; Khan, 1993; Rafiquddin, 1985; Hasan, 1894). One of the scientific studies revealed that the ethanolic seed extracts of *M. pruriens* have antidiabetic and weight reduction properties (Kareem, 1765).
12. **Karela (*Momorda charantia*):** *Karela* (bitter melon/pare) possess anti-microbial, hypoglycemic, and anti-carcinogenic properties. It is an effective hypoglycemic drug due to presence of *charantin* and insulin-like peptides (Majekodunmi et al., 2011; Beloin et al., 2005;

Grover and Yadav, 2004; Kumar et al., 2010; Virdia et al., 2003; Kahn and Flier, 2000). Numerous research studies have revealed that *Karela* improves insulin secretion and endorses new growth of beta cells (Shetty et al., 2005).

13. **Jamun (*Syzygium cumin*):** *Jamun* is one of the important medicinal plants which have the potential to treat DM, cancer, gout, and ischemic heart disease. Several studies investigated the antidiabetic activity of seed kernels of *Syzygium cumin* Linn against the streptozotocin-induced diabetic rats. All these studies reported a significant reduction in blood glucose levels (BGL) (Lotlikar and Rao, 1966; Ghani, 1998; Duke, 2002; Bhattacharya, 2008; Bhatia and Bajaj, 1972; Timbola et al., 2002).
14. **Kishneez (*Coriander sativum* L.):** *Tukhm Kishneez*, in Unani literature, is reported to possess numerous medicinal properties like anti-gastritis, anti-ulcer, sedative, anti-anxiety, and anti-stress activity. One of the research studies confirmed that *Kishneez* also possess antidiabetic potential against streptozocin induced wistar diabetic albino rats (Pandey and Khan, 2002).
15. **Kalonji (*Nigella sativa*):** It is also known as *Shooneez* in classical texts. *Ibn Sina*, in his encyclopedia *Al-Qanoon fit-Tibb*, mentioned *Kalonji* to possess numerous medicinal benefits. Ample research studies have also validated the significant reduction of the blood glucose level on intake of *Kalonji*, and hence it is strongly claimed to be a potent antidiabetic drug (Ahmed et al., 2018; Hamdan et al., 2019; Kaatabi et al., 2015).
16. **Gudmaar Booti (*Gymnema sylvestre*):** *Gudmaar booti* is a popular herb used in traditional system of medicine for diabetes, urinary infections, and respiratory ailments. Various scientific studies also confirm that *Gudmaar* possess potent antidiabetic and hypoglycemic effects (Daryabeygi et al., 2017; Makheswari and Sudarsanam, 2012).
17. **Satawar (*Asparagus racemosus*):** *Satawar* is an important medicinal plant of tropical and subtropical India. It is used for treatment of various gynecological and sexual disorders. Various studies have revealed the fact that *Satawar* possess potent antidiabetic and hypoglycemic property, and may be used as a novel antidiabetic compound or a dietary adjunct for the management of diabetes (Kodama et al., 2005; Hannan et al., 2012).

18. **Kachnaal (*Bauhinia variegate*):** In Unani classical texts, it is mentioned as a general tonic, astringent, and a remedy for skin diseases. Several research studies conducted to evaluate the anti-diabetic effect of *B. variegate*, and the results revealed a significant reduction in fasting and postprandial glucose levels in both types of diabetes (Akhtar and Shah, 1993; Nicolle et al., 2011; Patel and Mishra, 2011).
19. **Zanjabeel (*Zingiber officinale*):** *Zanjabeel, also called Adrak or Sounth in dry form,* is a popular cooking spice which possesses numerous medicinal properties (Mohan and Nandhakumar, 2014). Ample amount of research studies have demonstrated the *potent antidiabetic property of Z. officinale in* diabetic rats (Reza et al., 2017; Ibn Sina, 2010; Kato et al., 2006; Oliveira et al., 2005).
20. **Amla (*Emblica officinale*):** *Amla* is a dynamic herb used as food and medicine in Unani system of medicine. Various studies have been conducted to evaluate the antidiabetic activity, and results revealed a significant reduction in blood glucose level in type 2 diabetes (Rani et al., 2012; Kianbakht and Dabaghian, 2013; Akhtar et al., 2011; D'Souza et al., 2014).
21. **Tamar-Ul-Hind (*Tamarindus indica*):** *Tamar-ul-Hind* is a Persian word which means dates of India. Various research studies confirm its potent hypoglycemic and antidiabetic activities (Vikrant and Sharma, 2011; Parvin et al., 2013; Maiti et al., 2004).
22. **Gul-E-Gudhal (*Hibiscus rosasinensis*):** It is a well-known traditional herb, commonly known as Chinese hibiscus or China rose. It is widely used for various inflammatory conditions, gynecological disorders, and respiratory ailments. It can be used as a topical agent or as a supplement. A scientific study was conducted to validate its claim for antidiabetic property, and the reported results were encouraging for reducing the blood sugar level in diabetic rats (Maiti et al., 2005).
23. **Halela (*Terminalia chebula*):** Three types of *Halela* (*Halelasiyah, Halelazard,* and *Halelakabuli*) are mentioned in Unani classical texts which are used as a *Malayan* (laxative) in various compound formulations. A scientific research revealed that *Halela* possesses antidiabetic and reno-protective effects in diabetic rats (Mandade and Sreenivas, 2011).
24. **Gilo (*Tinospora cordifolia*):** *Gilo* is a popular medicinal plant, mentioned in classical texts, that exhibit antipyretic, analgesic,

anti-inflammatory, and antidiabetic potential (Nalamolu and Nammi, 2006; Kirtikar and Basu, 2003; Anonymous, 2003; Duke, 2002). Ample research studies also validate its safety and efficacy in lowering the blood sugar levels in diabetic rats (Kulkarni and Ansari, 2004; Stanely et al., 2000).

25. **Turbud (*Operculi naturpethum*):** *Turbud* is a well-known drug used as *Mushil* (purgative) and *Malayan* (laxative) in Unani line of treatment. Basically *Turbud Sufaid* (*Turbud* are of two types; *Sufaid* and *Siyah*) is commonly used for medicinal purposes while *Turbud Siyah* is not therapeutically recommended due to its emetic side effects. A scientific study was conducted to evaluate the efficacy of *Turbud Sufaid* in diabetic rats, and the results demonstrated that it possesses potent antidiabetic and hypoglycemic effects (Grover et al., 2000).

4.4.4.2 COMPOUND UNANI FORMULATIONS MENTIONED IN CLASSICAL TEXTS FOR MANAGEMENT OF DIABETES MELLITUS (DM)

1. ***Kushta Baiza-e-Murgh*:** Post Baizae Murgh (*Gallus gallus domesticus*), Salt, and Citerous lemon are chief ingredients of this formulation (Pulipaka et al., 2012) used for lowering blood sugar level in classical texts.
2. ***Kundri*:** It is a classical formulation mentioned for the treatment of DM. It has total 11 ingredients while Kunjad Siyah *(Sesamum indicum Linn),* Mur Makkai *(Commiphora Myrrha)*, Kundur (*Boswellia serrata* Roxb) are the chief ingredients of Kundri (Pulipaka et al., 2012), which having good results in management of DM and its associated symptoms.
3. ***Safoof-e-Sandal Ziabetus Wala*:** It is a classical formulation used for the treatment of DM (Hkm. Kabir-ud-din, 1967). Formulation has 10 ingredients in which Sandal Safed *(Santalum album),* Tukhme-Kahu *(Lactuca sativa),* Kateera *(Sterculia urens),* Tukhme Khurfa (*Portulaca oleraceae* Linn) and Samaghe Arabi *(Acacia arabica)* are the chief ingredients. The ingredients of this formulation having hypoglycemic activity in various research studies.
4. ***Safoof e Ziabetus*:** It is also a classical formulation used for the management of DM (Hkm. Kabir-ud-din, 1967). It contain Gilo khushk (*Tinospora cordifolia)* and Gudmaar booti *(Gymnema sylvestre)* which have potent antidiabetic activity.

Safoof-e-Hindi, Habb-e-Ziabetus, Qurs-e-Kafoor, Qurs-e-Tabasheer, Qurs-e-Ziabetus, Dawa-e-Ziabetus, Dawa-e-Ziabetus Sard and *Safoof-e-Ziabetus* (Kirmani et al., 2019) are various important formulations mentioned in classical text for management of DM.

4.5 CONCLUSION

Even though plenty of scientific work has been conceded regarding the etiology and pathophysiology of DM, and numerous mechanisms and associated theories have come to our understanding, the medical fraternity is unable to manage DM in effectual approach (preventive and curative aspect of diabetes). It is a decisive phase to consider whether the alternative medicine or traditional system of medicine may offer a better management potential towards all the health aspects (mental, physical, emotional, and spiritual) of diabetic patients. The Unani system of medicine is a comprehensive medical system, which meticulously deals with a variety of diseases. It offers preventive, therapeutic, and rehabilitative approach towards diabetes management. The Unani system considers diabetes as a *Sue Mizaj Haar* of various organs like kidney, liver, etc. Hence, the correction of dis-temperament by opposite *Mizaj* foods and medicines is an effective strategy for diabetes management. A holistic approach of Unani medicine also emphasizes on correction of *Asbab Sitta Zarooriyah* (six essential factors) which includes food and drinks, daily lifestyle, climate, resident, sleep, and wakefulness, balance in physical and mental rest and exertion and bowel regulation, which is deserted by contemporary system of medicine. Numerous Reims (*Tadaabeer*) which includes exercise, massage, steam bath, leeching, and *Hijama* therapy along with single and compound oral formulation is also offered in classical texts to manage DM successfully.

4.6.1 METHOD

Unani classical books and relevant articles are the major source of reference for compilation of this chapter. However, published scientific data from Google Scholar, Scopus, PubMed, and Science Direct were also utilized for literature appraisal.

KEYWORDS

- **diabetes**
- **hyperglycemia**
- ***sue mizaj***
- **unani pharmacopeia**
- ***Ziabetus shakkri***

REFERENCES

Ahmed, S. K., Chakrapani, C., Sampath, D., & Sunil, M., (2018). Evaluation of antidiabetic activity of fruit of *Coriandrum sativum*. Linn methanolic extract in streptozocin induced diabetic Wistar albino rats. *Int. J. Basic Clin. Pharmacol., 1*(7), 121–125.

Ahmed, T., & Anwar, M., (2009). Clinical importance of leech therapy. *Indian J. Trad. Knowledge, 8*, 443–445.

Akhtar, M. S., & Shah, M. V., (1993). Elemental constituents of antidiabetic screening of a folklore medicinal plant prescription. *Ind. J. Toxicol. Occup. Envt. Health, 2*, 46.

Akhtar, M. S., Ramzan, A., Ali, A., & Ahmad, M., (2011). Effect of amla fruit (*Emblica officinalis* Gaertn.) on blood glucose and lipid profile of normal subjects and type 2 diabetic patients. *Int. J. Food Sci. Nutr., 62*(6), 609–616.

Akhtar, M. W., Naseer, M., & Siddiqi, M., (2016). Anti-hyperglycemic effect of some Unani formulations in the management of diabetes mellitus type 2. *Int. J. Universal Pharmacy BioSci., 5*(3), 213–224.

Akhter, R., Mahabub-Uz-Zaman, M., Rahman, M. S., Khatun, M. A., Abdullah, A. M., Ahmed, N. U., & Islam, F., (2013). Comparative studies on antidiabetic effect with phytochemical screening of *Azadirachta indicia* and *Andrographis paniculata*. *IOSR J. Pharm. Biol. Sci., 5*(2), 122–128.

Anonymous, (2003). *The Wealth of India*. (Vol. I A-F, II G-Ph, III Pi-Z, pp. 207, 208, 33–35, 215, 216). New Delhi: Council of Scientific and Industrial Research.

Arumugam, S., Kavimani, S., Kadalmani, B., Ahmed, A. B. A., Akbarsh, M. A., & Raod, M. V., (2008). Antidiabetic activity of leaf and callus extracts of *Aegle marmelos* in rabbit. *Sci. Asia, 34*(3), 317–321.

Arya, V., Thakur, N., & Kashyap, C. P., (2011). Medicinal vigor of gul-e-ghafis (*Gentiana olivieri* Griseb): A review. *Unani Res., 1*(1), 12–17.

Banerjee, S. K., & Maulik, S. K., (2002). Effect of garlic on cardiovascular disorders: A review. *Nutr. J., 1*, 4–15.

Barthold, S. W., (2004). Unsung heroes in the battle against diabetes. *ILAR J., 45*(3), 227–230.

Beloin, N., Gbeassor, M., Akpagana, K., Hudson, J., De Soussa, K., Koumaglo, K., & Arnason, J. T., (2005). Ethnomedicinal uses of *Momordica charantia* (*Cucurbitaceae*) in Togo and relation to its phytochemistry and biological activity. *J. Ethnopharmacol., 96*(1, 2), 49–55.

Bhat, M., Kothiwale, S. K., Tirmale, A. R., Bhargava, S. Y., & Joshi, B. N., (2011). Antidiabetic properties of *Azardiracta indica* and *Bougainvillea spectabilis*: *In vivo* studies in murine diabetes model. *Evid Based Complement. Alternat. Med., 561625*.

Bhatia, I. S., & Bajaj, K. L., (1972). Tannins in black-plum (*Syzygium cumini* L.) seeds. *Biochem. J., 128*(1), 56P.
Bhattacharya, S. K., (2008). *Hand Book of Medicinal Plants* (5th edn., p. 338). Jaipur, Pointer Publication.
Brahmachari, G., (2004). Neem- an omnipotent plant: A retrospection. *Chem Bio Chem., 5*(4), 408–421.
CCRUM (2014). *Standard Unani Treatment Guidelines for Common Disease*. CCRUM: New Delhi.
Chaterjee, A., & Pakrashi, S. C., (1995). *The Treatise on Indian Medicinal Plants* (Vol. IV, pp. 207, 208). Publication & Information Directorate CSIR New Delhi.
D'souza, J. J., D'souza, P. P., Fazal, F., Kumar, A., Bhat, H. P., & Baliga, M. S., (2014). Antidiabetic effects of the Indian indigenous Fruit *Emblica officinalis* gaertn: Active constituents and modes of action. *Food Funct., 5*(4), 635–644.
Daryabeygi, K. R., Golzarand, M., Ghaffari, M. P., & Djafarian, K., (2017). *Nigella sativa* improves glucose homeostasis and serum lipids in type 2 diabetes: A systematic review and meta-analysis. *Complement Ther. Med., 35*, 6–13.
Dholi, S. K., Raparla, R., Mankala, S. K., & Nagappan, K., (2011). *In vivo* antidiabetic evaluation of neem leaf extract in alloxan-induced rats. *J. Appl. Pharma. Sci., 1*(4), 100–105.
Duke, J. A., (2002). *Handbook of Medicinal Herbs* (2nd edn., pp. 361–422). CRC Press: London.
Eidi, A., Eidi, M., & Esmaeili, E., (2006). Antidiabetic effect of garlic (*Allium sativum* L.) in normal and streptozotocin-induced diabetic rats. *Phytomed., 13*(9, 10), 624–629.
Eknoyan, G., & Nagy, J., (2005). A history of diabetes mellitus or how a disease of the kidneys evolved into a kidney disease. *Adv. Chronic Kidney Dis., 12*(2), 223–229.
El-Demerdash, F. M., Yousef, M. I., & El-Naga, N. I., (2005). Biochemical study on the hypoglycemic effects of onion and garlic in alloxan-induced diabetic rats. *Food Chem. Toxicol., 43*(1), 57–63.
Fröde, T. S., & Medeiros, Y. S., (2008). Animal models to test drugs with potential antidiabetic activity. *J. Ethnopharmacol., 115*(2), 173–183.
Ghani, N. M., (1998). *Khazainul Advia* (Vol. 2, pp. 240, 241). Idara Matbuat-i- Sulemani, Lahore (Pakistan).
Gray, A., & Flatt, P., (1997). Pancreatic and extra-pancreatic effects of the traditional antidiabetic plant, *Medicago sativa* (lucerne). *Brit. J. Nutr., 78*(2), 325–334.
Grover, J. K., & Yadav, S. P., (2004). Pharmacological actions and potential uses of *Momordica charantia*: A review. *J. Ethnopharmacol., 93*(1), 123–132.
Grover, J. K., Vats, V., & Rathi, S. S., (2000). Anti-hyperglycemic effect of *Eugenia jambolana* and *Tinospora cordifolia* in experimental diabetes and their effects on key metabolic enzymes involved in carbohydrate metabolism. *J. Ethnopharmacol., 73*(3), 461–470.
Gupta, V., & Keshari, B. B., (2013). *Withania coagulans* dunal. (Paneer doda): A review. *Int. J. Ayurved. Herbal Med., 3*(5), 1330–1336.
Hamdan, A., Idrus, H. R., & Mokhtar, M. H., (2019). Effects of *Nigella sativa* on type-2 diabetes mellitus: A systematic review. *Int. J. Environ. Res. Public Health, 16*(24), 4911.
Hannan, J. M., Ali, L., Khaleque, J., Akhter, M., Flatt, P. R., & Abdel-Wahab, Y. H., (2012). Anti-hyperglycaemic activity of *Asparagus racemosus* roots is partly mediated by inhibition of carbohydrate digestion and absorption, and enhancement of cellular insulin action. *Br. J. Nutr., 107*(9), 1316–1323.
Hasan, M., (1984). *Kitabul-Tauzihul-Advia* (pp. 160, 169, 197). Gulzar Mohammadi, Meerut.

Hattori, A., Yamada, N., Nishikawa, T., Fukuda, H., & Fujino, T., (2005). Antidiabetic effects of ajoene in genetically diabetic KK-A(y) mice. *J. Nutr. Sci. Vitaminol., 51*(5), 382–384.

Hemalatha, S., Wahi, A. K., Singh, P. N., & Chansouria, J. P. N., (2004). Hypoglycemic activity of *Withania coagulans* Dunal in streptozotocin-induced diabetic rats. *J. Ethnopharmacol., 93*(2, 3), 261–264.

Ibn, H. B., (2005). *Kitabul Mukhtarat Fil Tib* (Vol. III, pp. 326–328). CCRUM, New Delhi.

Ibn, S., (2007). In: Kantoori, S. G. H., (ed.), *Al Qanoon Fil Tib* (Vol. II, pp. 328, 353, 354, 1031–1033). Idara e Kitab us Shifa: New Delhi YNM.

Ibn, S., (2010). *Al Qanoon Fit Tib,* (1st edn., Vol. III, Part 2, pp. 263, 264, 783–785). (Arabic) 1408 Hijri. Jamia Hamdard, New Delhi.

Ibn, Z., (1986). *Kitabut Taiseer (Urdu translation)* (pp. 160–163). CCRUM, New Delhi.

Jakhetia, V., Patel, R., Khatri, P., Pahuja, N., Garg, S., Pandey, A., & Sharma, S., (2010). Cinnamon: A pharmacological review. *J. Adv. Sci. Res., 1*(2), 19–23.

Jani, D. K., Nesari, M. T., Vijayakumar, D., & Loriaux, L. D., (2006). Diabetes and the Ebers 33 papyrus: 1552 BC. *The Endocrinologist, 16*(2), 55–56.

Jha, K., Garg, A., Narang, R., & Das, S., (2015). Hirudotherapy in medicine and dentistry. *J. Clin. Diagn. Res., 9*(12), ZE05–ZE07.

Jurjani, A. H. I., (1996). *Zakhira Khawarzam Shahi*, (Vol. II, Part 6, pp. 540, 541). Idara Kitabul Shifa, New Delhi.

Kaatabi, H., Bamosa, A. O., Badar, A., Al-Elq, A., Abou-Hozaifa, B., Lebda, F., Al-Khadra, A., & Al-Almaie, S., (2015). *Nigella sativa* improves glycemic control and ameliorates oxidative stress in patients with type 2 diabetes mellitus: Placebo-controlled participant blinded clinical trial. *PLoS One, 10*(2), e0113486.

Kabiruddin, H. K. M., (1967). *Bayaz-e-Kabir, Part-II,* B.K (pp. 48–50, 62.). Daftar Almasih, Ballimaran, Delhi.

Kahn, B. B., & Flier, J. S., (2000). Obesity and insulin resistance. *J. Clin. Invest., 106*(4), 473–481.

Kamalakkannan, N., & Prince, P. S., (2005). The effect of *Aegle marmelos* fruit extract in streptozotocin diabetes: A histopathological study. *J. Herb Pharmacother., 5*(3), 87–96.

Kareem, N. M., (1965). *Makhzanul- Advia* (Urdu) (Vol. I, pp. 316, 317, 328, 329, 418, 419). Munshi Nawal Kishore. Kanpur.

Kato, A., Higuchi, Y., Goto, H., Kizu, H., Okamoto, T., Asano, N., Hollinshead, J., et al., (2006). Inhibitory effects of *Zingiber officinale* roscoe derived components on aldose reductase activity *in vitro* and *in vivo*. *J. Agric. Food Chem., 54*(18), 6640–6644.

Kavishankar, G. B., Lakshmidevi, N., Murthy, S. M., Prakash, H. S., & Niranjana, S. R., (2011). Diabetes and medicinal plants: A review. *Int. J. Pharm. Biomed. Sci., 2*(3), 65–80.

Kesari, A. N., Gupta, R. K., Singh, S. K., Diwakar, S., & Watal, G., (2006). Hypoglycemic and antihyperglycemic trichomonal, biochemical and toxicological activities of methanolic extract and some carbazole alkaloids isolated from the leaves of *Murraya koenigii* growing activity of *Aegle marmelos* seed extract in normal and diabetic rats. *J. Ethnopharmacol., 107*, 374–379.

Ketkar, A. Y., & Ketkar, C. M., (2004). Various uses of neem products. In: *The Neem Tree* (pp. 518–525). Weinheim, Germany: John Wiley & Sons.

Khan, A. M., Razaq, M. A., & Ahad, M. A., (1415). *Mojizul Mahsha (Arabi)* (pp. 378–387). Mujtabai: New Delhi, Hijri.

Khan, M. A., (1983). *Haziq*, (pp. 387–390). Jasem Book Depot, New Delhi.

Khan, M. A., (2006). *Romooze Aazm [Persian]*, (Farsi) (Vol. I, pp. 139–141). CCRUM, New Delhi.

Khan, N. M., (1993). *A Scientific Study of Some Unani Drugs Used for Improving the Male Sexual Function*. Dept. of Ilmul. Advia A.K.T.C A.M.U. Aligarh.

Kianbakht, S., & Dabaghian, F. H., (2013). Improved glycemic control and lipid profile in hyperlipidemic type 2 diabetic patients consuming *Salvia officinalis* L. leaf extract: A randomized placebo. Controlled clinical trial. *Complement. Ther. Med., 21*(5), 441–446.

Kirchhof, M., Popat, N., & Malowany, J., (2008). A historical perspective of the diagnosis of diabetes. *UWOMJ, 78*(1), 7–11.

Kirmani, A. N. I. K., Kabiruddin, H., Najibuddin, A., & Qandi, S., (2019). *Moalijat Sharah Asbab Urdu Hakimi Ilaj Unani Treat of Disease* (Vol. 3). Idara Kitabushshafa.

Kirtikar, K. R., & Basu, B. D., (1935). *Indian Medicinal Plants* (Vol. III, 2nd edn., pp. 1777, 1778). Periodical Expert book Agency: Delhi.

Kirtikar, K. R., & Basu, B. D., (2003). *Indian Medicinal Plants with Illustrations* (Vol. I, No. 2, pp. 77–80, 116–120, 172–174). Dehradun: Oriental Enterprises.

Kodama, T., Miyazaki, T., Kitamura, I., Suzuki, Y., Namba, Y., Sakurai, J., Torikai, Y., & Inoue, S., (2005). Effects of single and long-term administration of wheat albumin on blood glucose control: Randomized controlled clinical trials. *Eur. J. Clin. Nutr., 59*(3), 384–392.

Krishna, H. N. B., Sangha, R. B., Misra, N., & Pai, M. R. S. N. P., (2006). Antianxiety activity of NRANX-C, a polyherbal preparation in rats. *Indian J. Pharmacol., 38*(5), 330–335.

Kulkarni, P. H., & Ansari, S., (2004). *The Ayurvedic Plants* (pp. 221, 313). Delhi: Sri Satguru Publications.

Kumar, D. S., Sharathnath, K. V., Yogeswaran, P., Harani, A., Sudhakar, K., Sudha, P., & Banji, D., (2010). A medicinal potency of *Momordica charantia*. *Int. J. Pharma. Sci. Rev. Res., 1*(2), 95–100.

Kundlikrao, I. R., (2013). Antidiabetic drugs in Ayurveda. *Int. Res. J. Pharm., 4*(6), 21–24.

Liu, C. T., Wong, P. L., Lii, C. K., Hse, H., & Sheen, L. Y., (2006). Antidiabetic effect of garlic oil but not diallyl disulfide in rats with streptozotocin-induced diabetes. *Food Chem. Toxicol., 44*(8), 1377–1384.

Lotlikar, M. M., & Rao, M. R. R., (1966). Pharmacology of a hypoglycemic principle isolated from the fruits of *Momordica charantia* Linn. *Ind. J. Pharmacol., 28,* 129.

Maiti, R., Das, U. K., & Ghosh, D., (2005). Attenuation of hyperglycemia and hyperlipidemia in streptozotocin-induced diabetic rats by aqueous extract of seed of *Tamarindus indica*. *Biol. Pharm. Bull., 28*(7), 1172–1176.

Maiti, R., Jana, D., Das, U. K., & Ghosh, D., (2004). Antidiabetic effect of aqueous extract of seed of *Tamarindus indica* in streptozotocin-induced diabetic rats. *J. Ethnopharmacol., 92*(1), 85–91.

Majekodunmi, S. O., Oyagbemi, A. A., Umukoro, S., & Odeku, O. A., (2011). Evaluation of the antidiabetic properties of *Mucuna pruriens* seed extract. *Asian Pac. J. Trop. Med., 4*(8), 632–636.

Majoosi, A. A., (2010). In: Kantoori, G. H., (ed), *Kamilus Sana'ah*, (1st edn., pp. 154, 467, 472, 527). Idara Kitabus Shifa: New Delhi.

Makheswari, M. U., & Sudarsanam, D., (2012). Database on antidiabetic indigenous plants of Tamil Nadu, India. *Int. J. Pharma. Sci. Res., 3*(2), 287–293.

Mandade, R., & Sreenivas, S. A., (2011). Antidiabetic effect of aqueous ethanolic extract of *Hibiscus rosasinensis* L on streptozotocin-induced diabetic rats and the possible morphologic changes in the liver and kidney. *Int. J. Pharmacol., 7,* 363.

Maurya, R., Jayendra, A., Singh, A. B., & Srivastava, A. K., (2008). Coagulanolide, a withanolide from *Withania coagulans* fruits and anti-hyperglycemic activity. *Bioorg. Med. Chem. Lett., 18*(24), 6534–6537.

Ministry of Ayush, (2006). *National Formulary of Unani Medicine*. Ministry of Health and Family Welfare, New Delhi. Govt. of India Department of AYUSH.

Modak, M., Dixit, P., Londhe, J., Ghaskadbi, S., & Devasagayam, T. P., (2007). Indian herbs and herbal drugs used for the treatment of diabetes. *J. Clin. Biochem. Nutr., 40*(3), 163–173.

Mohamed, E. A. K., (2011). Antidiabetic, anti-hypercholesteremic and antioxidative effect of aloe vera gel extract in alloxan-induced diabetic rats. *Australian J. Basic Appl. Sci., 5*(11), 1321–1327.

Mohan, S., & Nandhakumar, L., (2014). Role of various flavonoids. hypotheses on novel approach to treat diabetes. *J. Med. Hypotheses Ideas, 8*, 1–6.

Nalamolu, K. R., & Nammi, S., (2006). Antidiabetic and renoprotective effects of the chloroform extract of *Terminalia chebula* Retz. seeds in streptozotocin-induced diabetic rats. *BMC Complement. Altern. Med., 6*, 17.

Narender, T., Shweta, S., Tiwari, P., Papi, R. K., Khaliq, T., Prathipati, P., Puri, A., et al., (2007). Antihyperglycemic and antidyslipidemic agent from *Aegle marmelos*. *Bioorg. Med. Chem. Lett., 17*(6), 1808–1811.

Narendhirakannan, R. T., Subramanian, S., & Kandaswamy, M., (2006). Biochemical evaluation of anti- diabetogenic properties of some commonly used Indian plants on streptozotocin-induced diabetes in experimental rats. *Clin. Exp. Pharmacol. Physiol., 33*(12), 1150–1157.

Nazamuddin, M., (2013). *Evaluation of Hypoglycemic Activity of Qurs Tabasheer in Experimentally Induced Diabetes in Animal Model*. Thesis submitted to Rajiv Gandhi University of Health Sciences (RGUHS), Bangalore, Karnataka, India.

Neelakantan, N., Narayanan, M., De Souza, R. J., & Van, D. R. M., (2014). Effect of fenugreek (*Trigonella foenum-graecum* L.) intake on glycemia: A meta-analysis of clinical trials. *Nutr. J., 13*(7), 1–11.

Nicolle, E., Souard, F., Faure, P., & Boumendjel, A., (2011). Flavonoids as promising lead compounds in type 2 diabetes mellitus: Molecules of interest and structure-activity relationship. *Curr. Med. Chem., 18*(17), 2661–2672.

Oliveira, A. C., Endringer, D. C., Amorim, L. A., Das, G. L., Brandão, M., & Coelho, M. M., (2005). Effect of the extracts and fractions of *Baccharis trimera* and *Syzygium cumini* on glycaemia of diabetic and non-diabetic mice. *J. Ethnopharmacol., 102*(3), 465–469.

Ozougwu, J. C., & Eyo, J., (2010). Studies of the antidiabetic activity of *Allium sativum* (garlic) aqueous extracts alloxan-induced diabetic albino rat. *Pharmacol. Online, 2*, 1079–1088.

Pandey, M., & Khan, A., (2002). Hypoglycemic effect of defatted seeds and water-soluble fiber from the seeds of *Syzygium cumini* (Linn.) skeels in alloxan diabetic rats. *Indian J. Exp. Biol., 40*(10), 1178–1182.

Parvin, A., Alam, M., Haque, M., Bhowmik, A., Ali, L., & Rokeya, B., (2013). Study of the hypoglycemic effect of *Tamarindus indica* Linn. seeds on non-diabetic and diabetic model rats. *Brit. J. Pharm. Res., 3*(4), 1094–1105.

Patel, D. K., Prasad, S. K., Kumar, R., & Hemalatha, S., (2012). An overview on antidiabetic medicinal plants having insulin mimetic property. *Asian Pac. J. Trop. Biomed., 2*(4), 320–330.

Patel, M. B., & Mishra, S. H., (2011). Hypoglycemic activity of c-glycosyl flavonoid from *Enicostemma hyssopifolium*. *Pharm Biol., 49*(4), 383–391.

Patil, P. R., Patil, S. P., Mane, A., & Verma, S., (2013). Antidiabetic activity of alcoholic extract of neem (*Azadirachta indica*) root bark. *Nat. J. Physiol. Pharm. Pharmacol., 3*(2), 142–146.

Pulipaka, S., Challa, S. R., & Pingili, R. B., (2012). Comparative antidiabetic activities of methanolic extract of *Operculina turpethum* stem and root against healthy and streptozotocin-induced diabetic rats. *Int. Curr. Pharma. J., 1*(9), 272–278.

Qarshi, M. H., (2011). *Jameul Hikmat* (Vol. II, pp. 864–870). Idara Kitabul Shifa, New Delhi.

Racovita, R. C., Peng, C., Awakawa, T., Abe, I., & Jetter, R., (2015). Very-long-chain 3-hydroxy fatty acids, 3-hydroxy fatty acid methyl esters and 2-alkanols from cuticular waxes of *Aloe arborescens* leaves. *Phytochem., 113*, 183–194.

Rafiquddin, M., (1985). *Kanjul Advia Mufrida* (1st edn., pp. 261–263, 337, 338, 567, 568). Muslim University Press. Aligarh.

Raghuram, T. C., Sharma, R. D., Sivakumar, B., & Sahay, B. K., (1994). Effect of fenugreek seeds on intravenous glucose disposition in non-insulin-dependent diabetic patients. *Phytother Res., 8*, 83–86.

Rahman, R., Pasha, S. A., Katosh, D. C., Siddiqui, K. M., Khan, M. A., & Jamil, S. S., (2013). *Unani System of Medicine: The Science of Health and Healing*. CCRUM, Department of AYUSH Ministry of Health & Family Welfare, Government of India, New Delhi 110020.

Ramirez, A. M., Perez, R. M., Garcia, E., & Garcia, F. E., (2020). Antidiabetic activity of *Aloe vera* leaves. *Evid. Based Complement. Alternat. Med.* Article ID 6371201.

Rani, M. P., Krishna, M. S., Padmakumari, K. P., Raghu, K. G., & Sundaresan, A., (2012). *Zingiber officinale* extract exhibits antidiabetic potential via modulating glucose uptake, protein glycation and inhibiting adipocyte differentiation: An *in vitro* study. *J. Sci. Food Agric., 92*(9), 1948–1955.

Rasool, M., & Varalakshmi, P., (2006). Suppressive effect of *Withania somnifera* root powder on experimental gouty arthritis: An *in vivo* and *in vitro* study. *Chem. Biol. Interact., 164*(3), 174–180.

Rengachary, S. S., Xavier, A., Manjila, S., Smerdon, U., Parker, B., Hadwan, S., & Guthikonda, M., (2008). The legendary contributions of Thomas Willis (1621–1675): The arterial circle and beyond. *J. Neurosurg., 109*(4), 765–775.

Samhita, S., (2008). In: Ambikadatta, S., (ed.), *Ayurveda Tattva Sandipika Purvardha-Sutra Sthana*, Chapter-33, Sloka-4th Kaviraja. Reprint Varanasi: Chaukhamba Sanskrit Sansthan.

Sanders, L. J., (2002). From Thebes to Toronto and the 21st century: An incredible journey. *Diabetes Spectrum, 15*, 56–60.

Sastri, B. N., (1950). *The Wealth of India* (First supplement series Vol. 1: A-Ci, p. 262). National Institute of Science Communication and Information Resources (NISCAIR), New Delhi – 110017.

Savona-Ventura, C., (2003). *The History of Diabetes Mellitus, A Maltese Perspective* (pp. 4–7). Available at http://staff.um.edu.mt/csav1/books/diabetes_hist.pdf (accessed on 18 December 2021).

Sharma, R., & Arya, V., (2011). A review on fruits having antidiabetic potential. *J. Chem. Pharma Res., 3*(2), 204–212.

Sheela, C. G., & Augusti, K. T., (1992). Antidiabetic effects of S-allyl cysteine sulphoxide isolated from garlic *Allium sativum* Linn. *Indian J. Exp. Biol., 30*(6), 523–526.

Shetty, A. K., Kumar, G. S., Sambaiah, K., & Salimath, P. V., (2005). Effect of bitter gourd (*Momordica charantia*) on glycaemic status in streptozotocin-induced diabetic rats. *Plant Foods Hum. Nutr., 60*(3), 109–112.

Singh, M., Kumar, N., Sood, S., Makkar, B., & Arora, V., (2010). Historical milestones in diabetes. *Australasian Med. J., 3*(13), 860–864.

Srinivasan, K., & Ramarao, P., (2007). Animal models in type 2 diabetes research: An overview. *Indian J. Med. Res., 125*(3), 451–472.

Stanely, P., Prince, M., & Menon, V. P., (2000). Hypoglycemic and Other Related Actions of *Tinospora cordifolia* Roots in Alloxan-Induced Diabetic Rats. *J. Ethnopharmacol., 70*(1), 9–15.

Taher, M., Majid, F., & Sarmidi, M. R., (2006). A Proanthocyanidin from *Cinnamomum Zeylanicum* stimulates phosphorylation of insulin receptor in 3t3-L1 Adipocytes. *Jurnal. Teknologi, 44*(F), 53–56.

Tanaka, M., Misawa, E., Ito, Y., Habara, N., Nomaguchi, K., Yamada, M., Toida, T., et al., (2006). Identification of five phytosterols from *Aloe vera* gel as antidiabetic compounds. *Biol. Pharm. Bull., 29*(7), 1418–1422.

Timbola, A. K., Szpoganicz, B., Branco, A., Monache, F. D., & Pizzolatti, M. G., (2002). A new flavonol from leaves of *Eugenia jambolana*. *Fitoterapia, 73*(2), 174–176.

Virdi, J., Sivakami, S., Shahani, S., Suthar, A. C., Banavalikar, M. M., & Biyani, M. K., (2003). Antihyperglycemic effects of three extracts from *Momordica charantia*. *J. Ethnopharmacol., 88*(1), 107–111.

WHO, (1999). *Definition, Diagnosis and Classification of Diabetes Mellitus and its Complications*. World Health Organization Department of Noncommunicable Disease Surveillance. Available at: http://apps.who.int/iris/bitstream/handle/10665/66040/WHO_NCD_NCS_99.2.pdf;jsessionid=C8CB413B3939181DF35E4EF3F524235C?sequence=1 (accessed on 18 December 2021).

CHAPTER 5

Omics Technology in Antidiabetic Plant Research

SEEMA SINGH,[1] CHARANJIT SINGH,[2] and ABDUL HADI[1]

[1]*Department of Botany, University of Kashmir, Hazratbal, Srinagar – 190006, Jammu and Kashmir, India, E-mail: seemakuniv@gmail.com (S. Singh)*

[2]*Consultant Medicine, Associated Hospital, Government Medical College, Baramulla – 193101, Jammu and Kashmir, India*

ABSTRACT

Diabetes mellitus (DM) is a metabolic disorder in which the ability to produce insulin is defective or there is resistance to the action of insulin at the cellular level culminating in raised levels of glucose in the blood. The option to look for alternative antidiabetics from plant origin arose as the conventional antidiabetic drugs were expensive and tagged with a variety of associated side effects. The potential of antidiabetic plants was explored and their active principles isolated which generally contained alkaloids, flavonoids, steroids, and organic acids. The antidiabetic plants not only delayed the diabetic complications but also proved to be rich sources of antioxidants which already had an established role to play in delaying the onset of this disease. Latest molecular techniques like omics help in providing in-depth knowledge about the active constituents of the medicinal plants, their characterization, pharmacokinetics, toxicological studies, and most importantly the standardization and quality control of the antidiabetic plant derivatives. Recent molecular techniques like DNA sequencing, DNA fingerprinting, and DNA microarrays help us in the identification of the biomedical resources

Antidiabetic Potential of Plants in the Era of Omics. Deepu Pandita, PhD, Anu Pandita, MSc, & Chander Bhanu, MBBS (Eds.)

and extend their help in differentiating simultaneous molecular effects of mixtures on various chemical agents. Thus, omics technology helps in the standardization and modernization of antidiabetic medicinal plants.

5.1 INTRODUCTION

'Omics' technology is a blanket term that encompasses high-end technologies namely genomics, transcriptomics, proteomics, and metabolomics. Genomics is the systematic description of a complete set of genes, their nucleotide sequences, organizations, functions, and their intra and interspecific interactions. Proteomics being more precise deals with the study of proteins and their complexes, their localizations, interactions, and post-translational modifications. Metabolomics is the in-depth analysis of the metabolites present within a cell, tissue, or organism. Transcriptomics is the study of the transcriptome that is generated by the genome, under some defined circumstances by the cell, using high-end technologies, such as microarray analysis. A comparative study of transcriptomes helps in the recognition of genes that are differentially expressed in specific cell populations or are the results of various types of treatments. Thus, omics technology gives us an in-depth knowledge of pharmacodynamics, pharmacokinetics, and toxicological characterization of active constituents of herbal plants (Pandey et al., 2016).

Since time immemorial, 'medicinal plants' have been considered as 'natural medicines.' Medicinal plants and their extracts have been used as reliable natural resources. Recent proteomic studies have highlighted the application of plants and their extracts as important sources in gaining information about the molecular basis of diseases such as diabetes. Intensive efforts are underway to create a path for the standardization of the plant resources and bioengineering of the metabolic pathways of medicinal plants. Over the past two decades, the worldwide prevalence of diabetes has risen alarmingly from an estimated 30 million cases in 1985 to 415 million cases in 2017. Based on current status, the IDF projects that around 642 million people will have diabetes by the year 2040. Even though both type 1 and type-2 diabetes are increasing the world over, type 2 diabetes is more prevalent among populations and is increasing more rapidly. Diabetes is a major cause of mortality. Recent estimates from the USA have estimated Diabetes as the 7th leading cause of death. IDF data suggests that around 5 million deaths have occurred worldwide in the age group of 20–79 years (14.5%). It was estimated that in 2015, $673 billion has been the global expenditure in relation to diabetes care. Up to 75% of individuals with diabetes are found in low or middle-income countries (Alvin et al., 2018).

5.2 PROTEOMICS IN PLANTS WITH ANTIDIABETIC POTENTIAL

The word 'proteomics' was introduced by Marc Wilkins in 1996 to define the complete genomic protein component. Proteomics is an important biotechnological tool to integrate genomic information and transcriptional activity with translational efficiency and enzyme activity. Proteins can be investigated for their availability, variable proteo forms due to post-translational modifications, and protein to protein interactions. Proteomics-based technologies are utilized for tracing various diagnostic markers, vaccine synthesis, understanding pathogenicity details, Varied signals and their responses in different expression formats, and correlation of various protein pathways in some conditions like Diabetes. Proteomics in plants is more challenging as the plant cells contain a rigid cellulosic cell wall, hence success depends primarily on cell wall disruption which in turn causes the release of proteins, secondarily presence of impurities other than proteins including lipids, polyphenols, organic acids, and a plethora of pigments interfering during isolation processes.

It is estimated that more than 800 plants belonging to different families may possess potential antidiabetic activity (Patel et al., 2012). Selected medicinal plants used as antidiabetics to increase insulin secretion and potentiate its action are *Acacia arabiaca* (Wadood et al., 1989), *Bergenia himalaica* Boriss (Hussain et al., 2004), *Cinnamomum zeylanicum* Blume (Verspohl et al., 2005), *Coriandrum sativum* L. (Gray et al., 1999), *Gymnema sylvestre* Retz. (Sugihara et al., 2000), *Vinca rosea* L. (Prabhakar et al., 2008). These plants directly act on the Ca^{2+} channels affecting the secretion of insulin. Changes in blood glucose concentrations trigger the release of insulin from the β cells. The secretion of insulin from granules (insulin stores) includes the closing of ATP-gated K+ channels and activation of voltage-gated Ca^{2+} channels (Jones et al., 1998). The medicinal plants which regulate the production and degradation of insulin are *Abies pindrow*, *Artemesia roxburghiana*, *Eugenia uniflora,* and *Ficus religiosa.*

Phytoconstituents such as alkaloids, flavonoids, glycosides, peptidoglycans, polysaccharides, carbohydrates, and amino acids also have hypoglycemic effects. Resveratrol a phytoalexin increases glucose uptake in diabetic rats mediated through nitric oxide (Penumathsa et al). *Tinospora cordifolia* contains Berberine which in turn exhibits hypoglycemic activities. Alkaloids such as Catharanthine, Vindoline, and Vindolinine are isolated from *Catharanthus roseus* L (Chattopadhyay, 1999) are known to reduce blood glucose levels (BGL). Medicinal plants such as *Ocimum sanctum* L)., *Alinia galanga* L. *Aloe vera* L. increase the insulin levels and show hypoglycemic effects (Mukherjee et al., 2006) *Momodica chatrantia* L. contains a steroidal saponin Charantin which possesses insulin like activity (Ng et al., 1986). β Sitosterol a steroid found in *Azadirachta indica* A. Juss., Andrographolide a diterpenoid steroid from *Andrographis paniculata* Nees exhibit hypoglycemic activities (Yu et al., 2003).

Insulin secretion from beta cells is increased by many plants like *Camellia sinensis* (green tea), *Capsicum frutescens*, *Ficus religiosa*, *Hibiscus rosa, swertia chirayata* and *Zingiber officinale*. *Vinca rosea* has an impact on saving the beta-cell mass of the pancreatic islets. *Ipomoea batata* has distinct hyperglycemic effects due to a reduction in insulin resistance at the peripheral level (Bhushan et al., 2010).

World over researchers through proteomics is assessing the interplay amongst ligands and their receptors in the molecular docking approach for designing newer molecules to treat diabetes (Uma and Sudarsanam, 2012). Docking studies on green tea flavonoids showed that Epicatechin is an insulin-mimetic agent. In *Peperomia pellucida* yohimbine exhibited reductase inhibition, hence indicating its potential use as an antidiabetic (Akhila et al., 2012). Scoraric acid D an isolate of *Scoparia dulcis* L. has hypoglycemic effects like alfa glucosidase inhibitors – Miglitol and Voglibose (Saikia et al., 2020). Flavonoids present in vegetarian food are bioactive components supporting healthy living (Zhu et al., 2019) conducted hepatic transcriptome and proteome analyses in diabetic mice to unravel the regulator mechanism of dietary avicularin (quercetin-3-O-α-arabinofuranoside), a flavonoid glycoside. Dietary avicularin treatment in high fat and STZ induced diabetes in mice, brought down the glycemic levels, triglycerides, low density lipoprotein, Hepatic Transaminases, and increased liver glycogen. Experimental results of RNA-Seq and iTRAQ confirmed the antidiabetic effects of avicularin in T2DM mice (Zhu et al., 2019). Ginsenoside an active compound in *Panax ginseng* has been used to treat diabetes (Cho, 2006; Xie et al., 2005). SELDI-TOF-MS techniques have been used to confirm the antidiabetic effects of Ginsenoside (Cho et al., 2006). Diabetes mellitus (DM) both type I and type II is a heterogeneous disease with cellular level alterations in the protein domains manifesting in some genes. These changes in proteins are measurable by various proteomic techniques like Mass spectrometry with LC-MS-MS and MALDI-TOF/TOF (Mukesh and Namita, 2013). *Cynodon dactylon* plant extracts have been used to study diabetes in experimental rats. It was found that certain proteins were up-regulated in rat livers with the help of proteomic biotechniques like 2D Electrophoresis and MALDI-TOF-MS. These proteomic approaches provide us with certain potential markers to better understand the molecular mechanisms associated with diabetes (Karthik et al., 2012).

Thus proteomics, especially in medicinal plants is proving to be a valuable tool in the pharmaceutical industry, where-in it plays a role in quality control of drugs, toxicological studies and standardization of formulations and decoctions. Proteomic assays have predominantly been used to explain the mechanism of action of some herbal preparations, characterization of herbal prescriptions and molecular diagnosis.

5.3 GENOMICS IN ANTIDIABETIC PLANTS

World over diabetes is among the leading causes of death By 2030 prevalence of diabetes may increase by 4.5%–5.0%. World Health Organization predictions indicate that the burden will shift on the developing countries. There have been various multidisciplinary approaches to control this disease, however herbal formulations owing to their lower cost and better safety profile are important components in the antidiabetic armamentarium. In Omics, genomic data will lead to the discovery of antidiabetic drugs from medicinal plants. Functional genomics reveals the functions of biological systems at different levels and helps us to analyze various life phenomena at the molecular level.

Various assay methods based on DNA polymorphism are available for identifying and isolating various herbal biomolecules. DNA chips with DNA sequencing were applied to various herbal plant samples for proper characterization. More recently another method "DNA barcoding" is being employed for characterization based on comparing sequences with standard DNA fragments. These methods are used for the identification of unknown species of plants by comparing the DNA barcode sequences from library sequences of known species (Lam et al., 2010). Unambiguous identification is a critical step for quality assurance, genetic diversity, phylogeny conservation of threatened species. Recent molecular techniques assist in the identification of medicinal plants. Polymerase chain reactions amplify nuclear and chloroplast DNA and the reaction products are analyzed by gel electrophoresis, sequencing, or hybridization with species-specific probes. The quality and presence of adulterants can be confirmed by genomic fingerprinting. The collaborative efforts of the molecular community in the generation of molecular "barcodes" of medicinal plants will one day make it possible to generate a unique barcode for every species on earth. Genomes of *Ocimum sanctum* L and *Momordica charantia* have been sequenced. The bitter gourd genome possesses a unique gene having some insulin-like activity (Polypeptide-p or p-insulin). Prospective genome sequencing projects will explore medicinal plants for various molecules possessing antidiabetic activity (Chakraborty, 2018).

North East (NE) India is known for its rich bioresources. It is home to numerous medicinal plants that have been used in traditional systems of medicine. Traditional resources need to be constantly improved by molecular breeding techniques using genic microsatellites or SSRs or expressed sequence tags (Sahu et al., 2014). Around 11 traditional plants having hypoglycemic activity have been worked up for identifying various microsatellite and gene-based Site-Specific Recombination markers. The selected antidiabetic plant were *Ananas comosus*, *Carica papaya*, *Catharanthus roseus* (Rasineni et al., 2010), *Citrus aurantium* (Sharma et al., 2008), *Cynodon dactylon*, *Dioscorea alata*, *Jatropha curcas*, *Mangifera indica*, *Morus indica*, *Nelumbo nucifera*, (Huralikuppi et al., 2006) and *Tinospora cordifolia* (Sangeetha et al., 2013; Jarald et al., 2008; Maithili et al., 2011). A host of medicinal plants with different biochemical properties possess antidiabetic potential. One such plant introduced in India *Costus pictus* a native of Mexico and also known as "Insulin plant" or "Spiral Ginger." Transcriptome sequencing of the green leaves of spiral ginger identified transcripts related to bixin biosynthesis and geranial biosynthesis pathways. They validated the presence of putative norbixin methyltransferase, a precursor of Bixin. The transcripts encoding these terpenoids are known to be Peroxisome Proliferator-Activated Receptor (PPAR) agonists and anti-glycation agents. Sequential extraction and high-performance liquid chromatography (HPLC) confirmed the presence of bixin in C. pictus methanolic extracts.

5.4 METABOLOMICS STUDY IN ANTIDIABETIC PLANTS

The systematic study of inimitable chemical fingerprints with a definite cellular process is called metabolomics. In metabolomics, an individual or a group of medicinal plant species from varied geographical zones can be analyzed for differences in the levels of metabolic responses. Once the collection of samples and preparation of the extract is completed a detailed analysis of the metabolite content and antihyperglycemic effects of plant samples is done. Finally, a relation between different metabolites and known bioactivities will be established. Metabolomics research has resulted in exploring biomarkers and analyzing the various metabolic disturbances during the disease. we can also get a detailed pathway of the mechanism of action of the various herbal medicines. Both *in vitro* and *in vivo* experiments were carried out successfully towards analyzing metabolic changes with the identification of the active compounds in antidiabetic medicinal plants(Hasanpour et al., 2020).

The random method is a detailed work out of various known and unknown metabolites. This method provides a broader overview of the metabolite contents, which is helpful in the study of viable biomarkers and the various metabolomic mechanisms of the disease. Various analytical techniques and data processing technologies also help in the acquisition of the varied metabolic pathways.

The targeted analysis lays focuses on a specific metabolic pathway wherein a small number of predefined active molecules are identified from the sample. The targeted analysis is applicable for molecular marker identification. *In vitro* experiments focus on the screening, identification, and characterization of the various bioactive compounds along with the enzyme inhibitory potential of medicinal plant extracts. During the preparation of medicinal plant extracts, advanced technologies like ultra-sonication and microwave methods are applied. Hasanpor et al. attempted the treatment of diabetes using the metabolomics approach in medicinal plants. Blood, urine, and fecal matter have also been used as samples in metabolomic studies. Multivariate statistical analysis is applied for the interpretation of a large amount of data extracted from various metabolomic studies (Hasanpour et al., 2020).

In vivo experiments conducted for establishing the antidiabetic potential of *Phyllanthus niruri* (Mediani et al., 2016) and Centella asiatica (Abas et al., 2016) were conducted on Streptozotocin-induced diabetic mice. In *P. niruri* the Phenolic compounds accompanied by lignans were the active biotic components that exhibited potential α-glucosidase activity (Patel et al., 2012). Mediani et al. reported glucose, choline, taurine, and creatinine as the potential biomarkers in diabetic mice. *P. niruri* increased the pyruvate content, and acetyl coenzyme A considerably reduced the serum glucose level alongside improving dyslipidemia (Mediani et al., 2016).

Ipomoea aquatica is another antidiabetic plant showing the presence of caffeic acid, catechin, chlorogenic acid, isochlorogenic acid A and C, and Quercetin derivatives. These derivatives can be further characterized by advanced mass spectroscopic analysis. In the metabolomic approach Sajak et al. further used NMR analysis to study the effect of these plant extracts.

Momordica charantia has also been used for lowering blood sugars in diabetics. Steroidal saponins like charantins, insulin like peptides and alkaloids are the active components responsible for the hypoglycemic effects of *M. charantia* (Ethan Basch et al., 2003). Chen et al. (2019) employed HPLC/MS to detect the differential metabolic response.

Li et al. employed different metabolomic techniques like lipidomics to analyze Kukoamine B and its effect on diabetic mice. They used the targeted approach for corroborating the influence of Kukoamine B, metformin, and rosiglitazone on the acylcarnitine dynamics. Rosiglitazone in combination with Kukoamine significantly increased long-chain acylcarnitines in diabetic mice.

Antidiabetic potential of the leaf extract of *Moringa oleifera* based on GCMS metabolic studies was investigated by (Khan et al., 2017). Both Puerarin and the chief ingredient of *Pueraria lobatae* Radix and berberine a major alkaloid of *Coptidis Rhizoma* show antidiabetic effects in diabetic rats.

Consumption of fresh salad dressing of *Centella asiatica* is prescribed in the treatment for DM. The active constituents being asiaticoside, asiatic acid, madecassoside, and madecassic acid. Oral treatment of Streptozotocin-induced obese diabetic rats with Asiatic acid lowered blood glucose, increased insulin production, and modulated the liver enzymes regulating glucose metabolism. Thus, studies on herbal drugs are centered on the mechanism of action, identifying biomarkers, and comparative analysis of genotypic disturbances leading to a diabetic phenotype. LC-MS based data often provide profiles of lipids while NMR based data reveals metabolites associated with TCA cycle, amino acid, and carbohydrate metabolism.

Hence the present-day herbal drug industry uses different omics techniques for standardization and quality control of herbal formulae, action mechanism in herbal drugs, and characterization of active constituents and detection of the molecular mechanisms to predict possible side effects and associated drug interactions. For future studies, focus on group size, sampling methodology, analytical procedures, and selection of statistical programs for appropriate data analysis is required.

TABLE 5.1 Some Antidiabetic Potential Medicinal Plants of Jammu and Kashmir State

SL. No.	Scientific Name	Monocot Family	Local Name	Ethno-Botanical Uses	Antidiabetic Potential
1.	*Acorus calamus L.*	Acoraceae	Sweet flag, Bach, Gora Bach, Vai Gander	*Acorus calamus* is shown to have anti-ulcerative, insecticidal anti-bacterial properties (Pandita et al., 2011). The ethanolic extract has anti-proliferative and immune-suppressant properties (Mehrotra et al., 2003). The rhizome is tonic and is used to cure stomach disorders. As an infusion, powder or as a whole for curing a cold, coughs, sore throat, toothache, rheumatism, and muscle pains (Melanie-Jayne et al., 2017). It acts as hepato-protective (Palani et al., 2011). *Acorus calamus* due to its active compound α asarone helps in overcoming noise-induced stress (Srikumar et al., 2005). The extract of *Acorus* prevents the oxidation of lipids by free radicals (Dharini et al., 2012). *Acorus* is known to have antifungal properties (Ghosh, 2006), antihelminthic properties (Merekar et al., 2011), prevents dementia as it is known to have β asarone as an AChE inhibitor (Mukherjee et al., 2007) and possesses antioxidant properties (Manikandan and Devi, 2005).	Plant extract at doses of 35–40 mg/Kg body weight showed the anti-hyperglycemic property in Streptozotocin treated rats. Diabetic rats when orally administered 185–200 mg/Kg bw of the extract for 20–21 days. After 20–21 days the blood glucose levels were lowered as compared to diabetic control (David et al., 2011).

TABLE 5.1 *(Continued)*

SL. No.	Scientific Name	Monocot Family	Local Name	Ethno-Botanical Uses	Antidiabetic Potential
2.	*Allium cepa L*	Liliaceae	Onion, Piyaz, Gandeh	The bulb is used in treating besting, headache, Nausea, vomiting, and muscle Sprain and headache (Kumar et al., 2015), skin boils, Hair fall and Diarrhea (Khanday et al., 2018).	Onion ether fractions showed significant hypoglycemic effects. Intake o *Allium cepa* (90–100 mg) decreased the fasting glucose in type I and type II diabetic persons. Fasting levels reduced to about 87–89 mg/dl in comparison to insulin (145 mg/dl) in type I diabetic persons and lower levels by 39–40 mg/dl as compared to glibenclamide (80–81 mg/dl) in type II diabetic persons (Imad et al., 2010)
3.	*Allium sativum L*	Liliaceae	Garlic, Lehsan	Garlic cloves are used to treat Influenza, Rheumatism (Kumar et al., 2015). Garlic constitutes amino acids cysteine and glutamine which acts as an anti-oxidant (Ayaz et al., 2007), cardiac problems, blood pressure regulation, hypoglycemic, reduces, and cholesterol levels. It is also known to possess antibacterial, antiviral, antifungal properties, thus strengthening the immune system.	The ethanolic extract on oral administration reduced the blood sugar levels. An extract of raw garlic at doses of 450–500 mg/Kg bw on intraperitoneal administration in STZ stimulated diabetic mice significantly reduced serum sugar levels. In alloxan stimulated diabetic mice 295–300 mg/Kg bw was the most effective dosage showing significant reductions in the glucose levels (Ozougwu et al., 2010).

TABLE 5.1 *(Continued)*

SL. No.	Scientific Name	Monocot Family	Local Name	Ethno-Botanical Uses	Antidiabetic Potential
4.	*Costus speciosus*	Zingiberaceae	Step Ladder Plant, Kushthain, Chemargatha	Possesses antibacterial, anti-inflammatory, antifungal, antioxidant, analgesic, antipyretic, hyperglycemic anti-diuretic, larvicidal, against stress and estrogenic activity. Rhizomes were used to cure pneumonia, rheumatism, dropsy, urinary diseases, jaundice mental disorders (Pawar et al., 2014).	The chief chemical components are diosgenin, curcumin, and curcuminoids. Eremanthin at doses of 5–20 mg/Kg body w for 60 days exhibited hypoglycemic effect in STZ mediated diabetic male Wistar mice. Rhizome ethanolic extract (190–200 mg/Kg) body weight shows a decline in blood glucose levels (Revathy et al., 2014).
5.	*Oryza sativa*	Poaceae	Rice, Dhan, Chol, Danaye, Tamul	Used in the cure of nutritional disorders, digestive system disorders, viral infections, Tonsillitis (Cabanting et al., 2016). The seed was used during dyspepsia-stasis, reflux. Brown rice extract has been used to treat diarrhea, indigestion, and nausea (Jamil et al., 2016).	Oral administration of an active component Oryzanol from rice bran oil at doses of 50–100 mg/Kg body weight in normal and STZ induced diabetic mice lowered blood sugar values. Extracts of purple rice delayed absorption of starch and sucrose absorption in diabetic rats. In humans, doses of 200 mg/Kg are known to prevent the progression of diabetic symptoms (Cho et al., 2012).

TABLE 5.1 *(Continued)*

SL. No.	Scientific Name	Monocot Family	Local Name	Ethno-Botanical Uses	Antidiabetic Potential
6.	*Setaria italic*	Poaceae	Italian millet, Koongni	It is fibrous and used as salad, Antihyperglycemic, and hypolipidemic	Aqueous extract of seeds at doses 300 mg/Kg body weight obtained a reduction of 70% in plasma sugar glucose readings and hence reported an efficient glycemic control, Reduced rates of HbA1C in diabetic treated mice in comparison to controls (Sireesha et al., 2011; Yallanki et al., 2011).
7.	*Smilax glabra Roxb.*	Smilacaceae	China Root, Sarsaparilla	The aerial tubers are used to cure boils and cysts, eczema anti-inflammatory and immune-modulatory (Shiyao et al., 2018).	Methanolic rhizome extracts at dosages of 95–100 mg/Kg bw around 4–5 hours of intraperitoneal administration reduced the blood sugars of both normal and diabetic rats. It is also known to suppress epinephrine-induced hyperglycemia in mice and enhanced insulin sensitivity (Fukunaga et al., 1991).
8.	*Triticum aestivum Linn.*	Poaceae	Wheat, Kanak	Floor is used to cure cancer patient (Alitheen et al., 2011), peptic ulcer (Ben-Arye, 2002), joint pain (Das, 2012). It used against toothache, foul mouth disorders, constipation, and skin diseases (Manisha Vats, 2011)	Ethanolic extracts demonstrated an antidiabetic effect in Streptozotocin stimulated diabetic mice. Freshly etiolated wheatgrass juice 30 ml/Kg shows a decline in blood glucose levels (Lalit et al., 2014).

TABLE 5.1 *(Continued)*

SL. No.	Scientific Name	Monocot Family	Local Name	Ethno-Botanical Uses	Antidiabetic Potential
9.	*Zea mays L.*	Poaceae	Maize, Makkai	Diuretic, against chronic cystitis, increase bile secretion, assists the elimination of water from tissues cardiac disorders and obesity (Muntean et al., 2016).	Corn silk extract when administered to Alloxan and Adrenaline induced hyperglycemic mice showed reduced levels of blood sugar and HbA1C. Insulin secretion being markedly improved in alloxan-stimulated hyperglycemic mice (Ranilla et al., 2009).
10.	*Aralia cachemirica Lindl*	Araliaceae	Aralia, Khoree	Rheumatism, arthritis, gastric problems, nephritis, lumbago, against inflammation, and diabetes mellitus.	Aqueous and alcoholic root extracts (245–250 mg/Kg)bw showed hypoglycemic effects in mice saturated with glucose, although the negligible effect was seen in normal fasting mice (Bhat et al., 2005).
11.	*Artemesia absinthium*	Asteraceae	Sweet Sagewort, Tethwan, Worm Wood	Stomach complaint, intestinal worm's liver problems, gastritis, jaundice, used to cure wounds and cuts, stasis, flatulent-dyspepsia, It also acts as an antifungal, antibacterial, antiparasitic, anti-ulcer, anti-cancer, neuroprotective, antidepressant, analgesic, antioxidant, and immunomodulator (Batiha et al., 2020).	Doses of 100–500 mg/Kg bw of plant extract on oral administration for 40–42 days exhibited hypoglycemia in normal and diabetic mice. The treated mice showed increased levels of HDL, liver glycogen, bodyweight, tolerance to glucose and decreased levels of triglycerides, total cholesterol, ALT, AST, Urea, Creatinine (Al-Shammy et al., 1994).

TABLE 5.1 *(Continued)*

SL. No.	Scientific Name	Monocot Family	Local Name	Ethno-Botanical Uses	Antidiabetic Potential
12.	*Bauhinia variegata* Linn.	Fabaceae	Orchid, Kachnara	It has antitumor, antimicrobial, antispasmodic, used as a tonic, treatment of cough and cold, anti-inflammatory, antigoitrogenic, and hepatoprotective properties.	The stem bark contains roseoside as its major constituent showing the increased release of insulin from β cells in culture. Bark extract (BVBE) was found to be acting like metformin (Frankish et al., 2010; Kumar et al., 2012).
13.	*Berberis aristata*	Berberidaceae	Indian Berberry, Kashmiri Kawdach	Treatment of ophthalmic diseases.	The root extracts in ethanol were studied in alloxan stimulated diabetic mice. A dosage of 71.42 and 100 mg/Kg bw depicted a reduction in levels of serum glucose. Levels of cholesterol along with triglyceride was enhanced in diabetic mice in relation to control mice (Semwal et al., 2009).
14.	*Bombax ceiba*	Malvaceae	Red silk cotton tree, kapok tree, Simbhal	Diarrhea, local antiseptic and hemostatic.	In Sprague Dawley rats, leaf extracts showed hypoglycemic activities at doses of 500 mg/Kg bw chiefly due to the presence of Shamimin a C-Flavonol glucoside. The bark extract containing triterpenoid compounds exhibited hypoglycemic effects in STZ stimulated and normal mice with doses of 600 mg/Kg bw (Saleem et al., 1999; Chetan et al., 2013).

TABLE 5.1 *(Continued)*

SL. No.	Scientific Name	Monocot Family	Local Name	Ethno-Botanical Uses	Antidiabetic Potential
15.	*Brassica nigra*	Brassicaceae	Black Mustard, Kali Aur, Kali Rayi	Possesses antioxidant, anti-inflammatory, antidiabetic, and hypolipidemic effects (Ali et al., 2015).	Diabetic rats when given aqueous extracts for 60 days showed a release of insulin from the pancreas resulting in decreased serum glucose levels. Aqueous extracts of seeds lead to an anti-hyperglycemic effect in streptozotocin-stimulated diabetic mice. It is also known to reduce the range of fasting glucose, HbA1C in diabetic rats after a month (Anand et al., 2009).
16.	*Brassica juncea*	Brassicaceae	Indian Mustard, Saetri Sarsu, Rai	Antimicrobial, antioxidant, anxiolytic, antifungal, goitrogenic.	Aqueous extract of the seeds at different dosages of 250–500 mg/Kg bw had a Hypoglycemic effect in diabetic rats as compared to normal. Methanolic extracts of leaves of *Brassica juncea* exhibited a dependency on doses and antihyperglycemic effect in glucose-induced hyperglycemic rats (Khan et al., 1995; Tripathi et al., 2001; Jung et al., 2009; Yokozawa et al., 2002)

TABLE 5.1 *(Continued)*

SL. No.	Scientific Name	Monocot Family	Local Name	Ethno-Botanical Uses	Antidiabetic Potential
17.	*Bryophyllm pinnatum Kurz*	Crassulaceae	Jakh Me Hayat, Air Plant, Miracle-Leaf	–	Extract of the whole plant caused a considerable decrease in the blood glucose levels of the fasting normal and fasted Streptozotocin-induced diabetic mice. When used with existing drugs such as Glibenclamide it enhanced the antidiabetic property of the drug (John, 2005).
18.	*Catharanthus roseus*	Apocynaceae	Vinca Periwinkle, Sadabahar	Antidiabetic, anti-cancer, hypotensive, and antiarrhythmic agent.	Leaves in an alcoholic considerably decreased the glucose ranges in control and Streptozotocin stimulated diabetic rats. The leaves in methanolic extracts affects the carbohydrate metabolism due to increased insulin secretion or prevention of harm caused by oxygen free radicals and enhanced glucose consumption (Benjamin et al., 1994; Gupta et al., 2009)
19.	*Coriandrum sativum L*	Apiaceae	Coriander, Dhaniya	Stomachic, diuretic, anti-convulsant, sedative, anti-mutagenic, anthelmintic.	The ethanolic extract of 200–400 mg/Kg bw of the seeds showed a considerable reduction in levels of serum glucose in Streptozotocin stimulated diabetic mice and it also increased the production of insulin from the pancreatic beta cells (Chitra et al., 1999; Nadeem et al., 2013).

TABLE 5.1 *(Continued)*

SL. No.	Scientific Name	Monocot Family	Local Name	Ethno-Botanical Uses	Antidiabetic Potential
20.	*Cucumis sativus* L.	Cucurbitaceae	Cucumber, Kheera	Antioxidant, antimicrobial, immunity booster, antidiabetic, antiaging.	Ethanolic extract of the fruit at different doses of 200–400/mg/Kg bw showed a considerable antihyperglycemic effect in Streptozotocin-induced rats in comparison to the standard drugs (Karthiyayini et al., 2009; Sharma et al., 2020).
21.	*Daucus carota* L.	Apiaceae	Carrot, Gajar	Daucus carota has shown the presence of properties such as hepatoprotective and antioxidant activity (Bishayee et al., 1995), Anticancer activity (Sparreboom et al., 2004), used to cure inflammation of the stomach, kidneys, and acts as analgestic (Porchezhian et al., 2000; Jasicka-Misiak et al., 2004).	Methanolic extract of seeds at different doses of 100, 200, and 300 mg/Kg body weight in association with the standard drug Glibenclamide on administration for 6 days decreased serum glucose levels. The male Swiss mice when orally overloaded with 80% ethanolic extracts showed an enhanced glucose tolerance (Swanstan et al., 1989).
22.	*Eucalyptus globulus*	Myrtaceae	Anukarpoor	Anti-Spasmodic (Modak et al., 2007), Anti-inflammatory (Nagpal et al., 2010), Antimalarial (Hardel et al., 2011), Anti-allergic, and antimalarial (Hardel et al., 2011).	Eucalyptus globulus plant aqueous extracts increase the secretion of insulin from pancreatic beta cells and also enhances the glucose consumption in the rat abdominal muscles. Alcoholic extract of eucalyptus globulus blockade on oral administration decreased serum glucose ranges and enhanced the serum insulin in diabetic mice only (Singh, 2011; Eidi et al., 2010).

TABLE 5.1 *(Continued)*

SL. No.	Scientific Name	Monocot Family	Local Name	Ethno-Botanical Uses	Antidiabetic Potential
23.	*Juglans regia* Linn	Juglandaceae	English walnut, Akhroot, Dhoon	Effective against skin diseases such as eczema, helminths, leucorrhea, sinusitis, diarrhea, stomach ache, arthritis, asthma, anorexia, diabetes mellitus, thyroid dysfunctions, cancers (Nael et al., 2011).	The extracts of *J. regia* containing the chief active compound as Juglone (5-hydroxy-1,4-naphthoquinone) decreased the blood glucose levels considerably. The aqueous extracts influence the glucose levels in the blood in type II diabetic persons. Its leaf extracts at doses of 90–100 mg// Kg bw daily twice for three months improved the glycemic regulation in type II diabetic persons (Amaral et al., 2004; Hosseini et al., 2014).
24.	*Marrubium vulgare* L.	Lamiaceae	Gand Soi	Yamaguchi et al. (2006) reported the anti-tumorigenicity effect of leaves to extract used tumors, cancer, and cystic fibrosis.	A single oral dosage of 495–500 mg/ Kg be for a period of nearly a month reduced the blood sugar ranges and enhances the insulin and tissue glycogen contents (Ramon et al., 1992).

TABLE 5.1 *(Continued)*

SL. No.	Scientific Name	Monocot Family	Local Name	Ethno-Botanical Uses	Antidiabetic Potential
25.	*Momordica charantia* Linn.	Cucurbitaceae	Bitter gourd, Karela	Leavespaste and decoction are used to cure menstrual disorders, fever, constipation, diarrhea, colic, infections, against worms and parasites, measles, hepatitis, and intestinal worms. Roots help in treating syphilis, rheumatism, boils, and ulcers (Agharkar, 1953; Jadhav, 2008).	Whole fruit extracts, both dried and fresh reduce blood sugar levels. This plant shows insulin-mimetic properties due to various active constituents, lectins, and insulin-like proteins. The aqueous extracts of unripe fruits liberate insulin from the β cells of obese hyperglycemic mice because of perturbations of membrane functions (Grover et al., 2002; Saxena et al., 2004).
26.	*Morus alba* Linn	Moraceae	Mulberry, Tul	Leaf extract was used against fever, sore throats, Eye irritation, headaches, dizziness, etc. The root bark is used in toothache and as a laxative. The fruit is used to cure premature graying and dizziness. Root bark is antitussive, antiasthmatic, sedative (Anonymous, 2001).	The ethanolic extracts of Morus alba showed better antidiabetic potential as compared to aqueous extract (Singh et al., 2005).

TABLE 5.1 *(Continued)*

SL. No.	Scientific Name	Monocot Family	Local Name	Ethno-Botanical Uses	Antidiabetic Potential
27.	*Nelumbo nucifera Gaertn*	Nymphaeaceae	Lotus, Kamal, Pamposh	Used to cure smallpox, pectoralgia, dysentery, leucoderma, spermatorrhea, cough, haematemesis, epistaxis, hemoptysis, haematuria, metrorrhagia, hyperlipidemia, fever, diarrhea, hyperdipsia, hepatopathy (Nishkruti et al., 2013).	The aqueous, alcoholic powdered extracts and also the sun-dried flowers of *Nelumbo* provide hypoglycemic effect in fasting normal albino rabbits. In STZ (streptozotocin) induced rabbits the rhizome and flower extracts showed significant antidiabetic property. The oral consumption of seed ash of Nelumbo at doses of around 195–200 mg/Kg bw for one month in STZ stimulated diabetic mice confirmed hypoglycemic activity of *Nelumbo*. The direct or indirect role played by the presence of considerable amounts of trace elements in insulin secretion cannot be ruled out (Huralikuppi et al., 1991).
28.	*Nymphaea stellata* Willd.	Nymphaeaceae	Blue Water Lily, Nippa	Cures menstrual irregularities, acts as anti-aging, rejuvenation (Yoo et al., 2006).	Flower extract of *Nymphaea* contains Nymphayol which not only protects the pancreatic beta cells against loss but induces insulin secretion. Ethanolic leaf extracts at doses of 100 and 200 mg/Kg bw decreased hyperglycemia. Flower extracts at doses of 300 mg/Kg bw cause a considerable reduction in the blood glucose levels in diabetic mice (Dhanabal et al., 2007; Sivasankari et al., 2010).

TABLE 5.1 *(Continued)*

SL. No.	Scientific Name	Monocot Family	Local Name	Ethno-Botanical Uses	Antidiabetic Potential
29.	*Saussurea lappa* Clarke	Asteraceae	Costus root, Kuth, Kutha, Kut	Antidiarrheal (Negi et al., 2013), Antiparasitic (Rhee et al., 1985), Anticonvulsant activity: (Gupta et al., 2009), Hypoglycemic (Upadhaya et al., 1993), Anti-hepatotoxic activity (Chen et al., 1995), Antimicrobial property (Khalid et al., 2011), Anti-inflammatory properties (Kim et al., 2012) Anticancerous (Hung et al., 2010).	Alcoholic root extracts of *Saussurea lappa*, when given to albino rats for a week, depicted the considerable Antidiabetic effect (Chaturvedi et al., 1993).
30.	*Tinospora cordifolia* Thunb.	Menispermaceae	Ghiloy	Immuno-modulator, reduce blood glucose levels, antioxidant, paste is applied to cure allergic problems, against inflammation, hypoglycemic (Choudhary et al., 2013).	The aqueous root extract when administered orally leads to a continuous decline in blood glucose levels (Dhaliwal, 1999).

Next-generation sequencing should also be done in order to investigate the medicinal properties of this plant. This will help in understanding the medicinal potential of plants and become a reservoir of resources for enterprising researchers involved in exploring new and promising treatments for Diabetes. Access to the experimental transcriptomic data in generic platforms will encourage worldwide genomic studies of related medicinal plants possessing potential for treatment of diabetes. A detailed description of the pharmacological and phytochemical portrayals of the antidiabetic plants in Jammu and Kashmir has been given by Pandita et al. (2015a, b). The details of other plants are enlisted in Table 5.1.

KEYWORDS

- **diabetes mellitus**
- **genomics**
- **omics**
- **proteomics**
- **transcriptomics**

REFERENCES

Abas, F. A., Khatib, V., Perumal, V., Suppaiah, A., Ismail, M., Hamid, K., & Shaari, N. L., (2016). Metabolic alteration in obese diabetes rats upon treatment with *Centella asiatica* extract. *J. Ethnopharmacol., 180*, 60–69.

Aderibigbe, A. O., Emudianughe, T. S., & Lawal, B. A., (1999). Antihyperglycaemic effect of *Mangifera indica* in rat. *Phytotherapy Res., 13*, 504–507.

Akhila, S., Aleykutty, N. A., & Manju, P., (2012). Docking studies on *Peperomia pellucidia* as antidiabetic drug. *International Journal of Pharmacy and Pharmaceutical Sciences, 4*(Suppl 4), 1–2.

Al- Shamany, L. A., Al-Khazraji, S. M., & Twajj, H. A. J., (1994). *Ethnopharmacol, 43*, 167–171.

Al- Snafi, A. E., (2015). The pharmacological importance of *Brassica nigra* and *Brassica rapa* grown in Iraq. *Journal of Pharmaceutical Biology, 5*(4), 240–253.

Alitheen, N. B., Oon, C. L., Keong, Y. S., Chuan, T. K., Li, H. K., & Yong, H. W., (2011). Cytotoxic effects of commercial wheatgrass and fiber towards human acute promyelocytic leukemia cells (HL60). *Pak. J. Pharm. Sci., 24*, 243–250.

Alvin, C., Powers, Kevin, D., Niswender, & Carmella, E. M., (2018). Diabetes mellitus. Diagnosis, classification and pathophysiology. In: Jameson, J. L., (ed.), *Harrison's Principles of Internal Medicine* (Vol. 2, pp. 2850–2859). Mc Graw Hill Education; New York.

Amaral, J. S., Seabra, R. M., Andrade, P. B., Valent, A. O. P., Pereira, J. A., & Ferreres, F., (2004). *J. Food Chem., 88*, 373–379.
Annadurai, R. S., Jayakumar, V., Mugasimangalam, R. C., et al., (2012). Next-generation sequencing and *de novo* transcriptome analysis of *Costus pictus* D. Don, a non-model plant with potent antidiabetic properties. *BMC Genomics, 13,* 663. https://doi.org/10.1186/1471-2164-13-663.
Aransiola, E. F., Daramola, M. O., Iwalewa, E. O., Seluwa, A. M., & Olufowobi, O. O., (2014). *International Journal of Biological, Food, Veterinary and Agricultural Engineering, 8*(1).
Aslam, B., Basit, M., Nisa, A. M., et al., (2017). Proteomics: Technologies and their applications. *Journal of Chromatographic Science., 55*(2), 182–196.
Attele, A. S., Zhou, X. Y. P., Wu, J. T. J. A., Zhang, L., Dey, L., Pugh, W., Rue, P. A., et al., (2002). Antidiabetic effects of Panax ginseng berry extract and the identification of an effective component. *Diabetes, 511*, 851–1858.
Ayaz, E., & Alpsoy, H. C., (2007). Sarimsak (*Allium sativum*) ve geleneksel tedavide kullanimi [Garlic (Allium sativum) and traditional medicine]. *Turkiye Parazitol. Derg., 31*(2), 145–149. Turkish. PMID: 17594659.
Batiha, G. E. S., Olatunde, A., El-Mleeh, A., Hetta, H. F., Al-Rejaie, S., Alghamdi, S., Zahoor, M., et al., (2020). Bioactive compounds, pharmacological actions, and pharmacokinetics of wormwood (*Artemisia absinthium*). *Antibiotics, 9*(6), 353. https://doi.org/10.3390/antibiotics9060353.
Bhat, Z. A., Ansari, S. H., Mukhtar, H. M., Naved, T., & Siddique, J. I., (2005). *Pharmazie, 60*, 712, 713.
Bhushan, M. S., Rao, C. H. V., Ojha, S. K., Vijaykumar, M., Verma, A., (2010). An analytical review of plants for antidiabetic activity with their phytoconstituent & mechanism of action. *Int. J. Pharm. Sci. Res., 1*(1), 29–46.
Cabanting, R. M., & Perez, L. M., (2016). An ethnobotanical study of traditional rice landraces (*Oryza sativa* L.) used for medical treatment in selected local communities of the Philippines. *J. Ethnopharmacol., 24;* 194, 767–773.
Chakraborty, P., (2018). Development of new antidiabetic drug from medicinal plant-genomic research. *J. Diabetol., 2*(1), 6, 7.
Chattopadhyay, R. R., (1999). A comparative evaluation of some blood sugar lowering agents of plant origin. *J Ethnopharmacol., 67*, 367–372.
Chaturvedi, P., Tripathi, P., Pandey, S., Singh, U., & Tripathi, Y. B., (1993). *Phytotherapy Research, 7*, 205–207.
Chen, J. L., Zheng, F., Wang, S., Huang, Z., Li, P., Zheng, S., Zhang, T., & Yi, H., (2019). Li. Metabolomics reveals the effect of zishen jiangtang pill, a Chinese herbal product on high-fat diet-induced type 2 diabetes mellitus in mice. *Front. Pharmacol., 10*(2019), 256.
Chitra, V., & Leelamma, S., (1999). *Food Chemistry, 67*(3), 229–231.
Cho, C. S., (2006). Some thoughts on Chinese medical paradiagram and review on the 2006 world congress on Chinese medicine. *J. Chin. Med.,1*(7), 391–400.
Cho, W. C., (2006). Research progress in SELDI–TOF–MS and its clinical applications. *Sheng Wu Gong Cheng Xue Bao, 22*(6), 871–876.
Cho, W. C., Yip, T. T., Chung, W. S., et al., (2006). Differential expression of proteins in the kidney, eye, aorta and serum of diabetic and non–diabetic rats. *J. Cell Biochem., 99*(1), 256–268.
Das, A., Raychaudhuri, U., & Chakraborty, R., (2012). Effect of freeze-drying and oven drying on antioxidant properties of fresh wheatgrass. *Int. J. Food Sci. Nutr., 63*, 718–721.

David, H. P., Rangachari, B., & Harshit, R. S., (2011). *Asian Pacific Journal of Tropical Biomedicine*, 1–9.

Denise, I. J., Gaspari, M., Greef, J. V. D., Robert, V. D. H., & Robert, V., (2005). Proteome analysis of the medicinal plant *Catharanthus roseus*. *Planta, 221*, 690–704.

Dhaliwal, K. S., (1999). *Method and Composition for Treatment of Diabetes*. US Patent, 5886029.

Dhanabal, S. P., Raja, M. K., Ramanathan, M., & Suresh, B., (2007). *Fitoterapia, 78*(4), 288–291.

Eidi, L., Eidi, M., Givian, R. M. H., & Abaspour, N. (2010). *Medical Sciences Journal of the Islamic Azad University, Tehran Medical Branch, 20*(1), 11–14.

Eleven, (2014). Traditional plants with potent antidiabetic properties. *OMICS A Journal of Integrative Biology, 18*(5).

Ethan, B., Steven, G., & Catherine, U., (2003). Bitter melon (*Momordica charantia*): A review of efficacy and safety. *American Journal of Health-System Pharmacy, 60*(4), 356–359.

Eun-Kyung, C., Kyung-Im, K., & Young-Ju, C., (2012). Jou*rnal of Life Science, 22*(1), 126–131.

Frankish, N., De Sousa, M. F., Mills, C., & Sheridan, H., (2010). *Planta Med., 76*(10), 995–997.

Fukunaga, T., Miura, T., Furuta, K., & Kato, A., (1997). *Biol. Pharm. Bull., 20*(1), 44–46.

Gray, A. M., & Flatt, P. M., (1999). Insulin-releasing and insulin-like activity of the traditional anti diabetic plant *Coriandrum sativum* (coriander). *Br. J. Nutr., 81*, 203–209.

Grover, J. K., Yadav, S., & Vats, V., (2002). *J. Ethnopharmacol., 81*(1), 81–100.

Hasanpour, M., Iranshahy, M., & Iranshahi, M., (2020). The application of metabolomics in investigating antidiabetic activity of medicinal plants. *Biomed. Pharmacother., 128*, 110263. doi: 10.1016/j.biopha.2020.110263. PMID: 32450525.

Hosseini, S., Jamshidi, L., Mehrzadi, Mohammad, K., Najmizadeh, A. R., Alimoradi, H., & Huseini, H., (2014). *J. Ethnopharmacol, 152*(3), 451–456.

Huralikuppi, J. C., Christopher, A. B., & Stephen, P. M., (1991). *Phytother. Res., 5*, 54–58.

Huralikuppi, J. C., Christopher, A. B., & Stephen, P. M., (2006). Antidiabetic effect of *Nelumbo nucifera* (Gaertn): Part I. Preliminary studies in rabbits. *Phytother. Res., 5*, 54–58.

Hussain, Z., Waheed, A., Qureshi, R. A., et al., (2004). The effect of medicinal plants of Islamabad and Murree region of Pakistan on insulin secretion from INS-1 cells. *Phytother. Res, 18*(1), 73–79.

Imad, M., Taj, E., Elhadi, M. A., & AbdElwahab, H. M., (2010). *Environ Health Insights, 4*, 71–77.

Jarald, E. E., Joshi, S. B., & Jain, D. C., (2008). Antidiabetic activity of aqueous extract and non polysaccharide fraction of *Cynodon* dactylon pers. *Ind. J. Exp. Biol, 46*, 660–667.

John, O. A. O., (2005). *Journal of Ethnopharmacology, 99*, 13–19.

Jones, P. M., & Persaud, S. J., (1998). Protein kinases, protein phosphorylation and the regulation of insulin secretion from pancreatic beta cells. *Endocr. Rev., 19*(4), 429–461.

Karthik, D., Ilavenil, S., Kaleeswaran, B., et al., (2012). Analysis of modification of liver proteome in diabetic rats by 2D electrophoresis and MALDI-TOF-MS. *Ind. J. Clin. Biochem, 27*, 221–230 https://doi.org/10.1007/s12291-012-0209-8.

Karthik, D., Ilavenil, S., Kaleeswaran, B., et al., (2012). Proteomic analysis of plasma proteins in diabetic rats by 2D Electrophoresis and MALDI–TOF–MS. *Appl. Biochem. Biotechnol., 166*, 1507–1519.

Karthiyayini, T., Rajesh, K., Kumar, K. L., Sahu, R., & Amit, R., (2009). *J. Biomed & Pharmacology, 2*(2), 351–355.

Khan, B. A., Abraham, A., & Leelamma, S., (1995). *Indian J. Biochem Biophys, 32*(2), 106–108.

Kumar, P., Baraiya, S., Gaidhani, S. N., & Gupta, M. D., (2012). *J. Pharmacol. Pharmacother, 3*, 64–66.

Lai, J., Li, R., Xu, X., Jin, W., Xu, M., & Zhao, H., (2010). Genome-wide patterns of genetic variation among elite maize inbred lines. *Nat. Genet., 42*(11), 1027–1030.

Maithili, V., Dhanabal, S. P., Mahendran, S., & Vadivelan, R., (2011). Antidiabetic activity of ethanolic extract of tubers of *Dioscorea alata* in alloxan-induced diabetic rats. *Ind. J. Pharmacol., 43*, 455–459.

Manisha, V., Harneet, S., & Satish, S., (2011). Phytochemical screening and antimicrobial activity of roots of *Murraya Koenigii* (Linn.) spreng. (Rutaceae). *Brazilian Journal of Microbiology, 42*, 1569–1573.

Mediani, A. F., Abas, M., Maulidiani, A., Khatib, C. P., Tan, I. S., Ismail, K., Shaari, A., & Ismail, N. L., (2016). Metabolic and biochemical changes in streptozotocin-induced obese-diabetic rats treated with *Phyllanthus niruri* extract. *J. Pharm. Biomed. Anal., 128*, 302–312.

Melanie-Jayne, R., Howes, R. F., & Peter, J. (2017). In: Bai-Yun, Z., & Kaicun, Z., (eds.), *Houghton Chapter Two – Effect of Chinese Herbal Medicine on Alzheimer's Disease* (pp. 29–56).

Modak, M., Dixit, P., Londhe, J., et al., (2007). Indians' herbs and herbal drugs used for the treatment of diabetes. *J. Clin. Biochem. Nutr., 40*, 163–173.

Mukesh, R., & Namita, P., (2013). Medicinal plants with antidiabetic potential: A review. *American Eurasian J. Agric. Environ. Sci., 13*(1), 81–94.

Mukherjee, P. K., Maiti, K., Mukherjiee, K., & Houghton, P. J., (2006). Leads from Indian medicinal plants with hypoglycemic potentials. *J. Ethnopharmacol., 106*, 1–28.

Nael, A. T., & Al-Wadaan, M. A., (2011). Utility and importance of walnut, Juglans regia Linn: A review. *African Journal of Microbiology Research, 5*(32), 5796–5805.

Narendra, K., Zishan, A. W., & Sandeep, D., (2015). Ethnobotanical study of the plants used by the local people of Gulmarg and its allied areas, Jammu & Kashmir, India. *Int. J. Curr. Res. Biosci. Plant Biol., 2*(9), 16–23.

Nishkruti, R. M., Ekta, P. P., Pragnesh, V. P., & Biren, S., (2013). *Nelumbo nucifera* (Lotus): A review on ethnobotany, phytochemistry and pharmacology. *Indian J. Pharm. Biol. Res., 1*(4), 152–167.

Pandey, R., Tiwari, R. K., & Shukla, S. S., (2016). Omics: A newer technique in herbal drug standardization and quantification. *J. Young Pharm, 8*(2), 76–81.

Pandita, D., Pandita, A., & Pandita, S., (2015). Pharmacological and phytochemical portrayal of monocotyledonous medicinal plants of Jammu and Kashmir abound with antidiabetic potential. *J. Microbiol. Biotech. Res., 5*(6), 28–33.

Pandita, D., Pandita, A., & Pandita, S., (2015). Pharmacological and phytochemical portrayal of dicotyledonous medicinal plants of Jammu and Kashmir abound with antidiabetic potential. *J. Microbiol. Biotech. Res., 5*(6), 16–27.

Patel, D. K., Prasad, S. K., Kumar, R., & Hemalatha, S., (2012). An overview on antidiabetic medicinal plants having insulin mimetic property. *Asian Pacific Journal of Tropical Biomedicine, 2*(4), 320–330.

Pawar, V. A., & Pawar, P. R., (2014). *Costus speciosus*: An important medicinal plant. *International Journal of Science and Research (IJSR), 3*(7).

Penumathsa, S. V., Thirunavukkarrasu, M., Zhan, et al., (2008). Resveratrol enhances GLUT -4 translocation to the caveolar lipid raft fractions through AMPK/AKT/Enos signaling pathway in diabetic myocardium. *J. Cell Mol. Med.*, 1582–1590.

Prabhakar, P. K., & Doble, M., (2008). A target based therapeutic approach towards diabetes mellitus using medicinal plants. *Curr. Diabetes Rev., 4*(4), 291–308. doi: 10.2174/157339908786241124. PMID:18991598.

Ramon, R. R., Alarcon-Aquilar, F., Lara-Lemus, A., & Flores- Saenz, J. L., (1992). *Arch. Med. Res, 23*(1), 59–64.

Ranilla, L. G., Apostolidis, E., Genovese, M. I., & Shetty, K., (2009). *J. Med. Food, 12*(4), 704–713.

Rasineni, K., Bellamkonda, R., Singareddy, S. R., & Desireddy, S., (2010). Antihyperglycemic activity of Catharanthus roseus leaf powder in streptozotocin-induced diabetic rats. *Pharmacognosy Res., 2*, 195–201.

Revathy, J., Abdullah, S. S., & Kumar, P. S., (2014). *Journal of Chemistry and Biochemistry, 2*(1), 13–22.

Sangeetha, M. K., Priya, C. D., & Vasanthi, H. R., (2013). Antidiabetic property of *Tinospora cordifolia* and its active compound is mediated through the expression of Glut-4 in L6 myotubes. *Phytomedicine, 20*, 246–248.

Saxena, A., & Vikram, N. K., (2004). *J. Altern. Complement Med, 10*(2), 369–378.

Semwal, B. C., Gupta, J., Singh, S., Kumar, Y., & Giri, M., (2009). *Int. J. of Green Pharmacy, 3:* 259–262.

Sharma, M., Fernandes, J., Ahirwar, D., & Jain, R., (2008). Hypoglycemic and hypolipidemic activity of alcoholic extract of citrus Aurantium in normal and alloxan-induced diabetic rats. *Pharmacology Online, 3*, 161–171.

Shiyao, H., Yiwei, Z., Jiayue, L., Lin, D., Jun, H., Dingbo, L., & Xueyan, F., (2018). Ethnomedicine, phytochemistry and pharmacology of *Smilax glabra*: An important traditional Chinese medicine. *The American Journal of Chinese Medicine, 46*(2), 261–297.

Sugihara, Y., Nogima, H., Matsuda, H., et al., (2000). Antihyperglycemic effects of gymnemic acid IV, a compound derived from *Gymnema sylvestre* leaves in streptozotocin-diabetic mice. *J. Asian Nat. Prod. Res., 2*, 321–327.

Swanstan-Flatt, S. K., Day, C., Flatt, P. R., Gould, B. J., & Bailey, C. J., (1989). *Diabetes Res., 10*(2), 69–73.

Uma, M. M., & Sudarsanam, D., (2012). A review on bioinformatics for diabetic mellitus. *International Journal of Pharma Sciences and Research, 3*(6), 389–395.

Verspohl, E. J., Bauer, K., & Neddermann, E., (2005). Antidiabetic effect of *Cinnamomum cassia* and *Cinnamomum*: Zeylanicum *in vivo* and *in vitro*. *Phytother. Res., 19*(3), 203–206.

Wadood, A., Wadood, N., & Shah, S. A., (1989). Effects of acacia arabica and *Caralluma edulis* on blood glucose levels of normal and alloxan diabetic rabbits. *J. Pak. Med. Assoc., 39*, 208–212.

Xie, J. T., Mchendale, S., & Yuan, C. S., (2005). Ginseng and diabetes. *Am J Chin Med., 33*(3), 397–404.

Yamaguchi, K., Liggett, J. L., Kim, N. C., & Baek, S. J., (2006). Anti-proliferative effect of horehound leaf and wild cherry bark extracts on human colorectal cancer cells. *Oncol. Rep., 15*, 275–281.

Yoo, M. J., Bell, C. D., Soltis, P. S., & Soltis, D. E., (2006). Divergence times and historical biogeography of *Nymphaeales*. *Syst. Bot., 30*, 693–704.

Yu, B. C., Hung, C. R., Chen, W. C., & Cheng, J. T., (2003). Antihyperglycemic effect of andragrapholide in streptozotocin –induced diabetic rats. *Planta Med., 69*, 1075–1079.

Zakir, H. K., Sumer, S., & Firdous, M., (2018). Medicinal plants used as folk medicine in Kashmir Himalaya. *IJRAR, 5*(3).

Zhu, X., Qiu, Z., Wouyang, W., Miao, J., et al., (2019). Hepatic transcriptome and proteome analyses provide new insights into the regulator mechanism of dietary avicularin in diabetic mice. *Food Research International, 125*, 108–570.

CHAPTER 6

Ethnobotanical Information of Medicinal Plants Used for Diabetes Mellitus in Seshachalam Biosphere Reserve, Andhra Pradesh, India

A. MADHUSUDHANA REDDY and C. NAGENDRA

*Yogi Vemana University, Kadapa – 516005, Andhra Pradesh, India,
E-mail: grassced@gmail.com (A. M. Reddy)*

ABSTRACT

Diabetes is a chronic disease with high or low blood glucose levels and its effects on insulin secretion. It gives major health issue of the people around the world especially in India. The disorder is now challenging in developed and developing countries to scientists. The diabetes if left unchecked and not medicated properly leads to death of patient. Ethnobotanical studies around the world are now greatly acknowledged in identifying antidiabetic plants. The ethnobotanical survey was undertaken with the help of 150 knowledgeable local rural and tribal community representatives in Seshachalam Biosphere Reserve for identifying plants curing diabetes. Seshachalam Biosphere Reserve also encompasses a major landscape of Eastern Ghats of Andhra Pradesh, lying between 13° 30' to 14° 11' 15" N and 79° 0' 30" to 79° 30' E and extended to an area of 1,840 sq.km. Our study revealed that about 68 plant species of the Biosphere Reserve is found to have antidiabetic therapeutic values. The present chapter deals with systematic enumeration of all the species used for diabetes by the local communities of Seshachalam Biosphere Reserve. The information led to the fair traditional botanical knowledge on herbal plants associated with the local communities and will be helpful in treating diabetes.

Antidiabetic Potential of Plants in the Era of Omics. Deepu Pandita, PhD, Anu Pandita, MSc, & Chander Bhanu, MBBS (Eds.)

6.1 INTRODUCTION

Diabetes is a chronic metabolic disorder represented by predisposition to the production of dramatically elevated blood glucose. The early known description of diabetes disorder dates back to 1,500 BC. In all traditional cultures viz Vedas, Chinese, Arabic, and Mediterranean cultures have been recorded about the diabetes mellitus (DM). Present day's the diseases are a major health problem in public they suffer with woozy health and die younger. Plants are playing an important role in our lives and have been an indispensable source of medicine since ancient times. Generally, plants have less toxicity, lesser side effects, and better compliance with low-cost of herbal drugs. It is a metabolic, non-communicable disease is the fourth largest killer in many developed and developing countries in the world. It posing a severe risk in the 21st century that has been (Wild et al., 2004) diagnosed. More than 30 million people suffered from diabetes around the universal. Life assurance can be halved with this disorder, particularly in developed countries where prevalence is growing and appropriate care is often not available. Life expectancy may be halved by this disease, especially in developed and developing countries where its occurrence is increasing and adequate treatment is often unavailable. Even in developed countries such as the USA, where sophisticated therapy is widely available, more demises are ascribed to diabetes than other diseases and motor vehicle accidents (WHO, 1985). In India, Ayurveda and several literatures revealed the use of plants in the action of various human diseases. India is rich in herbal and medicinal plants with a wealth of ideal geoclimatic conditions. It has well known and realistic knowledge of traditional herbal medicine. Herbal and medicinal plants grow abundantly in sufficient environmental conditions. An antidiabetic plant wealth of Seshachalam hills is not known fully. The current study incorporates the information about 68 antidiabetic plants from Seshachalam hills. An updated accepted name has been given to each species as per the Plants of the World Online (POWO) and International Plant Name Index (IPNI). Plant distribution followed based on flora of Andhra Pradesh (Pullaiah et al., 2018). Few plant names collected from Lankamala wildlife sanctuary (Reddy et al., 2015). Telugu names followed by flowering plants of Chittoor district of Andhra Pradesh (Madhava Chetty et al., 2008).

6.2 SIGNIFICANCE OF DIABETES

Etymologically, the word diabetes can be interpreted as "passing through." The term diabetes is derived from the Greek word ('dia' – through; 'betes' – pass). More accurately, it was described due to excessive quantity of sweetened taste, urine with an unusual smell, attended by excessive thirst, dry skin, intense fatigue, and general skinniness. Diabetes is defined as a type of chronic metabolic disorders characterized by hyperglycemia caused by insulin secretion deficiency and their intervention or both. High blood glucose and associated insulin shortage was diagnosed. It is a significant concern and co-morbidity of heart disorders. Diabetes cases are expected only 4% in 1995 and 5.4% will increase in 2025 around the worldwide (American Diabetes Association, 2010). The disease not only kills, it is a significant cause of adult sightlessness, kidney disease, gangrene, neuropathy, heart diseases and stroke.

6.3 TYPES OF DIABETES

Mellitus is based on the conditions at the time of analysis and a type of diabetes is assigned to the individual. A person cannot fit into a sole class and maybe given numerous classes.

6.3.1 TYPE I DIABETES

It is also called insulin-dependent diabetes or infantile-onset diabetes; total insulin deficiency is caused by β-cell death. It constitutes just 5–10% of the people suffered through diabetes (Atkinson et al., 2014). The major signs include high blood sugar, heavy thirst, repeated urination, elevated appetite, tiredness, and weight loss, slow wound healing, and blurred vision. Insulin injection is the only medication appropriate for this form of treatment. In United States total 1.8 million peoples suffered with Type-I diabetes in worldwide total 18 million peoples affected.

6.3.2 TYPE II DIABETES

Primarily it is called non-insulin reliant on diabetes which appeared owing to insulin opposing lead to dominant insulin secretive fault. Total 90–95% of people suffer from Type II diabetes (American Diabetes Association, 2010). Classic symptoms are constant appetite, frequent urination, and frequent thirst. Some signs include weight loss, burning of the scalp, and chronic infection. Significant chronic complications of type II diabetes include the acceleration of macro-vascular disease, retinopathy, renal disease, and renal disease neuropathy. Type II diabetes mellitus affects over 15 million people in the United States and nearly 150 million people throughout the world. In USA diabetes is seventh foremost cause of decease. NIDD therapy includes dietary changes, nutrition, exercise, and use of oral hypoglycemic specialists.

In 33% of patients, it can be very well managed by bucket diet, exercise, and weight reduction. The majority of the patients will require oral drugs that stimulate the pancreas to release more insulin or to extend the body's cell function to insulin.

6.3.2.1 IMPORTANT ORAL ANTIDIABETIC DRUG

Oral hypoglycemic agents are used in patients with Type II diabetes unresponsive to dietary regulation, or dietary control was initially effective but has now failed.

6.3.2.2 DIET IN DIABETES

One of the significant managements of diabetes is change in diet. Eating right helps to keep blood glucose levels under control, which helps from complications.

6.3.2.3 EXERCISE AND WEIGHT REDUCTION

Physical activity is significant influence in the instruction of blood sugar levels. Exercise will avoid the onset of NIDDM and control blood glucose for those already suffering from diabetes.

6.3.2.4 MINERAL SUPPLANTS

Nutritional supplements can also be useful in the treatment of diabetes. Several mineral supplements have been shown to be helpful to people with diabetes. Vitamin C, Vitamin E, Chromium, Magnesium, and Calcium are the most essential vitamins and supplements.

6.3.2.5 HERBAL REMEDIES

While different treatment approaches are possible for diabetes, normoglycemia, avoidance of diabetic complications and multiple pathological aspects are difficult to manage, except for the pancreatic transplantation process.

6.3.2.6 NEED SCOPE AND ALTERNATIVE REMEDIES

The modernization of the NIDDM lifestyle is a major health issue in developed countries. The high cost and low supply of existing treatments for many of India's pastoral population necessitates the production of indigenous and cheap botanical sources of antidiabetic and antihyperlipidemic raw or purified medicines.

6.3.3 GESTATIONAL DIABETES

Glucose intolerance is acclaimed with initiation or first gratitude in pregnancy period. Revisions suggest that a majority of cases of gestational diabetes tenacity with birth, determining whether or not the disorder continues following pregnancy and not ignoring the likelihood that unrecognized glucose tolerant could have begun concurrently with pregnancy (Atkinson et al., 2014).

There are numerous actions in which on antidiabetic medications might be used to promote beta-cell, pancreatic islet insulin release, restrict hormones that develop blood glucose levels (BGL), increase glycogen content, increase the sensitivity of insulin receptors, free radical rummaging, correct lipid, and protein metabolic disorders, increase the use of glucose in the tissue, resist lipid peroxidation and promote microcirculation in body (Bösenberg et al., 2008).

6.4 PHYTOCONSTITUENTS IN ANTIDIABETIC PLANTS

The medicinal plants have antidiabetic properties due to diverse forms of phytoconstituents types present. These constituents present in medicinal plants can be classified as:

- Imidazoline alkaloids generate glucose-dependent insulin secretion;
- Polysaccharides are reducing the glucose levels in blood and increase the serum insulin levels;
- Flavonoids are antidiabetic mediators that act by decreasing glucose levels, by dramatically letting down plasma cholesterol and triglyceride levels, and by-hepatic glucokinase action, by increasing insulin proclamation from beta-pancreatic islands (Chauhan et al., 2010);
- Dietary fibers help in reducing the concentration of glucose preoccupation and postprandial serum glucose;
- Secretion of insulin and saponins are induced by ferulic acid that causes release the insulin that prevents the glucose production levels in the bloodstream.

6.5 IMPORTANCE OF DIABETIC PLANTS

Presently certain herbal plants have been reported for beneficial in diabetes around the world, and it has been used as antidiabetic medicines. Although the availability of approved antidiabetic drugs in pharmaceutical stores, diabetes, and associated complications remained a major medical concern. The antihyperglycemic properties of these herbal plants are attributed to their potential to preserve the role of pancreatic tissues through inducing an improvement in insulin production or constraining intestinal glucose preoccupation or by facilitating the use of metabolites in insulin-dependent procedures. According to ethnobotanical knowledge records, some plants have antidiabetic ability. Literature reveals 400 plant species shows hypoglycemic activity it has been recorded, but exploration of new natural medicines from plants is still desirable because they produce compounds that show alternative and healthy effects on diabetes diseases. Many plants have antidiabetic impact. The active ingredients in several plants are used specifically as medicines, lead compounds or pharmacological managers.

6.6 STUDY AREA

The Sheshachalam hills are confined its distribution in Kadapa and Chittoor districts in Andhra Pradesh, India, and distributed very prominently in YSR Kadapa district and spreading the area beneath 13° 30' to 14° 11' 15" N and 79° 0' 30" to 79° 30' E and extended to 1,840 sq.km. The dominating forest type is tropical southern dry mixed deciduous forest (Champion and Seth, 1968). The area find diversity of wildlife as well as varied ecosystems and habitats, the area has been selected as the Seshachalam Biosphere Reserve in the month September 20, 2011. Seshachalam Biosphere Reserve contains world-famous blessed shrines widely known as Lord Venkateswara or Balaji. Seshachalam hills covers about the 365 theerthams contain periodic and perpetual waterfalls and helixes. In Andhra Pradesh state Thalakona is the highest waterfall with 270 feet. Seshachalam biosphere reserve forests are mainly divided in to 3 types they are they are dry deciduous mixed forest, moist deciduous forests, and scrubby forest type (Champion and Seth, 1968). Valuable *Pterocarpus santalinus* (Red sander) is abundantly growing in this region. Different tribal peoples are living in Seshachalam biosphere reserve these tribal communities are Yanadis and Chenchus. They are living from ancient times. The antidiabetic plants date gathered from tribal who are practices and experience in the use of knowledge. Information came in the form of vernacular names and plant specimens composed and known with local floras (Pullaiah et al., 2018; Madhava Chetty et al., 2008). Plants names are arranged in alphabetically and given each species information like scientific, vernacular names, family, and mode of action in Given Table 6.1 and morphology of some antidiabetic plants is shown in Figures 6.1 and 6.2.

FIGURE 6.1 Important antidiabetic plants of Tirumala Hills.

Memecylon umbellatum

Mucuna pruriens

Pterocarpus marsupium

Terminalia belirica

Terminalia chebula

Terminalia pallida

FIGURE 6.2 Important antidiabetic plants of Talakona Hills.

TABLE 6.1 List of Important Antidiabetic Plants

SL. No.	Scientific Name	Common Name	Family	Mode of Governance
1.	*Abrus precatorius* L.	Guravinda	Fabaceae	Daily two teaspoons of leaf juice given orally twice per day until recovery. This leaf extracts properly working against diabetes.
2.	*Acacia catechu* (L.f.) Willd.	Sandra chettu	Fabaceae	Heartwood extract used in diabetes treatment
3.	*Acacia nilotica* Delile	Nalla thumma	Fabaceae	Dry seed powder used as antidiabetic.
4.	*Adhatoda vasica* Nees	Addasaram	Acanthaceae	Leaf powder use in diabetes
5.	*Aegle marmelos* (L.) Corrêa	Bilva	Rutaceae	10 ml of young leaf juice combined with 4–5 drops of honey and given orally on an empty stomach twice per day (morning and evening) this will cure diabetes within a period of 3–4 weeks.
6.	*Ageratum conyzoides* L.	Pokabanthi	Asteraceae	Macerated form of the entire plant is taken 1 full spoon twice per day.
7.	*Andrographis paniculata* (Burm.f.) Wall. ex Nees	Nelavemu	Acanthaceae	4–8 leaves decocted in water and taken orally once per day after food or eaten younger leaves at daily up to cure.
8.	*Asparagus racemosus* Willd	Uttareni	Asparagaceae	Leaf powder using in diabetes treatment
9.	*Biophytum sensitivum* (L.) DC.	Pedda atipati	Oxalidaceae	Tender leaf extract is taken orally on early morning with empty stomach.
10.	*Bombax ceiba* L.	Buruga chettu	Malvaceae	Bark extract is using
11.	*Butea monosperma* (Lam.) Taub	Moduga	Fabaceae	10 ml fresh leaf juice is administered once daily from 6–10 days on an empty stomach. This is going to decrease blood sugar.

TABLE 6.1 *(Continued)*

SL. No.	Scientific Name	Common Name	Family	Mode of Governance
12.	*Caesalpinia bonduc* (L.) Roxb.	Gachakaya chettu	Fabaceae	The extract of seeds used in diabetes treatment.
13.	*Carissa carandas* L.	Pedda kalimi	Apocynaceae	The fruits were dried under shade and prepared powder. This powder material using in diabetes treatment.
14.	*Cassia fistula* L.	Rela chettu	Fabaceae	Fresh pods were collected washed with water and make it shade dry and make it powder. This pod extract is used in diabetic treatment.
15.	*Citrullus colocynthus* (L.) Schrad	Paparakaya	Cucurbitaceae	Dried leaves and fruits extract is used.
16.	*Corallocarpus epigaeu* Benth. and Hook.f.	Dhumpadonda	Cucurbitaceae	Rhizome powder use in diabetes treatment.
17.	*Cayratia trifolia* Domin	Gummediteega	Vitaceae	Empirically, it is used to cure various disorders and one of which is antidiabetic. The dried stem extract used as an antidiabetic agent.
18.	*Cipadessa baccifera* (Roth) Miq.	Ranaberi	Meliaceae	Fresh leaf extract is used in diabetic activity.
19.	*Cissampelos pareira* Linn.	Banka teega	Menispermaceae	Fresh twigs as well as leaf buds given orally for up to one week on an empty stomach. Strictly prohibited to eat sugar containing food during this course. Finally, this will decrease sugar levels.
20.	*Clerodendum serratum* (L.) Moon	Banala chettu	Lamiaceae	Tender leaf juice used in diabetes.

TABLE 6.1 *(Continued)*

SL. No.	Scientific Name	Common Name	Family	Mode of Governance
21.	*Cocculus hirsutus* (L.) Diels	Dussara teega	Menispermaceae	Aerial part extract is used
22.	*Costus speciosus* (J.Koenig) Sm.	Changalagadda	Costaceae	The leaves and rhizome extracts of this plant have been reporting to possess an antidiabetic in nature
23.	*Cryptostegia grandiflora* R.Br.	Garudapala	Apocynaceae	The aqueous extract of the aerial parts widely used in diabetic treatment.
24.	*Curculigo orchioides* Gaertn.	Nelathadi	Hypoxidaceae	Tuber extract is used as antidiabetic.
25.	*Decalepis hamiltonii* Wight and Arn.	Nannari teega	Asclepiadaceae	Fresh tubers dry in shade after make in to powder. This tuber extract used for cure.
26.	*Dioscorea bulbifera* L.	Adavidumpa	Dioscoreaceae	Fresh tubers powder used.
27.	*Dendrophthoe falcata* (L. f.)	Kukkanaluka	Loranthaceae	Whole plant is used as antidiabetic.
28.	*Ficus benghalensis* L.	Marri	Moraceae	In combination of 10 ml bark extract and 5 g of jaggery is effectively worked against diabetes. This should be taken orally once a day for up to two weeks its decrease blood sugar. Bark latex also effectively work against diabetes.
29.	*Ficus hispida* Linn	Kaakimedi	Moraceae	Mature fruits are consumed daily and stem bark is also used.
30.	*Ficus racemosa* Linn	Medi	Moraceae	40 g of boiled immature fruits and the same amount of fine rice given in normal meals up to two years will reduce sugar concentrations in urine.

TABLE 6.1 *(Continued)*

SL. No.	Scientific Name	Common Name	Family	Mode of Governance
31.	*Ficus religiosa* L.	Raavi	Moraceae	Bark extract is used.
32.	*Glycosmis pentaphylla* (Retz.) DC.	Tanaka	Rutaceae	Dried leaves powder using in diabetes treatment
33.	*Gmelina arborea* Rox. Ex. Sm.	Gummadi teku	Lamiaceae	Dried leaves powder used
34.	*Gmelina asiatica* L.	Chirugummadi	Lamiaceae	Decoction of the leaf extract taken orally for treating diabetes.
35.	*Gymnema sylvestre* (Retz.) R.Br. ex Sm.	Podapatri	Apocyanaceae	Tender leaves juice is taken orally on early morning with empty stomach until curing diabetes.
36.	*Hibiscus platanifolius* (Willd.) Sweet	Konda gogu	Malvaceae	Leaf extract using in diabetes.
37.	*Helicteres isora* L.	Thada chettu	Malvaceae	Root extract is taken orally one teaspoon full, or Daily early morning given root/ bark extract orally once a day up to two weeks on an empty stomach to cure sugar complaints.
38.	*Hemidesmus indicus* (L.) R.Br.	Sugandapala	Apocynaceae	Roots extracts were used in diabetic treatment.
39.	*Hemionitis arifolia* (Burm. f.) Moore	Ramabanam	Adiantaceae	Dried leaf powder used
40.	*Holarrehena pubescens* (Buch-Ham.) Wall	Munupala	Apocynaceae	Dried bark collected and prepared powder use in diabetic treatment.
41.	*Holostemma ada-kodien* Schult	Palajilledu	Apocynaceae	The Leaf extract used in diabetic treatment.

TABLE 6.1 *(Continued)*

SL. No.	Scientific Name	Common Name	Family	Mode of Governance
42.	*Hydrophila auriculata* (Schum.) Heine	Mulla godi	Acanthaceae	Leaf and stem extract are used in treating diabetes treatment.
43.	*Ichnocarpus frutescens* (L.) R.Br.	Nalla tiga	Apocynaceae	Tender leaf juice and its fruit in combination with 4 black pepper (Piper nigrum) given orally on morning before breakfast it will reduce sugar levels.
44.	*Kigelia Africana* (Lam.) Benth.	Enuguthondam	Bignoniaceae	The flowers were dried under shade and made in to powder. This powdered material used in diabetic treatment.
45.	*Madhuca indica* J.F.Gmel.	Ippa chettu	Sapotaceae	Decoction prepared with dried bark powder and this given orally for better results.
46.	*Mallotus philippensis* (Lam.) Muell-Arg	Kunkumapuvvu	Euphorbiaceae	The bark extract using in diabetes treatment.
47.	*Memecylon umbellatum* Burm.f.	Manjialli	Melastomataceae	Tender Leaf juice is taken orally for better results.
48.	*Mucuna pruriens* (L.) DC.	Duradhagondi	Fabaceae	Feeding of seed powder up to seven days it showing better results in blood glucose reduction.
49.	*Naravelia zeylanica* (L.) DC.	Gourikunthala	Rananculaceae	Fresh leaf juice used for better results.
50.	*Nelumbo nucifera* Gaertn	Thamara	Nymphaeaceae	Rhizome powder significantly reduce the diabetes.
51.	*Phyllanthus amarus* Schumach. and Thonn.	Nela usiri	Phyllanthaceae	Fresh leaf powder using in diabetic treatment.

TABLE 6.1 *(Continued)*

SL. No.	Scientific Name	Common Name	Family	Mode of Governance
52.	*Pterocarpus marsupium* Roxb.	Egisa	Fabaceae	Bark extract showing better results or heartwood soaked in water up to 24 hours and given 10 ml orally for 30 days this will give good results.
53.	*Pterocarpus santalinus* L.f.	Erra chandanam	Fabaceae	Bark extract used in diabetic treatment.
54.	*Salacia oblonga* Wall.	Garika musti	Celastraceae	Extract of the roots inhibited glucose level growth.
55.	*Semicarpus anacardium* L.f.	Nalla jeedi	Anacardiaceae	The dried stem barks were powdered with the grinder and used in this treatment.
56.	*Soymida febrifuga* (Roxb.) A.Juss.	Somidi chettu	Meliaceae	Bark extract is used
57.	*Sterculia urens* Roxb.	Erra poliki	Sterculiaceae	Gum used as antidiabetic
58.	*Strychnos nux-vomica* L.	Musti chettu	Loganiaceae	Seed extract using in diabetes treatment
59.	*Strychnus potatorum* L.f.	Chilla chttu	Loganiaceae	Fresh seed powder used as antidiabetic.
60.	*Swietenia mahagoni* L. Jacq.	Mahagani	Meliaceae	Bark extract using in diabetes.
61.	*Syzygium alternifolium* (Wight) Walp.	Mogi	Myrtaceae	Seed powder using in diabetes.
62.	*Terminalia arjuna* (Roxb.) Wight and Arn.	–	–	2–5 g dose of barks powder or macerated form with milk is taken 2 h after food for better results.
63.	*Terminalia bellirica* (Gaertn.) Roxb.	Thani chettu	Combretaceae	The fruits were collected and dried under 40°C after pericarp was powdered. This powder is used in diabetic treatment.

TABLE 6.1 *(Continued)*

SL. No.	Scientific Name	Common Name	Family	Mode of Governance
64.	*Terminalia chebula* Retz.	Nallakaraka	Combretaceae	Fresh mature Terminalia fruits were collected and dried after being made in to powder. This fruit powder using in diabetic treatment.
65.	*Terminalia pallida* Brandis	Karakkaya chettu	Combretaceae	Dried fruits powder used in diabetic treatment.
66.	*Tinospora cordifolia* Miers	Tippa teega	Menispermaceae	Bark extract and tender stem juice 15–20 ml with 2 drops of honey is given vocally twice a day for 15 days or 5 g of fresh stem powder and 3 g of pepper powder (Piper longum) given for one week by orally until cure.
67.	*Trichosanthes cucumerina* L. C	Adavi potlakaya	Cucurbitaceae	Aerial part extract used.
68.	*Ventilago madraspatana* Gaertn.	Surabhi	Rhamnaceae	Leaf juice is drastically decreasing the blood glucose.

6.7 CONCLUSION

Diabetes disease is a disorder originally characterized by a decrease of glucose homeostasis resultant from faults in insulin discharge, insulin action both resultant in damaged glucose breakdown and other energy-yielding fuels as lipids and proteins (Sivajothia et al., 2008). Several countries are presently facing significant changes in the number of people suffering from diabetes disorder. Total 30 million people grieved from diabetes around the world in 1985, and this quantity increased more than 170 million in 2000 it is reported by the World Health Organization (WHO). It is projected that this figure will growth more than 366 million in the year 2030 and that significantly it will increase in all developed countries and specially in between 45- and 64-years aged persons (Wild, 2004). Despite the prevalence of proven antidiabetic medicines on the pharmaceutical market, herbal plant treatments are used successfully to cure this disorder. Many conventional plant therapies are used worldwide for diabetes treatment. Plant based medications and herbal formulations are less toxic than pharmaceutical depend medicines.

In overall, there is a very slight biological understanding of particular modes of deed in the dealing of diabetes, although most plants have been found to contain compounds such as glycosides, alkaloids, terpenoids, flavonoids, etc., which are also thought to have antidiabetic impact. Research studies on natural therapies for diabetes disease are going worldwide and pose numerous problems not just to the doctors, but also to the researcher. To the best of our knowledge, ethnobotanical surveys conducted in Seshachalam biosphere reserve especially for antidiabetic plants. This present inventory therefore represents the available diabetic plants data.

KEYWORDS

- **diabetes mellitus**
- **ethnobotanical**
- ***Ocimum sanctum***
- **Seshachalam Biosphere Reserve**
- **World Health Organization**

REFERENCES

American Diabetes Association, (2010). Diagnosis and classification of diabetes mellitus. *Diabetes Care, 33*(Supplement 1), S62–S69.

Atkinson, M. A., Eisenbarth, G. S., & Michels, A. W., (2014). Type 1 diabetes. *The Lancet, 383*(9911), 69–82.

Bösenberg, L. H., & Van, Z. D. G., (2008). The mechanism of action of oral antidiabetic drugs: A review of recent literature. *Journal of Endocrinology, Metabolism and Diabetes of South Africa, 13*(3), 80–88.

Champion, H. G., & Seth, S. K., (1968). *A Revised Survey of the Forest Types of India*. Govt. of India Press, Delhi.

Chauhan, A., Sharma, P. K., Srivastava, P., Kumar, N., & Dudhe, R., (2010). Plants having potential antidiabetic activity: A review. *Der Pharmacia Letter, 2*(3), 369–387.

Harris, M. I., Eastman, R. C., Cowie, C. C., Flegal, K. M., & Eberhardt, M. S., (1997). Comparison of diabetes diagnostic categories in the US population according to 1997 American Diabetes Association and 1980–1985 World Health Organization diagnostic criteria. *Diabetes Care, 20*(12), 1859–1862.

Madhava Chetty, K., Sivaji, K., & Tulasi Rao, K., (2008). Flowering plants of Chittoor district, Andhra Pradesh, India. Published by *Students Offset Printers*, Tirupati, 61.

Pullaiah, T., & Karuppusamy, S., (2018). *Flora of Andhra Pradesh* (Vol. 3). Scientific Publishers.

Pullaiah, T., & Sri Rama, M. K., (2018). *Flora of Andhra Pradesh* (Vol. 2). Scientific Publishers.

Pullaiah, T., Ali, M. D., & Sandhya, R. S., (2018). *Flora of Andhra Pradesh* (Vol. 4). Scientific publishers.

Pullaiah, T., Chennaiah, E., & Sandhya, R. S., (2018). *Flora of Andhra Pradesh* (Vol. 1). Scientific Publishers.

Reddy, S. R., Reddy, A. M., & Babu, M. S., (2015). Traditional medicinal plants of Lankamalleswara Wildlife Sanctuary, Kadapa District, Andhra Pradesh, India. *American Journal of Ethnomedicine, 2*(6), 2348–9502.

Sandhya, S., Kumar, P. S., Vinod, K. R., Banji, D., & Kumar, K. (2010). Plants as potent antidiabetic and wound healing agents: A review. *Hygeia JD Med, 3*(1), 11–19.

Sivajothia, V., Dey, A., & Jayakar, B., (2008). *Iran. J. Pharm. Res., 7*, 53.

Vadivelan, R., Krishnan, R. G., & Kannan, R., (2019). Antidiabetic potential of *Asparagus racemosus* willd leaf extracts through inhibition of α-amylase and α-glucosidase. *Journal of Traditional and Complementary Medicine, 9*(1), 1–4.

Wild, S., Roglic, G., & Green, A., (2004). *Diabetes Care, 27*, 1047.

Wild, S., Roglic, G., Green, A., Sicree, R., & King, H., (2004). *Global prevalence of diabetes*: Estimates for the year 2000 and projections for 2030. *Diabetes Care, 27*(5), 1047–1053.

Part III
Mode of Action

CHAPTER 7

Overview of Steviol Glycosides: Action Toward Diabetes Control

PAYAL AGARWAL[1], RENITTA JOBBY[1], and PAMELA JHA[2*]

[1] *Amity Institute of Biotechnology, Amity University, Mumbai, Maharashtra, India*

[2] *Department of Biological Sciences, Sunandan Divatia School of Science, NMIMS Deemed to be University, Vile Parle (West), Mumbai 400 056., India, E-mail: pamelajha@gmail.com*

ABSTRACT

Numerous studies have reported *Stevia rebaudiana* to exhibit prompt effect against many physiological abnormalities including but not limited to, anti-hypersensitive, cardioprotective, anti-inflammatory, anticancer, antihyperglycemic. *Stevia rebaudiana* is recognized for its non-caloric sweetening properties and is a good candidate for developing new nutraceuticals treatment for diabetes. The bioactive phytochemical compounds extracted from *Stevia rebaudiana* are called steviol glycosides, a class of glycoconjugates that are revealed to be the most crucial for its activities against hyperglycemia. Steviol glycosides are administered orally, metabolized by distinct organs, and later eliminated by the body. They have been extensively studied for their response to high blood glucose levels (BGL). In this chapter, we aim to review the series of actions involved to achieve glucose homeostasis by steviol glycosides.

7.1 INTRODUCTION

Diabetes is a multi-faceted chronic ailment that influences an enormous fraction of the populace. It is characterized by an uncontrolled (and usually

Antidiabetic Potential of Plants in the Era of Omics. Deepu Pandita, PhD, Anu Pandita, MSc, & Chander Bhanu, MBBS (Eds.)

lasting) increase in blood glucose levels (BGL) – Hyperglycemia. The World Health Organization reports a drastic 5% elevation in mortality cases due to diabetes since 2016 (WHO – diabetes fact sheets, 2020). Occurrence of diabetes is accounted for two reasons: (i) islets of Langerhans of pancreatic β-cells fail to efficiently secrete insulin (a proteinaceous anabolic hormone responsible for lowering blood glucose level), as a consequence of anomalous immune response – this is type 1 diabetes which is formerly known as insulin-dependent diabetes mellitus or IDDM (Lucier et al., 2020); or (ii) development of insulin resistance by the body, where rise in blood glucose results from ineffective use of insulin by cells – this is type 2 diabetes, also referred to as non-insulin-dependent diabetes mellitus or NIDDM, and narrates the diabetic cases by 90% (Goyal et al., 2020).

Diabetes is routinely treated with diet therapy coupled with oral administration of insulin and non-insulin therapies like biguanides, thiazolidinediones, amylin analogs (Otto-Buczkowska et al., 2017), sulfonylureas (insulin secretagogues), alpha-glucosidase inhibitors (acarbose), dipeptidyl peptidase-4 inhibitors (sitagliptin, saxagliptin, alogliptin), dapagliflozin (sodium-glucose co-transporter-2 inhibitor) (Paik et al., 2019), and may have insulinotropic (Kalra et al., 2018) or inhibitory effect (Rosak et al., 2012) depending on the target molecule (Marín-Peñalver et al., 2016) with varying effectiveness (Garber et al., 2016). However, these drugs cannot attain a permanent, regular glucose homeostasis and are subject to many physiological side effects associated with them (AMBOSS).

Practitioners in Indian medical system have been composing original formulations of medicinal herbs to cure diabetes (Seth et al., 2004), since the earliest establishment of culture (Adhikari et al., 2018). This review aims to anthologize the antidiabetic actions of one of the most popular herbal supplements *Stevia rebaudiana* or “meethi patti” (sweet leaf) in Hindi. Natural compounds derived from plants have been vividly exploited in the pharmaceutical industries for their high effectiveness (Hosseini et al., 2015), uncomplicated synthesis (Altemimi et al., 2017), availability, less undesirable side effects (Veeresham, 2012), and cost-effectiveness (Payab et al., 2018; Kennedy et al., 2009). Shukia et al. (2000), compiled a comprehensive list of medicinal plants that have relevant antidiabetic effects.

7.2 MEDICINAL PLANTS

Phytochemicals are an evolutionary advantage of defense for plants to grapple with their sessile existence. These bioactive compounds have, however, been mined for their exclusive pleiotropic effects (Efferth et al., 2011) in the pharmaceutical research (Velu et al., 2018). The scientific community has been encouraged to perform an exhaustive research on identification of the medicinal plants (Firenzuoli et al., 2007; Wachtel-Galor et al., 2011) nesting this highly potential and biologically active array of compounds (Ahn, 2017). Phytochemicals were extensively investigated to demonstrate exceptional response as antiviral agents (Ghildiyal et al., 2020), indispensable anticancer lead (Singh et al., 2016), antimicrobial agents (Khatri et al., 2016), and antioxidants (Zhang et al., 2015).

There is an exhaustive list of medicinal plants that exhibit antidiabetic properties (Shukia et al., 2000). Some of these plants are enlisted along with the secondary metabolites involved and the mode of action (Table 7.1).

TABLE 7.1 Few Popular Plants with Antidiabetic Effect and Their Active Constituents

Medicinal Plant	Active Constituent	Phytochemical Classification	Mechanism of Action	References
Allium cepa (onion)	S-methyl cysteine sulfoxide	Phytoalexin	Promote glucose utilization	Kumari et al. (1995, 2002)
Allium sativum (garlic)	Garlic oil; allicin	Organosulfur compound	GLUT 4 expression; selective opening of SUR2	Liu et al. (2012); Dubey et al. (2012)
Annona squamosa (custard apple)	Quercetin-3-O-glucoside	Flavonoid	Modulation of insulin signaling	Davis et al. (2012); Panda et al. (2007)
Azadirachta Indica (Neem)	Nimbin	Triterpenoid	Islet regeneration/protection	Bhat et al. (2011)
Rubus suavissimus (Chinese Sweet tea)	Rubusoside	Steviol glycoside	Enhance insulin secretion	Ruili (2008)

7.3 *STEVIA REBAUDIANA*

Stevia rebaudiana (SrB), frequently referred to as "candy leaf," is a perennial, nutrient-rich, bushy shrub, belonging to the sunflower family – Asteraceae.

It is native to northeast Paraguay and neighboring parts of Brazil and Argentina. This seasonal plant is now cultivated worldwide for its sweet glycoside-rich leaves, that are claimed to be nearly 300 times sweeter than sucrose (Goyal et al., 2010), and thus used as a low-calorie natural sweetener (Savita et al., 2004). Countries like USA (Angelini et al., 2018), Europe (Tennant et al., 2017), Japan (Mitsuhashi et al., 1975), Canada (Nettleton et al., 2019), Korea (Gasmalla et al., 2015), India (Pal et al., 2015), and Argentina (Cosson et al., 2019) have worked towards increasing demand of purified steviol glycosides (discussed in Section 7.3.2) in response to surging rate of obesity and diabetic cases. Parts of South America, North America, and Southeast Asia have witnessed agricultural production of SrB irrespective of its limited natural habitat.

Thermal-stable traits (Šic et al., 2013) and impressive water retaining capacity (Campos et al., 2014) of SrB, by cause of essential amino acids present (Lemus-Mondaca et al., 2012), makes it very convenient to be included as an ingredient in recipes (Mogra et al., 2009). It can be used in addition to other protein supplements for strength training professions (Lima et al., 2019). Toxicological assessment of Stevia (Zhang et al., 2017) concludes its safe use as a food supplement. According to the FDA, the acceptable daily intake for purified SrB as food additives is 4 mg/kg of body weight. It is to be noted that the use of SrB is not limited as a sweetener. It has been explored for its mammoth therapeutic influence (illustrated in Figure 7.1). SrB is now known for its anticancer (Chen et al., 2018), anti-inflammatory (Kim et al., 2020), renoprotective (Rizwan et al., 2019), antibacterial (Theophilus et al., 2015), antihypertensive (Liu et al., 2003; Chan et al., 2000), anti-hyperlipidic, and antidiabetic attributes.

7.3.1 SECONDARY METABOLITES

Organic compounds that are not directly associated in the process of development, growth, and reproduction of an organism are called secondary metabolites. They are often indirectly derived from species-specific (Isah, 2019) primary metabolic pathways (Hussein et al., 2019). These substances, cultivated by plants, fungi, and bacteria, give its beneficiary a selective advantage to bolster their survivability and reproductive capabilities (Gupta et al., 2014). Secondary metabolites have found their rightful place in pharmaceutical industries as they continue to be explored for their unmatched pleiotropic benefits when used individually or as a mixture in medicines (Seca et al., 2019).

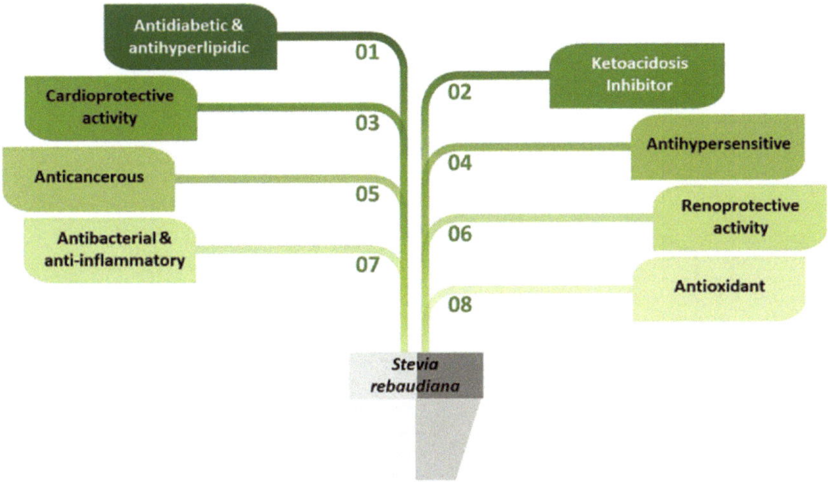

FIGURE 7.1 Pleiotropic effects of *Stevia rebaudiana.*

Secondary metabolites of *Stevia rebaudiana* are principally extracted from its leaves into a mixture commonly referred to as Stevia. Stevia tastes 30 to 150 times sweeter than sucrose or table sugar (Cardello et al., 1999), and is non-nutritive (Ashwell, 2015). The non-fermentable, heat, and pH stable active compounds present in Stevia are collectively called steviol glycosides (SG) (Ashwell, 2015). Glycosides are a class of glycoconjugates, that comprises a sugar group (glycone) bound to a non-sugar group (aglycone) via glycosidic linkage (N-linked or O-linked). It can be used as a storage strategy by plants, where, upon enzymatic degradation (Brito-Arias, 2016), the sugar component can free the active aglycone moiety for subsequent usage (against biotic or abiotic stress) (Erb et al., 2020).

Steviol (Ste) is the diterpene aglycone component of the glycosides (shown in Figure 7.2) present in the Stevia extract. It is readily metabolized by the body and thus plays a major role in the antidiabetic function of SG (Jeppesen et al., 2000). Among the glycosides present in *Stevia* (enlisted in Table 7.2) Stevioside (Stev) and Rebaudioside A (Reb A) are the most popular for their dominant composition and quality determinative factor respectively (Sharma et al., 2009). The difference in structure of the SG is discussed in Table 7.2. Structural discrepancies are attributed to the different mono/di/tri-saccharide, i.e., carbohydrate residues R1 and R2 present at positions C13 and C19, respectively.

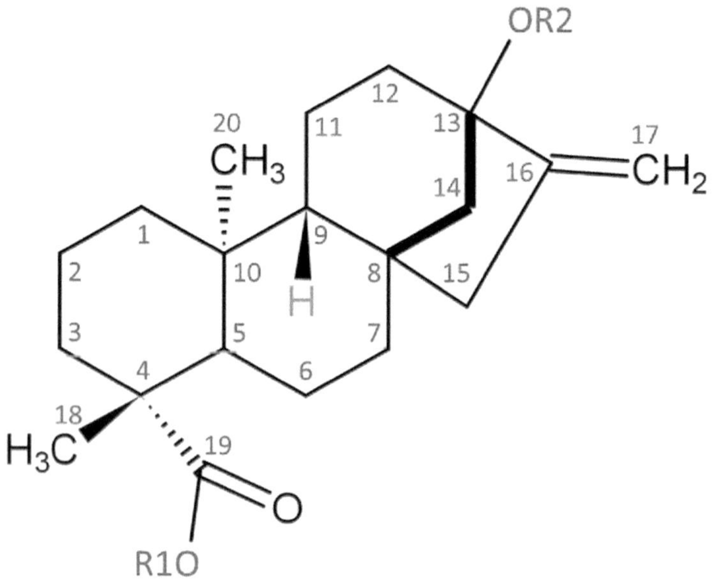

FIGURE 7.2 SG core structure with variable R1 and R2 residues.

TABLE 7.2 Major Steviol Glycosides Found in Dry Leaf Extract of *Stevia* (Kochikyan et al., 2006)

Compound	Abbreviation Used	R1	R2	Percentage (w/w) in Dry Leaf Extracts
Steviol	Ste	H	H	(Common backbone)
Stevioside	Stev	b-Glu	b-Glu-b-Glu (2→1)	4–14
Rebaudioside A	Reb A	b-Glu	b-Glu-b-Glu (2→1)-b-Glu (3→1)	2–4
Rebaudioside B	Reb B	H	b-Glu-b-Glu (2→1)-b-Glu (3→1)	Not reported
Rebaudioside C	Reb C	b-Glu	b-Glu-a-Rha(2→1)-b-Glu (3→1)	1–2
Rebaudioside D	Reb D	b-Glu-b-Glu (2→1)	b-Glu-b-Glu (2→1)-b-Glu (3→1)	0.2
Rebaudioside E	Reb E	b-Glu-b-Glu (2→1)	b-Glu-b-Glu (2→1)	0.2
Dulcoside A	Dul A	b-Glu	b-Glu-a-Rha(2→1)	0.4–0.7

7.3.1.1 BIOSYNTHESIS OF STEVIOL GLYCOSIDES

Biosynthesis of SG takes place in the leaves of *Stevia rebaudiana,* which are then transported to different parts of the plant. The precursors required for SG synthesis comes from primary metabolism pathways: glycolysis and photosynthesis. The whole process embroils 16 steps catalyzed by various enzymes present in the chloroplast, endometrium reticulum and the cytoplasm (Yadav et al., 2012). These steps can be presumptively divided in two stages: (a) production of Ste (the non-sugar backbone of SG) – 1st 11 steps that ends with the 1st committed step towards SG synthesis; and (b) stepwise glycosylation of Ste (loading of glycans) – final 5 steps catalyzed by a set of enzymes called glycosyltransferases. The biosynthesis pathway of SG is vividly discussed in Figure 7.3.

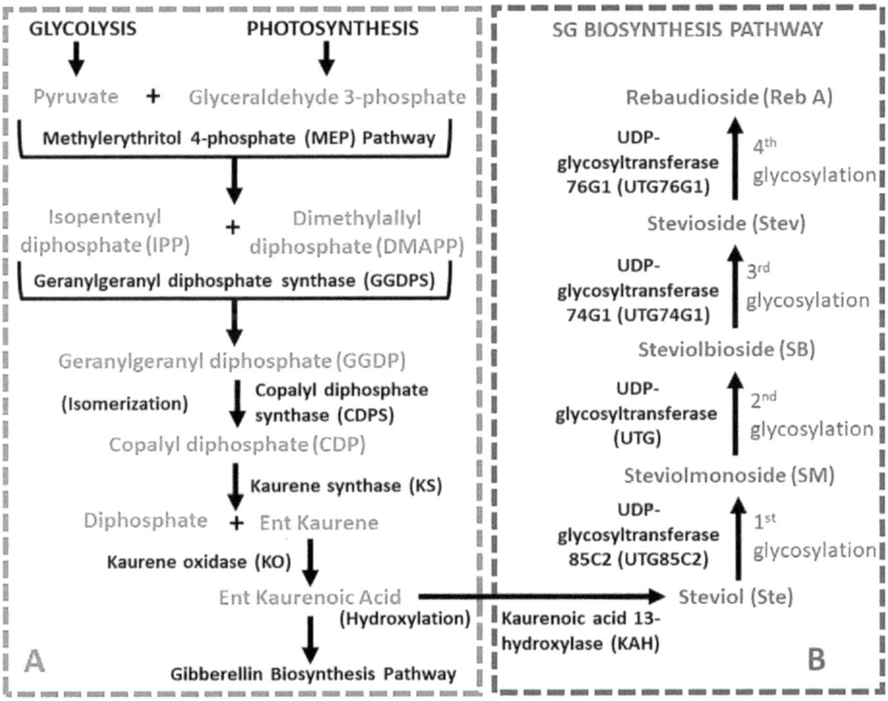

FIGURE 7.3 Biosynthesis pathway of steviol glycosides. (A) Production of steviol from precursors of glycolysis and photosynthesis. The MEP pathway occurs in plastids (until Kaurene production) and the endometrium reticulum (steviol formation); (B) stepwise glycosylation of Ste to produce different glycosides in the cytoplasm of each cell.

The first seven steps of stage A confirms to infamous Methylerythritol 4-phosphate (MEP) pathway that results in the formation of two isoprenoid precursors, namely, isopentenyl diphosphate (IPP) and dimethylallyl diphosphate (DMAPP) (Banerjee et al., 2014). Next four steps to steviol formation are also trivial to gibberellin biosynthesis. It includes the formation of kaurenoic acid from geranylgeranyl diphosphate (GDPP) with the help of two classes of enzymes: terpene cyclases (GGDPS, CDPS, KS) and cytochrome P450 monooxygenase (KO). Kaurenoic acid undergoes hydroxylation to produce Ste for biosynthesis of SG (Yamaguchi, 2008).

The SG biosynthesis begins by glycosylation of the hydroxyl group present at C13 of steviol, forming steviolmonoside (SM) – one glycan. The next step of glycosylation takes place at C2 of this previously added glycan, giving rise to steviolbioside (SB) – containing two glycans (Shibata et al., 1991). Third step of glycosylation occurs at the C19 carboxyl group of Ste backbone, producing Stev – three glycans. Stev can produce Reb A, by a fourth step of glycosylation at C3 of glucose, added to C13 during the first step of glycosylation (Brandle et al., 2007). Reb A thus contains four glycans.

7.3.2 DIABETY CONTROL BY SGS

7.3.2.1 METABOLIZATION AND ELIMINATION OF DIETARY SG

As SG enters the human gastrointestinal (GI) tract, enzymes secreted by various organs of the digestive system and the gut flora, completely convert SG to their backbone Ste. The glycosidic bond, that holds together both glucose and Ste units, can be cleaved by microbial enzymes (Magnuson et al., 2016). *Bacteroides* species, found in the human intestine, are capable of hydrolyzing SG with their β-glucosidase activity (Gardana et al., 2003). The sugar components thus freed, are not absorbed, and consequently does not commit to the energy requirements of the body. This can be explained by the hydrophilic nature of the *Stevia* mixture, possessing more than one/two glucose and rhamnose units (Rekha and Sharma, 2011). Hence, SG contributes to the non-caloric property of *Stevia rebuadiana* (Ashwell, 2015) mentioned earlier. The Ste component is readily absorbed by the body after being liberated from glycosides (Koyama et al., 2003). It undergoes a biotransformation reaction (enzymatic) in the liver which converts Ste to its activated form, i.e., steviol glucuronide (SVG) (steviol 19-O-â-D-glucopyranosiduronic acid) which is later eliminated by the body via urination (Geuns et al., 2006). This is briefly demonstrated in Figure 7.4.

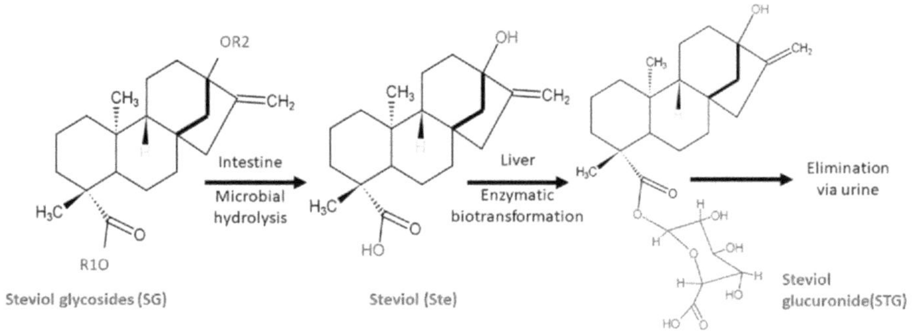

FIGURE 7.4 A proposed course of SG metabolism through different organs.
Source: Reprinted with permission from Geuns et al., 2006. © American Chemical Society.

7.3.2.2 MECHANISM OF SG ANTIDIABETIC ACTIVITY

Gu et al. (2019) demonstrated the *in vitro* stimulation of insulin secretion by SVG in a glucose- and dose-dependent trend, thus suggesting that SVG may play a key role in the antidiabetic effect of *Stevia rebaudiana*. Type 2 diabetes or NIDDM is characterized by an apparent rise in BGL (hyperglycemia) as a result of insulin insensitivity/resistance (Olokoba et al., 2012). The treatment for NIDDM aims to neutralize this abnormal, unhealthy glucose level in the blood. Chen et al. (2005) reported that Stevioside dose-dependently (0.5 mg/kg–5.0 mg/kg) increased blood insulin levels (reached max value within 90 minutes of Stev treatment) and decreased BGL (26.6 ± 3.1% for Stev dose of 5.0 mg/kg) in normal Wistar and NIDDM diabetic rats, respectively.

SG are professed to exhibit antidiabetic effect by two possible mechanisms:

- Stimulation of insulin secretion by β-pancreatic cells; and
- Promotion of glucose utilization by various cells in the body.

7.3.2.2.1 Stimulation of Insulin Secretion

Islets of Langerhans present in β-pancreatic cells is the site where insulin is produced in the body. These β cells are adept to detect and counter changes in blood glucose concentration by releasing reciprocal amounts of insulin into the bloodstream. Jeppesen et al. (2000) showcased a direct insulinotropic effect of SG diterpenes on the β cells.

Adenosine triphosphate (ATP)-sensitive potassium (K_{ATP}) channels, are acknowledged to assist in the regulation of glucose homeostasis by operating as metabolic sensors that are proficient at coupling metabolic status (β cell metabolism) to electrical activity (calcium entry). In absence of ATP, K_{ATP} channel is open and an efflux of K^+ ions is facilitated to preserve polarization of the cell. However, high ATP concentrations induce closing of the K_{ATP} channel (Craig et al., 2008). The ATP generated by glucose metabolism is considered to shut the K_{ATP} channel and therefore prevent it from maintaining the membrane potential of the cell (Ashcroft, 2005). The now-depolarized cell experiences an influx of calcium by voltage-gated calcium channels (VDCC), thereupon, insulin-containing vesicles are released (Minami et al., 2004). Classical drugs like sulfonylureas (insulin secretagogue) act by blocking the K_{ATP} channel to prompt insulin release from β cell (Ashcroft, 1996). However, this mechanism bears detrimental cardiovascular effects as a consequence of closed K_{ATP} channel in ventricular myocytes, which leads to the development of arrhythmic conditions (Morgan et al., 2014). Studies on Stevioside promoted their independence of K_{ATP} channel for stimulation of insulin secretion (Jeppesen et al., 200). The reduced K_{ATP} current by Reb A is not lasting and wears off as the glucose concentration drops (Abudula et al., 2008) and is visualized in Figure 7.5(A).

It has been shown that SG mediates insulin secretion by potentiating the activity of transient receptor potential cation channel (TRPM5) (Philippaert et al., 2017). TRPM5 is a calcium-activated channel that responds to increase in intracellular calcium concentration by enabling an influx of monovalent cations like K^+ and Na^+. Homologous to the activity of K_{ATP} channel, TRPM5 is also responsible for depolarization of β-pancreatic cells (Brixel et al., 2010). As a result of increase in membrane potential (positive), VDCC gets activated and permits the flow of calcium ions into the cell, consequently making insulin secretion feasible. Philippaert et al. (2017) illustrated the significance of TRPM5 for the insulinotropic action of SG. He later demonstrated that steviol was the key metabolite with TRPM5-potentiating effect (Philippaert et al., 2018). Stevioside is acknowledged as the only TRPM5-potentiating compound so far (Vennekens et al., 2017).

Glucagon is a major hormone that contravenes the action of insulin of reducing plasma glucose concentration. It is secreted by the α cells of pancreas and drives the increase of glucose levels in blood. β cells express Gcgrs genes that codes for a G-protein coupled receptor (GCGR) protein that is activated by glucagon (Huypens et al., 2000). When glucagon binds to GCGR, it activates a second messenger: phospholipase C. This secondary messenger catalyzes the splitting of PIP2 into diacylglycerol (DAG) and a sugar-phosphate called inositol 1,4,5-trisphosphate (IP3). The binding of IP3 to calcium channels on membranes of smooth endoplasmic reticulum (SER), triggers the release of stored calcium from SER to cytoplasm (Karp, 2008). This calcium can then contribute to insulin secretion. SVG was reported to upregulate the transcription of the Gcgr gene, thus contributing to the enhancement of insulin secretion from beta-cell (Gu et al., 2019). Glucagon-mediated insulinotropic bioaction of SG is briefly illustrated in Figure 7.5(B).

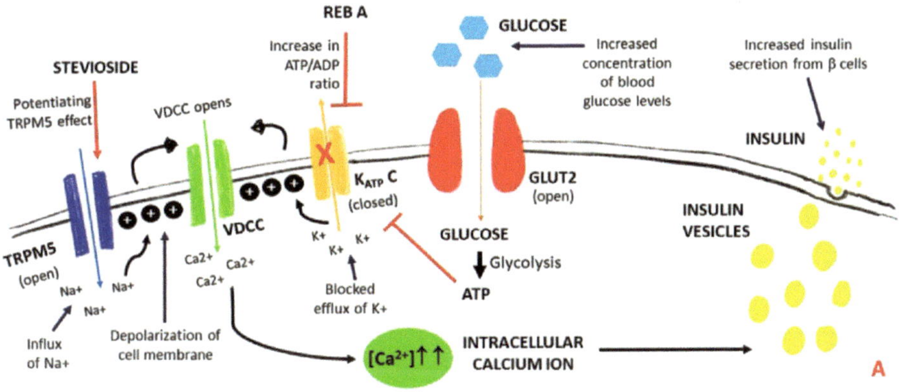

FIGURE 7.5(A) Insulinotropic effect of SG – Direct action of Stev and Reb α on β cells.

7.3.2.2.2 Promotion of Glucose Utilization

SG exhibits an insulin-mimicry action to increase glucose uptake. The principal mechanism of this effect is ergo to translocation of GLUT4 receptors to plasma membrane through the P13K-PKB/Akt pathway (Prata et al., 2017). Based on accumulated evidence, P13K-PKB/Akt pathway has been recognized as a key moderator of glucose metabolism in skeletal, adipose, and cardiac cells (Huang et al., 2018). Consequential actions of insulin-induced PI3K/Akt pathway, towards glucose homeostasis, is briefly illustrated in Figure 7.6.

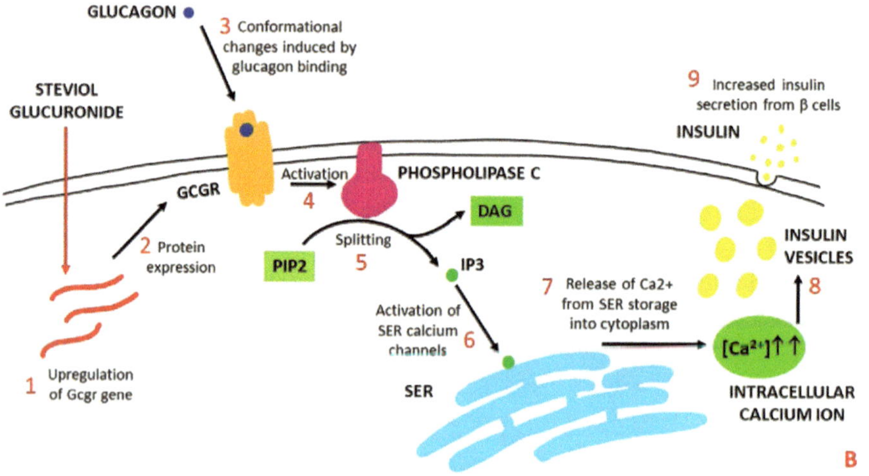

FIGURE 7.5(B) Insulinotropic effect of SG – Glucagon-mediated action of SVG on insulin secretion.

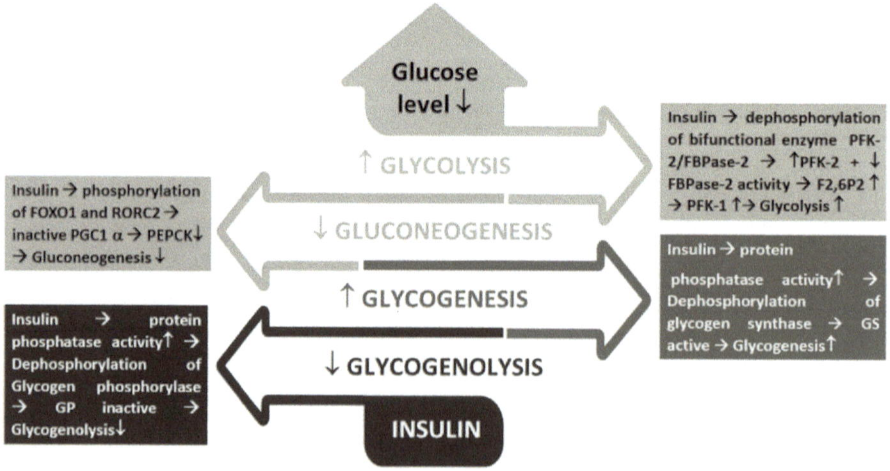

FIGURE 7.6 Action of insulin towards curtailment of blood glucose levels.

When insulin released from β cell into the bloodstream binds to insulin receptors (IR), tyrosine residues in insulin-receptor substrates (INS) might get phosphorylated. PI3K is a lipid kinase, containing two SH2 domains that specifically bind to the phosphorylated tyrosine residues. Phosphorylation of INS thus activates PI3-Kinase pathway. PI3K are turned on by many receptors, but only insulin receptors trigger GLUT4 translocation, in addition to other insulin-induced metabolic pathways. This is because of another kinase: a serine-threonine kinase PKB (also referred to as Akt), containing PH domain that interacts with PIP3 (a membrane-bound phosphoinositide) formed as a direct consequence of PI3K activation (Karp, 2008). PKB are reported to stimulate GLUT4 trafficking to cell surfaces in skeletal muscle cells (Wang et al., 1999). GLUT4, present in skeletal muscle cells, cardiomyocytes, and adipose cells, acknowledged for insulin-induced uptake of glucose in muscle cells and adipocytes. NIDDM develops a condition called insulin resistance, where cells stop responding to the presence of hormones. The apparent failure of GLUT4 translocation to cell membrane is speculated to be caused by defect in intracellular signaling or glucose toxicity from chronic hyperglycemia (Shepherd et al., 1999). One of the strategies to treat NIDDM is the prevention of insulin resistance. SG are potential ligands for insulin receptors on cells, promoting glucose influx in cells by mimicking insulin action (Prata et al., 2017). Hypoglycemic action of SG is briefly illustrated in Figure 7.7. Since the effect of SG seems to disappear with the decline in BGL, it is anticipated to prevent the development of hypoglycemia (Jeppesen et al., 2000).

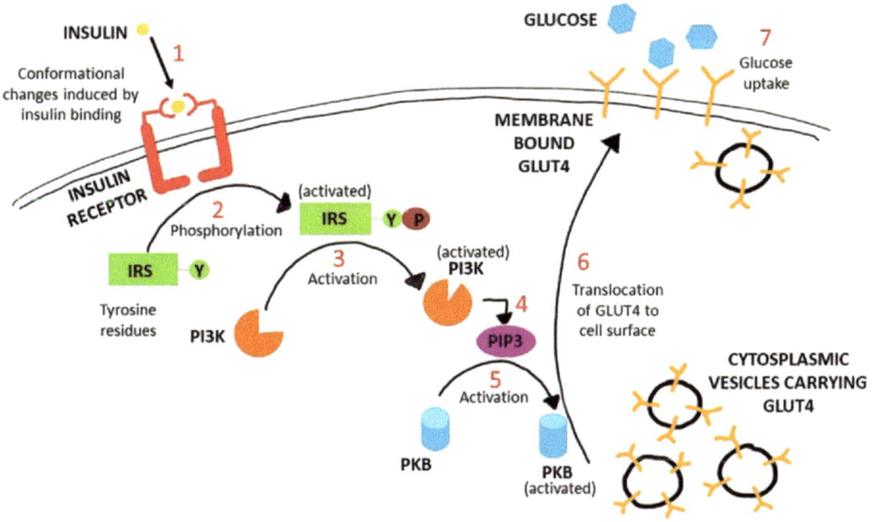

FIGURE 7.7 Insulin mimicking action of SG.

7.4 CONCLUSION

With surging obesity rates and diabetes cases, society is getting increasingly health-conscious. In addition to the rising popularity of plant-based remedies, it has become a necessity to consume healthy food, consequently avoiding side effects of synthetic supplements. In the meantime, we attempt to unveil the safe mechanism of one such candidate, *Stevia rebaudiana.* It can be concluded that steviol glycosides, phytochemicals extracted from *Stevia* leaves, may secure blood glucose homeostasis as a consequence of stimulating insulin secretion and promoting glucose utilization by the body. It curtails the existing blood glucose level without adding extra calories, and thus holds immense application as nutraceuticals. However, it is to be noted that not many toxicological studies have been performed on Stevia *rebaudiana.* It is essential to promote safe alternatives to pre-existing synthetic solutions with tremendous side effects. As a result, more scientific investigations can be designed that are centered on understanding cellular and/or molecular level toxicity of dietary steviol glycosides. Additionally, antidiabetic mechanisms of correspondence by steviol glycosides could be explored further. For instance, receptors specific to activated forms of steviol glycosides can be studied. There could be more clarity expected from future works aimed to analyze the detailed course of action by these bioactive compounds. The molecules that may be involved in the regulation/inhibition of steviol glycosides in humans, can be investigated.

KEYWORDS

- **blood glucose level**
- **diabetes**
- **insulinotropic effect**
- **secondary metabolite**
- **steviol**

REFERENCES

Abudula, R., Matchkov, V. V., Jeppesen, P. B., Nilsson, H., Aalkjaer, C., & Hermansen, K., (2008). Rebaudioside a directly stimulates insulin secretion from pancreatic beta cells: A glucose-dependent action via inhibition of ATP-sensitive k+-channels*. *Diabetes, Obesity and Metabolism, 10*(11), 1074–1085.

Adhikari, P., & Paul, S., (2018). History of Indian traditional medicine: A medical inheritance. *Asian Journal of Pharmaceutical and Clinical Research, 11*, 421. 10.22159/ajpcr.2017.v11i1.21893.

Ahn, K., (2017). The worldwide trend of using botanical drugs and strategies for developing global drugs. *BMB Reports, 50*(3), 111–116.

Altemimi, A., Lakhssassi, N., Baharlouei, A., Watson, D. G., & Lightfoot, D. A., (2017). Phytochemicals: Extraction, isolation, and identification of bioactive compounds from plant extracts. *Plants (Basel, Switzerland), 6*(4), 42.

Angelini, L. G., Martini, A., Passera, B., & Tavarini, S., (2018). Cultivation of *Stevia rebaudiana* bertoni and associated challenges. *Reference Series in Phytochemistry*, 35–85.

Ashcroft, F. M., (2005). ATP-sensitive potassium channelopathies: Focus on insulin secretion. *The Journal of Clinical Investigation, 115*(8), 2047–2058.

Ashcroft, F., (1996). Mechanisms of the glycaemic effects of sulfonylureas. *Hormone and Metabolic Research, 28*(09), 456–463.

Ashwell, M., (2015). Stevia, nature's zero-calorie sustainable sweetener: A new player in the fight against obesity. *Nutrition today, 50*(3), 129–134.

Banerjee, A., & Sharkey, T. D., (2014). Methylerythritol 4-phosphate (MEP) pathway metabolic regulation. *Nat. Prod. Rep, 31*(8), 1043–1055.

Barbieri, R., Coppo, E., Marchese, A., Daglia, M., Sobarzo-Sánchez, E., Nabavi, S. F., & Nabavi, S. M., (2017). Phytochemicals for human disease: An update on plant-derived compounds antibacterial activity. *Microbiol Res., 196*, 44–68.

Bhat, M., Kothiwale, S. K., Tirmale, A. R., Bhargava, S. Y., & Joshi, B. N., (2011). Antidiabetic properties of *Azadirachta indica* and *Bougainvillea spectabilis*: *In vivo* studies in murine diabetes model. *Evidence-Based Complementary and Alternative Medicine: ECAM, 2011*, 561625.

Brandle, J. E., & Telmer, P. G., (2007). Steviol glycoside biosynthesis. *Phytochemistry, 68*(14), 1855–1863.

Brito-Arias, M., (2016). Glycosides, synthesis, and characterization. In: *Synthesis and Characterization of Glycosides* (pp. 1–67). Springer, Boston, MA.

Brixel, L. R., Monteilh-Zoller, M. K., Ingenbrandt, C. S., Fleig, A., Penner, R., Enklaar, T., Zabel, B. U., & Prawitt, D., (2010). TRPM5 regulates glucose-stimulated insulin secretion. *Pflugers Archiv.: European Journal of Physiology, 460*(1), 69–76.

Campos, M., Barbosa-Martín, E., Matus-Basto, Á., Cabrera-Amaro, D., Murguía-Olmedo, M., Moguel-Ordonez, Y., & Betancur, D., (2014). Comparison of chemical and functional properties of *Stevia rebaudiana* (Bertoni) varieties cultivated in Mexican Southeast. *American Journal of Plant Sciences, 5*, 286–293.

Cardello, H., Da Silva, M., & Damasio, M., (1999). Measurement of the relative sweetness of stevia extract, aspartame and cyclamate/saccharin blend as compared to sucrose at different concentrations. *Plant Foods Hum. Nutr., 54*, 119–129.

Chan, P., Tomlinson, B., Chen, Y. J., Liu, J. C., Hsieh, M. H., & Cheng, J. T., (2000). A double-blind placebo-controlled study of the effectiveness and tolerability of oral stevioside in human hypertension. *British Journal of Clinical Pharmacology, 50*(3), 215–220.

Chen, J., Xia, Y., Sui, X., Peng, Q., Zhang, T., Li, J., & Zhang, J., (2018). Steviol, a natural product inhibits proliferation of the gastrointestinal cancer cells intensively. *Oncotarget, 9*(41), 26299–26308.

Chen, T., Chen, S., Chan, P., Chu, Y., Yang, H., & Cheng, J., (2005). Mechanism of the hypoglycemic effect of stevioside, a glycoside of *Stevia rebaudiana*. *Planta Medica, 71*(2), 108–113.

Cosson, P., Hastoy, C., Errazzu, L. E., Budeguer, C. J., Boutié, P., Rolin, D., & Schurdi-Levraud, V., (2019). Genetic diversity and population structure of the sweet leaf herb, *Stevia rebaudiana* B., cultivated and landraces germplasm assessed by EST-SSRs genotyping and steviol glycosides phenotyping. *BMC Plant Biology, 19*(1), 436.

Cowan, M. M., (1999). Plant products as antimicrobial agents. *Clinical Microbiology Reviews, 12*(4), 564–582.

Craig, T. J., Ashcroft, F. M., & Proks, P., (2008). How ATP inhibits the open K(ATP) channel. *The Journal of General Physiology, 132*(1), 131–144.

Davinelli, S., Sapere, N., Zella, D., Bracale, R., Intrieri, M., & Scapagnini, G., (2012). Pleiotropic protective effects of phytochemicals in Alzheimer's disease. *Oxidative Medicine and Cellular Longevity, 2012*, 386527.

Davis, J. A., Sharma, S., Mittra, S., Sujatha, S., Kanaujia, A., Shukla, G., Katiyar, C., et al., (2012). Antihyperglycemic effect of Annona squamosa hexane extract in type 2 diabetes animal model: PTP1B inhibition, a possible mechanism of action?. *Indian Journal of Pharmacology, 44*(3), 326–332.

Dubey, H., Singh, A., Patole, A. M., Tenpe, C. R., & Ghule, B. V., (2012). Allicin, a SUR2 opener: Possible mechanism for the treatment of diabetic hypertension in rats. *Revista Brasileira de Farmacognosia, 22*(5), 1053–1059.

Dyduch-Siemińska, M., Najda, A., Gawroński, J., Balant, S., Świca, K., & Żaba, A., (2020). *Stevia Rebaudiana* bertoni, a source of high-potency natural sweetener-biochemical and genetic characterization. *Molecules (Basel, Switzerland), 25*(4), 767.

Efferth, T., & Koch, E., (2011). Complex interactions between phytochemicals. The multi-target therapeutic concept of phytotherapy. *Current Drug Targets, 12*(1), 122–132.

Erb, M., & Kliebenstein, D. J., (2020). Plant secondary metabolites as defences, regulators, and primary metabolites: the blurred functional trichotomy. *Plant Physiol, 184*(1), 39–52.

Firenzuoli, F., & Gori, L., (2007). Herbal medicine today: Clinical and research issues. *Evidence-Based Complementary and Alternative Medicine: ECAM, 4*(Suppl 1), 37–40.

Garber, A. J., Abrahamson, M. J., Barzilay, J. I., Blonde, L., Bloomgarden, Z. T., Bush, M. A., et al., (2016). Consensus statement by the American association of clinical endocrinologists and American college of endocrinology on the comprehensive type 2 diabetes management algorithm-2016 executive summary. *Endocrine Practice: Official Journal of the American College of Endocrinology and the American Association of Clinical Endocrinologists, 22*(1), 84–113.

Gardana, C., Simonetti, P., Canzi, E., Zanchi, R., & Pietta, P., (2003). Metabolism of stevioside and rebaudioside A from *Stevia rebaudiana* extracts by human microflora. *Journal of agricultural and Food Chemistry, 51*(22), 6618–6622.

Gasmalla, M. A., Yang, R., & Hua, X., (2015). Extraction of rebaudioside-A by sonication from *Stevia rebaudiana* bertoni leaf and decolorization of the extract by polymers. *Journal of Food Science and Technology, 52*(9), 5946–5953.

Geisen, K., Végh, A., Krause, E., & Papp, J. G., (1996). Cardiovascular effects of conventional sulfonylureas and glimepiride. *Hormone and Metabolic Research = Hormon- und Stoffwechselforschung = Hormones et Metabolisme, 28*(9), 496–507.

Geuns, J. M., Buyse, J., Vankeirsbilck, A., Temme, E. H., Compernolle, F., & Toppet, S., (2006). Identification of steviol glucuronide in human urine. *Journal of Agricultural and Food Chemistry, 54*(7), 2794–2798.

Ghildiyal, R., Prakash, V., Chaudhary, V. K., Gupta, V., & Gabrani, R., (2020). Phytochemicals as antiviral agents: Recent updates. *Plant-Derived Bioactives: Production, Properties and Therapeutic Applications*, 279–295.

Goyal, R., & Jialal, I., (2020). *Diabetes Mellitus Type 2.* In: StatPearls [Internet]. Treasure Island (FL): StatPearls Publishing.

Goyal, S. K., Samsher, & Goyal, R. K., (2010). Stevia (*Stevia rebaudiana*) a bio-sweetener: A review. *International Journal of Food Sciences and Nutrition, 61*(1), 1–10.

Gu, W., Rebsdorf, A., Anker, C., Gregersen, S., Hermansen, K., Geuns, J. M., & Jeppesen, P. B., (2019). Steviol glucuronide, a metabolite of steviol glycosides, potently stimulates insulin secretion from isolated mouse islets: Studies *in vitro*. *Endocrinology, Diabetes & Metabolism, 2*(4).

Gupta, V. G., Schmoll, M., Herrera-Estrella, A., Upadhyay, R. S., Druzhinina, I., & Tuohy, M., (2014). *Biotechnology and Biology of Trichoderma* (pp. 125–137). Newnes.

Hosseini, A., & Ghorbani, A., (2015). Cancer therapy with phytochemicals: Evidence from clinical studies. *Avicenna Journal of Phytomedicine, 5*(2), 84–97.

Huang, X., Liu, G., Guo, J., & Su, Z., (2018). The PI3K/AKT pathway in obesity and type 2 diabetes. *International Journal of Biological Sciences, 14*(11), 1483–1496.

Hussein, R., & El-Anssary, A. A., (2019). Plants secondary metabolites: The key drivers of the pharmacological actions of medicinal plants. *Herbal Medicine.*

Huypens, P., Ling, Z., Pipeleers, D., & Schuit, F., (2000). Glucagon receptors on human islet cells contribute to glucose competence of insulin release. *Diabetologia, 43*(8), 1012–1019.

Isah, T., (2019). Stress and defense responses in plant secondary metabolites production. *Biological Research, 52*(1).

Jeppesen, P., Gregersen, S., Poulsen, C., & Hermansen, K., (2000). Stevioside acts directly on pancreatic β cells to secrete insulin: Actions independent of cyclic adenosine monophosphate and adenosine triphosphate-sensitive k+-channel activity. *Metabolism, 49*(2), 208–214.

Kalra, S., Bahendeka, S., Sahay, R., Ghosh, S., Md, F., Orabi, A., Ramaiya, K., et al., (2018). Consensus recommendations on sulfonylurea and sulfonylurea combinations in the management of type 2 diabetes mellitus – international task force. *Indian Journal of Endocrinology and metabolism, 22*(1), 132–157.

Karp, G., (2008). *Cell and Molecular Biology: Concepts and Experiments with Take Note Set* (5th edn., pp. 627–628). ISBN 978-0470-16961-2.

Kennedy, D. A., Hart, J., & Seely, D., (2009). Cost-effectiveness of natural health products: A systematic review of randomized clinical trials. *Evidence-Based Complementary and Alternative Medicine: ECAM, 6*(3), 297–304.

Khatri, S., Kumar, M., Phougat, N., Chaudhary, R., & Chhillar, A. K., (2016). Perspectives on phytochemicals as antibacterial agents: An outstanding contribution to modern therapeutics. *Mini-Reviews in Medicinal Chemistry, 16*(4), 290–308.

Kochikyan, V., Markosyan, A. A., Abelyan, L., ABalayan, A., & Abelyan, V. A., (2006). Combined enzymatic modification of stevioside and rebaudioside A. *Applied Biochemistry and Microbiology, 42*, 31–37.

Kowalczyk, M. C., Walaszek, Z., Kowalczyk, P., Kinjo, T., Hanausek, M., & Slaga, T. J., (2009). Differential effects of several phytochemicals and their derivatives on murine keratinocytes *in vitro* and *in vivo*: Implications for skin cancer prevention. *Carcinogenesis, 30*(6), 1008–1015.

Koyama, E., & Sakai, N., & Ohori, Y., & Kitazawa, K., & Izawa, O., & Kakegawa, K., & Fujino, A., & Ui, M., (2003). Absorption and metabolism of glycosidic sweeteners of stevia mixture and their aglycone, steviol, in rats and humans. *Food and Chemical Toxicology: An International Journal Published for the British Industrial Biological Research Association, 41*, 875–883.

Kumari, K., & Augusti, K. T., (2002). Antidiabetic and antioxidant effects of S-methyl cysteine sulfoxide isolated from onions (*Allium cepa* Linn) as compared to standard drugs in alloxan diabetic rats. *Indian Journal of Experimental Biology, 40*(9), 1005–1009.
Kumari, K., Mathew, B. C., & Augusti, K. T., (1995). Antidiabetic and hypolipidemic effects of S-methyl cysteine sulfoxide isolated from *Allium cepa* Linn. *Indian Journal of Biochemistry & Biophysics, 32*(1), 49–54.
Lemus-Mondaca, R., Vega-Gálvez, A., Zura-Bravo, L., & Ah-Hen, K., (2012). Response to the letter to editor regarding "*Stevia rebaudiana* bertoni, source of a high potency natural sweetener: A comprehensive review on the biochemical, nutritional and functional aspects extraction and safety of stevioside." *Food Chemistry, 135*(3), 1784.
Liakos, A., Karagiannis, T., Bekiari, E., Boura, P., & Tsapas, A., (2015). Update on long-term efficacy and safety of dapagliflozin in patients with type 2 diabetes mellitus. *Therapeutic Advances in Endocrinology and Metabolism, 6*(2), 61–67.
Lima, Y. C., Kurauti, M. A., Da Fonseca, A. G., Ferezini, J., Piovan, S., Malta, A., De Almeida, F., et al., (2019). Whey protein sweetened with *Stevia rebaudiana* bertoni (Bert.) increases mitochondrial biogenesis markers in the skeletal muscle of resistance-trained rats. *Nutrition & Metabolism, 16*, 65.
Liu, C. T., Hsu, T. W., Chen, K. M., Tan, Y. P., Lii, C. K., & Sheen, L. Y., (2012). The antidiabetic effect of garlic oil is associated with ameliorated oxidative stress but not ameliorated level of pro-inflammatory cytokines in skeletal muscle of streptozotocin-induced diabetic rats. *Journal of Traditional and Complementary Medicine, 2*(2), 135–144.
Liu, J. C., Kao, P. K., Chan, P., Hsu, Y. H., Hou, C. C., Lien, G. S., Hsieh, M. H., et al., (2003). Mechanism of the antihypertensive effect of stevioside in anesthetized dogs. *Pharmacology, 67*(1), 14–20.
Lucier, J., & Weinstock, R. S., (2020). *Diabetes Mellitus Type 1.* In: StatPearls [Internet]. Treasure Island (FL): StatPearls Publishing. Available from: https://www.ncbi.nlm.nih.gov/books/NBK507713/ (accessed on 18 December 2021).
Magnuson, B. A., Carakostas, M. C., Moore, N. H., Poulos, S. P., & Renwick, A. G., (2016). Biological fate of low-calorie sweeteners. *Nutr. Rev., 74*(11), 670–689. doi: 10.1093/nutrit/nuw032. PMID: 27753624.
Marín-Peñalver, J. J., Martín-Timón, I., Sevillano-Collantes, C., & Del Cañizo-Gómez, F. J., (2016). Update on the treatment of type 2 diabetes mellitus. *World Journal of Diabetes, 7*(17), 354–395.
Minami, K., Miki, T., Kadowaki, T., & Seino, S., (2004). Roles of ATP-sensitive K+ channels as metabolic sensors: Studies of Kir6.x null mice. *Diabetes, 53*(Supplement 3), S176–S180.
Mishra, P., & Singh, R., & Kumar, U., & Prakash, V., (2010). *Stevia rebaudiana* – A magical sweetener. *Global Journal of Biotechnology & Biochemistry, 5*, 62–74.
Mitsuhashi, H., Ueno, J., & Sumita, T., (1975). Yakugaku zasshi. *Journal of the Pharmaceutical Society of Japan, 95*(1), 127–130.
Modak, M., Dixit, P., Londhe, J., Ghaskadbi, S., & Devasagayam, T. P., (2007). Indian herbs and herbal drugs used for the treatment of diabetes. *Journal of Clinical Biochemistry and Nutrition, 40*(3), 163–173.
Mogra, R., & Dashora, V., (2009). Exploring the use of *Stevia rebaudiana* as a sweetener in comparison with other sweeteners. *J. Hum. Ecol., 25.*
Momtazi-Borojeni, A. A., Esmaeili, S. A., Abdollahi, E., & Sahebkar, A., (2017). A review on the pharmacology and toxicology of steviol glycosides extracted from *Stevia rebaudiana*. *Current Pharmaceutical Design, 23*(11), 1616–1622.

Morgan, C. L., Mukherjee, J., Jenkins-Jones, S., Holden, S. E., & Currie, C. J., (2014). Association between first-line monotherapy with sulphonylurea versus metformin and risk of all-cause mortality and cardiovascular events: A retrospective, observational study. *Diabetes, Obesity & Metabolism, 16*(10), 957–962.

Nettleton, J. E., Klancic, T., Schick, A., Choo, A. C., Shearer, J., Borgland, S. L., Chleilat, F., et al., (2019). Low-dose stevia (Rebaudioside A) consumption perturbs gut microbiota and the mesolimbic dopamine reward system. *Nutrients, 11*(6), 1248.

Olokoba, A. B., Obateru, O. A., & Olokoba, L. B., (2012). Type 2 diabetes mellitus: A review of current trends. *Oman Medical Journal, 27*(4), 269–273.

Otto-Buczkowska, E., & Jainta, N., (2017). Pharmacological treatment in diabetes mellitus type 1 – insulin and what else?. *International Journal of Endocrinology and Metabolism, 16*(1), e13008.

Paik, J., & Blair, H. A., (2019). Dapagliflozin: A review in type 1 diabetes. *Drugs, 79*(17), 1877–1884.

Pal, P. K., Kumar, R., Guleria, V., Mahajan, M., Prasad, R., Pathania, V., Gill, B. S., et al., (2015). Crop-ecology and nutritional variability influence growth and secondary metabolites of *Stevia rebaudiana* bertoni. *BMC Plant Biology, 15*, 67.

Panda, S., & Kar, A., (2007). Antidiabetic and antioxidative effects of Annona squamosa leaves are possibly mediated through quercetin-3-O-glucoside. *BioFactors (Oxford, England), 31*(3, 4), 201–210.

Payab, M., Hasani-Ranjbar, S., Aletaha, A., Ghasemi, N., Qorbani, M., Atlasi, R., Abdollahi, M., & Larijani, B., (2018). Efficacy, safety, and mechanisms of herbal medicines used in the treatment of obesity. *Medicine, 97*(1), e8825.

Philippaert, K., Pironet, A., Mesuere, M., Sones, W., Vermeiren, L., Kerselaers, S., Pinto, S., et al., (2017). Steviol glycosides enhance pancreatic beta-cell function and taste sensation by potentiation of TRPM5 channel activity. *Nature Communications, 8*(1).

Philippaert, K., Vennekens, R., & Light, P. E., (2018). Potentiation of TRPM5 with stevioside in the beta cells stimulates insulin secretion. *Diabetes, 67*(Supplement 1), 316-LB.

Prata, C., Zambonin, L., Rizzo, B., Maraldi, T., Angeloni, C., Vieceli, D. S. F., Fiorentini, D., & Hrelia, S., (2017). Glycosides from *Stevia rebaudiana* bertoni possess insulin-mimetic and antioxidant activities in rat cardiac fibroblasts. *Oxidative Medicine and Cellular Longevity, 2017*, 1–13.

Rekha, M., & Sharma, C. P., (2011). Nanoparticle mediated oral delivery of peptides and proteins. *Peptide and Protein Delivery*, 165–194.

Rizwan, F., Yesmine, S., Banu, S. G., Chowdhury, I. A., Hasan, R., & Chatterjee, T. K., (2019). Renoprotective effects of stevia (*Stevia rebaudiana* bertoni), amlodipine, valsartan, and losartan in gentamycin-induced nephrotoxicity in the rat model: Biochemical, hematological and histological approaches. *Toxicology Reports, 6*, 683–691.

Rosak, C., & Mertes, G., (2012). Critical evaluation of the role of acarbose in the treatment of diabetes: Patient considerations. *Diabetes Metab. Syndr. Obes., 5*, 357–367. doi: 10.2147/DMSO.S28340. PMID: 23093911; PMCID: PMC3476372.

Ruili, Q. (2008). Anti-inflammatory Activity of *Rubus suavissimus* S. Lee leaf extract in a C57BL/6J Mouse Model with LPS-induced Low-grade Chronic Inflammation. Thesis. pp. 64–68.

Savita, S., Sheela, K., Sunanda, S., Shankar, A., & Ramakrishna, P., (2004). *Stevia rebaudiana* – A functional component for food industry. *Journal of Human Ecology, 15*(4), 261–264.

Seca, A., & Pinto, D., (2019). Biological potential and medical use of secondary metabolites. *Medicines (Basel, Switzerland), 6*(2), 66.

Seon-Young, K., Mi-Jeong, J., & Hwangbo, M., Young-Doo, B., Tae-Young, J., Il-Je, C., & Seon-Young, J., (2013). Anti-inflammatory effect of *Stevia Rebaudiana* as a result of NF-??B and MAPK inhibition. *The Journal of Korean Oriental Medical Ophthalmology and Otolaryngology and Dermatology, 26*, 54–64.

Seth, S. D., & Sharma, B., (2004). Medicinal plants in India. *The Indian Journal of Medical Research, 120*(1), 9–11.

Sharma, M., Thakral, N. K., & Thakral, S., (2009). Chemistry and *in vivo* profile of ent-kaurene glycosides of *Stevia rebaudiana* bertoni: An overview. *Nat. Prod. Rad., 8*, 181–189.

Sharma, S., Walia, S., Singh, B., & Kumar, R., (2016). Comprehensive review on agro technologies of low-calorie natural sweetener stevia (*Stevia rebaudiana* bertoni): A boon to diabetic patients. *Journal of the Science of Food and Agriculture, 96*(6), 1867–1879.

Shepherd, P. R., & Kahn, B. B., (1999). Glucose transporters and insulin action--implications for insulin resistance and diabetes mellitus. *N. Engl. J., Med., 341*(4), 248–257.

Shibata, H., Sonoke, S., Ochiai, H., Nishihashi, H., & Yamada, M., (1991). Glucosylation of steviol and steviol-glucosides in extracts from *Stevia rebaudiana* bertoni. *Plant Physiology, 95*(1), 152–156.

Shukia, R., Sharma, S. B., Puri, D., Prabhu, K. M., & Murthy, P. S., (2000). Medicinal plants for treatment of diabetes mellitus. *Indian Journal of Clinical Biochemistry: IJCB, 15*(Suppl 1), 169–177.

Šic, Ž. J., Voća, S., Dobričević, N., & Ježek, D., & Bosiljkov, T., & Brncic, M., (2013). *Stevia rebaudiana* bertoni: A review of nutritional and biochemical properties of natural sweetener. *Agriculturae Conspectus Scientificus, 78*, 25–30.

Singh, S., Sharma, B., Kanwar, S. S., & Kumar, A., (2016). Lead phytochemicals for anticancer drug development. *Frontiers in Plant Science, 7*, 1667.

Tennant, D. R., & Bruyninckx, C., (2017). The potential application of European market research data in dietary exposure modelling of food additives. *Food Additives & Contaminants: Part A, 35*(3), 412–424.

Theophilus, P. A., Victoria, M. J., Socarras, K. M., Filush, K. R., Gupta, K., Luecke, D. F., & Sapi, E., (2015). Effectiveness of *Stevia Rebaudiana* whole leaf extract against the various morphological forms of borrelia burgdorferi *in vitro*. *European Journal of Microbiology & Immunology, 5*(4), 268–280.

Veeresham, C., (2012). Natural products derived from plants as a source of drugs. *Journal of Advanced Pharmaceutical Technology & Research, 3*(4), 200–201.

Velu, G., Palanichamy, V., & Rajan, A. P., (2018). Phytochemical and pharmacological importance of plant secondary metabolites in modern medicine. In: Roopan, S., & Madhumitha, G., (eds.), *Bioorganic Phase in Natural Food: An Overview*. Springer, Cham.

Vennekens, R., & Mesuere, M., & Philippaert, K., (2017). TRPM5 in the battle against diabetes and obesity. *Acta Physiologica*, 222.

Wachtel-Galor, S., & Benzie, I. F. F., (2011). Herbal medicine: An introduction to its history, usage, regulation, current trends, and research needs. In: *Herbal Medicine: Biomolecular and Clinical Aspects* (2nd edn.). Boca Raton (FL): CRC Press/Taylor & Francis. Chapter 1. Available from: https://www.ncbi.nlm.nih.gov/books/NBK92773/ (accessed on 18 December 2021).

Wang, Q., Somwar, R., Bilan, P. J., Liu, Z., Jin, J., Woodgett, J. R., & Klip, A., (1999). Protein kinase B/Akt participates in GLUT4 translocation by insulin in L6 myoblasts. *Molecular and Cellular Biology, 19*(6), 4008–4018.

Yadav, S. K., & Guleria, P., (2012). Steviol glycosides from stevia: Biosynthesis pathway review and their application in foods and medicine. *Critical Reviews in Food Science and Nutrition, 52*(11), 988–998.

Yamaguchi, S., (2008). Gibberellin metabolism and its regulation. *Annual Review of Plant Biology, 59*(1), 225–251.

Zhang, Q., Yang, H., Li, Y., Liu, H., & Jia, X., (2017). Toxicological evaluation of ethanolic extract from *Stevia rebaudiana* bertoni leaves: Genotoxicity and subchronic oral toxicity. *Regulatory Toxicology and Pharmacology: RTP, 86*, 253–259.

Zhang, Y. J., Gan, R. Y., Li, S., Zhou, Y., Li, A. N., Xu, D. P., & Li, H. B., (2015). Antioxidant phytochemicals for the prevention and treatment of chronic diseases. *Molecules (Basel, Switzerland), 20*(12), 21138–21156.

CHAPTER 8

Mechanism of Action of the Traditional and Modern Phyto-Biotics in the Treatment of Diabetes in the Current Era

K. VENKATA RATNAM,[1] LEPAKSHI MD. BHAKSHU,[2] and R. R. VENKATA RAJU[3]

[1]*Assistant Professor, Department of Botany, Rayalaseema University, Kurnool – 518007, Andhra Pradesh, India, E-mails: drvenkatapkd@gmail.com; kvratnam@rayalaseemauniversity.ac.in*

[2]*Assistant Professor, Department of Botany, PVKN Government College (Autonomous), Chittoor – 517002, Andhra Pradesh, India*

[3]*Department of Botany, Sri Krishnadeveraya University, Ananthapuramu – 515003, Andhra Pradesh, India*

ABSTRACT

Diabetes mellitus (DM) is a metabolic disorder and the third-leading root of morbidity after cardiovascular disease (CVD) and cancer besides obesity. According to "International Diabetes Federation," the number of diabetic patients may rise from 415 million to over 642 million or more by 2040. Natural products of plant origin have played an important role in diabetes therapy since ancient times, especially in Asia, India, and Africa. There are about 1,200 phytoconstituents isolated from medicinal plants scientifically proved its potential pharmacological activity/s, among them 400 phytoconstituents rich plant extracts manifest beneficial effects against type II diabetes. Mechanism of action of polyherbal formulations, medicinal plant extracts and bioactivities isolated from them includes by reducing blood glucose

Antidiabetic Potential of Plants in the Era of Omics. Deepu Pandita, PhD, Anu Pandita, MSc, & Chander Bhanu, MBBS (Eds.)

levels (BGL), effect on the insulin mechanism via inhibition of phosphoenolpyruvate carboxykinase and fructose-1, 6-bisphosphatase, regulation of glucose-6-phosphate dehydrogenase enzyme and antioxidant enzymes such as SOD, GPx, and catalase. The role of α-amylase (AAL) or α-glucosidase (AGL) inhibition mechanism in maintaining the DM also being discussed. Further, they attenuate pro-inflammatory markers via inhibiting the NFkB pathway, role of glucose transporter, PPAR γ agonist activity in *in vivo* (animal models) and *in vitro* (diabetic cell lines) models. The present study will discuss about the mechanism of action of popular herbal formulations, medicinal plant extracts and phytochemical components in animal and diabetic cell line models with special reference to insulin-mimetic effects in addition to carbohydrate metabolism and cytokine inhibition pathways. Further the study also focuses on the comparison of commercial synthetic and herbal drugs on its safety and effective usage.

8.1 INTRODUCTION

8.1.1 ROLE OF HERBALS IN DM AND TYPES OF PHYTO-THERAPEUTICS

Diabetes mellitus (DM) is one of the complex defect of metabolism, resulting from insufficient insulin synthesis/production or its failure to function. Diabetes can be classified into Insulin-dependent diabetes (Type I), which is because of insulin insufficiency owing to lacking functional beta cells and insulin independent (Type II) diabetes is a very common form which constitutes 90% among the diabetic population. The following cluster of symptoms are more prevalent:

- High profile of blood sugar;
- Abnormal thirst;
- Repeated urination or polyurea;
- Extreme or uneven carving correlated to weight loss;
- Clouded vision or cataract;
- Nausea and vomiting;
- Extreme weakness and tiredness;
- Irritation, discomfort, etc.

The traditional system of medicines succeeded in the prevention of diabetes and its complications either using mono or polyherbal formulas which are natural moreover with minimum side effects and affordable price. A plethora of herbals were scientifically evaluated and found highly beneficial in its prevention as well as treatment such as garlic, janmun, bitter guard, madhunashini, holy basil, fenugreek, aswagandha, *Phyllanthus amarus*, *Pterocarpus marsupium,* and guduchi (Modak et al., 2007). Prevention of oxidative processes at the molecular level were proved as an effect mechanism and evidenced by in the experimentally induced diabetic animal models as well as minimizing the severity of its complexity. The free radicals may initiate the destructive oxidation and trigger abnormalities in lipid biomarkers such as VLDL, LDL, and even HDL levels. The chains of oxidative processes effects on Lipid peroxidation which multiply oxidative stress in diabetic condition and lead to the generation of advanced glycation end products (AGEs) by non-enzymatic glycosylation might be causing abnormalities in cell and tissue metabolism. They are also involved in the enhanced vascular permeability through binding with the specific receptors of macrophages in both vascular channels. The resulting free radicals exert endothelial dysfunction. In addition, advanced glycosylated end products are formed around the nucleic acids and histone proteins leading to mutations and altered gene expression (Brownlee, 1996; Elgawish et al., 1999).

8.2 MECHANISM OF ACTION OF POLY HERBAL FORMULATIONS IN ANIMAL MODELS

8.2.1 POLY HERBAL FORMULATIONS

Diabecon, is an Ayurvedic polyherbal formulation of Himalaya herbals consists of about 10 medicinal herbs and the formulation showed its mechanism of action by increasing peripheral glucose utilization, hepatic, and muscle glucagon contents and c peptide level. It promoted damaged B cells and enhanced the capacity of B cell regeneration. Further, it modulated the glycated hemoglobin and lipid profile levels and normalized the micro-albuminuria (Mohan, 1998). Oral administration of Dianex (polyherbal formulation) at three doses ranging from 100 to 500 mg/kg body weight, significantly lowered glucose concentration levels in treated and normal rats for a period of six weeks (Mutalik et al., 2003).

Scientific investigations on the polyherbal formulation (DIASULIN) significantly controlled the glycemic index through increasing glucose metabolism through glycolysis and controlling gluconeogenesis. Further it significantly reduced the tissue lipids and membrane lipid peroxidation (Pari and Saravanan, 2004). Diabrid showed hypoglycemic activity and it is useful in maintain ace of body weight and blood pressure (Qadri et al., 2006). Diakyur exhibited significant hypoglycemic effect and anti-lipid peroxidative activity in experimental models (Joshi et al., 2007).

DRF/AY/5001 (polyherbal formulation) a product of Dabur Research Foundation, India, significantly inhibited epinephrine production in diabetic rats. The formulation is significantly normalized the glucose levels at 300 and 600 mg/kg bwt. concentrations. The enhancement in the ALP, SGPT, and SGOT, antioxidant enzymes such as GPx, CAT, and SOD, and glucose uptake by hemidiaphragm in serum (Mandlik et al., 2008). Hyperglycemia induced by epinephrine is mediated by the release of glucocorticoids, which affect glucose metabolism in the liver and also inhibit insulin release by the pancreas and detailed analysis is required to its mechanism. "Karnim Plus" is an indigenous polyherbal formulation of Unijules Life Sciences Ltd., Nagpur, India, exhibited glucose tolerance though it exerted insignificant effect on hypoglycemic activity. Moreover, it significantly ($P < 0.01$) affected cholesterol and triglycerides in diabetic rats, alleviating the lipid complications in some diabetic patients (Bangar et al., 2009).

A polyherbal formulation (DIHAR) made of the eight herbs on specific ratio has effectively normalized streptozotocin induced glucose and lipid levels with low levels of insulin to un-induced rats in addition to normalization of serum creatinine urea and improvement antioxidant effect (Patel et al., 2009). Pretreatment of 'Diashis' herbal formulation significantly increased the STZ-induced glucose metabolizing enzymes "hepatic hexokinase, glucose-6-phosphate dehydrogenase and glucose-6-phosphatase" activities and altered the diabetic profile. Moreover, it improved the stress related metabolites (Tushar et al., 2010). The herbal extracts of 10 effective medicinal plants has been formulated as DIASOL manufactured by Jyothy herbal laboratory, Chidambaram, India. It has been proved to control the blood plasma glucose concentration levels in diabetic rat models, which is very closer to standard drug, glibenclamide activity as it normalized the sugar level of plasma at 125 mg/kg bw given for 14 days to the rats of STZ-challenged diabetic models (Seevalen et al., 2010).

"Golyherb" polyherbal granules when administered orally effected significantly by lowering glucose concentrations of serum and enhanced glucose tolerance in diabetic rats (type 1). It altered lipids significantly and peroxidation of lipid along with improvement in antioxidant enzymes (Thakkar et al., 2010). Similarly, "Okudiabet" was effectively normalized the diabetic conditions alloxan-challenged diabetic animal models than standard drug glibenclamide (Ogbonnia et al., 2010).

Diabet herbal formulation made of six herbals namely *Curcuma longa*, *Coscinium fenestratum*, *Strychnos potatorum*, *Phyllanthus reticulatus*. *Tamarindus indica*, and *Tribulus terrestris,* exerted antidiabetic efficacy in glucose tolerance, while it is not effective hypoglycemic agent (Shrivastava et al., 2012). Polyherbal formulation made from *Glycosmis pentaphylla* (stem bark), *Tridax procumbens* and *Mangifera indica* (leaves) significantly reduced glucose concentration in fasting blood samples of Streptozotocin–Nicotinamide induced rats and showed normalized profile when compared with glibenclamide. Histopathological studies showed low level crowded body fluids and a slight reduction in the islets of Langerhans and normal β-cell population. It also reduced the damage of hepatocytes and necrosis in diabetic rat models at the tested concentrations (Ramesh et al., 2014).

8.2.2 POLY HERBAL FORMULATIONS IN IN-VITRO MODELS

Meshashringi, is a popular herbal traditional formulation of which have many pharmacological properties. Meshashringi and acarbose (standard drug) exhibited inhibitory effect on α-glucosidase/α-amylase and Dipeptidyl peptidase IV (DPP-IV) in *in vitro* models in the low concentrations (Sahu et al., 2016). Similarly, MD-1 (multi herbal formulation) reported for antidiabetic activity, having six herbs such as fruits of *Emblica officinalis* and *Eugenia jambolana,* aerial parts of *Phyllanthus amarus*, *Tinospora cordifolia*, and leaves of *Gymnema sylvestre* (ethanolic extract) exhibited inhibition effect on α-glucosidase activity strongly than the standard compound acarbose (Telapolu et al., 2018).

8.3 MECHANISM OF ACTION OF PLANT EXTRACTS IN ANIMAL MODELS OR AS ORAL ANTIDIABETIC DRUGS

8.3.1 EFFECT OF PLANT (SINGLE HERB) EXTRACTS IN ANIMAL MODELS

The commercially available synthetic drugs used in Diabetes might have limited applications and side effects which also became costly. The treatments used in traditional medicines that possess herbs might have many lead phyto-constituents that possess multiple beneficiary alleviating effects in addition to natural as well as affordable. Hence the people rely on the herbals for maintaining the metabolic disorders such as Diabetes (Saad et al., 2017). In this connection many researchers reported a good number of herbals used in diabetes as well as to reduce its complications. The following medicinal plants or their extracts were used frequently and proved the mechanism of its action.

Aqueous extract of *Scoparia dulcis* (sweet broom weed), reported for the antidiabetic activity and its action was attributed to secretion of insulin effect in separated islets of pancreas of mice evidenced by noteworthy reduction in the blood and also enhanced concentration of plasma insulin in streptozotocin treated rats. Moreover, it strongly reversed the levels of HDL, normalized the HMG-CoA reductase activity and atherogenic index in STZ-challenged rats indicating control over the lipid metabolism or dyslipidemia in STZ-challenged rats (Latha et al., 2004; Pari and Latha, 2006). The extracts of root and stem bark of *Anthocleista nobilis* at when given orally, altered significantly higher levels of fasting glucose levels in blood samples after treatment of weekdays in alloxan-induced rats (Philip et al., 2017; Ngwoke et al., 2018). Patel and Manigauha (2018) investigated the antidiabetic activity of petroleum ether and ethanol extracts, its fractions and water extracts of *Alangium salvifolium* leaves in streptozotocin (STZ) induced diabetic rat models significantly normalized the glucose and lipid profile. They conclude that polyphenol components and flavonoids present in F6 and F7 fractions impart the antidiabetic activity.

Aqueous extract of *Boswellia serrata* gum (200 and 400 mg/kg) investigated for antidiabetic effect and found significant effects on fasting blood glucose (FBG) and normalized symptoms of various organs and reduced streptozotocin-induced symptoms such as lymphocytic inflammation, anarchism of liver plates and necrosis of hepatocytes and in diabetic female rats (Azemi et al., 2012). Pretreatment of *Boerhavia diffusa* leaf extract at 200 mg/kg showed its antidiabetic mechanism by altering the hepatic enzymes levels experimental rats and its effect is stronger than the standard glibenclamide (Pari and Amarnath Satheesh, 2004). Oral administration of ethanolic extracts of *Momordica charantia* (bitter gourd) at 200 mg/kg dose exhibited significant antidiabetic efficacy in STZ challenged diabetic rats, by inhibiting the secretion of liver glycolytic enzymes (Shibib et al., 1993).

Administration of *Trigonella foenum-graecum* seed extracts strongly reduced in the glucose concentration in blood samples of untreated and alloxan treated diabetic rats (Khosla et al., 1995) in addition to improved glucose metabolism and modulated abnormal levels of creatinine kinase enzyme present in heart, skeletal, and liver tissues of diabetic experimental rats. The seeds also decreased renal and hepatic glucose-6-phosphatase and fructose-1, 6-bisphosphatase activity (Gupta et al., 1999). The extracts obtained from the hexane, methanol, and water of *Syzygium alternifolium* seeds were investigated on the diabetic rats induced by alloxan and concluded that the aqueous extract found efficient in the management of hyperglycemia highly significant than the organic fractions (Rao and Rao, 2001). Rama Mohan et al. (2010) observed the similar effect ethanol extract of leaves of *S. alternifolium* as an effective hypoglycemic activity when compared to aqueous and chloroform extracts in their scientific study.

8.3.2 MEDICINAL PLANTS HAVING INSULIN MIMETIC PROPERTY

Insulin, a key marker hormone secreted from the pancreas a gland located back of the stomach which controls glucose levels in the blood stream allowing the body to use glucose for energy. In case of Diabetes the function of insulin may be decreased or secreted in low levels or insulin resistance in the body leading to elevated glucose levels (Figure 8.1). Then insulin supplements might be given to avoid the hyperglycemic condition using the mimetic molecules which are similar to that of insulin, including raising the introduction of glucose into tissues, activating signaling molecules, inducing the expression of genes, and controlling metabolic processes. In this context, the traditional medicinal plants reported for the antidiabetic activity were studied for the mechanism and found the following medicinal plants possessing the mimetic effects were discussed. Some of the most effective plants are garlic, colocynth, *Ficus bengalensis*, *Gymnema sylvestre, Momordica charantia, Trigonella foenum greacum,* and were extensively investigated and characterized some lead phytoconstituents such as "epicatechin, epigallocatechin gallate, beta-pyrazol-1-yl alanine, cinchonain Ib, leucopelargonidin-3-O-alpha-L rhamnoside, leucocyanidin 3-O-beta-d-galactosyl cellobioside, roseoside, glycyrrhetinic acid, dehydrotrametenolic acid, strictinin, iso strictinin, pedunculagin, and christinin-A" reported to have significant insulin-mimetic effect in addition to antidiabetic activity (Patel et al., 2012).

8.3.3 HERBAL MEDICINES WITH THE INSULIN MIMETIC ACTIVITY

The extract from *Biophytum sensitivum* exerted elevation on the serum insulin, suggesting involvement of pancreas in 16-h fasted non-diabetic rabbits (Puri, 2001). Swerchirin (50 mg/kg) effectively controlled the blood glucose level and the histopathological studies evidenced that the enhanced insulin release from separated islets of the pancreas (Grover et al., 2002). A scientific study on antidiabetic efficiency of *Nigella sativa* seed oil, the hamster rodents were induced diabetes using Nicotine by intraperitoneally and streptozotocin intravenously after 15 minutes and observed the serum insulin levels using enzyme-immunoassay assays. Significant fall in the glucose levels correlated with insulin secretion from islets after treatment with the oil for four weeks suggesting insulinotropic property (Fararh et al., 2002) (Figure 8.2).

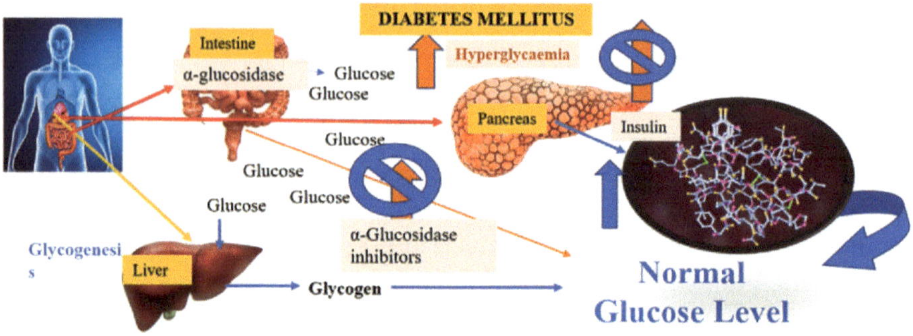

FIGURE 8.1 Network of carbohydrate metabolism in diabetes mellitus.

The antidiabetic efficiency of garlic and onion in experimentally induced diabetes in rat models were studied on the body weight, glucose levels in blood samples, plasma lipid markers and liver glycogen and proved that the onion extract and its components were more significant than the garlic. The other studies also indicated that garlic extracts improved insulin levels in the blood plasma (Grover et al., 2002; Mustafa et al., 2007; Kook et al., 2009).

The organic fractions of *Asparagus racemosus* root was exerted concentration dependent secretion of insulin in separated perfused rat pancreas/islet cells/clonal beta-cells through insulinotropic activity (Hannan et al., 2007). The alcoholic extract of Shathavri studied in STZ-challenged diabetic rat models at two doses, i.e., "200 and 400 mg/kg/b. wt." for 21 days exhibited noteworthy reduction in the glucose levels in blood samples, intake of fluid and substantial increased in the body weight. In addition, the extracts also exhibited improvement in the antioxidant enzymes in treated models (Ramachandran et al., 2011). The diabetic profile of molecular cascade has normalized effects were demonstrated with the extracts and isolated component glycyrrhetinic acid from the rhizome of *Radix glycyrrhizae* in the islets in addition to improvement in live cells along with enhanced insulin secretion (Ko et al., 2007). The extracts as well as Andrographolide isolated from *Andrographis paniculate* exhibited significant hypoglycemic effect in experimental animals and shown improvement in glucose metabolism (Gupta et al., 2007).

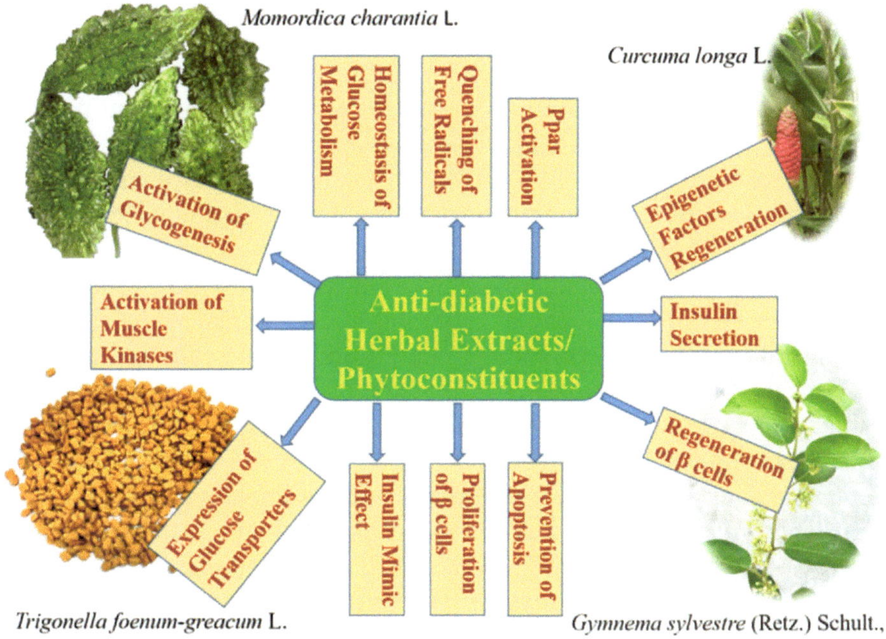

FIGURE 8.2 Possible mechanisms of antidiabetic (Type-II) effect of medicinal plants.

Four-week treatment with the extracts of *Capsicum frutescens* showed an enhancement effect in diabetic profile in STZ-induced rats fed with fatty diet inferred its insulin secretion effect (Islam et al., 2008). Similar effect were studied with the polypeptides of the roots of Ginseng (*Panax ginseng*) and Korean red ginseng (Bnouham et al., 2006; Kim, 2008) extracts of garlic (Islam et al., 2008) *Eriobotrya japonica* and its phytoconstituents such as procyanidin B-2, cinchonain Ib, chlorogenic acid and epicatechin (Qa'dan et al., 2009) aqueous extract of *Brassica nigra* seeds (Anand et al., 2009).

Cinnamaldehyde, one of the important constituents of *Cinnamon zeylanicum* exhibited insulinotropic effect through glucose transporter

(GLUT4) translocation in the experimental animals, especially in peripheral tissues (Anand et al., 2010). The extracts of *Annona muricata* reported the protection effect against the oxidative stress in the pancreatic tissues especially in streptozotocin-induced diabetic rat models. Whereas *Annona squamosa* (custard apple) exhibited antidiabetic activity through promoting insulin secretion in the pancreatic islets and reported in the increasing metabolizing effect of glucose levels in tissues and suppressing effects its release from the liver (Malviya et al., 2010). The aqueous extract of *Aegle marmelos* leaf reported as an effective glucose reducing agent in STZ-induced diabetic rat models in 15 days experiment, shown improvement in the diabetic carbohydrate and lipid profile and key enzymes activity along with normoglycemic index (Sharma et al., 2007). The extracts of *Alangium salvifolium* (Kshirsagar et al., 2010) and *Allium sativum* (garlic) were proved to exert the antihyperglycemic and antidiabetic activity through its insulin secretion effect in pancreatic cells (Ayodhya et al., 2010).

The extracts and oil of *Allium sativum* has noteworthy hypoglycemic effect in normal and diabetic rat/rabbit models, induced with alloxan, attributed to its insulin secretion property from parietal cells in the pancreas (Chauhan et al., 2010). Aqueous extract of roots of *Rubia cordifolia* exhibited notable glucose reducing activity in STZ-challenged rats in addition to many beneficial applications (Baskar et al., 2006). The seed extract of *Caesalpinia bonducella* expressed antihypoglycemic and antilipidemic effect in alloxan challenged diabetic rat models (Kannur et al., 2006) in addition the alcoholic extract showed an enhanced production of insulin (Ayodhya et al., 2010). Three fractions from the *Biophytum sensitivum* reported for the antidiabetic activity with the notable glucose reducing activity in STZ-challenged diabetic rats (Renuka and Saravanan, 2018). The effect has proved its enhanced insulin secretion from pancreatic beta cells in diabetic induced male rabbits (Ayodhya et al., 2010).

The water extract of punarnava leaves exhibited significant antidiabetic activity and normalized the alloxan effects and the activity of key enzymes of liver when treated 200 mg/kg and also noticed the oral glucose tolerance (Pari and Amarnath, 2004). The chloroform extract was reported to possess enhancement effect in the insulin sensitivity in addition to antihyperglycemic effect in STZ challenged diabetic rats (Malviya et al., 2010). The ethanolic extract prepared from leaves and stem bark of *Bougainvillea spectabilis* exhibited hypoglycemic activity and improved glucose uptake and glycogenesis in the liver along with insulin-mediated metabolism in type-1 diabetic rats induced with streptozotocin and alloxan respectively (Malviya et al., 2010; Sunil et al., 2012).

The antidiabetic activity *Cassia auriculata* leaf extracts (CLEt) in the rats induced with streptozotocin at the dose of 400 mg/kg daily for 15 days reported the significant enhancement in the activity of hepatic key enzymes of carbohydrate metabolic process. Moreover, the synthesis of glycogen in the target tissues were noticed (Gupta et al., 2010).

The extract prepared from fruit pulp of *Citrullus colocynthis* (300 mg/kg, p.o.) showed significant antidiabetic activity and correlated with the enhanced insulin level and reduced plasma glucose concentration levels in alloxan challenged diabetic rat models (Singh, 2011; Dallak et al., 2009a; Bnouham et al., 2006). *Coccinia indica* leaf extract exhibited insulin stimulatory effect from the beta cells in diabetic rats, and it revealed hypoglycemic activity, and the effect has been attributed to the insulinotropic or modulatory effect on the glucose metabolism (Singh, 2011).

The extract of *Cornus officinalis* reported enhancement in the GLUT4 transcription and promoted the growth of pancreatic islets and enhanced secretion of insulin from the pancreatic beta cells (Chen et al., 2008; Chauhan et al., 2010). Similar effect has also been studied with the acetone extracts of *Elephantopus scaber* with notable hypoglycemic effect, secretion of insulin and rebuilding of islets in the pancreas of STZ-treated rat models (Rao et al., 2010). extract of *Enicostemma littorale* exhibited antidiabetic effect in alloxan-induced rats and reported the glucose-induced insulin secretion by way of K+-ATP channel method (Maroo et al., 2002). *Ephedra distachya* alkaloid, L-ephedrine exerted an antihyperglycemic effect in diabetic mice attributed to secretion of insulin (Chauhan et al., 2010). Enhanced peripheral glucose metabolism in the experimental mouse model and enhanced insulin release from the pancreatic beta-cell line reported with the aqueous extract of *Eucalyptus* (Singh, 2011). The seed's extract of *Eugenia jambolana* tested on the isolated pancreatic islet cells of normal and diabetic animals was reported the enhanced insulin secretion in addition to inhibition effect on the activity of insulinase of liver and kidney (Bnouham et al., 2006; Modak et al., 2007; Singh, 2011).

The bitter fraction of *Aloe vera* exhibited glucose reducing effect in the rats along with the stimulation of the release of insulin from the beta cells of the pancreas (Singh, 2011). The seed diet of gum Arabic tree at 2 to 4 g/kg-b.wt. showed glucose reducing activity in *in vivo* study and rabbits through insulin secretion from the pancreas, which is a beneficial antidiabetic property (Singh, 2011). Banyan tree extract reported the secretion of elevated level of insulin in serum of tested animal models and normalization of diabetic profile along with the inhibitory effect of insulinase in liver and kidney (Bnouham et al., 2006; Singh, 2011). A "dimethoxy derivative of leucocyandin 3-O-beta-d-galactosyl cellobioside, glycoside of leucopelargonidin and Dimethoxy ether of leucopelargonidin-3-O-alpha-L rhamnoside" reported from *F. bengalensis* stem bark, reported the glucose reducing effect in blood in normal and treated rats through insulinomimetic activity (Bnouham et al., 2006; Ayodhya et al., 2010).

Gymnema sylvestre (madhunash) is one of the important antidiabetic medicinal herb of Indian traditional system, and it has been extensively studied for its glucose controlling effects in various experimental models. The alcoholic extract of its leaf reported for the stimulation and secretion of insulin in the islets of langerhans of beta cells in experimental rats as well as in beta cell lines pancreas. The administration of a aqueous extract of *G. sylvestre* at 400 mg/day concentration to insulin-dependent diabetes mellitus (IDDM) patients during the insulin therapy exhibited improvement in the management of blood glucose level in fasting in addition to the reduced demand for insulin supplementation (Grover et al., 2010) and it also necessitate in the proliferation and repairing of beta cells of pancreas in type II diabetic patients in addition to increased release of insulin in the serum (Bnouham et al., 2006). Gymnemic acid (dihydroxy gymnemic triacetate) is reported as one of the active principle involved in the release of insulin and regeneration and revitalization of the existing beta cells. The secretion insulin from the isolated islets of Langerhans in the in-vitro system without effecting its viability (Kaczmar, 1998; Saxena and Vikram, 2004; Rao et al., 2010). The root extracts of *Helicteres isora* obtained from the butanol fraction at a dose of 250 mg/kg reported antihyperglycemic effect in diabetic rats induced with alloxan which acted through insulin-sensitization (Venkatesh et al., 2003; Ayodhya et al., 2010).

Treatment of ethanol extract obtained from *Hibiscus* flower exhibited lenient activity, but noteworthy glucose-lowering effect, which was mainly due to insulin release at the rate of 14% for 21 days treatment by stimulation of pancreatic beta cells in streptozotocin challenged diabetic Wistar rats (Grover et al., 2002; Sachdewa and Khemani, 2003). The *Hordeum vulgare* tender seedlings reported the hypoglycemic and hyperinsulinemic effects which suggested the use of them as suitable cereals in diet for diabetes (Chauhan et al., 2010). *Lepechinia caulescens* reported to possess insulin-mimetic activity and significantly decreased glucose tolerance, whereas *Medicago sativa* aqueous extract evoked the stimulation in acute short term evaluation found that, insulin release in the pancreatic beta-cell line "BRIN-BD11" by 2.5–6.3 fold (Bnouham et al., 2006; Gray and Flatt, 1997).

The fruit juice of bitter gourd (*Momordica charantia*) showed significant hypoglycemic effect and enhanced insulin of plasma and noticed enhanced beta cells in diabetic rats. The constituents such as momordicin, charantin, proved to have insulin-mimetic activity through the "galactose-binding lectin" and "insulin-like protein" (Saxena and Vikram, 2004; Kaczmar, 1998). In addition, the aqueous extract of unripe fruits reported moderate induction of insulin secretion in cultured β-cell of mice suggested the insulinotropic effect through modulating property of membrane (Grover et al., 2002). It increases the rejuvenate partially dysfunctional cells or may initiate the improvement in the partially damaged cells which stimulates the production of insulin from the pancreas (Chauhan et al., 2010). In 20 days, experiment, ethanolic extract of *Ricinus communis* (Caster) seed, root, or leaf at 500 mg/kg, efficiently controlled the diabetic effects along with improved insulin in the blood which also reported normalized lipid markers and body weight of diabetic rats (Rao et al., 2002). *Syzygium cumini* is also familiar for using in diabetes and its fruit pulp is edible, when tested for antidiabetic effect in the diabetic rats (STZ triggered) even in short-term oral treatment, exhibited hypoglycemic effect through stimulating insulin mediated by the protection from the insulinase activity (Grover et al., 2002).

Pulverized seed powder of *Mucuna pruriens* reported the antidiabetic and hypoglycemic activity at tested doses in non-treated or diabetic rabbits (induced by alloxan) has been attributed the insulin secretion, in addition seeds containing a few trace minerals such as manganese, zinc, etc., which is beneficial effect in the management of sugar levels in diabetes (Grover et al., 2002; Ayodhya et al., 2010). Flavonoid characterized form Guinea plum plant showed glucose-lowering effect in diabetic animal models (Rao et al., 2010). Jiangtangsu, a phytoconstituent of *Prunella vulgaris* reported for significant hypoglycemic activity in diabetic mice which acts through repairing the β-cells and produces the insulin (Chauhan et al., 2010). The polysaccharide content of *Radix rehmanniae* rhizome reported stimulation of insulin and glucose-lowering activity in experimental rats (untreated and streptozotocin treated) along with the diminution of glycogen (Chauhan et al., 2010).

The extracts of *Salvia lavandifolia* and *Selaginella tamariscina* extract showed significant antidiabetic activity through stimulation insulin secretion in the experimental rats from the beta cells of the pancreas (Bnouham et al., 2006). Oral treatment of hexane fraction obtained from *Swertia chirayita* at

"250 mg/kg," notable reduced sugar levels in blood samples and enhanced plasma insulin levels without effecting glycogen of liver in normal rats in addition secreting insulin from beta cells for 28 days (Bnouham et al., 2006). Insulin release from the beta-cell line, i.e., BRIN-BD11 has been demonstrated using aqueous extracts (Thomson et al., 2014). Decoction prepared from thorny burnet "*Sarcopoterium spinosum*" showed an insulinotropic effect and enhanced glucose uptake in the liver, in addition to stimulation of insulin release (Rao et al., 2010). The WE of *Tabernanthe iboga* exerted its antidiabetic activity with insulinotropic compounds might be involved the controlling "K^+ ATP" and enhancing the calcium influx mediated through "voltage-sensitive Ca^{2+} channels." (Souza et al., 2011). Similarly, *Tinospora crispa* extract exhibited antihyperglycemic activity through stimulation of insulin secretion mediated by controlling Ca^{2+} concentration of beta-cells (Bnouham et al., 2006).

The WE of *Teucrium polium* (TP) enhanced insulin secretion by stimulating the pancreas, this may be due to initiation and "re-initiation of islets" of experimental rats (Esmaeili and Yazdanparast, 2004; Yazdanparast et al., 2005). This activity has attributed to the apigenin reported from the methanol fraction (Mirghazanfari et al., 2010). In *in vivo* experiment, oral administration of *Tribulus terrestris* (gokshura) extract significantly modulated the alloxan-induced elevated blood glucose in experimental rats (Chauhan et al., 2010) and similar effect in rabbits (Amina El-Shaibany et al., 2015).

The *Trigonella foenum-graecum* seeds (Fenugreek) reported to controlling carbohydrate metabolism in reduced insulin cases (Kaczmar, 1998). Whereas in another study, fenugreek seeds evoked "glucose-induced insulin secretion" (Saxena et al., 2004) whereas, the seed powder, extracts, gum, and leaves showed anti-hyperglycemic effect by delaying "gastric emptying due to the high fibrous nature and by suppressing the carbohydrate metabolizing enzymes in addition to improvement in insulin production (Gupta et al., 2001; Chauhan et al., 2010).

8.4 MECHANISM OF ACTION OF PHYTOCONSTITUENTS IN ANIMAL MODELS

Antidiabetic effects of "scoparic acid D, diterpenoid" isolated from ethanolic "*S. dulcis,*" was investigated in STZ challenged diabetic rats. Scoparic acid D was administered orally in "three different doses, i.e., 10, 20, and 40 mg/kg b. wt." during the experimental period. Treatment with "20 mg mg/kg body weight," significantly affected BGL (blood glucose levels) (Latha et al., 2009). Berberine is an alkaloid, naturally occurring in the genus *Berberes* of family Berberidaceae. Plants containing berberine were practiced to treat cold, fever, lung infections, metabolic illness, and influenza traditionally (Fabricant and Farnsworth, 2001). Timosaponin B2 (TB-2), is a steroid saponin isolated from *Anemarrhena asphodeloide* (Meng et al., 1999). Combined therapy of both berberine and TB-2 at 100 mg/kg/day displayed antidiabetic efficiency by reducing FBGL non-fasting and fasting BGL and showed beneficial role in OG oral glucose test than metformin a standard drug control after a six week experiment in non-obese diabetic Goto-Kakizaki rats (Tian et al., 2019).

Chemical components isolated from traditional medicinal plant sources enhances insulin secretion/increased insulin production are Epigallocatechin-3-gallate (isolated from *Camellia sinensis*) (Yun et al., 2006; Wolfram et al., 2006), Capsaicin (from *Capsicum annuum*) (Akiba et al., 2004; Gram et al., 2007; Islam and Choi, 2008), "Curcumin" (from turmeric rhizomes) (Na et al., 2011; Meghana et al., 2007; El-Moselhy et al., 2011; Hussain, 2002), Genistein (from *Glycine max*) (Choi et al., 2008; Fu et al., 2010), "Quercetin" (Hii and Howell, 1985) ginkgolides from *Ginkgo biloba* (Pinto et al., 2009), Gurmarin, betaine, choline, trimethylamine from *Gymnema sylvestre* (Sugihara et al., 2000; Preuss et al., 2004), (–) Hydroxycitric acid from *Gymnema sylvestre* and Garcinia cambogia (Preuss et al., 2004; Hayamizu, 2003), ferulic acid from *Curcuma longa* (Ohnishi et al., 2004), leucine, isoleucine from *Aloe vera* (Ajabnoor et al., 1990).

Amino acids, amines, and carboxylic acid derivatives enhances insulin secretion/increased insulin production and glycogen synthesis are lectins, mistletoe lectin I, II, III, viscotoxin B, cycliton from *Viscum album* (Adaramoye et al., 2012; Eno et al., 2008), furfural, caprylic acid from Agaricus campestris (Manohar et al., 2002; Gray and Flatt, 1998) and bis DEHP from *C. auriculata* (Abesundara et al., 2004). Glycosides that enhances insulin secretion/increased insulin production and glycogen synthesis are gymnemic acid, gymnemosides from madhunashini (Sugihara et al., 2000; Preuss et al., 2004), Vin α-ginsenoside R3 from American ginseng (Vuksan et al., 2000), Scopolin roscoside II, astragalin, and skimming from Mulberry (Gulubova and Boiadzhiev, 1975), C-glycosides from *Trigonella foenum graecum* (Gupta et al., 1999; Kluwer, 1999), Momordin, momordicine, charantin from *Momordica charantia* (Sarkar et al., 1996).

Alkaloids, a group of naturally occurring nitrogen-containing plant compounds showed their mechanism of action of antidiabetic efficiency by targeting on carbohydrate digestion/absorption and glucose transport are trigonelline, carpaine, and gentianine compounds from fenugreek (Khosla et al., 1995), achyranthine, betaine, and β-ecdysone from prickly chaff (Akhtar and Iqbal, 1991), Aegelin, marmesin, marmelosin from *Aegle marmelos* (Kamalakkannan and Prince, 2005; Ponnachan et al., 1993).

Amino acids and carboxylic acid derivatives showed their antidiabetic activity by targeting carbohydrate digestion/absorption and glucose transport, are 4-Hydroxyisoleucine, n-hydroxyisoleucine from *Trigonella foenum graecum* (Hui et al., 2009; Khosla et al., 1995), Nitrosamines from *Areca catechu* (Mannan et al., 2000), ethyl brevifolin carboxylate and brevifolin carboxylic acid from Bhumyamalaki *(P. amarus)* (Ali ct al., 2006), Carbohydrates targeting carbohydrate metabolism, insulin secretion and glucose transport are glucomannan, cellulose, mannose from *Aloe vera,* caryophylline from *Ocimum sanctum, Syzygium aromaticum* and *Protein-bound polysaccharide from Alpinia galanga, Aloe vera* and *Ocimum sanctum* (Van de Venter et al., 2008), gum, pectin/fibers and mucilaginous fiber from fenugreek, sweet orange and Ivy guard (Kar et al., 2003; Nandini et al., 2003), D-threitol, D-arabinitol, palmitic acid from *Hericium erinaceus* (Khan et al., 2003; Liang et al., 2013), cinnzeylanin, D-glucan, L-arabino-D-xylan, and cinnzeylanol from Cinnamon (Solomon and Blannin, 2007), Mucopolysaccharide from *Opuntia* spp. and Inulin, lagavulin from *Taraxacum officinale* (Godard et al., 2010).

Caffeine solution (0.01%) significantly reduced animal body weight, fat level and lowered insulin resistance in 90% pancreatectomized diabetic rats for in 84 days experimental duration. It also enhanced glucose-stimulated insulin release and increased beta-cell number (Park et al., 2007). Flavonoid-rich fraction obtained from Indian kino tree showed pancreatic beta-cell regranulation. Epicatechin isolated from flavonoid rich fraction of Indian kino tree was found to be increasing insulin release or by regenerating the beta cell (Saxena et al., 2004; Modak et al., 2007).

The antidiabetic effect of amentoflavone (AME) was assessed in the "high fat diet/STZ induced diabetic in mice model" for a period of 8 weeks. The component exhibited its antidiabetic activity through modulating the glucose metabolism enzymes, the expression of pAkt, Akt, and GLUT4 immunoreactivity. It also lowered the levels of glucose, bad lipid markers and glucagon and enhanced good cholesterol and insulin levels. Further AME enhanced the activities of enzyme levels such as phosphofructokinase-1, pyruvate kinase and glucokinase and inhibited the activities of glucose-6-phosphatase, glycogen synthase kinase-3 and phosphoenolpyruvate carboxykinase. AME modulated superoxide dismutase associated with reduced TBARS formation. AME exhibited antidiabetic efficiency by regulating glucose and lipid metabolism via modulating antioxidant system/activating PI3K/Akt pathway (Chengfu et al., 2019).

8.5 A-GLUCOSIDASE/AMYLASE INHIBITORS AS ANTIDIABETICS

Management of Diabetes is one of the important aspects through good diet practices as well as using the oral drugs such as alpha-glucosidase inhibitors prescribed by the physicians. Scientific reports states that acarbose (a natural compound) did not show any significant toxicity, except gastrointestinal problems associated with loose stools, flatulence or abdominal discomfort have also been reported. The higher incidence of these symptoms in US (about 40%), next in Germany (about 25%) and about 17% in Asian trials (Rosak and Mertes, 2012). Hence the diabetic patients were relying on the natural products for the maintenance of normal glucose levels using traditional practices. It is beneficial particularly, non-insulin-dependent diabetes mellitus to reduce glucose concentration after meal, which can decrease the intake of glucose through slowing of the assimilation of carbohydrates and ultimately the flux of glucose levels was regulated by the α-glucosidase/ amylase inhibitors (Figure 8.1).

The important classes of secondary metabolites isolated from medicinal plant sources are alkaloids, flavonoids, anthocyanins, terpenoids, phenolic components and glycosides, which are able to attenuate the activity of the enzymes through binding α-glucosidase or carbohydrate metabolizing enzymes (Patel et al., 2012; Rosak and Mertes, 2012). Many scientists were able to trace out the mode of action of natural products and discussed as follows. "Quercetin 3-O-β-d-xylopyranosyl (1"' to 2") – β-d-galactopyranoside," a flavonoid glycoside isolated from devil tree *(Alstonia scholaris)* strongly inhibited maltase with "IC_{50} of 1.96 mM" (Anurakkum et al., 2007). Methanol extract obtained from *Adhatoda vasica* significantly inhibited RI AGL (rat intestinal α-glucosidase) and isolated "Vasicine and Vasicinol," showed a significant sucrose inhibitory effect with IC_{50} values 125 and 250 μM, respectively (Gao et al., 2008). Umbellactam A, B, and C constituents characterized from *Piper umbellatum* showed average α-glucosidase inhibitory activity (Tabopda et al., 2008). Three ellagitannins, i.e., chebulinic acid, chebulanin, and chebulagic acid separated by bioassay-guided fractionation technique exhibited potent maltase inhibitory activity with at 690, 97, and 36 μM, respectively from the fruits of *Terminalia chebula* (Gao et al., 2007).

Two active constituents, i.e., "(-)-3-O-galloylcatechin and (-)-3-O-galloylepicatechin" isolated from *Bergenia ciliata* proved AGI effect in rat intestine (Bhandari et al., 2008). Adisakwattana et al. (2009) investigated AGI effect of "anthocyanin, cyanidin-3-galactoside" against intestinal sucrose. *Curcuma longa* (turmeric) is an important source of curcumin, demethoxycurcumin, and bisdemethoxycurcumin were investigated effect on α-glucosidase suppression activity in in vitro studies, used as spice in traditional Indian dishes (Du et al., 2006). A new pentacyclic triterpenoids, "3b-acetoxy-16b-hydroxybetulinic acid" form "*Fagara tessmannii*" (Mbaze et al., 2007) and Segetalic acid isolated from roots of "*Gypsophila oldhamiana.*" Luo et al. (2008) demonstrated the AG/amylase suppression activity. Bio-active flavonoids, i.e., "1,3-dihydroxy-2-methylanthraquinone, 1,2-dihydroxyanthraquinone and 1-hydroxy-2-methylanthra quinone" from *Rubia cordifolia* using bioassay-guided fractionation method reported its therapeutic potential of α-glucosidase inhibition (Kang et al., 2009). The extracts of *Ochna obtusata* leaves (Gourigari et al., 2016) and *Dendrophthoe falcata* (Bhakshu et al., 2017) were reported the inhibition effect of α-glucosidase activity which are used for diabetes in the traditional medicines.

8.6 MECHANISM OF ACTION OF POLY HERBAL FORMULATIONS IN DIABETIC CELL LINE MODELS

The cell line provides an important and reliable *in vitro* model for the pharmacological studies before testing on the human and for many diseases, a particular line has been developed and most of the trials were correlated with the experimental outputs. Hence the present study also focused on the research work done so far using the cell lines and analyzed as follows.

The cell lines such as L6 and 3T3L1 are well known cell line models to investigate glucose uptake/absorption in *in vitro* models (Gould et al., 1989). Ethanol extract of MD-1 polyherbal formulation of six medicinal plants enhanced glucose assimilation similar to control cells in a concentration-dependent manner. Further, the formulation significantly inhibited lipid accumulation than standard pioglitazone (PGZ) in 3T3L1 preadipocytes. The formulation showed its mechanism of action by up-regulating the transcription factor peroxisome proliferator-activated receptor (PPARγ), a marker of regulator of adipocyte activation (Telapolu et al., 2018).

8.7 MECHANISM OF ACTION OF CHEMICAL COMPONENTS IN CELL LINE MODELS

Scoparic acid D (SA–D), at 20 µg/ml concentration exerted the insulinotropic effect when pretreatment with SA-D isolated from *Scoparia dulcis* has significantly altered the STZ-induced cytotoxicity and NO production in rat insulinoma cells (Latha et al., 2009). Cytopiloyne, a polyyne group secondary metabolite, isolated from *Bidens Pilosa* (orfic), exhibited concentration dependent antidiabetic activity through enhancing the enhanced transcription and insulin levels in "rat insulinoma RIN-m5F cells." Protein kinase C, diacyl-glycerol in addition to calcium, may be involved to enhance insulin release and production (Chang et al., 2013).

Natural components isolated from medicinal plants enhances insulin secretion/increased insulin production in animal/human cell lines are cytopiloyne, isolated from *Bidens orfic* (Ubillas et al., 2000; Chang et al., 2007, 2013; Bartolome et al., 2013; Lai et al., 2015). Epigallocatechin-3-gallate from tea plant (Han, 2003; Fu and Koo, 2006; Wolfram et al., 2006; Yun et al., 2006; Landis-Piwowar et al., 2007; Cai and Lin, 2009; Suh et al., 2010), Curcumin from *Curcuma longa* (Best et al., 2007; Chuengsamarn et al., 2012; Meng et al., 2013), Genistein (from *Glycine max*, Ohno et al., 1993; Sorenson et al., 1994; Persaud et al., 1999; Elliott et al., 2002; Liu et al., 2006; Gilbert and Liu, 2013), Gymnemic acids from *Gymnema sylvestre* (Sugihara et al., 2000; Daisy et al., 2009), "Nymphayol" from *Nymphaea forficata,* (Subash-Babu et al., 2009), "Ginsenoside" of *Panax ginseng* (Xie et al., 2005; Park et al., 2008; Chen et al., 2012), Berberine from *Rhizoma coptidis* (Leng et al., 2004; Lee et al., 2006; Yin et al., 2008; Zhou et al., 2008; Vuddanda et al., 2010; Dong et al., 2012). Quercetin which is a very common flavonoid isolated from divergent plant sources (Youl et al., 2010; Cho et al., 2012).

Chemical constituent, 4-Hydroxybenzoic acid isolated from *Pandanus odorus* significantly enhanced hepatic insulin and glycogen levels at a dose of "5 mg/kg in healthy rats" (Bnouham et al., 2006). Antidiabetic efficiency of stevioside was investigated in *in vitro* models using "isolated mouse islets and clonal β-cells (INS-1)." It was observed that the component showed antihyperglycemic, insulin secretion and prevented glycogen buildup activity in diabetic GK rat models (Jeppesen et al., 2000). Jeppesen et al. (2002) and Jung et al. (2006) demonstrated INS stimulating effect of stevioside and steviol of *Stevia rebaudiana* from β-cells.

Amino acid, 4-Hydroxyleucine of fenugreek seeds enhanced glucose provoked secretion of INS in animals and human beings (Brocaet et al., 1999; Haeri et al., 2009; Grover et al., 2002; Modak et al., 2007). Allicin or "S-allyl cysteine sulfoxide (SACS)," provoked insulin secretion and enhanced antioxidant potential of beta cells isolated from rat (Bnouham et al., 2006; Modak et al., 2007). The extract and the active principle, roseoside of *Bauhinia orficate,* exhibited significant insulin stimulation activity in INS-1 insulin-secreting cell line (Frankish et al., 2010).

Glycosides that enhances insulin secretion/increased insulin production/growth of pancreatic β cells and glycogen synthesis are gomordin, momordicine, charantin from *Momordica charantia* (Chao et al., 2003), Vin α-ginsenoside R3 from American ginseng (Vuksan et al., 2000), Astragalin, scopolin, orficat, roscoside II from *Morus alba* (Gulubova and Boiadzhiev, 1975), active principles isolated from bitter guard, i.e., "momordin, momordicine, charantin, momorcharaside A and B, momorcharin A and B" (Sarkar et al., 1996), isocucurbitacin B and cucurbitacin B from *Helicters isora* and momordin-a, luffin-a from *Luffa cylindnica* (Lemus et al., 1999), Kotalanol, salacinol characterized from *Salacia reticulata, S. oblonga* (Huang et al., 2008), Arbutin, eriolin from *Arctostaphylos uvaursi* (Moon et al., 2007), colocynthin, citrullol, colosynthetin, elaterin, elatericin B, from *Citrullus colocynthis* (González-Tejero et al., 2008; Ziyyat et al., 1997), Leucocyanidin, pelargonidin from *Ficus benghalensis* (Singh et al., 2009; Cherian et al., 1995; Kumar et al., 1994) and Taraxacin from *Taraxacum officinale* (Broadhurst et al., 2000).

Flavonoids that enhances insulin secretion/increased insulin production/rebirth of pancreatic β cells and glycogen synthesis are chrysin, isoquercitrin from Morus alba (Roman-Ramos et al., 1995), Flavanone glycosides, i.e., myrciacitrins I, II, and myrciaphenones A and B characterized from *Myrcia multiflora* and α-Cephalin, myricetin-3'-glucoside, ambrettolide from *Abelmoschus moschatus* (Chattopadhyay, 1999; Ngueyem et al., 2009), *Citrus* flavonoids, hesperidin, naringin from *Camellia sinensis* (Jung et al., 2004), apigenin quercetrin, quercetin, apigenin-7-O-glucoside and rutin, from *Urtica dioica, Bauhinia varigtla, Ginkgo biloba* (Hussain et al., 2004; Jellin et al., 1999), Naringenin from *Camellia sinensis* (Waltner-Law et al., 2002), α-Terpineol, hexanol from *Agaricus campestris* (Gray and Flatt, 1998; Rankin et al., 2008) from grape seed (Pinent et al., 2004), Amarogentin, swerchirin, chirantin, gentiopicrin from *Swertia chirayita* (Van de Venter et al., 2008), Tribulusamides A and B, "kaempferol-3-β-D-(6'P-coumaroyl)glucoside" and "kaempferol-3-glucoside" from Gokshura fruits (Cooper et al., 2003;

Kirtikar and Basu, 1998), Shamimin from *Biophytum sensitivum* (Puri and Baral, 1998; Puri, 2001), Leucopelargonidin, dulcitol from *Casearia esculenta* (Prakasam et al., 2005), matteuorien, matteuorienin, matteuorienate A, B, C from *Matteuccia orientalis* (Shane-McWhorter, 2001).

Bioactive components showing their antidiabetic efficiency through by increasing pancreatic beta-cell regeneration or by inhibiting the pancreatic beta-cell apoptosis are kinsenoside from *Anoectochilus roxburghii* (Zhang et al., 2007), Epigallocatechin-3-gallate from *Camellia sinensis* (Han, 2003; Suh et al., 2010; Yun et al., 2006), Conophylline from *Ervatamia microphylla* (Saito et al., 2012; Umezawa et al., 1996, 2003; Ogata et al., 2004; Kawakami et al., 2010; Kodera et al., 2009). Saponins showed antidiabetic effect by enhancing growth of pancreatic β-cells, insulin release are "stigmasterol, quercitol, gymnemic acid IV" from Gumar leaves (Sugihara et al., 2000; Preuss et al., 2004), Quinquenoside L3 and L9 from American ginseng (Vuksan et al., 2000), Andrographolide from *Andrographis paniculata* (Yu et al.., 2003), 3-O-β-D-glucopyranoside from *Myrtus communis* (Alipour et al.., 2014), 3-Hepatadecanone, 8-hexadecenoic acid hexadecenoic acid from *Asparagus adscendens* (Mathews et al., 2006), Ginsenosides Rg2, panaxan type components of *Panax ginseng* (Ma et al., 2008; Attele et al., 2002), Lactucain C from *Lactuca indica* (Hou et al., 2003), e-Glucoside, mangiferin, salacinol, kotalanol, epigallocatechin from *Salacia reticulata, Salacia oblonga* (Krishnakumar et al., 1999), Allo-aromadendrene, T-cadinol, α-gurjunene, β-eudesmol, β-ubebene, aromadendrene from *Artemisia pallens* (Ruikar et al., 2011), Sotolon [3-hydroxy-4,5-dimethyl-2(5H)-furanone], Trigonellin from *Trigonella foenum-graecum* (Khosla et al., 1995), "Ursolic acid, mulberrofuran-U" from *Morus insignis, Myrtus communis* (Basnet et al., 1993) and "muinol, azorellanol, mulin-11,3-dien-20-oic-acid, mulinolic acid" from *Azorella compacta* (Borquez et al., 2007; Fuentes et al., 2005).

Alkaloids, group of naturally occurring nitrogenous compounds showed their mechanism of action of antidiabetic efficiency by targeting on glucose/carbohydrate metabolism are berberine from *Tinospora cordifolia* (Singh et al., 2003), "Sotolon [4,5-dimethyl-3-hydroxy-2(5H)-furanone], trigonelline, gentianine, carpaine" from fenugreek seeds (Hui et al., 2009), β-ecdysone, betaine, and achyranthine, from *Achyranthus aspera* (Akhtar and Iqbal, 1991). Peptidoglycans involved in glucose transport, carbohydrate digestion and absorption are fenugreekine from *Trigonella foenum graecum* (Khosla et al., 1995), Gluten, taraxacerin from *Taraxacum officinale* (Hussain et al., 2004; Yarnell and Abascal, 2009) and Glucosamines from *Aloe vera* (Ajabnoor, 1990).

Polyphenol and its derivatives showed its glucose-lowering effect by regulating transport of glucose, metabolism of carbohydrates and its absorption, are curcumin, turmerone, germacrone, zingiberene from Curcuma longa (Kar et al., 2003; Zhang et al., 2013), Ellagic acid and its derivatives from "*Potentilla candican, Phyllanthus niruri, Caesalpinia ferrea, Arctostaphylos uvaursi*" (Ueda et al., 2004), "Ellagic acid, corosolic acid, 4-hydroxybenzoic acid, 3-O-methylprotocatechuic acid, caffeic acid, p-coumaric acid, kaempferol" from *Lagerstroemia speciosa, Acacia Arabica* (Naisheng et al., 2008) carvacrol, linalool from Ocimum sanctum (Hannan et al., 2006; Broadhurst et al., 2000), and Mangiferin from *Salacia species* (Yoshikawa et al., 2001). Saponins showed their antidiabetic mechanism in glucose transport and carbohydrate metabolism are Diosgenin from Trigonella foenum-graecum (Khosla et al., 1995), Kotalagenin-16-acetate, diterpene, triterpenes from Salacia oblonga, *Croton cajucara* (Krishnakumar et al., 1999; Silva et al., 2001).

Phytoconstituents showed their beneficial effects in alloxan/STZ induced diabetic animal model through modulating lipid/glucose metabolism are Soy isoflavones (genistein, orficat) from *Glycin max, Curcuma longa* (Howes et al., 2003; Mezei et al., 2003), Silymarin, silybin, silychristin, silidianin from *Silybum marianum* (Huseini et al., 2006), Tinosporine, cordifolide, tinosporide, cordifole, columbin from *Tinospora cordifolia, Tinospora crispa* (Hui et al., 2009; Kar et al., 2003; Van de Venter et al., 2008; Noor and Ashcroft, 1998), "S-methyl cysteine sulfoxide, S-allyl cysteine sulfoxide" from *Allium cepa* (Roman-Ramos et al., 1995; Kumari and Augusti, 2007).

Flavonoids that showed Antidiabetic efficiency through by quenching free radicals generated through oxidative stress are epigallocatechin-gallate, gallocatechin, epicatechin, (+) catechin, (-) epicatechin from tea leaves, pomegranate, *Satureja khuzestanica, Bauhinia forficta* (Hii and Howell, 1984; Waltner-Law et al., 2002; Vessel et al., 2003; Li et al., 2008), Kaempferol, isorhamnetin from *Ginkgo biloba* (Jellin et al., 1999), tribulusamides A and B, kaempferol-3-beta-D-(6'P-coumaryl)glucoside, kaempferol-3-glucoside from *Tribulus terrestris* (Cooper et al., 2003; Kirtikar and Basu, 1998), ferulic acid from *Curcuma longa* (Ohnishi et al., 2004) and catharanthine, vindoline, vindolinene, vinblastine, vincristine from *Catharanthus roseus, Vinca roesa* (Chattopadhyay, 1999; Jarald et al., 2008; Kar et al., 2003). Bioactive flavonoid glycosides, i.e., isostrictinin, strictinin, and pedunculagin isolated from guava plant, have been clinically proved as potential antidiabetic drugs, attributed to enhanced acceptability to INS (Chauhan et al., 2010).

8.8 MECHANISM OF ACTION OF PLANT EXTRACTS IN DIABETIC CELL LINE MODELS

Rat insulinoma cells (RINm5F) are well known and standardized cell line model for pancreatic β-cells used to determine the efficiency of herbal drugs and bioactive components on cytokine activation in vitro models. Pretreatment of water extract of *Scoparia dulcis* effectively altered the STZ-induced cytotoxicity and NO secretion in "rat insulinoma cell line" (Latha et al., 2004). The AE of *Agrimony eupatoria* (Rosaceae) provoked insulin secretion during pretreatment on BRIN-BD11 β-cell lines (Bnouham et al., 2006). Christinin-A, a saponin glycoside rich butanol fraction obtained from *Zizyphus spinachristi* leaves, potentially initiated glucose challenged insulin release in normal rats (Abdel-Zaher et al., 2005). After four weeks of treatment, the extract significantly enhanced serum insulin and cAMP content in the pancreas in diabetic rats (Bnouham et al., 2006).

Treatment of primary rat islets of Sprague-Dawley rats with different concentrations of berberine, i.e., 1, 3, 10, and 30 μmol/L, enhanced Glucose-stimulated insulin release in concentration-dependently. The component did not exhibit any toxicity on rat islets at 10 μmol/L concentrations. Further, the component significantly up-regulated the levels of mRNA and hepatic nuclear factor 4 alpha (HNF4α) in concentration-dependently. It also enhanced the secretion levels of Glucokinase in rat islets (Wang et al., 2008). In another study carried out by Ko et al. (2005), berberine at 50 μM concentration exhibited significantly sensitized insulin secretion in 3T3-L1 adipocytes, and it also enhanced glucose-activated insulin secretion and multiplication in Min6 cells through by enhanced insulin/insulin-like growth factor-1 signaling pathway. Dichloromethane – methanol (equal ratio) extract obtained from leaves and twigs of *Catharanthus roseus*, significantly increased the STZ induced lower levels of hepatic enzymes, i.e., malate dehydrogenase, succinate dehydrogenase, glycogen synthase and glucose 6-phosphate-dehydrogenase, at 500 mg/kg in a weekdays treatment. Further the extract normalized TBARS in diabetic rats indicating antioxidative mechanism (Singh et al., 2001).

The antidiabetic efficacy and its mechanism of water extract obtained from *Brachylaena elliptica* (a South African medicinal plant) was investigated using multiple activities in different cell line models, i.e., make use of glucose in HepG2/L6 cell lines, storage of triglycerides in 3T3-L1, INS-1 cell multiplication and glucose metabolism, NO secretion in RAW macrophages using standard procedures. The extract exhibited low toxicity and significantly enhanced glucose uptake in HepG2 cells, which was significantly good than the standard compound berberine. But the extract did not display any significant glucose metabolism activity in L6 cells and INS-1 cells, uptake glucose in muscle and triglycerides storage in 3T3-L1 cells. The effect of *Brachylaena elliptica* extract evaluated for the glucose metabolizing capacity on the HepG2 cells and L6 myotubes, INS-1 cells, in addition to INS-1 proliferation, triglyceride accumulation in 3T3-L1 and Nitric Oxide production in RAW macrophages and found effective in glucose metabolism in HepG2 cells whereas no significant effect on the glucose and triglycerides in the tested samples and summarized as the effect of the selected extract is not related to insulin-mimetic activity. The extract displayed a proliferative effect on INS-1 cells at 25 μg/ml when compared to the negative control. The plant also displayed dose-dependent reduction in NO production in RAW cells and also with weak significant inhibition of CYP3A4 activity (Sagbo et al., 2018). Glycosides "momordin, momordicine, charantin, momorcharaside A and B, momorcharin A and B" isolated from *Momordica charantia* (Chao et al., 2003), Epigallocatechin gallate (a major component of tea plant) enhanced streptozotocin-induced lower levels of INS and protected oxidative damages in diabetic rats. It further increased insulin secretion levels in rats fed with high fat diet for a period of 15 days. Further, these phytoconstituents regulated insulin and insulin-independent AMPK pathways regulate the translocation of GLUT4 vesicles to plasma-membrane resulting in stimulation of glucose uptake in skeletal muscle (Islam and Choi, 2007; Figure 8.1).

8.9 FUTURE PROSPECTS

A patent has claimed by Mitra et al. (2009), from, Himalaya Global Holdings Ltd, India on the experimental details of the extracts of roots from *Cipadessa baccifera* using possible mechanisms and proved improvement in the type-II diabetes.

Since the diabetes mellitus (DM) incidence appeared in middle aged and above, the management of glucose level in the bloodstream and became an important part of general health. Hence medications along with the physical exercise also boost. The severe conditions leading to many complications making sickness in patients and rushing to the diabetologists and the treatment became more complication. In addition, the people are not able to recognize the early stages of diabetes due to ignorance or lack of awareness. The treatment with traditional medicines were explained and usage of herbs in single or combinations were acting effectively which are proved by several researchers at extract or molecular level. The modern medicines also proving its effectiveness even it reached to implantation of beta cells or pluripotent stem cells which may not be affordable especially in developing countries. Hence the efforts towards development of herbal based treatments or research may be enhanced which have multifaceted benefits (e.g., Antioxidant, Antimicrobial, Hepatoprotective properties, etc.,) may influence on the diabetic complications and helps in normalization.

8.10 CONCLUSION

The medicinal plant extracts or its components which possess the activating or inhibitory roles as in the above mechanism act a key role in the discovery/development of effective therapeutic agents to treat Diabetes. Many diabetic formulations effectively control the Diabetes are yet to prove its molecular mechanism to develop safe medications. Extensive research is required to progress in the development of medications not only to maintain the sugar levels in the bloodstream and also to regain the juvenile potency of the organs affected. More over the traditional plants provide an additional advantage to decrease the metabolic oxidative stress, which might be the source for the development of a chain of free radicals and leading to degenerative disorders. Hence the researchers have to find out the effective medicine at an affordable cost using the high throughput methods. The present study is expected to pay attention among the scientific community to take necessary initiation towards the development of safe medicaments for diabetes for the betterment of humankind.

KEYWORDS

- **cascades of diabetic target molecules**
- **diabetes mellitus**
- **herbal medicine**
- **role of key enzymes in maintaining diabetes**
- ***Tinospora crispa***

REFERENCES

Abdel-Zaher, A. O., Salim, S. Y., Assaf, M. H., & Abdel-Hady, R. H., (2005). *Teucrium polium* antidiabetic activity and toxicity of *Zizyphus spina-christi* leaves. *J. Ethnopharmacol., 101*(1–3), 129–138.

Abesundara, K. J. M., Mastui, T., & Matsumoto, K., (2004). α-Glucosidase inhibitory activity of some Sri Lanka plant extracts, one of which, *Cassia auriculata*, exerts a strong antihyperglycemic effect in rats comparable to the therapeutic drug acarbose. *J. Agric. Food Chem., 52*, 2541–2545.

Adaramoye, O., Amanlou, M., Habibi- Rezaei, M., et al., (2012). Methanolic extract of African mistletoe (*Viscum album*) improves carbohydrate metabolism and hyperlipidemia in streptozotocin-induced diabetic rats. *Asian Pac. J. Trop. Med., 5*, 427–433.

Adisakwattana, S., Charoenlertkul, P., & Yibchok-Anun, S., (2009). α-Glucosidase inhibitory activity of cyanidin-3-galactoside and synergistic effect with acarbose. *J. Enz. Inhib. Med. Chem., 24*, 65–69.

Ajabnoor, M. A., (1990). Effect of aloes on blood glucose levels in normal and alloxan diabetic mice. *J. Ethnopharmacol., 28*, 215–220.

Akhtar, M. S., & Iqbal, J., (1991). Evaluation of the hypoglycaemic effect of *Achyranthes aspera* in normal and alloxan-diabetic rabbits. *J. Ethnopharmacol., 31*, 49–57.

Akhtar, M. S., (1992). Hypoglycemic activities of some indigenous medicinal plants traditionally used as antidiabetic drugs. *J. Pak. Med. Assoc., 42*, 271–277.

Akiba, Y., Kato, S., Katsube, K. L., et al., (2004). Transient receptor potential vanilloid subfamily 1 expressed in pancreatic islet *β* cells modulates insulin secretion in rats. *Biochem. Biophy. Res. Commu., 321*(1), 219–225.

Ali, H., Houghton, P. J., & Soumyanath, A., (2006). α-Amylase inhibitory activity of some Malaysian plants used to treat diabetes; with particular reference to *Phyllanthus amarus*. *J. Ethnopharmacol., 107*, 449–455.

Alipour, G., Dashti, S., & Hosseinzadeh, H., (2014). Review of pharmacological effects of *Myrtus communis* L. and its active constituents. *Phytother. Res., 28*, 1125–1136.

Anand, P., Murali, K. Y., Tandon, V., Murthy, P. S., & Chandra, R., (2010). Insulinotropic effect of cinnamaldehyde on transcriptional regulation of pyruvate kinase, phosphoenolpyruvate carboxykinase, and GLUT4 translocation in experimental diabetic rats. *Chem. Biol. Interact., 186*(1), 72–81.

Anand, P., Murali, Y. K., Tandon, V., Murthy, P. S., & Chandra, R., (2009). Insulinotropic effect of aqueous extract of *Brassica nigra* improves glucose homeostasis in streptozotocin induced diabetic rats. *Exp. Clin. Endocrinol. Diabetes, 117*(6), 251–256.

Anurakkun, N. J., Bhandari, M. R., & Kawabata, J., (2007). α-Glucosidase inhibitors from Devil tree (*Alstonia scholaris*). *Food Chem., 103*, 1319–1323.

Attele, A. S., Zhou, Y. P., Xie, J. T., et al., (2002). Antidiabetic effects of *Panax ginseng*. *Diabetes, 51*, 1851–1858.

Ayodhya, S., Kusum, S., & Anjali, S., (2010). Hypoglycaemic activity of different extracts of various herbal plants Singh. *Int. J. Ayurveda Res. Pharm., 1*(1), 212–224.

Azemi, M. E., Namjoyan, F., Khodayar, M. J., Ahmadpour, F., Padok, A. D., & Panahi, M., (2012). The antioxidant capacity and antidiabetic effect of *Boswellia serrata* triana & planch aqueous extract in fertile female diabetic rats and the possible effects on reproduction and histological changes in the liver and kidneys. *Jundishapur J. Nat. Pharm. Prod, 7*(4), 168–175.

Bangar, O. P., Jarald, E. E., Asghar, S., & Ahmad, S., (2009). Antidiabetic activity of a polyherbal formulation (Karnim plus). *Inter. J. Gree. Pharm.*, 211–214.

Bartolome, A. P., Villase˜nor, I. M., & Yang, W. C., (2013). *Bidens pilosa* L. (*Asteraceae*): Botanical properties, traditional uses, phytochemistry, and pharmacology. *Eviden. Based Comple. Altern. Med.*, 51. Article ID 340215.

Basnet, P., Kadota, S., & Terashima, S., (1993). Two new 2-arylbenzofuran derivatives from hypoglycaemic activity-bearing fractions of *Morus insignis*. *Chem. Pharm. Bull. (Tokyo), 41*, 1238–1243.

Best, L., Elliott, A. C., & Brown, P. D., (2008). Curcumin induces electrical activity in rat pancreatic β-cells by activating the volume-regulated anion channel. *Biochem. Pharmacol., 73*(11), 1768–1775.

Bhakshu, L. M., Ratnam, K. V., Suresh, B. M. V., & Raju, R. R. V., (2017). Evaluation of *Dendrophthoe falcata* leaf extracts for antioxidant, antidiabetic and anticholenesterase activities using *in vitro* assays. In: Ramachandra, B., (ed.), Proceedings of the National Seminar on *Modern Trends in Chemistry Research* (pp. 93–105). Roshan Publications, Vishakapatnam A.P., India.

Bhandari, M. R., Anurakkun, N. J., Hong, G., & Kawabata, J., (2008). α-Glucosidase and α-amylase inhibitory activities of Nepalese medicinal herb pakhanbhed (*Bergenia ciliata* haw.) *Food Chem., 106*, 247–252.

Bnouham, M., Ziyyat, A., Mekhfi, H., Tahri, A., & Legssyer, A., (2006). Medicinal plants with potential antidiabetic activity-a review of ten years of herbal medicine research (1990–2000). *Int. J. Diabetes Metab, 14*, 1–25.

Borquez, J., Loyola, L. A., Morales, G., et al., (2007). Azorellane diterpenoids from *Laretia acaulis* inhibit nuclear factor-kappa B activity. *Phytother. Res., 21*, 1082–1086.

Broadhurst, C. L., Polansky, M. M., & Anderson, R. A., (2000). Insulin-like biological activity of culinary and medicinal plant aqueous extracts *in vitro*. *J. Agric. Food Chem., 48*, 849–852.

Broca, C., Gross, R., Petit, P., Sauvaire, Y., Manteghetti, M., Tournier, M., et al., (1999). 4-hydroxyisoleucine: Experimental evidence of its insulinotropic and antidiabetic properties. *Ame. J. Physiol., 277*(4 Pt 1), E617–E623.

Brownlee, M., (1996). Advanced protein glycosylation in diabetes in diabetes and ageing. *Ann. Rev. Med., 46*, 223–234.

Cai, E. P., & Lin, J. K., (2009). Epigallocatechin gallate (EGCG) and rutin suppress the glucotoxicity through activating IRS2 and AMPK signaling in rat pancreatic *β* cells. *J. Agri. Food Chem., 57*(20), 9817–9827.

Chang, C. L. T., Chang, S. L., Lee, Y. M., et al., (2007). Cytopiloyne, a polyacetylenic glucoside, prevents type 1 diabetes in nonobese diabetic mice. *J. Immunol., 178*(11), 6984–6993.

Chang, C. L. T., Liu, H. Y., Kuo, T. F., et al., (2013). Antidiabetic effect and mode of action of cytopiloyne." *Eviden. Based Comple. Altern. Med.*, 13. Article ID 685642.

Chao, C. Y., & Huang, C. J., (2003). Bitter gourd (*Momordica charantia*) extract activates peroxisome proliferator-activated receptors and upregulates the expression of the acyl CoA oxidase gene in H4IIEC3 hepatoma cells. *J. Biomed. Sci., 10*, 782–791.

Chattopadhyay, R. R., (1999). A comparative evaluation of some blood sugar lowering agents of plant origin. *J. Ethnopharmacol., 67*, 367–372.

Chauhan, A., Sharma, P. K., Srivastava, P., Kumar, N., & Duehe, R., (2010). Plants having potential antidiabetic activity: A review. *Der. Pharm. Lett., 2*(3), 369–387.

Chen, C. C., Hsu, C. Y., Chen, C. Y., & Liu, H. K., (2008). *Fructus corni* suppresses hepatic gluconeogenesis related gene transcription, enhances glucose responsiveness of pancreatic beta-cells, and prevents toxin induced beta-cell death. *J. Ethnopharmacol., 117*(3), 483–490.

Chen, F., Chen, Y., Kang, X., Zhou, Z., Zhang, Z., & Liu, D., (2012). Anti-apoptotic function and mechanism of ginseng saponins in Rattus pancreatic beta-cells. *Biolo. & Pharmaceu. Bull., 35*(9), 1568–1573.

Chengfu, S., Chuanbin, Y., Man, G., Yingying, K., Peipei, Y., Xiaolan, W., Min, L., et al., (2019). Antidiabetic activity and potential mechanism of amentoflavone in diabetic mice. *Molecules, 24*(11), 2184.

Cherian, S., Sheela, C. G., & Augusti, K. T., (1995). Insulin sparing action of leucopelergonidin derivative isolated from *Ficus bengalensis* Linn. *Indian J. Exp. Biol., 33*, 608–611.

Cho, J. M., Chang, S. Y., Kim, D. B., Needs, P. W., Jo, Y. H., & Kim, M. J., (2012). Effects of physiological quercetin metabolites on interleukin-1β-induced inducible NOS expression. *J. Nutri. Biochem., 23*(11), 1394–1402.

Choi, M. S., Jung, U. J., Yeo, J., Kim, M. J., & Lee, M. K., (2008). Genistein and daidzein prevent diabetes onset by elevating insulin level and altering hepatic gluconeogenic and lipogenic enzyme activities in non-obese diabetic (NOD) mice. *Dia. Metabol. Res. Rev., 24*(1), 74–81.

Chuengsamarn, S., Rattanamongkolgul, S., Luechapudiporn, R., Phisalaphong, C., & Jirawatnotai, S., (2012). Curcumin extract for prevention of type 2 diabetes. *Diabetes Care, 35*(11), 2121–2127.

Cooper, E. J., Hudson, A. L., Parker, C. A., et al., (2003). Effects of the beta-carbolines, harmane and pinoline, on insulin secretion from isolated human islets of Langerhans. *Eur. J. Pharmacol., 482*, 189–196.

Daisy, P., Eliza, J., & Mohamed, F. K. A. M., (2009). A novel dihydroxy gymnemic triacetate isolated from *Gymnema sylvestre* possessing normoglycemic and hypolipidemic activity on STZ induced diabetic rats. *J. Ethnopharmacol., 126*(2), 339–344.

Dallak, M., Al-Khateeb, M., Abbas, M., Elessa, R., Al-Hashem, F., Bashir, N., et al., (2009a). *In vivo*, acute, normo-hypoglycemic, antihyperglycemic, insulinotropic actions of orally administered ethanol extract of *Citrullus colocynthis* (L.) schrab pulp. *Ame. J. Biochem. Biotechnol., 5*(3), 119–126.

Dong, H., Wang, N., Zhao, L., & Lu, F., (2012). Berberine in the treatment of type 2 diabetes mellitus: A systemic review and meta-analysis. *Eviden. Based Comple. Alter. Med.* Article ID 591654, 12.

Du, Z. Y., Liu, R. R., Shao, W. Y., Mao, X. P., Ma, L., Gu, L. Q., et al., (2006). α-Glucosidase inhibition of natural curcuminoids and curcumin analogs. *Eur. E. Med. Chem., 14*, 213–218.

Eidi, A., Eidi, M., & Esmaeili, E., (2006). Antidiabetic effect of garlic (*Allium sativum* L.) in normal and streptozotocin-induced diabetic rats. *Phytomedicine., 13*(9, 10), 624–629.

Elgawish, A., Glomb, M., Friendlander, M., & Monnier, V. M., (1999). Involvement of hydrogen peroxide in collagen cross-linking by high glucose *in vitro* and *in vivo*. *J. Biol. Chem., 271*, 12964–12971.

Elliott, J., Scarpello, J. H. B., & Morgan, N. G., (2002). Differential effects of genistein on apoptosis induced by fluoride and pertussis toxin in human and rat pancreatic islets and RINm5F cells. *J. Endocrin., 172*(1), 137–143.

El-Moselhy, M. A., Taye, A., Sharkawi, S. S., El-Sisi, S. F. I., & Ahmed, A. F., (2011). The antihyperglycemic effect of curcumin in high fat diet fed rats. Role of TNF-alpha and free fatty acids. *Food Chem. Toxicol., 49*(5), 1129–1140.

Eno, A. E., Ofem, O. E., Nku, C. O., et al., (2008). Stimulation of insulin secretion by *Viscum album* (mistletoe) leaf extract in streptozotocin-induced diabetic rats. *Afr. J. Med. Sci., 37*, 141–147.
Esmaeili, M. A., & Yazdanparast, R., (2004). Hypoglycaemic effect of *Teucrium polium*: Studies with rat pancreatic islets. *J. Ethnopharmacol., 95*(1), 27–30.
Fabricant, D. S., & Farnsworth, N. R., (2001). The value of plants used in traditional medicine for drug discovery. *Environ. Heal. Perspect.*, (Suppl. 109), 69.
Fararh, K. M., Atoji, Y., Shimizu, Y., & Takewaki, T., (2002). Insulinotropic properties of *Nigella sativa* oil in streptozotocin plus nicotinamide diabetic hamster. *Res. Vet. Sci., 73*(3), 279–282.
Fararh, K., Atoji, Y., Shimizu, Y., & Takewaki, T., (2003). Isulinotropic properties of *Nigella sativa* oil in Streptozotocin plus nicotinamide diabetic hamster. *Res. in Veter. Sci., 73*, 279–282.
Frankish, N., De Sousa, M. F., Mills, C., & Sheridan, H., (2010). Enhancement of insulin release from the beta-cell line INS-1 by an ethanolic extract of *Bauhinia variegata* and its major constituent roseoside. *Planta Med., 76*(10), 995–997.
Fu, Y., & Koo, W. W. L., (2006). EGCG protects HT-22 cells against glutamate-induced oxidative stress. *Neurotoxi. Res., 10*(1), 23–29.
Fu, Z., Zhang, W., Zhen, W., et al., (2010). Genistein induces pancreatic beta-cell proliferation through activation of multiple signaling pathways and prevents insulin-deficient diabetes in mice. *Endocrinology, 151*(7), 3026–3037.
Fuentes, N. L., Sagua, H., Morales, G., et al., (2005). Experimental antihyperglycemic effect of diterpenoids of *Laretia acaulis* and *Azorella compacta* Phil (Umbelliferae) in rats. *Phytother. Res., 19*, 713–716.
Gao, H., Huang, Y. N., Gao, B., Li, P., Inagaki, C., & Kawabata, J., (2008). Inhibitory effect on α-glucosidase by *Adhatoda vasica* nees. *Food Chem., 108*, 965–972.
Gao, H., Huang, Y. N., Xu, P. Y., & Kawabata, J., (2007). Inhibitory effect on α-glucosidase by the fruits of *Terminalia chebula* retz. *Food. Chem., 105*, 628–634.
Gilbert, E. R., & Liu, D., (2013). Antidiabetic functions of soy isoflavone genistein: Mechanisms underlying its effects on pancreatic β-cell function. *Food Funct., 4*(2), 200–212.
Godard, M. P., Ewing, B. A., Pischel, I., et al., (2010). Acute blood glucose lowering effects and long-term safety of opun DiaTM supplementation in prediabetic males and females. *J. Ethnopharmacol., 130*, 631–634.
González-Tejero, M. R., Casares-Porcel, M., Sánchez-Rojas, C. P., et al., (2008). Medicinal plants in the Mediterranean area: Synthesis of the results of the project Rubia. *J. Ethnopharmacol., 116*, 341–357.
Gould, G. W., Derechin, V., James, D. E., Tordjman, K., Ahern, S., Gibbs, E. M., Lienhard, G. E., & Mueckler, M., (1989). Insulin stimulated translocation of the HepG2/erythrocyte type glucose transporter expressed in 3T3 L1 adipocytes. *J. Biol. Chem., 264*, 2180–2184.
Gourigari, T. R., Lepakshi, B. M., Kamsala, R. V., & Raju, R. R. V., (2016). Evaluation of anticholinergic, antidiabetic and antioxidant activity of leaf extracts of *Ochna obtusata* DC using *in vitro* assays. *Inter. J. Pharm. Pharmaceut. Sci., 8*(6), 82–87.
Gram, D. X., Ahr′en, B., Nagy, I., et al., (2007). Capsaicin-sensitive sensory fibers in the islets of Langerhans contribute to defective insulin secretion in Zucker diabetic rat, an animal model for some aspects of human type 2 diabetes. *Euro. J. Neurosci., 25*(1), 213–223.

Gray, A. M., & Flatt, P. R., (1997). Pancreatic and extra-pancreatic effects of the traditional antidiabetic plant, *Medicago sativa* (Lucerne). *Br. J. Nut., 78*, 325–334.

Gray, A. M., & Flatt, P. R., (1998). Insulin-releasing and insulin-like activity of *Agaricus* campestris (mushroom). *J. Endocrinol., 157*, 259–266.

Grover, J. K., Yadav, S., & Vats, V., (2002). Medicinal plants of India with antidiabetic potential. *J. Ethnopharmacol., 81*(1), 81–100.

Gulubova, R., & Boiadzhiev, T. S., (1975). Morphological changes in the endocrine pancreas of the rabbit after the administration of a *Morus alba* extract. *Eksp. Med. Morfol., 14*, 166–171.

Gupta, A., Gupta, R., & Lal, B., (2001). Effect of *Trigonella foenum-graecum* (fenugreek) seeds on glycaemic control and insulin resistance in type 2 diabetes mellitus: A double blind placebo controlled study. *J. Asso. Physici. India., 49*, 1057–1061.

Gupta, D., Raju, J., & Baquer, N. Z., (1999). Modulation of some gluconeogenic enzyme activities in diabetic rat liver and kidney: Effect of antidiabetic compounds. *Indian J. Expt. Biol., 37*, 196–199.

Gupta, R., Bajpai, K. G., Johri, S., & Saxena, A. M., (2007). An overview of Indian novel traditional medicinal plants with antidiabetic potentials. *Afr. J. Tradit. Complement. Altern. Med., 5*(1), 1–17.

Gupta, S., Sharma, S. B., Usha, R. S., Bansal, S. K., & Prabhu, K. M., (2010). Elucidation of mechanism of action of *Cassia auriculata* leaf extract for its antidiabetic activity in streptozotocin induced diabetic rats. *J. Med. Food, 13*(3), 528–534.

Haeri, M. R., Izaddoost, M., Ardekani, M. R., Nobar, M. R., & White, K. N., (2009). The effect of fenugreek 4-hydroxyisoleucine on liver function biomarkers and glucose in diabetic and fructose-fed rats. *Phytother. Res., 23*(1), 61–64.

Han, M. H., (2003). Epigallocatechin gallate, a constituent of green tea, suppresses cytokine-induced pancreatic β-cell damage. *Exper. Molec. Med., 35*(2), 136–139.

Hannan, J. M., Marcnah, L., Au, L., et al., (2006). *Ocimum sanctum* leaf extracts stimulate insulin secretion from perfused pancreas, isolated islets and clonal pancreatic β-cells. *J. Endocrinol., 189*, 127–136.

Hannan, J. M., Marenah, L., Ali, L., Rokeya, B., Flatt, P. R., & Abdel-Wahab, Y. H., (2007). Insulin secretory actions of extracts of *Asparagus racemosus* root in perfused pancreas, isolated islets and clonal pancreatic beta-cells. *J. Endocrinol., 192*(1), 159–168.

Hayamizu, K., (2003). Effect of *Garcinia cambogia* extract on serum leptin and insulin in mice. *Fitoterapia., 74*, 267–273.

Hii, C. S. T., & Howell, S. L., (1985). Effects of flavonoids on insulin secretion and 45Ca2+ handling in rat islets of Langerhans. *J. Endocri., 107*(1), 1–8.

Hou, C. C., Lin, S. J., Cheng, J. T., et al., (2003). Hypoglycemic dimeric guianolides and a lignan glycoside from *Lactuca indica*. *J. Nat. Prod., 66*, 625–629.

Howes, J. B., Tran, D., Brillante, D., et al., (2003). Effects of dietary supplementation with isoflavones from red clover on ambulatory blood pressure and endothelial function in postmenopausal type 2 diabetes. *Diabetes Obes. Metab., 5*, 325–332.

Huang, T. H., He, L., Qin, Q., et al., (2008). Salacia oblonga root decreases cardiac hypertrophy in Zucker diabetic fatty rats: Inhibition of cardiac expression of angiotensin II type 1 receptor. *Diabetes Obes. Metab., 10*, 574–585.

Hui, H., Tang, G., & Go, V. L., (2009). Hypoglycemic herbs and their action mechanisms. *Chin. Med., 12*, 4–11.

Huseini, H. F., Larijani, B., Heshmat, R., et al., (2006). The efficacy of *Silybum marianum* (L.) gaertn (silymarin) in the treatment of type 2 diabetes: A randomized, double-blind, placebo-controlled clinical trial. *Phytother. Res., 20*, 1036–1039.

Hussain, H. E. M. A., (2002). Hypoglycemic, hypolipidemic and antioxidant properties of combination of Curcumin from *Curcuma longa*, Linn, and partially purified product from *Abroma augusta*, Linn. in streptozotocin induced diabetes. *Indian J. Clin. Biochem., 17*(2), 33–43.

Hussain, Z., Waheed, A., Qurshi, R. A., et al., (2004). The effect of medicinal plants of Islamabad and Murree region of Pakistan on insulin secretion from INS-1 cells. *Phytother. Res., 18*, 73–77.

Islam, M. S., & Choi, H., (2007). Green tea, antidiabetic or diabetogenic: A dose response study. *Biofactors., 29*(1), 45–53.

Islam, M. S., & Choi, H., (2008). Dietary red chilli (*Capsicum frutescens* L.) is insulinotropic rather than hypoglycemic in type 2 diabetes model of rats. *Phytother. Res., 22*(8), 1025–1029.

Jarald, E. E., Sheeja, E., Motwani, S., et al., (2008). Comparative evaluation of antihyperglycaemic and hypoglycaemic activity of various parts of *Catharanthus roseus* Linn. *Res. J. Med. Plant., 2*, 10–15.

Jellin, J. M., Batz, F., & Hitchens, K., (1999). *Pharmacist's Letter/Prescriber's Letter Natural Medicines Comprehensive Database.* Stockton, CA: Therapeutic Research Faculty.

Jeppesen, P. B., Gregersen, S., Alstrup, K. K., & Hermansen, K., (2002). Stevioside induces antihyperglycaemic, insulinotropic and glucagonostatic effects *in vivo*: Studies in the diabetic Goto-Kakizaki (GK) rats. *Phytomedicine, 9*(1), 9–14.

Jeppesen, P. B., Gregersen, S., Poulsen, C. R., & Hermansen, K., (2000). Stevioside acts directly on pancreatic beta cells to secrete insulin: Actions independent of cyclic adenosine monophosphate and adenosine triphosphate-sensitive K+-channel activity. *Metabolism, 49*(2), 208–214.

Joshi, C. S., Priya, E. S., & Venkataraman, (2007). Hypoglycaemic antilipidperoxidative effects of a polyherbal formulation, Diakyur in experimental animal models. *J. Health Sci., 53*(6), 734–739.

Jung, M., Park, M., Lee, H. C., Kang, Y. H., Kang, E. S., & Kim, S. K., (2006). Antidiabetic agents from medicinal plants. *Curr. Med. Chem., 13*(10), 1203–1218.

Jung, U. J., Lee, M. K., Jeong, K. S., et al., (2004). The hypoglycemic effects of hesperidin and naringin are partly mediated by hepatic glucose regulating enzymes in C57BL/KsJ-db/db mice. *J. Nutri., 134*, 2499–2503.

Kaczmar, T., (1998). Herbal support for diabetes management. *Clin. Nutri. Insights, 6*(8), 1–4.

Kamalakkannan, N., & Prince, P. S. M., (2005). The effect of *Aegle marmelos* fruit extract in streptozotocin diabetes: A histopathological study. *J. Herb. Pharmacol., 5*, 87–96.

Kang, W., Zhang, L., & Song, Y., (2009). Alpha-glucosidase inhibitors from *Rubia cordifolia. Zhongguo Zhong Yao Za Zhi., 34*(9), 1104–1107.

Kannur, D. M., Hukkeri, V. I., & Akki, K. S., (2006). Antidiabetic activity of *Caesalpinia bonducella* seed extracts in rats. *Fitoterapia, 77*(7, 8), 546–549.

Kar, A., Choudhary, B. K., & Bandyopadhyay, N. G., (2003). Comparative evaluation of hypoglycaemie activity of some Indian medicinal plants in alloxan diabetic rats. *J. Ethnopharmacol., 84*, 105–108.

Kawakami, M., Hirayama, A., Tsuchiya, K., Ohgawara, H., Nakamura, M., & Umezawa, K., (2010). Promotion of β-cell differentiation by the alkaloid conophylline in porcine pancreatic endocrine cells. *Biomed. and Pharmacothe., 64*(3), 226–231.

Khan, M. A., Tania, M., Liu, R., et al., (2003). *Hericium erinaceus*: An edible mushroom with medicinal values. *J. Complement. Integr. Med., 10*(1), 1–6.

Khosla, P., Gupta, D. D., & Nagpal, R. K., (1995). Effect of *Trigonella foenum-graecum* (fenugreek) on blood glucose in normal and diabetic rats. *Indian J. Physiol. Pharmacol., 39*, 173, 174.

Khosla, P., Gupta, D. D., & Nagpal, R. K., (1995). Effect of *Trigonella foenum-graecum* (fenugreek) on serum lipids in normal and diabetic rats. *Indian J. Physiol. Pharmacol., 27*, 89–93.

Kim, K., & Kim, H. Y., (2008). Korean red ginseng stimulates insulin release from isolated rat pancreatic islets. *J. Ethnopharmacol., 120*(2), 190–195.

Kirtikar, K. R., & Basu, B. D., (1991). *Indian Medicinal Plants* (Vol. 1). Bishen Singh Mahendra Pal Singh.

Kluwer, W. C., (1999). *The Review of Natural Products by Facts and Comparisons*. St Louis, MO: Facts & Comparisons®.

Ko, B. S., Choi, S. B., Park, S. K., Jang, J. S., Kim, Y. E., & Park, S., (2005). Insulin sensitizing and insulinotropic action of berberine from *Cortidis rhizoma. Biol. Pharm. Bull., 28*(8), 1431–1437.

Ko, B. S., Jang, J. S., Hong, S. M., Sung, S. R., Lee, J. E., Lee, M. Y., et al., (2007). Changes in components, glycyrrhizin and glycyrrhetinic acid, in raw *Glycyrrhiza uralensis* fisch, modify insulin sensitizing and insulinotropic actions. *Biosci. Biotechnol. Biochem., 71*(6), 1452–1461.

Kodera, S., Yamada, Y., Yamamoto, Y., et al., (2009). Administration of conophylline and betacellulin-*δ*4 increases the *β*-cell mass in neonatal streptozotocin-treated rats. *Endocrine Journal, 56*(6), 799–806.

Kook, S., Gun-Hee, K., & Choi, K., (2009). The antidiabetic effect of onion and garlic in experimental diabetic rats: Meta-analysis. *J. Med. Food., 12*, 552–560.

Krishnakumar, K., Augusti, K. T., & Vijavammal, P. L., (1999). Hypoglycaemic and anti-oxidant activity of *Salacia oblonga* wall extract in streptozotocin induced diabetic rats. *Indian J. Physiol. Pharmacol., 43*, 510–514.

Kshirsagar, R. P., Darade, S. S., & Takale, V., (2010). Effect of *Alangium salvifolium* (Alangiaceae) on dexamethasone induced insulin resistance in rats. *J. Pharm. Res., 3*(11), 2714–2716.

Kumar, R. V., & Augusti, K. T., (1994). Insulin sparing action of leucocyanidin derivative isolated from *Ficus bengalensis* Linn. *Indian J. Biochem. Biophys., 31*, 73–76.

Kumari, K., & Augusti, K. T., (2007). Lipid lowering effect of S-methyl cysteine sulfoxide from *Allium cepa* Linn. in high cholesterol diet fed rats. *J. Ethnopharmacol., 109*, 367–371.

Lai, B. Y., Chen, T. Y., Huang, S. H., et al., (2015). *Bidens pilosa* formulation improves blood homeostasis and *β*-cell function in men: A pilot study. *Eviden. Based Comple. Altern. Med.*, 1–5. Article ID 832314.

Landis-Piwowar, K. R., Huo, C., Chen, D., et al., (2007). A novel prodrug of the green tea polyphenol (–)-epigallocatechin-3-gallate as a potential anticancer agent. *Cancer Research, 67*(9), 4303–4310.

Latha, M., Pari, L., Ramkumar, K. M., Rajaguru, P., Suresh, T., Dhanabal, T., Sitasawad, S., & Bhonde, R., (2009). Antidiabetic effects of scoparic acid D isolated from *Scoparia dulcis* in rats with streptozotocin-induced diabetes. *Natur. Prod. Res., 23*(16), 1528–1540.

Latha, M., Pari, L., Sitasawad, S., & Bhonde, R., (2004). Insulin secretagogue activity and cytoprotective role of the traditional antidiabetic plant *Scoparia dulcis* (Sweet Broomweed), *Life Sciences, 75*(16), 2003–2014.

Lee, Y. S., Kim, W. S., Kim, K. H., et al., (2006). Berberine, a natural plant product, activates AMP-activated protein kinase with beneficial metabolic effects in diabetic and insulin-resistant states. *Diabetes, 55*(8), 2256–2264.

Lemus, I., Garcia, R., Dclvillar, E., et al., (1999). Hypoglycemic activity of four plants used in Chilean popular medicine. *Phytother. Res., 13*, 91–94.

Leng, S. H., Lu, F. E., & Xu, L. J., (2004). Therapeutic effects of berberine in impaired glucose tolerance rats and its influence on insulin secretion. *Acta Pharmacologica Sinica., 25*(4), 496–502.

Li, Y., Qi, Y., Huang, T. H., et al., (2008). Pomegranate flower: A unique traditional antidiabetic medicine with dual PPAR-alpha/-gamma activator properties. *Diabetes Obes. Metab., 10*, 10–17.

Liang, B., Guo, Z., Xie, F., et al., (2013). Antihyperglycemic and antihyperlipidemic activities of aqueous extract of *Hericium erinaceus* in experimental diabetic rats. *BMC Complement. Altern. Med., 13,* 253.

Liu, D., Zhen, W., Yang, Z., Carter, J. D., Si, H., & Reynolds, K. A., (2006). Genistein acutely stimulates insulin secretion in pancreatic beta-cells through a cAMP-dependent protein kinase pathway. *Diabetes, 55*(4), 1043–1050.

Luo, J. G., Ma, L., & Kong, L. Y., (2008). New triterpenoid saponins with strong α- glucosidase inhibitory activity from the roots of *Gypsophila oldhamiana. Bio-org. Med. Chem., 16*, 2912–2920.

Ma, S. W., Benzie, I. F., Chu, T. T., et al., (2008). Effect of *Panax ginseng* supplementation on biomarkers of glucose tolerance, antioxidant status and oxidative stress in type 2 diabetic subjects: Results of a placebo-controlled human intervention trial. *Diabetes Obes. Metab., 10*, 1125–1127.

Malviya, N., Jain, S., & Malviya, S., (2010). Antidiabetic potential of medicinal plants. *Acta Pol. Pharm., 67*(2), 113–118.

Mandlik, R. V., Desai, S. K., & Naik, S. R., (2008). Antidiabetic activity of polyherbal formulation (DRF/AY/5001). *Indian J. Exp. Biol., 46*, 599–606.

Mannan, N., Boucher, B. J., & Evans, S. J. W., (2000). Increased waist size and weight in relation to consumption of *Areca catechu* (betel-nut); a risk factor for increased glycaemia in Asians in East London. *Bri. J. Nutr., 83*, 267–275.

Manohar, V., Talpur, N. A., Echard, B. W., et al., (2002). Effects of a water-soluble extract of maitake mushroom on circulating glucose/insulin concentrations in KK mice. *Diabetes Obes. Metab., 4*, 43–48.

Maroo, J., Vasu, V. T., Aalinkeel, R., & Gupta, S., (2002). Glucose lowering effect of aqueous extract of *Enicostemma littorale* Blume in diabetes: A possible mechanism of action. *J. Ethnopharmacol., 81*(3), 317–320.

Mathews, J. N., Flatt, P. R., & Abdel-Wahab, Y. H., (2006). *Asparagus adscendens* (Shweta musali) stimulates insulin secretion, insulin action and inhibits starch digestion. *Bri. J. Nutrt., 95*, 576–581.

Mbaze, L. M., Poumale, H. M., Wansi, J. D., Lado, J. A., Khan, S. N., Iqbal, M. C., et al., (2007). α-Glucosidase inhibitory pentacyclic triterpenes from the stem bark of *Fagara tessmannii* (Rutaceae). *Phytochem., 68*, 591–595.

Meghana, K., Sanjeev, G., & Ramesh, B., (2007). Curcumin prevents streptozotocin-induced islet damage by scavenging free radicals: A prophylactic and protective role. *Euro. J. Pharmacol., 577*(1–3), 183–191.

Meng, B., Li, J., & Cao, H., (2013). Antioxidant and anti-inflammatory activities of curcumin on diabetes mellitus and its complications. *Curr. Pharmaceu. Desig., 19*(11), 2101–2113.

Meng, Z. Y., Zhang, J. Y., Xu, S. X., & Sugahara, K., (1999). Steroidal saponins from *Anemarrhena asphodeloides* and their effects on superoxide generation. *Planta Med., 65*, 661–663.

Mezei, O., Banz, W. J., Steger, R. W., et al., (2003). Soy isoflavones exert hypoglycemic and hypolipidemic effects through the PPAR pathways in obese Zucker rats and murine RAW 264.7 cells. *J. Nutr., 133*, 1238–1243.

Mirghazanfari, S. M., Keshavarz, M., Nabavizadeh, F., Soltani, N., & Kamalinejad, M., (2010). The effect of *Teucrium polium* L. extracts on insulin release from in situ isolated perfused rat pancreas in a newly modified isolation method: The role of Ca^{2+} and K^+ channels. *Iran Biomed. J., 14*(4), 178–185.

Mitra, S. K., Saxena, E., & Babu, U. V., (2009). *Novel Antidiabetic Herbal Composition, Method for Preparing the Same and Use Thereof.* Patent, US20090238900A1, United States.

Modak, M., Dixit, P., Londhe, J., Ghaskadbi, S., & Paul, A. D., T., (2007). Indian herbs and herbal drugs used for the treatment of diabetes. *J. Clin. Biochem. Nutr., 40*(3), 163–173.

Mohan, V., (1998). Evaluation of diabecon (D-400) as an antidiabetic agent- a double-blind placebo-controlled trial in NIDDM patients with secondary failure to oral drugs. *Ind. J. Clin. Pract., 9*(8), 18.

Moon, Y. H., Nam, S. H., Kang, J., et al., (2007). Enzymatic synthesis and characterization of arbutin glucosides using glucansucrase from *Leuconostoc mesenteroides*. *Appl. Microbiol. Biotechnol., 27*, 559–567.

Mustafa, S. S. S., Eid, N. I., Jafri, S. A., El-Latif, H. A. A., & Ahmed, H. M. S., (2007). Insulinotropic effect of aqueous ginger extract and aqueous garlic extract on the isolated perfused pancreas of streptozotocin induced diabetic rats. *Pak. J. Zool., 39*(5), 279–284.

Mutalik, S., Sulochana, B., Devi, U. P., & Udupu, N., (2003). Preliminary studies on acute and sub-acute toxicity of an antidiabetic herbal preparation, Dianex. *Ind. J. Exp. Biol., 41*, 316–320.

Na, L. X., Zhang, Y. L., Li, Y., et al., (2011). Curcumin improves insulin resistance in skeletal muscle of rats. *Nutri. Metabol. Cardio. Diseas., 21*(7), 526–533.

Naisheng, B., Kan, H., Roller, M., et al., (2008). Active compounds from *Lagerstroemia speciosa*, insulin like glucose uptake-stimulatory/inhibitory and adipocyte differentiation-inhibitory activities in 3T3-L1 cells. *J. Agric. Food Chem., 56,* 11668–11674.

Nandini, C. D., Sambaiah, K., & Salimath, P. V., (2003). Dietary fibres ameliorate decreased synthesis of heparan sulphate in streptozotocin induced diabetic rats. *J. Nutri. Biochem., 14*, 203–210.

Ngueyem, T. A., Brusotti, G., Caccialanza, G., et al., (2009). The genus *Bridelia*: A phytochemical and ethnopharmacological review. *J. Ethnopharmacol., 124*, 339–349.

Ngwoke, K. G., Akwagbulam, A. G., Erhirhie, E. O., Ajaghaku, D. L., Okoye, F., & Esimone, C. O., (2018). Antioxidant, anti-inflammatory, analgesic properties, and phytochemical characterization of stem bark extract and fractions of *Anthocleista nobilis*. *Pharmacog. Res., 10*(1), 81–87.

Noor, H., & Ashcroft, S. J. H., (1998). Insulinotropic activity of *Tinospora crispa* extract: Effect on β-cell Ca2+ handling. *Phytother. Res., 12*, 98–102.

Ogata, T., Li, L., Yamada, S., et al., (2004). Promotion of *β*-cell differentiation by conophylline in fetal and neonatal rat pancreas. *Diabetes, 53*(10), 2596–2602.

Ogbonnia, S. O., Mbaka, G. O., Adekunle, A., Anyika, E. N., Gbolade, O. E., & Nwakakwa, N., (2010). Effect of a poly-herbal formulation, okudiabet, on alloxan- induced diabetic rats. *Agri. and bio. J. North America, 1*(2), 139–145.

Ohnishi, M., Matuo, T., Tsuno, T., et al., (2004). Antioxidant activity and hypoglycemic effect of ferulic acid in STZ-induced diabetic mice and KK-Ay mice. *Biofactors., 21*, 315–319.

Ohno, T., Kato, N., Ishii, C., et al., (1993). Genistein augments cyclic adenosine 3–5-monophosphate (cAMP) accumulation and insulin release in min6 cells. *Endocri. Res., 19*(4), 273–285.

Pari, L., & Amarnath, S. M., (2004). Antidiabetic activity of *Boerhaavia diffusa* L.: Effect on hepatic key enzymes in experimental diabetes. *J Ethnopharmacol., 91*(1), 109–113.

Pari, L., & Latha, M., (2006). Antihyperlipidemic effect of *Scoparia dulcis* (Sweet Broomweed) in streptozotocin diabetic rats. *J. Med Food, 9*(1), 102–107.

Pari, L., & Saravanan, R., (2004). Antidiabetic effect of diasulin, a herbal drug, on blood glucose, plasma insulin and hepatic enzymes of glucose metabolism in hyperglycaemic rats. *Diabetes Obes. Metab., 6*(4), 286–292.

Park, S., Ahn, I. S., Kwon, D. Y., Ko, B. S., & Jun, W. K., (2008). Ginsenosides Rb1 and Rg1 suppress triglyceride accumulation in 3T3-L1 adipocytes and enhance *β*-cell insulin secretion and viability in min6 cells via PKA-dependent pathways. *Biosci. Biotech. Biochem., 72*(11), 2815–2823.

Park, S., Jang, J. S., & Hong, S. M., (2007). Long-term consumption of caffeine improves glucose homeostasis by enhancing insulinotropic action through islet insulin/insulin-like growth factor-1 signaling in diabetic rats. *Metabolism, 56*(5), 599–607.

Patel, A. K., & Manigauha, A., (2018). Antioxidant and antidiabetic activity of isolated flavonoids from *Alangium salvifolium* leaves extracts. *Inter. J. Gre. Pharm., 12*(2), 82–90.

Patel, D. K., Prasad, S. K., Kumar, R., & Hemalatha, S., (2012). An overview on antidiabetic medicinal plants having insulin mimetic property. *Asia. Paci. J. Trop. Biomed., 2*(4), 320–330.

Patel, S. S., Shah, S. S., & Goyal, R. K., (2009). Antihyperglycemic, antihyperlipidemic and antioxidant effects of Dihar, a polyherbal ayurvedic formulation in streptozotocin induced diabetic rats. *Indian J. Exp. Biol., 47*(7), 564–570.

Persaud, S. J., Harris, T. E., Burns, C. J., & Jones, P. M., (1999). Tyrosine kinases play a permissive role in glucose-induced insulin secretion from adult rat islets. *J. Mole. Endocrin., 22*(1), 19–28.

Philip, F. U., Joshua, E. A., Helen, N. A., & Charles, O. N., (2017). Antidiabetic activity and phytoconstituents of *Anthocleista nobilis. J. Bioequiv. Availab., 9*(Suppl 5).

Pinent, M., Blay, M., Blade, M. C., et al., (2004). Grape seed-derived procyanidins have an antihyperglycemic effect in streptozotocin induced diabetic rats and insulin mimetic activity in insulin-sensitive cell lines. *Endocrinology, 145*, 4985–4990.

Pinto, M. D. S., Kwon, Y. I., Apostolidis, E., et al., (2009). Potential of *Ginkgo biloba* L. leaves in the management of hyperglycemia and hypertension using *in vitro* models. *Bioresour. Technol., 100*, 6599–6609.

Ponnachan, P. T., Paulose, C. S., & Panikkar, K. R., (1993). Effect of leaf extract of *Aegle marmelos* in diabetic rats. *Indian J. Exp. Biol., 31*, 345–347.

Prakasam, A., Sethupathy, S., & Pugalendia, K. V., (2005). Antiperoxidative and antioxidant effects of *Casearia esculenta* root extract in streptozotocin induced diabetic rats. *Yale J. Biol. Med., 78*, 15–23.
Preuss, H. G., Bagchi, D., Bagchi, M., et al., (2004). Effects of a natural extract of (-)-hydroxycitric acid (HCA-SX) and a combination of HCA-SX plus niacin-bound chromium and *Gymnema sylvestre* extract on weight loss. *Diabetes Obes. Metab., 6*, 171–180.
Puri, D., & Baral, N., (1998). Hypoglycemic effect of *Biophytum sensitivum* in the alloxan diabetic rabbits. *Indian J. Physiol. Pharmacol., 42*, 401–406.
Puri, D., (2001). The insulinotropic activity of a Nepalese medicinal plant *Biophytum sensitivum*: Preliminary experimental study. *J. Ethnopharmacol., 78*(1), 89–93.
Qa'dan, F., Verspohl, E. J., Nahrstedt, A., Petereit, F., & Matalka, K. Z., (2009). Cinchonain Ib isolated from *Eriobotrya japonica* induces insulin secretion *in vitro* and *in vivo*. *J. Ethnopharmacol., 124*(2), 224–227.
Qadri, N. M., Rehman, Z., & Shireen, K., (2006). Evaluation of antidiabetic activity of diabrid, a herbal formulation in type-II diabetic patients. *J. Chem. Soc. Pak., 28*(3), 281–283.
Rama, M. A., Prasad, K. V. S. R. G., & Sharm, J. A. R. P., (2010). Hypoglycemic and antihyperglycemic activity of *Syzygium alternifolium* (wt.) walp. leaf extracts in normal and diabetic rats. *International Journal of Drug Development Research, 2*(1), 27–32.
Ramachandran, V., Mandal, D., Payyavala, U., Sangai, P. D., Muthureddy, N., Ataraj, S., Shanish, A., & Kannan, E., (2011). Hypoglycemic, antioxidant and hypolipidemic activity of *Asparagus racemosus* on streptozotocin-induced diabetic in rats. *Advan. Appl. Sci. Res., 2*(3), 179–185.
Ramesh, P. R., Vijaya, C., & Parasuraman, S., (2014). Antidiabetic activity of polyherbal formulation in streptozotocin –nicotinamide induced diabetic Wistar rats. *J. Trad. Comple. Med., 4*(2), 108–117.
Rankin, J. W., Andreae, M. C., Oliver, C. C. Y., et al., (2008). Effect of raisin consumption on oxidative stress and inflammation in obesity. *Diabetes Obes. Metab., 10*, 86–96.
Rao, B. K., & Rao, C. H., (2001). Hypoglycemic and antihyperglycemic activity of *Syzygium alternifolium* (Wt.) walp. seed extracts in normal and diabetic rats. *Phytomedicine: International Journal of Phytotherapy and Phytopharmacology, 8*(2), 88–93.
Rao, M. U., Sreenivasulu, M., Chengaiah, B., Reddy, K. J., & Chetty, C. M., (2010). Herbal medicines for diabetes mellitus: A review. *Int. J. Pharm. Tech. Res., 2*(3), 1883–1892.
Renuka, C., & Saravanan, K., (2018). Antidiabetic potential of *Biophytum sensitivum* whole plant extracts in STZ induced diabetic rats. *Inter. J. Sci. and Engine. Res, 9*(6), 72–78.
Roman-Ramos, R., Flores-Saenz, J. L., & Alarcon- Aguilar, F. L., (1995). Anti-hyperglycemic effect of some edible plants. *J. Ethnopharmacol., 48*, 25–32.
Rosak, C., & Mertes, G., (2012). Critical evaluation of the role of acarbose in the treatment of diabetes: Patient considerations. *Diabetes Metab. Syndr. Obes., 5*, 357–367.
Ruikar, A. D., Khatiwora, E., & Ghayal, N. A., (2011). Studies on aerial parts of *Artemisia pallens* wall for phenol, flavonoid and evaluation of antioxidant activity. *J. Pharm. Bioallied. Sci., 2*, 302–305.
Saad, B., Zaid, H., Shanak, S., & Kadan, S., (2017). *Anti-diabetes and Anti-Obesity Medicinal Plants and Phytochemicals*. Springer International Publishing AG. doi: 10.1007/978-3-319-54102-5_5.
Sachdewa, A., & Khemani, L., (2003). Effect of *Hibiscus rosa sinensis* Linn. ethanol flower extract on blood glucose and lipid profile in streptozotocin induced diabetes in rats. *J. Ethnopharmacol., 89*, 61–66.

Sagbo, I. J., Van De, V. M., Koekemoer, T., & Bradley, G., (2018). *In vitro* antidiabetic activity and mechanism of action of *Brachylaena elliptica* (Thunb.) DC. *Eviden. Based Comple. Altern. Med.,* 13. Article ID 4170372.

Sahu, P. K., Prasad, P., & Roy, A., (2016). Screening of *in vitro* antidiabetic activity of herbal formulation meshashringi. *UK J. Pharmaceu. Biosci., 4*(6), 75–79.

Saito, R., Yamada, S., Yamamoto, Y., et al., (2012). Conophylline suppresses pancreatic stellate cells and improves islet fibrosis in Goto-Kakizaki rats. *Endocrinology, 153*(2), 621–630.

Sarkar, S., Pranava, M., & Marita, R., (1996). Demonstration of the hypoglycemic action of *Momordica charantia* in a validated animal model of diabetes. *Pharmacol. Res., 33,* 1–4.

Saxena, A., & Vikram, N. K., (2004). Role of selected Indian plants in management of type 2 diabetes: A review. *J. Altern. Complement. Med., 10*(2), 369–378.

Seevalen, S. H. B., Debasish, S., Vethambur, B., & Tajudeen, K., (2010). Evaluation on safety and efficacy of a polyherbal antidiabetic formulation-DIASOL. *Asian Pac J. Mol. Biol. Biotech., 18*(1), 59–61.

Shane-McWhorter, L., (2001). Biological complementary therapies: A focus on botanical products in diabetes. *Diabetes Spectr., 14,* 199–208.

Sharma, A. K., Agarwal, V., Sharma, S., Chauhan, B., Sharma, A. D., & Punia, R., (2011). Antidiabetic effect of bark of *Alangium salvifolium* in alloxan-induced diabetic rats. *J. Glob. Phar. Technol., 3,* 26–32.

Sharma, B., Santosh, K., & Satapathi, P. R., (2007). Hypoglycemic and hypolipidemic effect of *Aegle marmelos* (L.) leaf extract on streptozotocin induced diabetic mice. *Inter. J. Pharmacol., 3,* 444–452.

Shibib, B. A., Khan, L. A., & Rahman, R., (1993). Hypoglycemic activity of *Coccinia indica* and *Momordica charantia* in diabetic rats: Depression of the hepatic gluconeogenic enzymes glucose-6-phosphatase and fructose-1, 6- biphosphatase and elevation of liver and red-cell shunt enzyme glucose-6-phosphate dehydrogenase. *Biochem. J., 292,* 267–270.

Shrivastava, S., Lal, V. K., & Pant, K. K., (2012). Polyherbal formulations based on Indian medicinal plants as antidiabetic phytotherapeutics. *Phytopharmacol., 2*(1), 1–15.

Silva, R. M., Santos, F. A., Rao, V. S., et al., (2001). Blood glucose and triglyceride lowering effect of trans-dehydrocrotonin, a diterpene from *Croton cajucara* benth in rats. *Diabetes Obes. Metab., 3,* 452–456.

Singh, L. W., (2011). Traditional medicinal plants of Manipur as antidiabetics. *J. Med. Plant Res., 5*(5), 677–687.

Singh, R. K., Mehta, S., Jaiswal, D., et al., (2009). Antidiabetic effect of *Ficus bengalensis* aerial roots in experimental animals. *J. Ethnopharmacol., 123,* 110–114.

Singh, S. N., Vats, P., Suri, S., et al., (2001). Effect of an antidiabetic extract of *Catharanthus roseus* on enzymic activities in streptozotocin induced diabetic rats. *J. Ethnopharmacology, 76*(3), 269–277.

Singh, S. S., Pandey, S. C., Srivastava, S., et al., (2003). Chemistry and medicinal properties of *Tinospora cordifolia* (Guduchi). *Indian J. Pharmacol., 35,* 83–91.

Solomon, T. P., & Blannin, A. K., (2007). Effects of short-term cinnamon ingestion on *in vivo* glucose tolerance. *Diabetes Obes. Metab., 9,* 895–901.

Sorenson, R. L., Brelje, T. C., & Roth, C., (1994). Effect of tyrosine kinase inhibitors on islets of Langerhans: Evidence for tyrosine kinases in the regulation of insulin secretion. *Endocrinology, 134*(4), 1975–1978.

Souza, A., Mbatchi, B., & Herchuelz, A., (2011). Induction of insulin secretion by an aqueous extract of *Tabernanhte -iboga* Baill. (Apocynaceae) in rat pancreatic islets of Langerhans. *J. Ethnopharmacol., 133*(3), 1015–1020.

Subash-Babu, P., Ignacimuthu, S., Agastian, P., & Varghese, B., (2009). "Partial regeneration of beta-cells in the islets of Langerhans by nymphayol a sterol isolated from *Nymphaea stellata* (Willd.) flowers. *Bio-org. Med. Chem., 17*(7), 2864–2870.

Sugihara, Y., Nojima, H., Matsuda, H., et al., (2000). Antihyperglycemic effects of gymnemic acid IV, a compound derived from *Gymnema sylvestre* leaves in streptozotocin-diabetic mice. *J Asian Nat. Prod. Res., 2*, 321–327.

Sugihara, Y., Nojima, H., Matsuda, H., Murakami, T., Yoshikawa, M., & Kimura, I., (2000). Antihyperglycemic effects of gymnemic acid IV, a compound derived from *Gymnema sylvestre* leaves in streptozotocin-diabetic mice. *J. Asia. Nat. Prod. Res., 2*(4), 321–327.

Suh, K. S., Chon, S., Oh, S., et al., (2010). Prooxidative effects of green tea polyphenol (−)-epigallocatethin-3-gallate on the HIT-T15 pancreatic beta cell line. *Cell Biology and Toxicology, 26*(3), 189–199.

Sunil, J. Y., Kumar, M. S., & Khan, Y., (2012). Hypoglycemic activity of *Bougainvillea spectabilis* stem bark in normal and alloxan-induced diabetic rats. *Asian Pacific J. Trop. Biomed.,* S919–S923.

Tabopda, T. K., Ngoupayo, J., Liu, J., Mitaine-Offer, A. C., Tanoli, S. A., Khan, S. N., et al., (2008). Bioactive aristolactam from *Piper umbellatum. Phytochem., 69*, 1726–1731.

Telapolu, S., Kalachavedu, M., Punnoose, A. M., & Bilikere, D., (2018). MD-1, a poly herbal formulation indicated in diabetes mellitus ameliorates glucose uptake and inhibits adipogenesis – an *in vitro* study. *BMC Comple. Altern. Med., 18,* 113.

Thakkar, N. V., & Patel, J., (2010). Pharmacological evaluation of – Glyoherb: A polyherbal formulation on streptozotocin- induced diabetic rats. *Int. J. Diabetes Dev. Ctries., 30*(1), 1–7.

Thomson, H., Ojo, O. O., Flatt, P., & Abdelwahab, Y., (2014). Antidiabetic actions of aqueous bark extract of *Swertia chirayita* on insulin secretion, cellular glucose uptake and protein glycation. *J. Exp. Integr. Med., 4,* 268.

Tian, X., Liu, F., Li, Z., Lin, Y., Liu, H., Hu, P., Chen, M., et al., (2019). Enhanced antidiabetic effect of berberine combined with timosaponin B2 in goto-kakizaki rats, associated with increased variety and exposure of effective substances through intestinal absorption. *Front. Pharmacol, 10*, 19.

Tushar, K. B., Debasis, D., Kausik, C., Kazi, M. A., & Debidas, G., (2010). Effect of *Diashis*, a polyherbal Formulation in streptozotocin-induced diabetic male albino rats. *Int. J. Ayurveda Res., 1*(1), 18–24.

Ubillas, R. P., Mendez, C. D., Jolad, S. D., et al., (2000). Antihyperglycemic acetylenic glucosides from *Bidens pilosa. Planta Medica, 66*(1), 82, 83.

Ueda, H., Kawanishi, K., & Moriyasu, M., (2004). Effects of ellagic acid and 2-(2,3,6-trihydroxy-4-carboxyphenyl) ellagic acid on sorbitol accumulation *in vitro* and *in vivo*. *Biol. Pharm. Bull., 27*, 1384–1387.

Umezawa, K., Hiroki, A., Kawakami, M., et al., (2003). Induction of insulin production in rat pancreatic acinar carcinoma cells by conophylline. *Biomed. Pharmacother., 57*(8), 341–350.

Umezawa, K., Taniguchi, T., Toi, M., et al., (1996). Growth inhibition of K-ras-expressing tumors by a new vinca alkaloid, conophylline, in nude mice. *Drugs under Exp. Cli. Res, 22*(2), 35–40.

Van De, V. M., Roux, S., Bungu, L. C., et al., (2008). Antidiabetic screening and scoring of 11 plants traditionally used in South Africa. *J. Ethnopharmacol., 119*, 81–86.

Venkatesh, S., Dayanand, R. G., & Madhava, R. B., (2003). Antihyperglycemic activity of *Helicteres isora* roots in alloxan-induced diabetic rats. *Pharmaceu. Biol., 41*(5), 347–350.

Vessal, M., Hemmati, M., & Vasei, M., (2003). Hypoglycemic effects of quercetin in streptozocin-induced diabetic rats: Comparative biochemistry and physiology. *Toxicol. Pharmacol., 135*, 357–364.

Vuddanda, P. R., Chakraborty, S., & Singh, S., (2010). Berberine: A potential phytochemical with multispectrum therapeutic activities. *Expert Opinion on Investigational Drugs, 19*(10), 1297–1307.

Vuksan, V., Sievenpiper, J. L., Koo, V. Y., et al., (2000). American ginseng (*Panax quinquefolius* L.) reduces postprandial glycemia in nondiabetic subjects and subjects with type 2 diabetes mellitus. *Arch. Intern. Med., 160*, 1009–1013.

Waltner-Law, M. E., Wang, X. L., Law, B. K., et al., (2002). Epigallocatechin gallate: A constituent of green tea, represses hepatic glucose production. *J. Biol. Chem., 277*, 34933–34940.

Wang, Z. Q., Lu, F. E., Leng, S. H., Fang, X. S., Chen, G., Wang, Z. S., et al., (2008). Facilitating effects of berberine on rat pancreatic islets through modulating hepatic nuclear factor 4 alpha expression and glucokinase activity. *World J. Gastroenterol., 14*(39), 6004–6011.

Wolfram, S., Raederstorff, D., Preller, M., et al., (2006). Epigallocatechin gallate supplementation alleviates diabetes in rodents. *J. Nutrition, 136*(10), 2512–2518.

Xie, J. T., Wang, C. Z., Wang, A. B., Wu, J., Basila, D., & Yuan, C. S., (2005). Antihyperglycemic effects of total ginsenosides from leaves and stem of *Panax ginseng*. *Acta Pharmacol. Sin., 26*(9), 1104–1110.

Yarnell, E., & Abascal, K., (2009). Dandelion (*Taraxacum officinale* and *T. mongolicum*). *Integ. Med., 8*, 35–38.

Yazdanparast, R., Esmaeili, M. A., & Helan, J. A., (2005). *Teucrium polium* extract effects pancreatic function of streptozotocin diabetic rats: A histopathological examination. *Iran Biomed. J., 9*(2), 81–85.

Yin, J., Xing, H., & Ye, J., (2008). Efficacy of berberine in patients with type 2 diabetes mellitus. *Metab. Clini. Experi., 57*(5), 712–717.

Yoshikawa, M., Nishida, N., Shimoda, H., et al., (2001). Polyphenol constituents from *Salacia* species: Quantitative analysis of mangiferin with alpha glucosidase and aldose reductase inhibitory activities. *Yakugaku Zasshi, 121*, 371–378.

Youl, E., Bardy, G., Magous, R., et al., (2010). Quercetin potentiates insulin secretion and protects INS-1 pancreatic *β*-cells against oxidative damage via the ERK1/2 pathway. *Br. J. Pharmacol, 161*(4), 799–814.

Yu, B. C., Hung, C. R., Chen, W. C., et al., (2003). Antihyperglycemic effect of andrographolide in streptozotocin induced diabetic rats. *Planta Med., 69*, 1075–1079.

Yun, S. Y., Kim, S. P., & Song, D. K., (2006). Effects of (–)-epigallocatechin-3-gallate on pancreatic beta-cell damage in streptozotocin- induced diabetic rats. *Euro. J. Pharmacol., 541*(1, 2), 115–121.

Zhang, D., Fu, M., Gao, S. H., et al., (2013). Curcumin and diabetes: A systematic review. *Evid. Based Complement. Alternat. Med.,* 636053.

Zhang, Y. H., Cai, J. Y., Ruan, H. L., Pi, H. F., & Wu, J. Z., (2007). Antihyperglycemic activity of kinsenoside, a high yielding constituent from *Anoectochilus roxburghii* in streptozotocin diabetic rats. *J. Ethnopharmacol., 114*(2), 141–145.

Zhou, L., Wang, X., Shao, L., et al., (2008). Berberine acutely inhibits insulin secretion from *β*-cells through 3,5-cyclic adenosine 5-monophosphate signaling pathway. *Endocrinology, 149*(9), 4510–4518.

Ziyyat, A., Lcgssyer, A., Mekhfi, H. R., et al., (1997). Phytotherapy of hypertension and diabetes in oriental Morocco. *J. Ethnopharmacol, 58*, 45–54.

CHAPTER 9

Molecular Insights of Active Plant-Based Antidiabetic Drug Molecules

PRAVALLIKA,[1] B. S. SUPREETHA,[1] V. VEENA,[1] J. SELVAKUMAR,[2] and HARIKRISHNAN[3]

[1]*Department of Biotechnology, School of Applied Sciences, REVA University, Rukmini Knowledge Park, Kattigenahalli, Yelahanka Post, Bengaluru – 5600064, Karnataka, India, Tel.: 080-66226622, Fax: 66226622, E-mails: veena.v@reva.edu.in; btveenavijaykumar@gmail.com (V. Veena)*

[2]*Department of Chemistry, Wright State University, 3640 Colonel Glenn HWY, Dayton, OH45435, USA*

[3]*Department of Chemistry, School of Arts and Sciences, Vinayaka Mission Research Foundation-Aarupadai Veedu (VMRF-AV) Campus, Paiyanoor, Chennai – 603104, Tamil Nadu, India, E-mail: harich4@gmail.com*

ABSTRACT

An array of diabetic controlling plant-based molecules has been reported as a source for antidiabetic chemotherapeutic strategies. Diabetes referred as metabolic and chronic endocrine disorder with severe complications. Conventional medicines include injectable/oral insulin, sulfonylurea, biguanidine, glucosidase inhibitors and glinides. Upon the detailed study on the biology of various types of diabetes, new specific targeted drug candidates are developed. These may be insulin in inhalation, mimetics of incretin, analogs of amylin, analogs of gastric-inhibitory polypeptides (GIPs), peroxisome proliferator-activated receptor (PPAR) inhibitors and dipeptidyl peptidase 4 (DPP 4). The major drawback of these drugs is incurable in nature,

Antidiabetic Potential of Plants in the Era of Omics. Deepu Pandita, PhD, Anu Pandita, MSc, & Chander Bhanu, MBBS (Eds.)

high cost and undesirable adverse effects. However, plant-based molecules have more advantages that include cost-effective, durable, and minimum side effects. Major plant families that are used in traditional knowledge with proven to diabetic treatment includes *Asteraceae, Araliaceae, Cucurbitaceae, Lamiaceae, Leguminosease, and Rosaceae.* Major proven diabetic management plants include *Momordica charantia, Citrullus colocynthis, Allium sativum, Trigonella species, Gymnema sylvestre* and *Ficus bengalensis*. Phytoconstituents that are responsible for diabetic management from these plants includes flavonoids, terpenes, polyphenolics, alkaloids, tannins, and saponins. Exploring the novel targets for antidiabetic agents such as glucagon-like peptide 1 (GLP 1), sodium-glucose co-transporter 2 (SGLT 2) and DPP 4 have paved way to target the diabetes at various levels. The molecular, physiological, pharmacological, and toxicological research have proven to be various diabetic management levels such as prophylactic, therapeutic index, mode of actions, targets, efficacy, and safety application seems to be promising. In this chapter, we emphasis on the various class of plant-based molecules that specifically targets the molecular multitargets in diabetics management from preclinical and clinical trials with mode of actions. These plant-based molecules have different targets and multitargets to overcome the drug resistance, incurable, toxic, side-effects, efficacy, and high-cost drugs.

9.1 INTRODUCTION

Diabetes mellitus (DM), a chronic endocrine metabolic disease that is caused due to defect in insulin dynamics. It is characterized by the body's inability to transform sugar into energy that causes elevated blood sugar levels referred as hyperglycemia. Due to multisystem involvement, it leads to serious complications that include cardiovascular, nervous, renal, and other body systems (Unnikrishnan et al., 2016; International Diabetes Federation, 2015). As DM seems to be the disorders associated with the metabolism of carbohydrate, proteins, and fats. It leads to increase in volume of blood and out of urine (polyuria) resulting in heavy water loss in the body and finally results in increased thirst (polydipsia). It is been estimated that diabetes cases might increase to 366 million in 2030 in mostly from developing countries. It became as a serious problem spreading rapidly all over the world (Kumar et al., 2017; Wild et al., 2004; Eleazu et al., 2016). Life style changes that include regular exercise and proper diet management can be preventive measures for DM (Whiting et al., 2011; Gwatidzo et al., 2017).

Diabetes are of two type based on the deficient in insulin production (Type 1) and resistance to insulin (Type 2). Type 1 is reported because of degeneration of β-cell in the pancreas so that body completely stops insulin production. Hence, it is insulin-dependent is characterized by impaired production of insulin finally leads to deficiency of insulin in the body. In this case, patients must take insulin daily through injections. However, in type 2 diabetic conditions there is insufficient usage of even production of insulin in normal levels. Hence, it is insulin resistance condition is because of impaired functioning of β-cells of pancreas. The complications of these types of diabetes majorly involves kidney, nervous, cardiovascular, body defense system and body systems in the body that results in disease severity (Sarayani et al., 2014; Hasan et al., 2015; Chaudhury et al., 2017). Commonly, the majority of people are known to be having type-2 than type-1 diabetes (Wilcox et al., 2005). However, type-1 are generally treated by insulin, but the combination of treatment is needed for type-2 diabetes.

9.1.1 MOLECULAR SIGNALING IN VARIOUS DIABETES

Deep understanding of glucose homeostasis regulation in the body have led to elucidation of the mode of action of antidiabetic agents. It involves two hormones, insulin produced by pancreatic β-cells and glucagon that is produced by islets of Langerhans of pancreas that act antagonistic to regulate the concentration of glucose in blood (Campbell et al., 2015). Glucagon secreted during fasting state increases the blood glucose level while excess of glucose in blood is reduced by the action of insulin (Buckingham et al., 2013). These hormones regulate and maintain normal blood glucose levels (BGL). However, insufficient production of insulin or insulin functions or the body's response to insulin might lead to disturbance in diabetes. Insulin, anabolic peptide hormone that binds with insulin receptor on cell surface of certain cell types (Boucher et al., 2014). Binding of insulin with insulin receptor have known to activate the insulin signaling pathway. Insulin binds to insulin receptor through the ligand-binding domain, which induces the conformational changes in receptor and activates the RTK domain (Mayer et al., 2007; Gutmann et al., 2018). This leads to phosphorylation of three tyrosine residues of β-subunits of RTK. Phosphorylation activates the domain of Src homology 2 (SH 2) or domain of phospho-tyrosine binding (PTB)

(Saltiel et al., 2001; Kaneko et al., 1999). These results in the expression of glucose transporter type 4 (GLUT 4), a glucose transporting proteins, activation of target enzymes and other necessary transcription factors (Petersen et al., 2018).

In a healthy individual, after ingestion of meals the insulin is produced from β pancreatic cells. Insulin acts on the multiple tissues and organs of the body (Morigny et al., 2016; Haeusler et al., 2017). The five major target organs of insulin actions are liver, muscle tissues, adipose tissues, pancreas, and intestine (Kumar et al., 2014; Bharti et al., 2013) (Figure 9.1). Action of insulin on these targets organs results in the lowering of the glucose in blood and restore the normal level of glucose in blood.

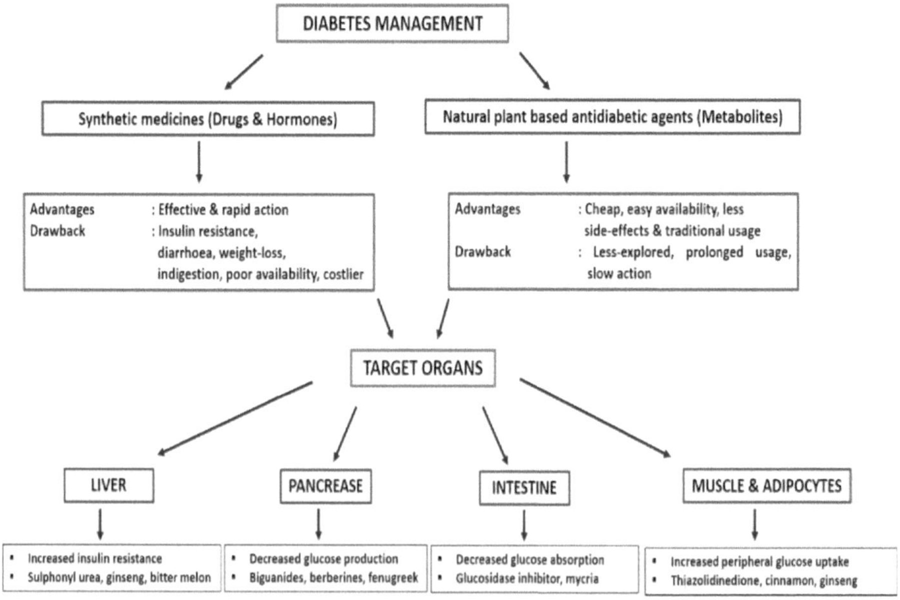

FIGURE 9.1 Diabetic management through synthetic and natural products with their mode of action on various organs, advantages, and drawbacks.

In the liver, insulin promotes the induction of glycogen synthesis from the glucose through activation of glycogen synthase by glycogenesis to reduce high levels of blood glucose. In addition, in liver cells, insulin is also known to inhibit the phospho-enol pyruvate carboxykinase (PEPCK) and glucose 6 phosphatase (G6Pase) enzymes (Quinn et al., 2005). These enzymes are key regulatory enzymes in the process of gluconeogenesis. Insulin is also known to inhibit the glycogen phosphorylase (GP) of glycogenolysis pathway that produce glucose and regulate the glucose homeostasis. In liver cells, the GLUT-2 transporters that is involved in glucose uptake are highly expressed but they are insulin-insensitive.

In muscles, the insulin promotes the high amount of GLUT-4 transporter translocation on the cell surface to increase the glucose uptake by the cells (Stöckli et al., 2011). Insulin is also known to increases the glycolysis and glycogenesis through the inhibition of glycogenolysis and proteolysis in muscle cells (DeFronzo et al., 2009; Cartee, 2015). These activities ultimately lower the blood glucose level, and muscle cells are plays critical roles in controlling the postprandial high levels of blood glucose.

In adipose tissues, the adipocytes express the GLUT-4 transporters that are insulin sensitive. Thereby, insulin increases the high glucose uptake and increase the glucose metabolism through glycolysis. In addition, insulin is also known to increase the lipogenesis and inhibits lipid degradation that is necessary for balancing the serum lipid level in the body (Pataky et al., 2018).

In these cells, the GLUT-4 transporters are known to be sequestered and storage of vesicles referred as GLUT-4 storage vesicles (GSV) (Miinea et al., 2004). These GSVs are translocated out the surface of the cell membrane through exocytosis in presence of insulin. This process happens through phosphatidylinositol 3 kinase (PI3-K) activation. Activated SH 2 domain present on the regulatory p85 subunit of PI3-K now bind with tyrosine-phosphorylated IRS that activate p110 catalytic domain of PI3-K. Then, the PI3-K that have activated now is recruited on the cell membrane where p110 is converted to phosphatidylinositol 4, 5 biphosphate (PIP-2) to phosphatidylinositol-3,4,5-triphosphate (PIP-3). Thereby, the activity of PIP-3 is upregulates the 3-phosphatidylinositol-dependent kinase-1 (PDK-1) that phosphorylates to activate the protein kinase-B (Akt). Further activated Akt can phosphorylate Rab GTPase activating protein AS160, that further inactivate a class of small G-proteins, Rab that are found on the GSVs to suppress the GTPase activity. As a result, production of GTP is increased to activate the Rab. Now, Rab regulates GSV is activated to translocate GLUT 4 expression on the cell surface (Pataky et al., 2018).

9.1.2 *CONVENTIONAL DIABETES TREATMENT*

Insulin that are oral and injectable, sulfonylureas, biguanides, inhibitors of glucosidase and glinides are commonly used as conventional treatment for diabetics (Figure 9.1). Other advanced approaches are stem cell therapy, nanotechnology, and gene therapy. The major limitations associated with conventional medicines are gastrointestinal, hematopoietic, cardiovascular, nervous system, urinary, and skin problems. Combination of herbs and antidiabetic medications results in positively enhanced antidiabetic effects. Indirect or extrapancreatic either mechanisms or organs approach is the current focus of research other than the conventional approach.

Depending on the method of administration, antidiabetic drugs are injectable and inhaled or oral-based. Recent approaches lead to develop a new kind of therapeutics that includes analogs of amylin, analogs of gastric-inhibitory polypeptides (GIPs) and incretin mimetics. Several synthetic antidiabetic molecules are being reported with different mechanism of action (Diamant et al., 2003). Sulfonylurea and meglitinide are some secretagogues of insulin and act to induce the secretion of insulin in β-cells of the pancreas. Bi-guanides and thiazolidinedione act as insulinomimetics agents (Loke et al., 2009; Nanjan et al., 2018). Acarbose has inhibitory activities in α-glucosidase on the intestinal cells to reduce the digestion process and reduce the absorption of dietary glucose (Chiasson et al., 2002). Recent classes of antidiabetic agents are reported to exhibit anti-hyperglycemic potential that includes incretin hormones GLP 1 and GIPs. These molecules regulate the appetite, satiety, reduce the glucagon secretion and induce the production of insulin. Entero-endocrinal cells present in a tract of gastro-intestine produces these molecules. However, they are quickly degraded using (DPP 4) in the GI tract (Davidson et al., 2008; Seino et al., 2010). Hence, the analogs and agonists receptor of GLP 1, GIPs, and DPP 4 are major targets for intervention of therapeutic development for type 2 diabetics.

Inhibitors of sodium-glucose cotransporter (SGLT) seems to next novel class of oral medicine to treat the diabetes. These SGLT 2 proteins in tubules of renal proximal in kidney are involved in reabsorption of glucose in blood. Hence, inhibiting the SGLT 2 aid in facilitating the removal of glucose in urine. SGLT 1, glucose transporters (GLUT) are needed for intestinal glucose absorption and inhibit SGL 1 can delay the glucose absorption (Musso et al., 2019; Cefalo et al., 2019). However, the dual inhibitor of SGLT-1/2, sotaglifozin has reported for controlling he blood glucose and representing efficient for both diabetic treatment (Campbell et al., 2008; Hsia et al., 2018; Rieg et al., 2018). Most widely insulin secretagogues and sensitizers are recommended. Biguanide, metformin is commonly used for type 2 diabetics for obese patient as first-line therapy (Briones et al., 2016). It increases insulin sensitivity through enhancement in uptake of glucose by muscles, Further, it also suppresses the gluconeogenesis in the liver cells (Wanner et al., 2018).

Currently, the conventional therapies for type 2 diabetes aid in regulating the BGL but they have undesirable side-effects. Insulin and secretagogues of insulin will reduce the blood glucose resulting in a sudden hypoglycemic condition. Sulfonylureas and thiazolidinediones results in over-weight, biguanides, inhibitors of α-glucosidase and incretin are associated with adverse side-effects in gastrointestinal effects (Samuel et al., 2012; Madiraju et al., 2018; Swinnen et al., 2018). Hence, alternative effective therapeutic strategies with least side effects are in need.

9.2 PLANT-BASED DIABETES MANAGEMENT

Traditionally, the medicinal plants are the main source for the treatment of several disease conditions. Researchers have future prospective for prevention and treatment strategies for diabetes (Sarker et al., 2006; Cragg et al., 2014). Ethnomedicine are of higher values that are not produced through synthetic routes. Various phytoconstituents that include flavonoid, alkaloid, amino-acids, amine, derivatives of carboxylic acid, anthranoid, carbohydrates, glycol-sides, minerals, vitamins, inorganic compounds, peptidoglycans, saponins, polyphenol, and their derivatives have been candidate molecules for diabetics control and treatment (Oh et al., 2015; Desai et al., 2015). Globally, more than 1,200 plants are recognized have antidiabetic potentials. Most commonly plants are from the families such as *Fabaceae, Asteraceae, and Lamiaceae* (Wang et al., 2014).

9.2.1 MEDICINAL PLANTS FOR DIABETES

Traditional medicinal systems throughout the world uses the plants are used from long decades. Now, plants continue to be new remedies for mankind (Baldwa et al., 1977). Several plants serve as the basis for the drug discovery and development for treatment of diabetes from > 3,500 years (Khanna et al., 1981; Yibchok-Anun et al., 2006; Bharti et al., 2016).

Most important families having antidiabetic activity were *Asteraceae and Myrtaceae.* Species most frequently used were *Syzygium cumini and Sphagneticola trilobata.* These plants have shown promising results mainly *Syzygium cumini.* Aloe vera having anti-hyperglycemic effect shown reduced blood glucose level (Kang et al., 2012; Ismail et al., 2017). Oleanolic acid of quinoa yogurt beverage can be used as a dietary food supplement for antidiabetic activities. Quinoa from *Chenopodium quinoa* having bioactive compounds constituting secondary metabolites are having different activities like antidiabetic, anti-inflammatory, and anti-cholesterol (Hui et al., 2009; Grover et al., 2000; Alarcon-Aguilara et al., 1998). The traditional knowledge with proven to diabetic treatment includes *Asteraceae, Araliaceae, Cucurbitaceae, Lamiaceae, Leguminoseae, Liliaceae, Moraceae, and Rosaceae.* Largely, they have proven for diabetic management plants include *Citrullus colocynthis, Momordica charantia, Gymnema sylvestre, Allium sativum, Trigonella foenum greacum* and *Ficus bengalensis.* These plants contain phytoconstituents include polyphenols, flavonoids, terpenoids, alkaloids, coumarins, quinones, tannins, saponins, carbohydrates, cardiac glycosides and minerals that are having natural hypoglycemic functions (Rotshteyn et al., 2004; van de Venter et al., 2008). There are some other phytoconstituents like tannins, glycosides, minerals having antidiabetic effect through hypoglycemic property, insulin release activity, uptake of glucose, glucose utilization and also inhibiting the intestinal absorption (Bharti et al., 2018; Luo et al., 1998). Though they have beneficial aspects, they also have some risks and can affect the liver, kidney, and may also cause neurological disturbances.

9.2.2 PHYTOCONSTITUENTS OF ANTIDIABETIC MOLECULES

Traditionally, plant have various mechanism of action to exhibit the antidiabetic potential. The mechanisms include regulating the digestion and absorption of carbohydrates, regulating the glycolytic pathway and Krebs' cycle, inhibiting the gluconeogenesis, regulating the HMP shunt, repressing the glycogenolysis process, inhibiting the activity of α-glucosidase and glucose absorption, increase the secretion of insulin and potentiates its actions, potentiates insulin action by inactivation of insulinase and activate the PI3K (Figure 9.2; Table 9.1). Hypoglycemic potentials from plants commonly include alkaloids, polysaccharides, saponins, ferulic acids, flavonoids, imidazdine compounds, sulfur-containing compounds, glycosides, peptidoglycans, phenolics, and polyphenols (Bharathi et al., 2013, 2016; Luo et al., 1998).

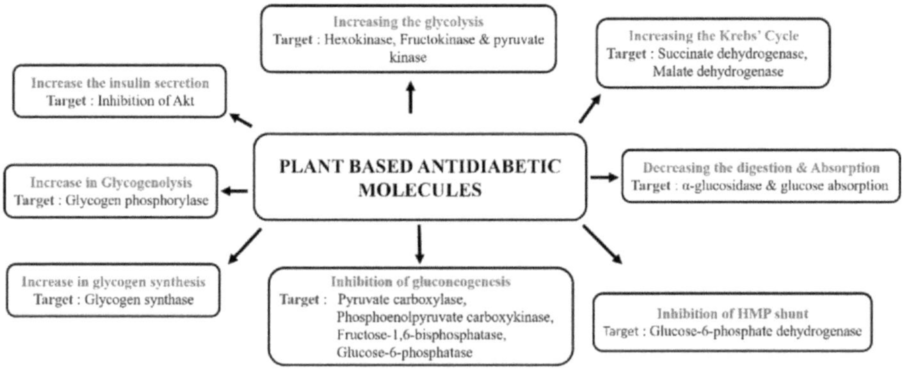

FIGURE 9.2 Plant-based antidiabetic molecules with their molecular mechanism of action and their target proteins.

9.2.3 MOLECULAR MECHANISM OF ACTIVE PLANT-BASED ANTIDIABETIC MOLECULES

9.2.3.1 PHYTOMOLECULES THAT REGULATE THE DIGESTION AND ABSORPTION OF CARBOHYDRATES

In the diet, the carbohydrates, especially starch and complex carbohydrates are converted to simple sugars that includes glucose, fructose, and maltose by digestion process that happens from mouth to colon. Firstly, the dietary starch is converted to oligosaccharides by salivary α-amylase and pancreatic carbohydrate degrading enzymes. Secondly, the α-glucosidase membrane-bound enzyme present in the small intestine further converts oligosaccharides to simple sugars. Hence, the α-glucosidase inhibition is another critical and effective treatment for diabetics that can reduce the absorption of glucose into blood (Hui et al., 2009; Khosla et al., 1995; Akhtar et al., 1991; Mannan et al., 2000; Ali et al., 2006).

Various phytomolecules that are reported to inhibit α-glucosidase that includes sotolon (4,5-dimethyl-3-hydroxy-2(5H)-furanone), trigonelline, gentianine, carpaine, betaine, achyranthine, β-ecdysone, castanospermine, epifagomine, fagomine, Nitrosamines, Brevifolin carboxylic acid, ethyl-brevifolin-carboxylate, glucomannan, caryophyllene, protein-bound polysaccharide, cellulose, mannose, D-threitol, D-arabinitol, palmitic acid, L-arabino-D-xylan, cinnzeylanin, cinnzeylanol, D-glucan, inulin, laevulin, fenugreekine, glucosamine, curcumin, turmerone, germacrone, zingiberene, ellagic acid and its derivatives, corosolic-acid, 4-hydroxy-benzoic acid, 3-O-methylprotocatechuic acid, caffeic acid, p-coumaric acid, kaempferol, wedelolactone, dimethyl wedelolactone, kotalagenin-16-acetate, diterpene, and triterpenoids.

TABLE 9.1 Antidiabetic Phytoconstituents and Their Mechanism of Actions

SL. No.	Mode of Action	Molecular Targets	Phytoconstituents	References
1.	Increasing glycolysis pathway and Krebs cycle	Hexokinase, fructokinase, pyruvate kinase, succinate dehydrogenase, malate dehydrogenase	S-methyl cysteine sulfoxide, S-allyl cysteine sulfoxide, Leucopelargonidin, dulcitol, Aegelin, marmesin, marmelosin, Ginsenosides Rg2, panaxan A, B, C, D, E, Lactucain C, e-Glucoside, mangiferin, salacinol, kotalanol, epigallocatechin, Kotalagenin-16-acetate, diterpene, triterpens, Mallic acid, chlorogenic acid	Kumari et al. (2007); Roman-Ramos et al. (1995); Prakasam et al. (2005) Hou et al. (2003); Krishnakumar et al. (1999); Silva et al. (2001); Wadood et al. (1989)
2.	Inhibition of gluconeogenesis	Pyruvate carboxylase, phosphoenolpyruvate carboxykinase, fructose-1,6-bisphosphatase, glucose-6-phosphatase	Camphor, eugenol, trans-βocimene, geraniol, α-pinene, limonene, p-cymene, 1,8-cineole, thujone, Epigallocatechin-gallate, gallocatechin, epicatechin, (+) catechin, (−) epicatechin, Leucopelargonidin, dulcitol, (+) Catechin, (−) epicatechin, chiorogenic acid, liquiritigenin, isoliquiritigerin, Momordin, momordicine, charantin, Tinosporine, cordifolide, tinosporide, cordifole, columbin	Hii et al. (1984); Waltner-Law et al. (2002); Vessal et al. (2003); Li et al. (2008); Prakasam et al. (2005); Grover et al. (2000); Wadood et al. (1989); Van de Venter et al. (2008); Chao et al. (2003); Sarkar et al. (1996); Hui et al. (2008)
3.	Inhibition of HMP shunt	Glucose-6-phosphate dehydrogenase	e-Glucoside, mangiferin, salacinol, kotalanol, epigallocatechin, Kotalagenin-16-acetate, diterpene, triterpens, Mallic acid, chlorogenic acid	Prakasam et al. (2005); Hou et al. (2005); Krishnakumar et al. (1999); Silva et al. (2001); Wadood et al. (1989)

TABLE 9.1 *(Continued)*

SL. No.	Mode of Action	Molecular Targets	Phytoconstituents	References
4.	Increase in glycogen synthesis	Glycogen synthase	Aegelin, marmesin, marmelosin, Allicin, apigenin, alliin, polypeptide-P, lectins, mistletoe lectin I, II, III, viscotoxin B, cycliton, Furfural, caprylic acid, Bis (2-ethyl hexyl) phthalate (DEHP), Momordin, momordicine, charantin, Momorcharaside A and B, momorcharin A and B, Cucurbitacin B, isocucurbitacin B, Kotalanol, salacinol, Arbutin, eriolin, Citrullol, colocynthin, elaterin, elatericin B, colosynthetin, Leucocyanidin, pelarogonidin, Cytrus bioflavonoids (hesperidin, naringin, Amarogentin, swerchirin, chirantin, gentiopicrin	Kamalakkannan et al. (2005); Ponnachan et al. (1993); Gholap et al. (2004); Kumar et al. (2014); Chao et al. (2003); Adaramoye et al. (2012); Eno et al. (2008); Gray et al. (1999); Abesundara et al. (2004); Sarkar et al. (1996); Lemus et al. (1999); González-Tejero et al. (2008); Ziyyat et al. (1997); Moon et al. (2007); Singh et al. (2003); Cherian et al. (1994); Kumar et al. (2014); Jung et al. (2004)
5.	Inhibition of glycogenolysis	Glycogen phosphorylase	Cytrus bioflavonoids (hesperidin, naringin),	Jung et al. (2004)

TABLE 9.1 *(Continued)*

SL. No.	Mode of Action	Molecular Targets	Phytoconstituents	References
6.	Inhibition of digestion and absorption of glucose	Inhibition of α-glucosidase and glucose absorption	Sotolon [4,5-dimethyl-3-hydroxy-2(5H)-furanone], trigonelline, gentianine, carpaine compounds, Betaine, achyranthine, βecdysone, Castanospermine, epifagomine, fagomine, Nitrosamines, Brevifolin carboxylic acid, ethyl brevifolin carboxylate, Glucomannan, Caryophyllene, Protein-bound polysaccharide, Cellulose, mannose, D-threitol, D-arabinitol, palmitic acid, L-arabino-D-xylan, cinnzeylanin, cinnzeylanol, D-glucan, Inulin, laevulin, Fenugreekine, Glucosamines, Curcumin, turmerone, germacrone, zingiberene, Ellagic acid and its derivatives, corosolic acid, 4-hydroxybenzoic acid, 3-O-methylprotocatechuic acid, caffeic acid, p-coumaric acid, kaempferol, Wedelolactone, dimethyl wedelolactone, Kotalagenin-16-acetate, diterpene, triterpens	Hui et al. (2009); Khosla et al. (1995); Akhtar et al. (1991); Mannan et al. (2000); Ali et al. (2006); Van de Venter et al. (2008); Khosla et al. (1995); Naisheng et al. (2008); Ananthi et al. (2003); Hannan et al. (2006); Broadhurst et al. (2000)

TABLE 9.1 *(Continued)*

SL. No.	Mode of Action	Molecular Targets	Phytoconstituents	References
7.	Increase the insulin secretion	Inhibition of Akt signaling pathway	Harmine, pinoline, Gurmarin, betaine, choline, trimethylamine, (–) Hydroxycitric acid, Castanospermine, epifagomine, fagomine, Leucine, isoleucine, alanin, Polypeptide-P, Lectins, mistletoe lectin I, II, III, viscotoxin B, cycliton, Furfural, caprylic acid, Bis (2-ethyl hexyl) phthalate (DEHP), Aloin, barbaloin, isobarbaloine, aloetic acid, aloe-emodin, emodin, cinnamic acid, crysophanic acid, Vicine, Torachrysone, toralactone, rhein, alaternin, Camphor, eugenol, trans-βocimene, geraniol, α-pinene, limonene, p-cymene, 1,8-cineole, thujone, Glucomannan, Caryophylline, Vin α-ginsenoside R3, Momordin, momordicine, charantin, Momorcharaside A and B, momorcharin A and B, Cucurbitacin B, isocucurbitacin B, Momordin-a, luffin-a, Kotalanol, salacinol, Arbutin, eriolin, Citrullol, colocynthin, elaterin, elatericin B, colosynthetin, Leucocyanidin, pelarogonidin, Taraxacin, Chrysin, isoquercitrin, Myrciaphenones A and B, myrciacitrins I and II, α-Cephalin, myricetin-3'-glucoside, ambrettolide, Flavanols, flavones, flavanones, Quercetin, quercetrin, apigenin, rutin, apigenin-7-O-glucoside, Naringenin, α-Terpineol, hexanol, Shamimin, Leucopelargonidin, dulcitol, Matteuorien, matteuorienin matteuorienate A, B, C., Glucosamines	Sugihara et al. (1991); Preuss et al. (1991); Hayamizu et al. (1991); Akhtar et al. (1991); Hannan et al. (2006); Broadhurst et al. (2000); Chattopadhyay et al. (1999); Liu et al. (2008); Waltner-Law et al. (2002).

9.2.3.2 PHYTOMOLECULES REGULATING GLYCOLYTIC PATHWAY AND KREBS' CYCLE

Plant based alkaloids from numerous medicinal plants have been reported to possess antidiabetic potentials. Glycolysis, metabolic pathway of sugars converted to glucose than glucose through a series of ten reactions. Among them, three regulatory reactions that are catalyzed by glucokinase, phosphor-fructokinase, and pyruvate kinase respectively. Some of the molecules that inhibit the glycolysis and Krebs cycle. Alkaloids have inhibitors of enzymes in glycolysis, namely glucokinase, phospho-fructokinase, and pyruvate kinase. Berberine, an alkaloid isolated have reported for antidiabetic activity through increasing the glucokinase and phospho-fructokinase activities that results in enhanced glucose transportation and its utilization (Singh et al., 2003).

Tricarboxylic acid (Krebs) cycle is the centralized energy-producing process that occurs in the matrix of mitochondria. The pyruvate produced in glycolysis gets converted to carbon dioxide and water through acetyl-Co-A that produces nicotinamide adenine dinucleotide (NADH). During seven enzymatic reactions of the Krebs cycle, the succinate dehydrogenases and malate synthase are known to be regulated by the plant metabolites. Similarly, the natural inhibitors derived from medicinal plants for glycolysis and Krebs cycle includes S-methyl cysteine-sulfoxide, S-allyl cysteine-sulfoxide, leucopelargonidin, dulcitol, aegelin, marmesin, marmelosin, ginsenosides-Rg 2, panaxan-A to E, lactucain-C, e-glucoside, mangiferin, salacinol, kotalanol, epigallocatechin, kotalagenin-16-acetate, diterpene, triterpens, Mallic acid and chlorogenic acid that have been isolated from *Tinospora cordifollia* (Singh et al., 2003), *Barberisaristata sp., Alium sepa* (Kumar et al., 2014), *Tinospora cordifollia, Tinospora crispa* and *Camellia sinensis*.

9.2.3.3 PHYTOMOLECULES THAT INHIBITS THE GLUCONEOGENESIS

Gluconeogenesis is the major multi-step metabolic pathway that occurs in liver and kidney cells. Gluconeogenesis pathway converts pyruvate or/and other three carbon metabolites such as lactate and alanine are converted to glucose. In the gluconeogenesis, the seven reaction are reversible that involves four important enzymes namely pyruvate carboxylase, PEPCK, fructose 1,6-bisphosphatase and G6Pase. Hence, four enzymes of gluco are the major targets for the phytomolecules to act as antidiabetic properties through which glucose formation in the body can be regulated (Waltner-Law et al., 2002; Vessal et al., 2003; Li et al., 2004, 2008; Prakasam et al., 2005). Some of the plant molecules reports that include sotolon-[4,5-dimethyl3-hydroxy-2(5H)-furanone], carpaine, trigonelline, and gentianine isolated from *Trigonella foenum graecum* have reported to decrease the fructose-1,6-bisphosphatase activity to exhibit antidiabetic (Singh et al., 2003).

The phytomolecules that are reported to target and inhibit these enzymes includes camphor, eugenol, trans-β-ocimene, thujone, geraniol, α pinene, limonene, p cymene, 1,8-cineole, epigallocatechin-gallate, gallocatechin, epicatechin, (+) catechin, (−) epicatechin, leucopelargonidin, dulcitol, (+) catechin, (−) epicatechin, chiorogenic acid, liquiritigenin, isoliquiritigerin, momordin, momordicine, charantin, tinosporine, cordifolide, tinosporide, cordifole, and columbin. These metabolites are reported to be isolated from plants that includes *Momordica charantia, Taraxacum officinale, Citrullus colocynthis, Arctostaphylos uvaursi, Salacia reticulate, Salacia oblonga, Luffa cylindnica* and *Momordica charantia* (Grover et al., 2000; van de Venter et al., 2008; Chao et al., 2003; Sarkar et al., 1996; Hui et al., 2009).

Glycogen synthase enzyme catalysis the addition and progressive lengthening of glycogen chain. In the diabetics, there is improper conversion of glycogen that causes high glucose levels in blood. Activation and increasing the glycogen synthase activity is another important mechanism exhibited by phytochemicals. Some phytomolecules known to increase the activity of glycogen synthase of gluconeogenesis includes aegelin, marmesin, marmelosin, allicin, apigenin, alliin, polypeptide-P, lectins, mistletoe lectin I, II, III, viscotoxin-B, cycliton, furfural, caprylic acid, bis-(2-Ethylhexyl)phthalate (DEHP), momordin, momordicine, charantin, momorcharaside-A/B, momorcharin-A/B, cucurbitacin-B, isocucurbitacin-B, inulin, laevulin, fenugreekine, glucosamine, curcumin, turmerone, germacrone, zingiberene, ellagic acid derivatives, corosolic acid, 4-hydroxybenzoic acid, 3 O methylprotocatechuic acid, caffeic acid, p-coumaric acid, kaempferol, wedelolactone, dimethyl wedelolactone, kotalagenin-16-acetate, diterpene, and triterpenoids. These molecules are reported in the plants such as *Aegle marmelose, Curcuma longa, Dioscorea esculenta, Momordica charantia, Murraya koenigii, Ocimum sanctum, Piper betle and Coccinia indica* (Prabhakar et al., 2008).

9.2.3.4 PHYTOMOLECULES THAT INHIBIT OF HMP SHUNT

Hexose monophosphate (HMP) shunt, an alternative pathway for oxidation of glucose other than glycolysis. It is an anabolic pathway that produce pentose sugar (ribose-5-phosphate) and NADPH from six molecules of glucose. The key regulatory enzyme glucose 6 phosphate dehydrogenase (G6PDH) of HMP pathway is implicated in diabetes. G6PDH inhibitors of the HMP shunt pathway that are reported includes glucosamine, curcumin, turmerone, germacrone, zingiberene, ellagic acid and its derivatives, corosolic acid, 4-hydroxybenzoic acid, 3-O-methylprotocatechuic acid, caffeic acid, p-coumaric acid, kaempferol, Wedelolactone, dimethyl wedelolactone, kotalagenin-16-acetate, diterpene, triterpens, e-Glucoside, mangiferin, salacinol, kotalanol, epigallocatechin, kotalagenin-16-acetate and diterpene. These are isolated from *Aegle marmelose, Coscinium fenestratum Colebr, Gongronema latifolium, Momordica charantia, Murraya koenigii, Ocimum sanctum, Aconitum carmichaeli, Coccinia indica, Colocassia esculenta, Catharanthus roseus, Trigonella foenum graecum, Momordica charantia* (Khan et al., 1995; Prabhakar et al., 2008).

9.2.3.5 INHIBITION OF GENERATED FREE RADICALS

In diabetes, the excess of glucose has been reported to increase the glycation of membrane proteins and lipid to generate reactive oxygen species (Wanders et al., 2000). The ROS is a defense mechanism but have deleterious effects to induce secondary complications. In the body, the several enzymes such as superoxide dismutase, catalase, reduced glutathione (GSH) and glutathione S transferases are involved in reducing the ROS generated (van de Venter et al., 2008). The phytomolecules such as catharanthine, vindoline, and vindolinine have reported to activate free radical scavenging enzymes (Jarald et al., 2008; Chattopadhyay et al., 1999). Thereby, antioxidant activities of the molecule the adverse effects of ROS. Similarly, the vinblastine and vincristine obtained in *Vinca rosea* are known to activate free radical scavenging enzymes.

9.2.3.6 PHYTOMOLECULES THAT INHIBIT THE GLYCOGENOLYSIS

Glycogenolysis is the pathway through which glycogen is hydrolyzed to glucose catalyzed by GP. Few plant metabolites that include Cytrus bioflavonoids (hesperidin, naringin) have been reported in various plants such as *Aegle marmelose, Murraya koenigii, Ocimum sanctum* and *Brassica juncea* (Jung et al., 2004; Prabhakar et al., 2008). The leaf extracts obtained from *Azadirachta indica* have reported for reduction of glucose and induced glycogenolysis in rabbit that was induced diabetics (Jung et al., 2004).

9.2.3.7 PHYTOMOLECULES THAT PROMOTES INCREASE THE INSULIN PRODUCTION

Insulin, a peptide hormone produced by β pancreatic cells that regulates the glucose in blood through anabolic process for conversion of glucose to glycogen synthesis, triacylglycerols, and proteins. Majority of botanicals have known to stimulate the production of insulin in β pancreatic cells to increase the expression of insulin gene and simultaneously inhibit the insulin degradation. Type I diabetics patients due to other complications of genetic, autoimmune, environmental, or viral infections have benefits with such botanicals. The β-carbolines derivatives and pinoline isolated from *Tribulus terrestris* have reported to promote insulin secretion by pancreatic cells through regeneration. Secretion of insulin from secretary granules and inhibiting the insulin degradation is also another important mechanism to exhibit antidiabetic effects. Most of type 1 patients dependent on exogenous insulin in order to maintain the glucose level at normal range. The β-carbolines isolated from *Tribulus terrestris* have been reported to promote secretion of insulin from pancreatic β-cells. Marmelon, aegelin, and marmesin isolated from *Aegle marmelos* have reported for regeneration of pancreatic β-cells, thereby increasing the insulin secretions. Many phytomolecules that includes harmine, pinoline, gurmarin, betaine, choline, trimethylamine, (–)-hydroxycitric acid, castanospermine, epifagomine, fagomine, leucine, isoleucine, alanin, polypeptide-P, lectins, mistletoe lectin I, II, III, viscotoxin B, cycliton, furfural, caprylic acid, DEHP, aloin, barbaloin, isobarbaloine, aloetic acid, aloe-emodin, emodin, cinnamic acid, crysophanic acid, vicine, torachrysone, toralactone, rhein, alaternin, camphor, eugenol, trans-βocimene, geraniol, α-pinene, limonene, p-cymene, 1,8-cineole, thujone, glucomannan, caryophylline, vin-α-ginsenoside R3, momordin, momordicine, charantin, Momorcharaside A/B, momorcharin A/B, cucurbitacin B, isocucurbitacin B, momordin-a, luffin-a, Kotalanol, salacinol, Arbutin, eriolin, Citrullol, colocynthin, elaterin, elatericin B, colosynthetin, Leucocyanidin, pelarogonidin, Taraxacin, chrysin, isoquercitrin, myrciaphenones A/B, myrciacitrins I/II, α-Cephalin, myricetin-3'-glucoside, ambrettolide, flavanols, flavones, flavanones, quercetin, apigenin, rutin, apigenin-7-O-glucoside, naringenin, α-terpineol, hexanol, shamimin, leucopelargonidin, dulcitol, matteuorien, matteuorienin matteuorienate A-C and glucosamines.

Peroxiosome proliferator-activated receptors (PPARs), a set of transcription factors that are suggested to involve in the multiplication of peroxisomes of cells. These peroxisomes are involved in the proteins, carbohydrates, and lipid metabolism. PPARs are of three types namely α, β/ǿ and ɣ. The α-PPARs are involved in the metabolism of lipids, while β-PPARs has been implicated in diseases such as cancer, infertility, and dyslipidemia but PPARs are involved in the insulin resistance. Dehydroglyasperin-C, dehydroglyasperin-D, glyasperin-B, glycycoumarine, glycyrin, and glyasperin-D derived from Glycyrrhiza uralensis and neolgnan dehydroeugenol derived from Syzygium aromaticum have reported to inhibit the ɣ-PPARs. The ɣ-PPARs are most important target for curcumin and 6-gingerol that mediate the signaling through cyclic AMP for alteration of cellular functions.

9.2.3.8 PHYTOMOLECULES THAT INCREASES THE GLYCOGEN SYNTHESIS

The plant metabolites that are known to increase in glycogen synthesis through the inhibition of glycogen synthase. Some reported metabolites includes aegelin, marmesin, marmelosin, allicin, apigenin, alliin, polypeptide-P, Lectins, mistletoe lectin I, II, III, viscotoxin-B, cycliton, Furfural, caprylic acid, Bis (2-ethyl hexyl) phthalate (DEHP), momordin, momordicine, charantin, momorcharaside A and B, momorcharin A and B, cucurbitacin-B, isocucurbitacin B, kotalanol, salacinol, arbutin, eriolin, citrullol, colocynthin, elaterin, elatericin-B, colosynthetin, leucocyanidin, pelarogonidin, Cytrus bioflavonoids (hesperidin, naringin, amarogentin, swerchirin, chirantin, and gentiopicrin. Some plants that have been reported in plants such as *Aegle marmelos, Tribulus terrestris, Caralluma edulis, Syzygium cumini, Acacia Arabica*, etc.

Other mechanism for antidiabetic potential reported by the plant metabolites includes Insulin secretion, regeneration of pancreatic β cells, Decrease glycosuria and AGEs, Glucose transport, Cholesterol synthesis, insulinonematic activity, lipid, and glucose metabolism, PPAR activation and HMG Co-A suppression. These miscellaneous targets of the antidiabetic drugs have a potential role in the advanced stages of diabetes.

9.3 CONCLUSION

Several scientific reports prove that plants have significant medicinal values which seems to be an alternative and effective strategies in the discovery of novel therapeutic drugs for management of diabetics. Commonly recommended compounds currently in market, metformin acts as anti-hyperglycemic compound was discovered from *Galega ofcinalis*. Several research reports from antibiotics to anticancer drugs or the backbone of the potent drugs was derived from plant sources. However, several plant-based antidiabetic drugs are in clinical trials. Scientific investigation through the evidence-based medicinal research have been helpful in discovery of novel highly efficient natural products of for the diabetic management is needed in order to address the multifaceted mechanistic control of hyperglycemic conditions. Further, the molecular basis of diabetics has discovered various molecular targets for the diabetics and its implications. Thereby, the targeted therapy for diabetes seems to be the essential and attractive area of research.

KEYWORDS

- **clinical outcome**
- **mode of action**
- **molecular targets**
- **multi-targets**
- **plant-based antidiabetics**
- **polyphenols**
- **preclinical studies**

REFERENCE

Abesundara, K. J. M., Mastui, T., & Matsumoto, K., (2004). α-glucosidase inhibitory activity of some Sri Lanka plant extracts, one of which, *Cassia auriculata*, exerts a strong antihyperglycemic effect in rats comparable to the therapeutic drug acarbose. *J. Agric. Food Chem., 52*, 2541–2545.

Adaramoye, O., Amanlou, M., HabibiRezaei, M., et al., (2012). Methanolic extract of African mistletoe (Viscum album) improves carbohydrate metabolism and hyperlipidemia in streptozotocin-induced diabetic rats. *Asian Pac. J. Trop. Med., 5*, 427–433.

Akhtar, M. S., & Iqbal, J., (1991). Evaluation of the hypoglycaemic effect of *Achyranihes aspera* in normal and alloxan-diabetic rabbits. *J. Ethnopharmacol., 31*, 49–57.

Alarcon-Aguilara, F. J., Roman-Ramos, R., Perez-Gutierrez, S., et al., (1998). Study of the antihyperglycemic effect of plants used as antidiabetics. *J. Ethnopharmacol., 61*, 101–110.

Ali, H., Houghton, P. J., & Soumyanath, A., (2006). α-Amylase inhibitory activity of some Malaysian plants used to treat diabetes; with particular reference to *Phyllanthus amarus*. *J Ethnopharmacol., 107*, 449–455.

Ananthi, J., Prakasam, A., & Pugalendi, K. V., (2003). Antihyperglycemic activity of *Eclipta alba* leaf on alloxan-induced diabetic rats. *J. Biol. Med., 76*, 97–102.

Baldwa, V. S., Bhandari, C. M., Pangaria, A., & Goyal, R. K., (1977). Clinical trial in patients with diabetes mellitus of an insulin-like compound obtained from plant source. *Upsala J. Med. Sci., 82*(1), 39–41.

Bharti, S. K., Krishnan, S., & Gupta, A. K., (2013). *Herbal Formulation to Combat Type 2 Diabetes Mellitus*. Germany: LAMBERT Academic Publishing.

Bharti, S. K., Krishnan, S., & Kumar, A., (2016). Phytotherapy for diabetes mellitus: Back to nature. *Minerva Endocrinol., 41*, 143–146.

Bharti, S. K., Krishnan, S., Kumar, A., & Kumar, A., (2018). Antidiabetic phytoconstituents and their mode of action on metabolic pathways. *Ther. Adv. Endocrinol. Metab., 9*(3), 81–100.

Boucher, J., Kleinridders, A., & Kahn, C. R., (2014). Insulin receptor signaling in normal and insulin-resistant states. *Cold Spring Harb. Perspect. Biol., 6*(1), a009191.

Briones, R. M., Sarmah, A. K., & Padhye, L. P., (2016). A global perspective on the use, occurrence, fate and effects of antidiabetic drug metformin in natural and engineered ecosystems. *Environ. Pollut., 219*, 1007–1020.

Broadhurst, C. L., Polansky, M. M., & Anderson, R. A., (2000). Insulin-like biological activity of culinary and medicinal plant aqueous extracts in-vitro. *J. Agric. Food Chem., 48*, 849–852.

Buckingham, B., Beck, R. W., Ruedy, K. J., Cheng, P., Kollman, C., et al., (2013). Effectiveness of early intensive therapy on $-cell preservation in type 1 diabetes. *Diabetes Care, 36*, 4030–4035.

Campbell, J. E., & Drucker, D. J., (2015). Islet α cells and glucagon—Critical regulators of energy homeostasis. *Nat. Rev. Endocrinol., 11*(6), 329–338.

Campbell, R. K., & White, J. R., (2008). More choices than ever before: Emerging therapies for type 2 diabetes. *Diabetes Educ., 34*(3), 518–534.

Cartee, G. D., (2015). Mechanisms for greater insulin-stimulated glucose uptake in normal and insulin-resistant skeletal muscle after acute exercise. *Am. J. Physiol. Endocrinol. Metab., 309*(12), E949–959.

Cefalo, C. M. A., Cinti, F., Mofa, S., Impronta, F., Sorice, G. P., Mezza, T., et al., (2019). Sotaglifozin, the first dual SGLT inhibitor: Current outlook and perspectives. *Cardiovasc Diabetol., 18*(1), 20.

Chao, C. Y., & Huang, C. J., (2003). Bitter gourd (*Momordica charantia*) extract activates peroxisome proliferator-activated receptors and upregulates the expression of the acyl CoA oxidase gene in H4IIEC3 hepatoma cells. *J. Biomed. Sci, 10*, 782–791.

Chattopadhyay, R. R., (1999). A comparative evaluation of some blood sugar lowering agents of plant origin. *J. Ethnopharmacol., 67*, 367–372.

Chaudhury, A., Duvoor, C., Dendi, V. S. R., Kraleti, S., Chada, A., et al., (2017). Clinical review of antidiabetic drugs: Implications for type 2 diabetes mellitus management. *Front. Endocrinol., 8*. 10.3389/fendo.2017.00006.

Cherian, S., Sheela, C. G., & Augusti, K. T., (1995). Insulin sparing action of leucopelergonidin derivative isolated from *Ficus bengalensis* Linn. *Indian J. Exp. Biol., 33*, 608–611.

Chiasson, J. L., Josse, R. G., Gomis, R., Hanefeld, M., Karasik, A., & Laakso, M., (2002). Acarbose for prevention of type 2 diabetes mellitus: The STOP-NIDDM randomized trial. *Lancet, 359*(9323), 2072–2077.

Cragg, G. M., Grothaus, P. G., & Newman, D. J., (2014). New horizons for old drugs and drug leads. *J. Nat. Prod., 77*(3), 703–723.

Davidson, J. A., Parente, E. B., & Gross, J. L., (2008). Incretin mimetics and dipeptidyl peptidase-4 inhibitors: Innovative treatment therapies for type 2 diabetes. *Arq. Bras. Endocrinol. Metabol., 52*(6), 1039–1049.

DeFronzo, R. A., & Tripathy, D., (2009). Skeletal muscle insulin resistance is the primary defect in type 2 diabetes. *Diabetes Care, 32.*

Desai, S., & Tatke, P., (2015). Charantin: An important lead compound from *Mormodica charantia* for the treatment of diabetes. *J. Pharmacogn. Phytochem., 3*(6), 163–166.

Diamant, M., & Heine, R., (2003). Thiazolidinediones in type 2 diabetes mellitus. *Drugs, 63*(13), 1373–1406.

Eleazu, C. O., Eleazu, K. C., & Iroaganachi, M. A., (2016). Effect of cocoyam (Colocasia esculenta), unripe plantain (Musa paradisiaca) or their combination on glycated hemoglobin, lipogenic enzymes and lipid metabolism of streptozotocin-induced diabetic rats. *Pharm. Biol., 54*, 91–97.

Eno, A. E., Ofem, O. E., Nku, C. O., et al., (2008). Stimulation of insulin secretion by Viscum album (mistletoe) leaf extract in streptozotocin-induced diabetic rats. *Afr. J. Med. Med. Sci., 37*, 141–147.

Gholap, S., & Kar, A., (2004). Hypoglycaemic effects of some plant extracts are possibly mediated through inhibition in corticosteroid concentration. *Pharmazie, 59*, 876–878.

González-Tejero, M. R., Casares-Porcel, M., Sánchez-Rojas, C. P., et al., (2008). Medicinal plants in the Mediterranean area: Synthesis of the results of the project Rubia. *J Ethnopharmacol., 116*, 341–357.

Gray, A. M., & Flatt, P. R., (1999). Insulin-releasing and insulin-like activity of the traditional antidiabetic plant *Coriandrum sativum* (coriander). *Br. J. Nutr, 81*, 203–209.

Grover, J. K., & Vats, V., (2000). Shifting paradigm from conventional to alternate medicine: An introduction on traditional Indian medicine. *Asia Pac. Biotech News, 5*, 28–32.

Gutmann, T., Kim, K. H., Grzybek, M., Walz, T., & Coskun, Ü., (2018). Visualization of ligand-induced transmembrane signaling in the full-length human insulin receptor. *J. Cell Biol., 217*(5), 1643–1649.

Gwatidzo, S. D., & Williams, J. S., (2017). Diabetes mellitus medication use and catastrophic healthcare expenditure among adults aged 50+ years in China and India: Results from the WHO study on global AGEing and adult health (SAGE). *BMC Geriatrics, 17.*

Haeusler, R. A., Mcgraw, T. E., & Accili, D., (2017). Biochemical and cellular properties of insulin receptor signaling. *Nat. Rev. Mol. Cell Biol., 19*(1), 31.

Hannan, J. M., Marcnah, L., Au, L., et al., (2006). *Ocimum sanctum* leaf extracts stimulate insulin secretion from perfused pancreas, isolated islets and clonal pancreatic β-cells. *J. Endocrinol., 189*, 127–136.

Hasan, S. S., Clavarino, A. M., Mamun, A. A., & Kairuz, T., (2015). A comparative drug utilization study of the treatment of diabetes in Malaysia and Australia. *Aust. Med. J., 8*, 179–188.

Hii, S. C. T., & Howell, S. L., (1984). Effects of epicatechin on rat islets of Langerhans. *Diabetes, 33*, 291–296.

Hou, C. C., Lin, S. J., Cheng, J. T., et al., (2003). Hypoglycemic dimeric guianolides and a lignan glycoside from *Lactuca indica*. *J. Nat. Prod., 66*, 625–629.

Hsia, D. S., Grove, O., & Cefalu, W. T., (2018). An update on SGLT2 inhibitors for the treatment of diabetes mellitus. *Curr. Opin. Endocrinol Diabetes Obes., 24*(1), 73–79.

Hui, H., Tang, G., & Go, V. L., (2009). Hypoglycemic herbs and their action mechanisms. *Chin Med., 12*, 4–11.

International Diabetes Federation, (2015). *IDF Diabetes Atlas* (7th edn.). Brussels: International Diabetes Federation.

Ismail, H. F., Majid, F. A. A., & Hashim, Z., (2017). Eugenia polyantha enhances adipogenesis via CEBP-α and adiponectin overexpression in 3T3-L1. *Chem. Eng. Trans., 56*(1), 1117–1122.

Jarald, E. E., Sheeja, E., Motwani, S., et al., (2008). Comparative evaluation of antihyperglycaemic and hypoglycaemic activity of various parts of *Catharanthus roseus* Linn. *Res. J. Med. Plant, 2*, 10–15.

Jung, U. J., Lee, M. K., Jeong, K. S., et al., (2004). The hypoglycemic effects of hesperidin and naringin are partly mediated by hepatic glucose regulating enzymes in C57BL/KsJ-db/db mice. *J. Nutr., 134*, 2499–2503.

Kamalakkannan, N., & Prince, P. S. M., (2005). The effect of *Aegle marmelos* fruit extract in streptozotocin diabetes: A histopathological study. *J. Herb. Pharmacol, 5*, 87–96.

Kaneko, K., Shirotani, T., Araki, E., Matsumoto, K., Taguchi, T., Motoshima, H., et al., (1999). Insulin inhibits glucagon secretion by the activation of PI3-kinase in In-R1-G9 cells. *Diabetes Res. Clin. Pract., 44*(2), 83–92.

Kang, C., Lee, H., Jung, E. S., Seyedian, R., Jo, M., Kim, J., et al., (2012). Saffron (*Crocus sativus* L.) increases glucose uptake and insulin sensitivity in muscle cells via multipathway mechanisms. *Food Chem., 135*(4), 2350–2358.

Kar, A., Choudhary, B. K., & Bandyopadhyay, N. G., (2003). Comparative evaluation of hypoglycaemie activity of some Indian medicinal plants in alloxan diabetic rats. *J. Ethnopharmacol., 84*, 105–108.

Khan, B. A., Abraham, A., & Leelamma, S., (1995). Hypoglycemic action of *Murraya koenigii* (curry leaf) and *Brassica juncea* (mustard): Mechanism of action. *Indian J. Biochem. Biophys., 32*(2), 106–108.

Khanna, P., Jain, S. C., Panagariya, A., & Dixit, V. P., (1981). Hypoglycemic activity of polypeptide-p from a plant source. *J. Nat. Prod., 44*(6), 648–655.

Khosla, P., Gupta, D. D., & Nagpal, R. K., (1995). Effect of *Trigonella foenum* graecum (Fenugreek) on serum lipids in normal and diabetic rats. *Indian J. Physiol. Pharmacol., 27*, 89–93.

Krishnakumar, K., Augusti, K. T., & Vijavammal, P. L., (1999). Hypoglycaemic and anti-oxidant activity of Salacia oblonga wall extract in streptozotocin-induced diabetic rats. *Indian J. Physiol. Pharmacol., 43*, 510–514.

Kumar, A., Bharti, S. K., & Kumar, A., (2014). Type 2 diabetes mellitus: The concerned complications and target organs. *Apollo Med., 11*, 161–166.

Kumar, A., Bharti, S. K., & Kumar, A., (2017). Therapeutic molecules against type 2 diabetes: What we have and what are we expecting? *Pharmacol. Rep., 69*, 959–970.

Kumari, K., & Augusti, K. T., (2007). Lipid-lowering effect of S-methyl cysteine sulfoxide from Allium cepa Linn in high cholesterol diet-fed rats. *J. Ethnopharmacol., 109*, 367–371.

Lemus, I., Garcia, R., Delvillar, E., et al., (1999). Hypoglycemic activity of four plants used in Chilean popular medicine. *Phytother. Res., 13*, 91–94.

Li, Y., Qi, Y., Huang, T. H., et al., (2008). Pomegranate flower: A unique traditional antidiabetic medicine with dual PPAR-alpha/-gamma activator properties. *Diabetes Obes. Metab., 10*, 10–17.

Liu, K. Z., Li, J. B., Lu, H. L., et al., (2004). Effects of *Astragalus* and saponins of *Panax notoginseng* on MMP-9 in patients with type 2 diabetes. *Macroangiopathy, 29*, 264–266.

Loke, Y. K., Singh, S., & Furberg, C. D., (2009). Long-term use of thiazolidinediones and fractures in type 2 diabetes: A meta-analysis. *Can. Med. Assoc. J., 180*(1), 32.
Luo, J., Fort, D. M., Carlson, T. J., et al., (1998). *Cryptolepis sanguinolenta*: An ethnobotanical approach to drug discovery and the isolation of a potentially useful new antihyperglycaemic agent. *Diabet. Med., 15*, 367–374.
Madiraju, A. K., Qiu, Y., Perry, R. J., Rahimi, Y., Zhang, X. M., Zhang, D., et al., (2018). metformin inhibits gluconeogenesis by a redox-dependent mechanism *in vivo*. *Nat. Med., 24*(9), 1384–1394.
Mannan, N., Boucher, B. J., & Evans, S. J. W., (2000). Increased waist size and weight in relation to consumption of Areca catechu (betel-nut); a risk factor for increased glycaemia in Asians in East London. *Br. J. Nutr., 83*, 267–275.
Mayer, J. P., Zhang, F., & Dimarchi, R. D., (2007). Insulin structure and function. *Pept. Sci., 88*(5), 687–713.
Miinea, C. P., Sano, H., Kane, S., Sano, E., Fukuda, M., Peranen, J., et al., (2005). AS160, the Akt substrate regulating GLUT4 translocation, has a functional rab GTPase-activating protein domain. *Biochem. J., 391*(Pt 1), 87–93.
Moon, Y. H., Nam, S. H., Kang, J., et al., (2007). Enzymatic synthesis and characterization of arbutin glucosides using glucansucrase from *Leuconostoc mesenteroides*. *Appl. Microbiol. Biotechnol., 27*, 559–567.
Morigny, P., Houssier, M., Mouisel, E., & Langin, D., (2016). Adipocyte lipolysis and insulin resistance. *Biochimie., 125*, 259–266.
Musso, G., Gambino, R., Cassader, M., & Paschetta, E., (2019). Efficacy and safety of dual SGLT 1/2 inhibitor sotaglifozin in type 1 diabetes: Meta-analysis of randomized controlled trials. *BMJ., 365*, l1328.
Naisheng, B., Kan, H., Roller, M., et al., (2008). Active compounds from Lagerstroemia speciosa, insulin-like glucose uptake-stimulatory/inhibitory and adipocyte differentiation-inhibitory activities in 3T3-L1 cells. *J. Agric. Food Chem., 56*, 11668–11674.
Nanjan, M. J., Mohammed, M., & Kumar, B. R. P., (2018). Chandrasekar MJN. Thiazolidinediones as antidiabetic agents: A critical review. *Bioorg. Chem., 77*, 548–567.
Noor, H., & Ashcroft, S. J. H., (1998). Insulinotropic activity of *Tinospora crispa* extract: Effect on β-cell Ca^{2+} handling. *Phytother. Res, 12*, 98–102.
Oh, Y. S., (2015). Plant-derived compounds targeting pancreatic beta cells for the treatment of diabetes. *Evid. Based Complement Altern. Med., 2015*, 629863.
Pataky, M. W., Yu, C. S., Nie, Y., Arias, E. B., Singh, M., Mendias, C. L., et al., (2019). Skeletal muscle fiber type-selective effects of acute exercise on insulin-stimulated glucose uptake in insulin-resistant, high-fat-fed rats. *Am. J. Physiol. Endocrinol. Metab., 316*(5), E695–706.
Petersen, M. C., & Shulman, G. I., (2018). Mechanisms of insulin action and insulin resistance. *Physiol. Rev., 98*(4), 2133–2223.
Ponnachan, P. T., Paulose, C. S., & Panikkar, K. R., (1993). Effect of leaf extract of *Aegle marmelose* in diabetic rats. *Indian J. Exp. Biol., 31*, 345–347.
Prabhakar, P. K., & Doble, M. A., (2008). Target-based therapeutic approach towards diabetes mellitus using medicinal plants. *Current Diabetes Reviews, 4*, 291–308.
Prakasam, A., Sethupathy, S., & Pugalendia, K. V., (2005). Antiperoxidative and antioxidant effects of *Casearia esculenta* root extract in streptozotocin-induced diabetic rats. *Yale J. Biol. Med., 78*, 15–23.

Quinn, P. G., & Yeagley, D., (2005). Insulin regulation of PEPCK gene expression: A model for rapid and reversible modulation. *Curr. Drug Targets Immune Endocr. Metabol. Disord., 5*(4), 427–437.

Rieg, T., & Vallon, V., (2018). Development of SGLT1 and SGLT2 inhibitors. *Diabetologia, 61*(10), 2079–2086.

Roman-Ramos, R., Flores-Saenz, J. L., & AlarconAguilar, F. L., (1995). Anti-hyperglycemic effect of some edible plants. *J. Ethnopharmacol., 48*, 25–32.

Rotshteyn, Y., & Zito, S. W., (2004). Application of modified *in vitro* screening procedure for identifying herbals possessing sulfonylurea-like activity. *J. Ethnopharmacol., 93*(2, 3), 337–344.

Saltiel, A. R., & Kahn, R., (2001). Insulin signaling and the regulation of glucose and lipid metabolism. *Nature, 414*(6865), 799–806.

Samuel, V. T., & Shulman, G. I., (2012). Mechanisms for insulin resistance: Common threads and missing links. *Cell, 148*(5), 852–871.

Sarayani, A., Rashidian, A., & Gholami, K., (2014). Low utilization of diabetes medicines in Iran, despite their affordability (2000–2012): A time-series and benchmarking study. *BMJ Open,* 005859.

Sarkar, S., Pranava, M., & Marita, R., (1996). Demonstration of the hypoglycemic action of *Momordica charantia* in a validated animal model of diabetes. *Pharmacol. Res., 33*, 1–4.

Sarker, S. D., Latif, Z., & Gray, A. I., (2006). *Natural Product Isolation* (2nd edn., pp. 1–26). Totowa: Humana Press Inc.

Seino, Y., Fukushima, M., & Yabe, D., (2010). GIP and GLP-1, the two incretin hormones: Similarities and differences. *J. Diabetes Invest., 1*(1, 2), 8–23.

Silva, R. M., Santos, F. A., Rao, V. S., et al., (2001). Blood glucose and triglyceride-lowering effect of trans-dehydrocrotonin, a diterpene from *Croton cajucara* benth in rats. *Diabetes Obes. Metab., 3*, 452–456.

Singh, R. K., Mehta, S., Jaiswal, D., et al., (2009). Antidiabetic effect of Ficus bengalensis aerial roots in experimental animals. *J. Ethnopharmacol., 123*, 110–114.

Stöckli, J., Fazakerley, D. J., & James, D. E., (2011). GLUT4 exocytosis. *J. Cell Sci., 124*, 4147.

Swinnen, S. G., Hoekstra, J. B., & DeVries, J. H., (2009). Insulin therapy for type 2 diabetes. *Diabetes Care, 32* (Suppl 2), S253–259.

Ugochukwu, N. H., & Babady, N. E., (2003). Antihyperglycemic effect of aqueous and ethanolic extracts of *Gongronema latifolium* leaves on glucose and glycogen metabolism in livers of normal and streptozotocin-induced diabetic rats. *Life Sci., 73*(15), 1925–1938.

Unnikrishnan, R., Anjana, R. M., & Mohan, V., (2016). Diabetes mellitus and its complications in India. *Nat. Rev. Endocrinol., 12*, 357–370.

Van De, V. M., Roux, S., Bungu, L. C., Louw, J., Crouch, N. R., Grace, O. M., et al., (2008). Antidiabetic screening and scoring of 11 plants traditionally used in South Africa. *J Ethnopharmacol., 119*(1), 81–86.

Vessal, M., Hemmati, M., & Vasei, M., (2003). Hypoglycemic effects of quercetin in streptozocin-induced diabetic rats: Comparative biochemistry and physiology. *Toxicol. Pharmacol., 135*, 357–364.

Wadood, A. N., & Shah, S. A., (1989). Effects of *Acacia arabica* and *Caralluma edulis* on blood glucose levels of normal and alloxan diabetic rabbits. *J. Pak. Med. Assoc., 39*, 208–212.

Waltner-Law, M. E., Wang, X. L., Law, B. K., et al., (2002). Epigallocatechin gallate: A constituent of green tea, represses hepatic glucose production. *J. Biol. Chem., 277*, 34933–34940.

Wanders, R. J. A., Grunsven, E. G., & Jansen, G. A., (2000). Lipid metabolism in peroxisomes: Enzymology, functions and dysfunction of the fatty acid α and β oxidation system in humans. *Biochem. Soc. Trans., 28*, 141–148.

Wang, H., Kan, W., Cheng, T., Yu, S., Chang, L., & Chuu, J., (2014). Differential antidiabetic effects and mechanism of action of charantin-rich extract of Taiwanese *Momordica charantia* between type 1 and type 2 diabetic mice. *Food Chem. Toxicol., 69*, 347–356.

Wanner, C., & Marx, N., (2018). SGLT2 inhibitors: The future for treatment of type 2 diabetes mellitus and other chronic diseases. *Diabetologia, 61*(10), 2134–2139.

Whiting, D. R., Guariguata, L., Weil, C., & Shaw, J., (2011). IDF diabetes atlas: Global estimates of the prevalence of diabetes for 2011 and 2030. *Diabetes Res. Clin. Pract., 94*, 311–321.

Wilcox, G., (2005). Insulin and insulin resistance. *Clin. Biochem. Rev., 26*(6), 19–39.

Wild, S., Roglic, G., Green, A., Sicree, R., & King, H., (2004). Global prevalence of diabetes: Estimates for the year 2000 and projections for 2030. *Diabetes Care, 27*, 1047–1053.

Yatoo, M. I., Saxena, A., Gopalakrishnan, A., Alagawany, M., & Dhama, K., (2017). Promising antidiabetic drugs, medicinal plants and herbs: An update. *Int. J. Pharmacol., 13*, 732–745.

Yibchok-Anun, S., Adisakwattana, S., Yao, C. Y., Sangvanich, P., Roengsumran, S., & Hsu, W. H., (2006). Slow acting protein extract from fruit pulp of *Momordica charantia* with insulin secretagogue and insulinomimetic activities. *Biol Pharm Bull., 29*(6), 1126–1131.

Ziyyat, A., Lcgssyer, A., Mekhfi, H. R., et al., (1997). Phytotherapy of hypertension and diabetes in oriental Morocco. *J. Ethnopharmacol., 58*, 45–54.

Part IV

Recent Biotechnological Advancements of Antidiabetic Plants and Plant-Based Anti-Diabetes Drugs and Future Perspectives with Genome Editing Technologies

CHAPTER 10

Molecular Breeding and Application of Molecular Markers for Improvement of Antidiabetic Medicinal Plants

SAMEENA SHAIK,[1] ANAND KUMAR,[1] GUNTI MALLIKARJUNA,[1] GURU LAKSHMI KOLA,[1] P. CHANDRA OBUL REDDY,[2] and A. CHANDRA SEKHAR[1]

[1]*Molecular Genetics and Functional Genomics Laboratory, Department of Biotechnology, School of Life Sciences, Yogi Vemana University, Kadapa – 516005, Andhra Pradesh, India, E-mails: chandrasekhar9@yahoo.com; acsekhar@yogivemanauniversity.ac.in (A. C. Sekhar)*

[2]*Plant Molecular Biology Laboratory, Department of Botany, School of Life Sciences, Yogi Vemana University, Kadapa – 516005, Andhra Pradesh, India*

ABSTRACT

Diabetes mellitus (DM) is a common chronic metabolic disorder characterized by the destruction of pancreatic β cells or reduced insulin activity/secretion, which has a significant effect on all age group peoples around the world. About 20% of the global population is known to get affected by this disease as per the World Health Organization (WHO) globally it is projected that about 336 million people will be affected by 2030. Many associated adverse effects have been found with the formulations that are used currently to treat this disease. Globally, the use of plants and their parts as source of medicines for both preventive and curative purposes is an ancient tradition. The quality of herbal medicine depends on the authenticity of the

Antidiabetic Potential of Plants in the Era of Omics. Deepu Pandita, PhD, Anu Pandita, MSc, & Chander Bhanu, MBBS (Eds.)

raw material. A lot of adulteration has been attributed to increased global demand for raw herbal drugs. The primary source of adulteration is in the plant resource, mainly due to the misidentification of phenotypically similar plant species from the same genus. Several physical-chemical methods such as chromatography and spectroscopy-based methods are widely used to detect the adulteration in the herbal powders, however these methods are influenced by genetic, developmental stage and environmental factors. Similarly, medicinal property of any plant species depends on the production of phytochemical compounds or secondary metabolites. Thus, there is a need for the development of alternative methods. The medicinal plants interact with their environment that determincs their active constituents. Molecular markers analysis deals with the identification and exploitation of naturally occurring DNA sequence polymorphisms; these markers can be utilized to detect the adulteration, accurate identification of plant species and improvement of medicinal plants for their active ingredients through Marker Assisted Selection (MAS).

10.1 INTRODUCTION

Medicinal plants have been utilized all over the world from times unknown, to cure the human and animal diseases and maintains the health. Use of medicinal plants in those days are utmost importance. In India medicinal plants industry is growing at the rate of 7–15% annually, and it may be 20% in the future, which shows a huge requirement of medicinal plants in the upcoming time. The use of medicinal plants has been in vogue since ancient times in India, which is described in ancient books like Ayurveda and charaksamhita. Medicinal plants are known for their secondary metabolites with healing/curing medicinal properties. The natural medicines are much better and safer than synthetic drugs. In recent years use of natural medicines has increased tremendously, leading to a rapid growth of phytopharmaceutical industries. Diabetes mellitus (DM) is characterized metabolic disorder with hyperglycemic conditions that results from defects in insulin secretion, insulin action or both. Diabetes are majorly three types: type 1, type 2, and gestational DM. Deficiency of insulin production by β pancreatic cells is the main causative for Type 1 diabetes, while obesity and initial phase of progressive insulin resistance are for Type 2 diabetes. Sustained hyperglycemic condition is a common symptom of both forms of diabetes. Destitational diabetes shows the degree of glucose intolerance during pregnancy, irrespective of glycemic status after delivery. However, it is disappeared after delivery. It affect 3 to

10% of pregnancies. Tiredness, increased thirst, and uncontrollable frequent urination are the common symptoms of diabetes. As described by Zaid et al. (2008), it is a combination of metabolic disorders but not a single disease. It is associated with secondary complications including diabetic retinopathy (DR), which is assessed by glycated hemoglobin levels as well with a wide range of microvascular complications (Priyia, 2016; Hiller et al., 1988). Dietary fiber is known for its ability to reduce the incidence of diabetes.

About 20% of the global population is affected by diabetes, with a prediction of about 336 million people by 2030 as reported by the World Health Organization (WHO). It is assumed that 75 to 80% of the world population will be rely on herbal medicines for fundamental and primary health care mainly due to their cultural and traditional acceptability and better compatibility with the human body and less side effects. India is the second largest exporter of herbal medicine with less than 2% global market share. Rich traditional knowledge and untapped herbal occurrence create a novel untapped potential for molecular targets using combined OMICS and systems sciences (JagajjitSahu et al., 2014). India is a rich in bioresources, molecular breeding strategies facilitate the implementation of these bioresources and improve the herbal plants qualities. Molecular markers have been proved as a boon in plant breeding strategies for genetic diversity studies as well for managing the plant resources (Ford-Lloyd et al., 1997). Molecular-level identification of plant species has more meaningful rather than phenotypic based identification, which was more environmental depended (Powell et al., 1996). In the last decade, various types of molecular markers have been developed and established for authentication of different medicinal plants species. DNA based molecular markers are specific because of its unique genetic sequence (Alireza, 2013). Traditional breeding strategies are being employed for the genetic identification, cultivation as well improvement of herbal plants which has their own limits. While, the advent of molecular markers and their applications for the identification of the genuine herbal genetic resources has revolutionized the breeding strategies for the herbal plant improvement (Leela et al., 2009). Occurrence/extend of polymorphism attribute to its genome size as well for the diversity that a species contain. Clonal selection has been proposed as a pathogenetic mechanism in various chronic disease such as diabetes. Application of molecular markers, and Real-Time PCR technique make the life of a conventional breeder easy over for elite germplasm identification as well their clonal propagation (Dipendra, 2017). Molecular markers analysis deals with the identification and exploitation of naturally occurring DNA sequence variations; these markers can be utilized to detect the adulteration, accurate identification of plant species and improvement of medicinal plants through Marker Assisted Selection (MAS). Power of herbal plants depend of the bioactive content of phytochemicals/ secondary metabolites found. Recent outburst of the pandemics and their control by the herbal based components highlight the power of herbal plants. Bioactive components are known for their capacity to modulate the metabolic procedures resulting the better wellbeing (Marchetti et al., 2019; Chang et al., 2019). Thus, there is a need for the development of alternative methods.

10.2 ANTIDIABETIC PLANTS

There are plenty of plant based products available or consumed to deal with DM. With about 400 antidiabetic plant combinations, 700 formulations/ recipes are available to combat the diabetes (Sivani et al., 2017). There is a sudden increase in the percent utilization of the herbal based medicines in last decade or so across the globe. Well documented traditional knowledge and recorded evidence of practiced knowledge on herbal based products has a positive impact on the traditional Indian Ayurveda that has the capacity to capitalize the worldwide market (Thatte, 2009). Advanced resource-poor regions of developing countries depend majorly on medicinal plants or plant based products for diabetes management (Rawat et al., 2013). Most of the antidiabetic plants belongs to the family of *Rosaceae, Moraceae, Liliaceae, Araliaceae, Euphorbiaceae, Cucurbitaceae, and Asteraceae*:

1. **Bitter Gourd (*Momordica charantia* L.):** It is a popular and highly consumable vegetable belongs to Cucurbitaceaethe family and originated in eastern Asia. It is traditionally cultivated as a medicinal plant and vegetable for the treatment of diabetics because of its medicinal and nutritional properties. It is widely cultivated across all regions of India, South America, Malaysia, and Africa, China (Minraj et al., 1993). Vernacularly this is known as karela. Because of its bitter taste, this is also called bitter gourd. Polypeptide-p, an insulin analog containing plant parts are used in the traditional medicines of India, which are used in treatments of diabetes (Wang et al., 2014). Inhibitory capacity of bitter melon on a-glucosidase activity to suppresses postprandial hyperglycemia has been reported (Ubenso et al., 2007). Bitter gourd has the capacity to manage almost all secondary symptoms of diabetes viz., polydipsia, polyphagia, and polyuria. It can also reduce the urinary excretion of excess glucose

without admitting insulin (Karunanayake et al., 1984). Ethanolic extract of bitter guard lowered blood sugar levels fasted. *M. charntia* contains two compounds one is inhibit hexokinase activity and other glucose uptake (Meir et al., 1985). Bitter guard has the healing property on the pancreas by helping it to rejuvenate the beta cells and this by increasing the ability of the pancreas to produce insulin (Shetty et al., 2005). Bitter guard contains common active ingredient and phytonutrients such as polypeptide and hypoglycemic agent called Charantin, which has a capacity to increase glycogen synthesis in adipose and liver cells while glucose uptake in liver. Further free radical scavenging activity of the bitter guard has the capacity to act as an antioxidant agent and support production of healthy insulin levels.

2. ***Costus pictus***: Commonly known as insulin plant or spiral ginger belongs to family costaceae. This is originated in Mexico and introduced in India in the year 2000. These plant widely occur in Kerala. Local people traditionally use leaves of these plants for diabetes. Therefore, this is popularly called insulin plant (Ahamad et al., 2009). Antioxidative properties of this plants strongly support or correlate with its antidiabetic property (Majumdar et al., 2012). *C. pictus* is an herbaceous monocot, rhizomatous, perennial, branched grow in shady conditions. George et al. (2007), reported that fresh leaves contains 18 chemical compounds identified by using GC-MS. Chromatogram methods showed that, ether fraction contained major herbal ingredient, bis (2'-ethyl hexyl)-1, 2-benzene dicarboxylate (59.04%) and acid fraction contained major component such as haxadecanoic acid (44.53%) and 4, 8, 12, 16-tetra methyl heptadecan 4-olide (27.86%) which are known for their antidiabetic properties.
3. ***Catharanthus roseus***: Commonly known as Madagascar periwinkle and vernacularly known as sadabahar, belongs to apocynaceae family (Ranjan et al., 2008). These plant contains very high medicinal values because of this property it was known to control diabetes. Flowers are different in colors such as rose, pink, blue, and lavender. Plant attain the height up to 30–100 cm. It is cultivated for its antibiotic properties and alkaloids, which have antidiabetic activities. Extracts of roots and shoots are useful in diabetes. Variety of oral hypoglycemic agents are available that can use for diabetes treatment along with insulin. But none of them stand in the market due to their added side effects long-term microvascular and macrovascular

complications (Piedrola et al., 2001). In traditional Indian medication decoction of some leaves of these plants maintain the glucose levels in blood. Toosang et al. (2013) reported that alkaloids isolated from Catharanthus roseus leaves show high ameliorative effect in diabetes. *C. roseus* leaves are known to contain about 150 alkaloids. Four alkaloids were isolated from leaves are vindoline I, vindolidine II, vindolicine III and vindolinine IV, in which vindolinine IV shows high effect in mitigation of diabetes effect. The methanolic extracts of this plant shows good antidiabetic activity, while alcoholic extracts shows promising antihyperglycemic activities without significant changes in the body weight in alloxan-induced hyperglycemic rats (Mohammed et al., 2010). Methanolic extract of *catharanthusroseus* leaves contains flavonoids and terpenes.

4. ***Gymnema sylvestre*:** It belongs to the family *asclepiadaceae*. It is originated from India and Sri Lanka. It has a strong, thick, and woody climber which is used for the treatment of diabetes and decoction of leaves is useful in sore throat infection. Leaves are sweet in taste (Magda et al., 2011). Mastication *G. Sylvestre* (20 mg/day) leaves for about 20 to 60 days helps in normalizing blood sugar levels through pancreatic beta-cell repaired or regeneration in type 2 diabetes (Shanmugasundaram et al., 1990). It is one of the popular antidiabetic plant in Asia. By supplementing *gymnemasylyestre* leaves raised insulin levels in diabetic patient. AFLP marker has been applied to detect genetic divergence of progeny of *Gymnema sylvestre* from the parent (Magda et al., 2011).
5. ***Allium sativum* (Garlic):** It is a well-known popular spice in India and other counties. Garlic belongs to the family *Liliaceae*. This is species of onion genus and it grows up to 1.2 m in height. Garlic has versatile medicinal properties used since ancient times. Garlic is commercially available in the form of oil, powdered, and capsules which are used for medicinal purposes, Such as lowering the blood glucose level and improves the lipid profile (Elkayam et al., 2003). They are important dietary supplements. Ethanol extract of garlic is known to regulated the blood-sugar level and normalize the activity of both glucose-6-phosphatase and liver hexokinase. It also helps to elevate liver glycogen, serum insulin and free amino acids. Eidi et al. (2005) revealed that glibnclamide drug stimulates beta cells to release insulin and lower the glucose levels, but garlic extract is more effective than glibenclamide. Fresh garlic extracts decreases the

sugar cholesterol levels and improves the lipid profiles (Knipschild et al., 1989). Garlic has multi medicinal properties that prevent from bacterial and fungal infections, digestion, gastric problem, antioxidants, and enhance the immune system.

6. ***Eugenia jambolana* or *Syzygiumcumini* (Jamun):** It belongs to the family myrtaceae. This is branched, woody, a large tree commonly found in India, Thailand, Philippines, and other countries. Its fruit is berry like violate or black color. This is a popular fruit in India commonly known as jamun. This fruits have multiple medicinal properties. It is useful in liver protection, diarrhea, and dysenteries, sore throat. It has anti-cancer properties. Extract of jamun seed kill larva of *Aedesaegypti* and show antibacterial properties also. *E. jambolana* is a higher source of vitamin A and vitamin C. It has been widely used since ancient times for the treatment of diabetes. Fruit extract enhance and stimulate the secretion of insulin and suppress the insulinase activity from liver and kidney and decrease cholesterol level in body.
7. ***Cinnamomum zeylanicum*:** It is a member of the family Lauraceae and originated from Sri Lanka. This is widely used across East Asia and Europe. It is commonly known as Cinnamon. This is one of the spice, aromatic, and condiment flavoring, obtained from the inner bark of several trees from several species of the genus Cinnamomum. It contains mainly cinnamaldehyde an volatile oil ingredient that has antidiabetic properties that help lower down the glucose level in the blood. Cinnamon decreases total plasma sugar level and improves insulin sensitivity. It also significantly reduced gastric emptying and profoundly decreased postprandial glycemic response. Cinnamon aqueous extract is known for its potential antidiabetic property that regulates uncoupling protein-1 (UCP-1) that in turn helps in translocation of GLUT4 in the muscle and adipose tissues.

10.3 MOLECULAR MARKER TYPES

Molecular breeding is regarded as an important tool for the improvement of many crop species and it is used for varietal identification, seed purity and genetic distance analysis. Marker assisted backcrossing mediated genome wide polymorphism identifications and tagging are the major applications. Markers are used to solve the problems by distinguishing the actual material from the substitutes and superior drugs. A variety of molecular markers can be developed and used for the construction of genetic maps using linkage analysis. Genome based techniques has done by molecular markers techniques. The efficiency of molecular markers depends on their ability to detect polymorphism; Molecular markers are also used for the identification of hypoglycemic agents in plants which prospects antidiabetic properties. Molecular markers are relatively less time consuming and no need of huge population.

Basic molecular markers are grouped into three types on the basis of:

- Mode of gene action (dominant or co-dominant markers);
- Methods of detection (hybridization or non-hybridization based molecular markers);
- Mode of transmission (paternal and maternal organelle inheritance, bi-parental or maternal nuclear inheritance).
- Based on the type of methodology used for authentication of polymorphism. They are:
- Non-hybridization or PCR molecular markers;
- Sequence based molecular markers.

1. **Hybridization-Based Molecular Markers:** Markers produced by the technique of hybridization is known as hybridization based molecular markers. The difference in the lengths of DNA fragments generated by specific restriction enzymes from genomic DNAs of two or more individuals of a species is known as hybridization. These markers reveal polymorphism after some distance away (Barendse et al., 1994):

   i. **Restriction Fragment Length Polymorphism (RFLPs):** This is the first markers that were used to identify differences in a genetic sequence by using restriction enzyme. These markers cut the double stranded DNA of highly specific restriction site. It is an important tool for plant genome mapping to digest the PCR products of regions with sequences used restriction enzymes. Biswas used RFLP markers to validate some medicinal species such as *panax, fritillaria, pallidiflora, Atractylodes* species and others from their impurities. RFLP method involves application of southern hybridization. Polymorphism levels of RFLP are very low. Pourmohammad (2014) famed *citrus maxima, C. citrus and C. medica* from each other by using RFLP markers

ii. **Variable Number Tandem Repeat (VNTR):** These are the important source of RFLP markers which are used in linkage mapping of diploid genomes. Variable tandem repeats find its application in forensic science by DNA fingerprinting and the CODIS databases. And are rarely used for plant identification (Biswas et al., 2014). In the past few years for the development of VNTR repeats evidence has been collected in a huge spectrum of pathological states.

iii. **Microarray or Chip Hybridization:** Microarray is a collection of thousands of microscopic spots of DNA molecules arranged in an array. Each microscopic spot contains the 10^{12} picomoles of know sequence. Microarray is also known as DNA chip or biochip. Microarray hybridization can be used for confirmation of Plant-based medicines. DNA identification of specific medicinal materials is an efficient and cost effective way to analyze highly processed medicine and will assist in observing their quality and legality.

2. **PCR-Based Molecular Markers:** These are second-generation markers which involves in in-vitro amplification of particular DNA sequences with the help of particular primers and a DNA polymerase enzyme. The amplified fragments are separated by electrophoresis and identified the bands by using different methods. These markers are perfect for the identification of medicinal plant species. These DNA based molecular markers are helps in the improvement of medicinal plants. The genetic information is unique in DNA based molecular markers for every species because of its physiology conditions, environmental factors, and age:

   i. **Amplified Fragment Length Polymorphism (AFLP):** This is PCR based markers associated with RFLP. Peters J. used AFLP markers for the construction of a physical map of the *Arabidopsis* genome. It has been used for authentication and studying genetic diversity of Chinese medicines (Yuan et al., 2003). AFLP markers generates highly reliable and reproduce data, no prior DNA knowledge requires for genotyping and detects high polymorphism with in a single primer. In *Chlorophytum* *borivilianum* used AFLP markers for the assessment of genetic relation between individuals (NirajTripathi et al., 2012).

ii. **Random Amplification of Polymorphic DNA (RAPD):** These markers are very advantageous because it exhibits higher polymorphism, requirement of little quantity of DNA, technical simplicity, rapidity, no requirement of prior DNA sequences. Fu (1999) used RAPD markers to differentiate the samples of *Codonopsis pilosula, Andrographis paniculata and, Allium schoenoprasum L.* which was collected from various geographical regions. Padiya et al. used RAPD markers were used for the authentication of *Lymnophila heterophylla*as medicinal plant form Odisha region in India. These markers are used as a complementary tool for systematic relationship studies and in quality control. These markers are very helpful for differentiating the genetic relationship and diversity patterns between medicinal plants. RAPD markers determines the active ingredients of the medicinal plants, *Ledebouriella seseloides* and *Atractylodes macrocephala, Astragalus membanaceus.*

iii. **Simple Sequence Repeats/Short Tandem Repeat/Microsatellite Polymorphism (SSR/STR):** These markers are short repeat of any nucleotide. This PCR based marker, easily detectable. SSR maker provide highly polymorphism. This is also known as short tandem repeat or microsatellite and sometimes referred as simple sequence length polymorphism (SSLP). Amplification of SSR generate unique pattern and specific profile for identification of individual. SSR marker revealed 12 polymorphism locus in *Rhodiola rosea*. SSRs variations play an important role of genomic diversity that ease the study of genetic differences in human health and diseases.

iv. **Inter Simple Sequence Repeat (ISSR) Marker:** It was developed by Zietkiewicz. This is PCR based molecular marker located between two identical repeated regions (SSR region). ISSR is a simple and no need for prior knowledge of DNA sequencing. This is dominant and highly produced polymorphism. Application of ISSR markers helps to detect elite plants based on genotype and improve the herbal plants properties through plant molecular breeding program. These markers are relatively most recent in molecular technology. Domyati et al. reported that *B. aegyptiaca* shows highest number of amplicons and lowest in *Z. simplex*.

v. **Sequence-Characterized Amplified Regions (SCAR) Marker:** This marker was first developed by Paran and Michelmore in 1993. This is more reproducible, locus specific and co-dominant marker. SCAR markers are easy, reliable, and advantageous for authentication of medicinal plant used in the preparation of herbal medicine. SCARs can be used as physical landmarks in the genome and genetic marker. These markers are more informative for genetic mapping and also for comparative mapping

3. **Sequence-Based Molecular Markers:** These are PCR based sequence markers which is amplified by PCR techniques to detect the other genomic sequences whose location is already mapped. These are followings:

i. **Single Nucleotide Polymorphisms (SNPs):** These markers are used for the identification of single nucleotide change or indel at any specific sequences. These markers are used in plant and animal genomes, mainly in agricultural breeding works for linkage and genetic map constructions. Coles et al. (2005) used *chenopodium quina* wild variety to identify SNP by expressing the libraries. Immediate applications of SNPs are genetic mapping experiments, germplasm introduction and evolutionary relationships in *chenopodium*. SNPs are the last molecular markers. Due to their occurrence across the human genome have become important application for the development of genetic markers for human diseases.

ii. **Molecular Markers Applications with Special Reference to Antidiabetic Plants:**

- Molecular markers technology facilitates an independent way to express the medicinal plant materials.
- Use of molecular markers are generally referred to as biochemical constituents, including phenotype and genotype.
- Medicinal plants which contains active ingredients and intricate molecules that have further developed and analyzed accurately by using molecular marker technologies.
- Biomarkers is a constant, easily available for screening the DR in the population and used for the need of future treatment.

- Complete genome sequences for Arabidopsis and rice, explore in genomic resources for various models such as Lycopersicon, Medicago, and populous has opened new door for medicinal plant breeding by comparative genomics (Gilmore s et al., 2003).
- Application of molecular markers technologies also lies in the area of plant biology like systematics, population genetics, evolutionary biology and conservation genetics, genomics, detection of the wild ancestors of domestic species and the development of geographic profile of genetic diversity (Sharma et al., 2019).
- It is considered that monogenic form of diabetes is the metabolic and molecular abnormalities that induces diabetes potential for drug therapy for diabetes subtype (Robert, 2010).
- By using various molecular markers selection of elite germplasm helps to improve the quality of herbal plants.
- Molecular plant breeding involves some modern tools like quantitative trait loci (QTL) and marker-assisted selection (MAS) by using molecular markers based on their phenotypic and genetic diversity.
- In medicine, a molecular marker can be used as a substance that is introduced in an organism to detect diseases.

10.4 CONCLUSION AND FUTURE PERSPECTIVE

Prospective future of medicine will direct to connect genotype to phenotypes and distinguish common characters. Drugs designing and therapy will directly depend upon metabolic and genetic profile linked with diseases. For this challenge anew long-read sequencing approaches are needed. Advancement of molecular technologies has made easy and comfortable plant breeding strategies. By using these technologies, antidiabetic plants can be improved rapidly. In Molecular plant breeding, genetic polymorphism information is used to develop healthcare through gene therapy and used for design the new drugs. The microsatellite markers developed by Jagajjitsahu et al. (2014) can be considered as principal depository for upcoming refinement of the yield through marker assisted breeding and also for characterization of new variants with antidiabetic properties.

KEYWORDS

- **antidiabetic**
- **association**
- **medicinal plants**
- **molecular marker**
- **phytochemicals**

REFERENCES

Ahmad, B. A., & Kumar, A. R., (2009). *In vitro* propagation of monocot (*Costus pictus* D. don) an antidiabetic medicinal plant. *Journal of Agricultural Technology*.

Ahmed, M. F., Kazim, S. M., Ghori, S. S., Mehjabeen, S. S., Ahmed, S. R., Ali, S. M., & Ibrahim, M., (2010). Antidiabetic activity of *Vinca rosea* extracts in alloxan-induced diabetic rats. *Int. J. Endocrin.*

Ayer, D. K., (2017). Breeding for quality improvement in turmeric (*Curcuma longa* L.). *Advances in Plants & Agriculture Research., 6*(6), 201–204.

Barendse, W., Armitage, S. M., Kossarek, L. M., Shalom, A., Kirkpatrick, B. W., Ryan, A. M., Clayton, D., et al., (1994). A Genetic linkage map of the bovine genome. *Nat. Gene., 6*, 227–235.

Biswas, K., & Biswas, R., (2014). DNA Molecular markers based authentication of herbal drugs. *IJPR*S.

Coles, N., Cole, M. C., Christensen, S., Jellen, E., & Maughan, P., (2005). Development and use of an expressed sequence tag library in quinoa (*Chenopodium Quinoa* Willd.) for the discovery of single nucleotide polymorphism. *Plant Science, 168*(2), 439–447.

Domyati, F. M., Rania, Y. A. A., Edris, S., Mansour, A., Sabir, G., & Bahieldin, A., (2010). Molecular markers associated with genetic diversity of some medicinal plants in Sinai. *J. Med. Plants Res., 5*(2), 200–210.

Eidia, A., Eidib, M., & Esmaeilia, E., (2006). Antidiabetic effect of garlic (*Allium sativum* L.) in normal and streptozotocin-induced diabetic rats. *Phytomedicine, 13*(9, 10), 624–629.

Elkayam, A., Mirelman, D., Peleg, E., & Rosenthal, T., (2003). The effects of allicin on weight in fructose-induced hyperinsulinemic, hyperlipidemic, hypertensive rats. *American Journal of Hypertension, 16*(12), 1053–1056.

Ford, L. B. V., Jackson, M. T., & Newbury, H. J., (1997). Molecular markers and the management of genetic resources in seed gene banks: A case study of rice. *J. CAB International., 19*(5), 108.

Fu, R., Wang, J., Zhang, Y., Wang, Z., & Shaw, P. C., (1999). Differentiation of medicinal *Codonopsis* species from adulterants by polymerase chain reaction-restriction fragment length polymorphism. *Planta Med., 65*(7), 648–650.

George, A., Thankamma, A., Remadevi, V. K., & Fernande, I. A., (2007). Phytochemical investigation of insulin plant (Costuspictus). *Asian Journal of Chemistry.*

Gilmore, S., & Peakall, R., (2003). Isolation of microsatellite markers in *Cannabis sativa* L. In *Fibre Crop Varieties: Molecular Ecology Notes*. Wiley Online Library.

Hiller, R., Sperduto, R. D., Podgor, M. J., Ferris, F. L., & Wilson, P. W., (1988). Diabetic retinopathy and cardiovascular disease in type II diabetics. The Framingham heart study and the Framingham eye study. *American J. Epidemiology, 128*(2), 402–409.

karunanayake, E. H., Welihinda, J., Sirimanne, S. R., & Sinhadorai, G., (1984). Oralhypoglycaemic activity of some medicinal plants of Sri Lanka. *Journal of Ethnopharmacology, 11*(2), 223–231.

Kleijnen, J., Knipschild, P., & Ter, R. G., (1989). Garlic, onions and cardiovascular risk factors. A review of the evidence from human experiments with emphasis on commercially available preparations. *BJCP*.

Majumdar, M., & Prachi, S. P., (2017). Antibacterial, antioxidant and antiglycation potential of *Costus pictus* from southern region India. *Asian J. Plant Sci. Res.*

Marchetti, L., Pellati, F., Graziosi, R., Brighenti, V., Pinetti, D., & Bertelli, D., (2019). Identification and determination of bioactive phenylpropanoid glycosides of *Aloysia polystachya* (Griseb. et Moldenke) by HPLC-MS. *Journal of Pharmaceutical and Biomedical Analysis, 166*, 364–370.

Meir, P., & Yaniv, N., (1985). A Vitro study on the effects of *Mormodica charantia* on glucose uptake and glucose metabolism in rats. *Planta Med., 51*(1), 12–16.

Miniraj, N., Prasanna, K. P., & Peter, K. V., (1993). Bitter gourd *Momordi caspieces* genetic improvement of vegetable plants. *Genetic Improvement of Vegetable Crops*, 239–246.

Mukesh, R., & Namita, R., (2013). Medicinal plants with antidiabetic potential. *J. Agric. Environ. Sc.*

Osmana, M. A., Dhawanb, S. S., Bahlb, J. R., Darokarb, M. P., & Khanujac, S. P. S., (2011). AFLP marking and polymorphism among progenies of *Gymnema sylvestre*: An important medicinal plant of India. *Natur. Prod. Comunic.*

Padiya, R., Sanjay, K., & Banerje, (2013). Garlic as an antidiabetic agent: Recent progress and patent reviews. *Recent Patents on Food, Nutrition & Agriculture, 5*(2), 105–127.

Peters, J. L., Constandt, H., Neyt, P., Cnops, G., Zethof, J., Zabeau, M., & Gerats, T., (2001). A physical amplified fragment length polymorphism map of *Arabidopsis*. *Plant Physiology*.

Piedrola, G., Novo, E., Escobar, F., & García-Robles, R., (2001). White blood cell count and insulin resistance in patients with coronary artery disease. *Annales D'endocrinologie, 62*(1), 7–10.

Pourmohammad, A., (2013). Application of molecular markers in medicinal plant studies. *Agriculture and Environment., 5(1)*, 80–90.

Powell, W., Morgante, M., Andre, C., Hanafey, M., Voge, J., Antoni, S. T., & Rafalski, A. (1996). The comparison of RFLP, RAPD, AFLP and SSR (microsatellite) markers for germplasm analysis. *Molecular Breeding., 2*, 225–238.

Pusparajah, P., Han, L. L., & Abdul, K. K., (2016). Molecular markers of diabetic retinopathy. *Front. Physiol.*

Robert, S. J., (2010). Editorial: Genetic markers in diabetes mellitus: The need and promise for specific drug therapies in defined subtypes of diabetes patients. *Therapy, 7*, 309–312. 10.2217/thy.10.41.

Sahu, J., priyabrata, S., Choudhury, M. D., Dehury, B., Madhumita, B., Mahendrakumar, M., & Anupam, D. T., (2014). Rediscovering medicinal plants potential with OMICS. *J. Int. Bio., 18*(5).

Saravanamuttu, S., & Sudarsanam, D., (2012). Antidiabetic plants and their active ingredients. *IJPSR, 3*(10), 3639–3650.

Shanmugasundaram, E. R. B., Rajeswarf, G., Baskaran, K., Rajesh, K. B. R., Radha, K. S., & Kizar, A. B., (1990). Use of *Gyiwvema sylvestre* leaf extract in the control of blood glucose in insulin-dependent diabetes mellitus. *Journal of Ethnopharmacology, 30*(3), 281–294.

Shanmugasundaram, E. R., Gopinath, K. L., Radha Shanmugasundaram, K., & Rajendran, V. M., (1990). Possible regeneration of the islets of Langerhans in streptozotocin-diabetic rats given *Gymnema sylvestre* leaf extracts. *Journal of Ethnopharmacology, 30*(3), 265–279.

Robert S. J., (2010). Editorial: Genetic markers in diabetes mellitus: The need and promise for specific drug therapies in defined subtypes of diabetes patients. *Therapy, 7*, 309–312. 10.2217/thy.10.41.

Shaw, R. K., Acharya, L., & Mukherjee, A. K., (2009). Assessment of genetic diversity in a highly valuable medicinal plant *Catharanthus roseus* using molecular markers. *Crop Breeding and Applied Biotechnology, 9*, 52–59.

Shetty, A. K., Suresh, G., Sambaiah, K. K., & Salimath, P. V., (2005). Effects of bitter guard on glycemic status in streptozotocin induced diabetic rats. *Plant Foods Hum. Nutr., 60*, 109–112.

Soon, H. T., Chung, Y. L., Hazrina, H., Aditya, A., Mohammadjavad, P., Won, F. W., Shiau-Chuen, C., et al., (2013). Antidiabetic and antioxidant properties of alkaloids from *Catharanthus roseus* (L.) G. don. *Molecules, 18*(8), 9770–9784.

Tatikonda, L., Wani, S. P., Kannan, S., Beerelli, N., Sreedevi, T. K., Hoisington, D. A., et al., (2009). AFLP-based molecular characterization of an elite germplasm collection of *Jatropha curcas* L. a biofuel plant. *Plant Scie., 176*(4), 505–513.

Thatte, U., (2009). Still in search of a herbal medicine. *Indian J. Pharmacol., 41*(1), 1–3.

Tripathi, N., Chouhan, D. S., Saini, N., & Tiwari, S., (2012). Assessment of genetic variations among highly endangered medicinal plant *Bacopa monnieri* (L.) from central India using RAPD and ISSR analysis. *3 Biotech, 2*, 327–336.

Uebanso, T., Arai, H., Taketani, Y., Fukaya, M., Yamamoto, H., Mizuno, A., Uryu, K., et al., (2007). Extracts of *Momordi cacharantia* suppress postprandial hyperglycemia in rats. *Journ. Nutr. Sci. Vit., 53*(6), 482–488.

Wang, L., Waltenberger, B., Pferschy, W. E. M., Blunder, M., Liu, X., Malainer, C., Blazevic, T., et al., (2014). Natural product agonists of peroxisome proliferator-activated receptor-gamma (PPARg). *Biochemical Pharmacology, 92*(1), 73–89.

Yuan, M., & Hong, Y., (2003). Heterogeneity of Chinese medical herbs in Singapore assessed by fluorescence AFLP Analysis. *The American Journal of Chinese Medicine, 31*(5), 773–779.

Zaid, H., Antonescu, C. N., Randhawa, V. K., & Amira, K., (2008). Insulin action on glucose transporters through molecular switches, tracks and tethers. *Biochem. J., 413*(2), 201–215.

CHAPTER 11

Biotechnologically Engineered Transgenic Medicinal Plants: Exploration of Antidiabetic Properties

PRIYA SUNDARRAJAN[1] and LEKHA BHAGTANEY[2]

[1]*Associate Professor, Department of Life Science and Biochemistry and Director, Caius Research Laboratory, St. Xavier's College (Autonomous), Mumbai – 400001, Maharashtra, India, E-mail: priya.s@xaviers.edu*

[2]*Sunandan Divatia School of Science, NMIMS, Vile Parle (West), Mumbai – 400056, Maharashtra, India, E-mail: lekhabhagtaney@gmail.com*

ABSTRACT

Diabetes mellitus (DM) arises due to impaired insulin secretion or activity and hence can be categorized as a complex metabolic disorder of numerous etiologies. If not well managed, it can lead to serious health consequences due to disturbances in carbohydrate, protein, and lipid metabolism. From ancient times, it is known that medicinal plants have certain biologically active substances that exhibit antidiabetic, anticancer, antimicrobial, and other such properties. This therapeutic potential of medicinal plants arises from its secondary metabolites, which play an important role in planning of drug development approach towards treatment. A number of techniques can be utilized to increase the production of these bioactive compounds. Biotechnology-based transgenic approaches are one of the recently used methods to enhance the *in vitro* production of such compounds. For successful transformation of plants, the following parameters need to be taken into consideration viz. gene constructs, vectors, transformation technique and integration of the transgene.

Antidiabetic Potential of Plants in the Era of Omics. Deepu Pandita, PhD, Anu Pandita, MSc, & Chander Bhanu, MBBS (Eds.)

There can be two types of gene transfer approach in plants: *Agrobacterium tumefacians* Ti plasmid vector-mediated gene transfer and direct gene transfer. Each method has its own advantages and disadvantages. For the screening of these products, there are two types of studies: *in vitro* studies which typically include antidiabetic assays and *in vivo* studies which include diabetes induced animal models. The studies conducted provide information on the antidiabetic properties of these transgenic plants, such as the increase in sensitivity towards insulin, effect on beta-pancreatic cells, insulin-like effects and inhibitors of enzymes facilitating hyperglycemia in the body. Further metabolite profiling can be used to understand which specific components in these medicinal plants contribute to the antidiabetic property. Identifying these specific active components can help in providing an optimal and more efficient medicinal preparation. Biotechnology-based transgenic approaches were found to be safe and effective for secondary metabolite production and can be used for therapeutic purposes.

11.1 INTRODUCTION

Diabetes is a condition related to carbohydrate metabolism, which is defined by chronic hyperglycemia caused due to insulin deficiency or impairment in insulin function. According to WHO, around 422 million people worldwide are suffering from diabetes, out of which the majority of the people are living in low-and middle-income countries. The number of deaths due to diabetes is 1.6 million. The incidence of the disease has been increasing steadily over the past few decades. Access to affordable treatment is most important to people living with diabetes. The target now is to cease the rise in diabetes and obesity by the year 2025.

Since time immemorial, plants have been used as therapeutic agents to combat a variety of diseases. The main agents responsible for the therapeutic action of plants are secondary metabolites. Many studies have been focusing on enhancing the production of these bioactive compounds through various biotechnological tools. Latest developments in plant tissue culture and genetic modification studies have helped in developing more efficient ways to increase the yield of pharmaceutically important molecules. In comparison to traditional plantation methods, plant-tissue culture provides a better alternative for large-scale production as the regulated supply of bioactive compounds does not depend on plant availability (Khan et al., 2009).

Biotechnology-based approaches in medicinal plants with such as tissue culture-based regeneration, callus-mediated organogenesis, regeneration through somatic embryogenesis, calli suspension culture and heterologous biosynthesis methods have been discussed in brief. This chapter also explores in depth the utilization of transgenesis as a method to enhance the yield of bioactive compounds having important antidiabetic properties with a few examples of studies conducted on the same. Genome sequencing studies provide information regarding the different biosynthetic pathways of secondary metabolites, genome annotations and evolutionary framework of the medicinal plant, thereby helping in the development of new and better therapy. Further, a few important bioactive compounds showing potent activity against diabetes along with the *in vitro* and *in vivo* studies used to check the efficacy of these molecules have been discussed. Metabolite profiling is utilized to identify the specific components that are responsible for the antidiabetic property in these medicinal plants to achieve optimal medicinal preparation.

11.2 DIABETES MELLITUS (DM): PATHOPHYSIOLOGY AND MANAGEMENT

Diabetes is characterized by impaired glucose control and involves multiple organ systems. Insulin and glucagon are the two hormones that keep up the blood glucose homeostasis. The two hormones namely glucagon and insulin are secreted by the cells (α-cells and β-cells) of pancreas, namely Islets of Langherhans. The β-cells produce insulin which maintains the levels of blood glucose, post-prandial or otherwise. α-cells secrete glucagon during fasting state which raises the blood glucose levels (BGL). Normal blood glucose or normoglycemia is achieved through the action of both these hormones (Lankatillake et al., 2019). Diabetes is also related to oxidative stress where there exists an instability in the balance between the generation of cellular reactive oxygen species (ROS) and the body's natural antioxidant mechanism to counter it. ROS cause lipid peroxidation, protein oxidation and nucleic acid degradation, retardation of enzyme activity, and leads to cell death due to activation of the apoptotic pathway. Therefore, ROS are extremely harmful by-products of cells (Gothai et al., 2016).

DM can be classified into Type 1 diabetes mellitus (T1DM) and Type 2 diabetes mellitus (T2DM). In T1DM, there is loss of β-cells caused by autoimmune reactions leading to deficiency of insulin. This condition is more often seen in children and young adults. Because of this, regular insulin supply is of utmost importance to maintain BGL. β-cell destruction is faster in infants and children than in adults. Thus, children experience ketoacidosis whereas adults exhibit modest to severe hyperglycemia and ketoacidosis in response to stress or infection (Tiwari, 2015). T2DM is majorly a lifestyle disorder and is also caused due to genetic predisposition. The prevalence of T2DM is higher than T1DM. In patients suffering from T2DM, insulin resistance and gradual degradation of β-cells is observed. There is a need for abnormally high levels of insulin for maintenance of a quantifiably regular biological response in insulin resistance. When cells and tissues become resistant to normal insulin levels, the β-pancreatic cells increase the levels of insulin to keep up the regular levels of blood glucose resulting in hyperinsulinemia. Chronic stimulation of β-cells to produce high levels of insulin results in debilitation and dysregulation of β-cells, finally leading to deficiency of insulin due to β-cell failure. Insulin resistance precedes T2DM by 10–20 years (Lankatillake et al., 2019).

11.2.1 TREATMENT WITH SYNTHETIC DRUGS

The treatment regimen of diabetes is dependent on the seriousness of hyperglycemia, risks of hypoglycemia, hepatic, and renal functions, body mass index (BMI), self-regulation of BGL and cost of treatment. Treatment of T1DM includes insulin injections, islet neogenesis associated protein (INGAP) peptide therapy which aims to increase islet cell regeneration, triggering secretion of insulin via Glucagon-like Peptide (GLP) analogs like exenatide and liraglutide, inhibition of dipeptidyl peptidase-4 (DPP-4) by Sitagliptin. Treatment of T2DM includes enhancement of insulin secretion by sulfonylureas and repaglinide, an increase in the activity of insulin in fat and muscle by troglitazone, promotion of insulin activity in liver tissue by metformin and delay in the absorption of carbohydrate by miglitol and acarbose. Combinational therapy is also provided in certain cases like insulin-sulfonylureas for reduced daily requirement of insulin, insulin-metformin for minimizing weight gain during therapy and insulin-troglitazone for reduced insulin requirement and improved maintenance of BGL (Tiwari, 2015).

11.2.2 SIDE EFFECTS DUE TO SYNTHETIC DRUGS

Usage of synthetic drugs comes along with some unavoidable side effects. Insulin and sulfonylureas treatment. There have been reports of prominent side effects caused due to the use of synthetic drugs such as drug resistance, adverse effects and even toxicity. Insulin and sulfonylurea can cause weight gain and metformin (a biguanide) can induce hypoglycemia, digestive problems like diarrhea, nausea, and sometimes lactic acidosis. Thiazolidinedione has been reported to cause weight gain (Tiwari, 2015). Sulfonylurea or insulin in combination with sitagliptin or linagliptin can cause severe hypoglycemic condition. The various side effects of DPP-4 inhibitors include headache, nausea, common cold and allergic reactions. α-glucosidase inhibitors can cause side effects like flatulence and diarrhea (Hasan et al., 2018).

11.2.3 MEDICINAL PLANTS USED FOR DIABETES TREATMENT

Since there is the presence of such severe side effects with current synthetic drugs, newer, more natural therapies for diabetes are being sought out. Many novel compounds isolated from medicinal plants have shown potent hypoglycemic effects that display equal or even better activity than the synthetic treatment. Nevertheless, several plant bioactive compounds have not been well characterized. Many medicinal plants show the presence of important classes of compounds like terpenoids, flavonoids, phenolics, and others that have shown a decrease in BGL (Tiwari, 2015). Some medicinal plants that have been traditionally used to treat diabetes along with their active constituents have given in Table 11.1.

TABLE 11.1 Some Medicinal Plants and Their Active Constituents Displaying Important Antidiabetic Properties

Plant	Active Constituents Against Diabetes	References
Camellia sinensis	Catechins	Salehi et al. (2019)
Curcuma longa	Curcumin	Lankatillake et al. (2019)
Morus alba	Morin	Lankatillake et al. (2019)
Coptidis rhizoma	Berberine	Ga and Mahalingam (2016)
Embelia ribes	Embelin	Ga and Mahalingam (2016)
Panax ginseng	Ginsenoside	Ga and Mahalingam (2016)
Momordica charantia	Charantin	Lankatillake et al. (2019)
Tinospora cordifolia	Palmatine, Berberine	Ga and Mahalingam (2016); Salehi et al. (2019)
Cinnamomum zeylanicum	Cinnamaldehyde	Chakraborty (2018)
Allium sativum	Allicin	Chakraborty (2018)

Several studies have reported the promising activity of these bioactive agents for the treatment of diabetes. More than a single bioactive compound extract, a combination of multiple extracts has shown even better effects against diabetes. Combination extract therapies demonstrate enhanced activities of multiple bioactive compounds in single therapies, thereby decreasing the drug load given to the patients (Gothai et al., 2016).

11.3 ENHANCED PRODUCTION OF SECONDARY METABOLITES IN PLANTS USING BIOTECHNOLOGY AS A TOOL

Biotechnology is a powerful tool implemented to improve the quality and production of industrially important plants. The biggest advantage of using biotechnology is the controlled working conditions which mean that the plants are unaffected by the natural environmental changes. Other advantages of using *in vitro* plant cell and organ culture include faster production of superior quality plant material, easy extraction of bioactive compounds, efficient production of valuable compounds, easily scalable production process and pathogen-free plant-derived product. For the production of secondary metabolite, the parts of plants that are commonly used are calli, roots, shoots, plant cell suspension cultures. In case of requirement of a specific metabolite which is produced in specialized plant tissue, induction of cell differentiation becomes necessary. Thus, the type of culture used depends on the required secondary metabolite (Kowalczyk et al., 2020). In general, biotechnology-based approaches in plants are as in subsections.

11.3.1 TISSUE CULTURE-BASED REGENERATION

Numerous methods can be used for *in vitro* regeneration, however, the most common method used is micropropagation. Genetically identical plants are obtained on propagation from existing meristems. For medicinal plant-based studies, regeneration from root and shoot meristems have provided more successful results. An increase in the rate of multiplication and pathogen-free plant product are some of the advantages of the micropropagation technique. Several studies have reported that plant growth regulators like auxins and cytokinins have various effects on the growth of medicinal plants (Tripathi and Tripathi, 2005).

For instance, the plant *Ajuga bracteosa* produces an array of metabolites that have many pharmacological applications. The extract of this plant has a therapeutic effect against many conditions like diabetes, gastrointestinal disorders, cancer, and inflammation. A study was conducted to develop an efficient method for the direct regeneration of shoot on *Ajuga bracteosa* by the use of leaf and stem explants. Thiadiazuron (TDZ) is proven to affect endogenous auxin and cytokinin production thereby facilitating organogenesis. TDZ also amplifies the yield of therapeutically important plant metabolites. Hence, TDZ was utilized as a plant growth regulator. This study reported a high shoot induction frequency for stem explants by utilizing TDZ in the culture medium (Ali et al., 2018).

11.3.2 CALLUS-MEDIATED PRODUCTION OF SECONDARY METABOLITES

Callus formation occurs at the site of injury in the plant to seal the wound and for water retention. In the laboratory, callus formation occurs on artificial media containing auxin and cytokinin which are plant growth regulators. Callus formation followed by organogenesis is achieved by the differential application of the known plant growth regulators under controlled conditions in the culture medium. Regeneration or organogenesis by callus tissue results in plants having genetically variable cells. For secondary metabolite production, callus cultures have importance (Tripathi and Tripathi, 2005).

Ledebouria revoluta is a medicinal plant extensively utilized in pharmaceutical industries. The scaly bulbs have antioxidant properties. This is effective against diabetes as hyperglycemia induces the formation of free radicals from the auto-oxidation of glucose (Bajaj and Khan, 2012). A study was performed using *Ledebouria revoluta* in order to improve the quality of the plant components to meet the high demand of the industry. The bulb scale of *Ledebouria revoluta* was utilized as an explant for callus formation. 2,4-dichlorophenoxyacetic acid, a synthetic auxin, played an essential role in de-differentiation of organized tissue of explant. For shoot induction, the cytokinin concentration was reduced to a minimum and auxin was completely withdrawn from the medium. Kinetin (KIN) and 6-(c,c-dimethylallylamino) purine (2iP) are the two cytokinins which were successful in inducing bulblets of *Ledebouria revoluta* from the calli (Haque et al., 2018).

Important bioactive compounds are often produced utilizing calli suspension cultures. To obtain specific bioactive compounds, Medicinal Plant Cell Suspension Cultures (MPCSC) are used in which single cells or differently sized aggregates of cells are grown in an agitated liquid medium. To increase the synthesis of the desired compound, the gene constructs are designed and inserted accordingly (Kowalczyk et al., 2020).

Clausena harmandiana is a medicinal plant that contains important bioactive compounds known as carbazole alkaloids which are responsible for the therapeutic activity. Carbazoles identified from this plant have exhibited elevated glucose uptake activity (Noipha et al., 2010). Therefore, it is a prospective candidate in diabetes therapy. A study was performed to optimize the production of two carbazole alkaloids clausine K and 7-methoxymukonal using callus suspension culture. Their results suggested that five to six-week-old cultures produced maximum levels of clausine K and 7-methoxymukonal. This culture was used to investigate the impact of different factors on the production of carbazole alkaloids. Effects of precursors such as sucrose and tryptophan, elicitors like yeast extract and light (purple light conditions and dark conditions) could improve the production of carbazole alkaloids (Kitisripanya et al., 2018).

11.3.3 REGENERATION THROUGH FORMATION OF SOMATIC EMBRYOS

The development of embryo-like structures from somatic cells without gametic fusion is called somatic embryogenesis. The advantages of this method include abundant production of specific plant material, genetic diversity, and transformation. A combination of the growth regulators auxins and cytokinins is used to encourage somatic embryogenesis (Ebrahimi et al., 2018).

A study was performed on the plant *Hypoxis hemerocallidea* for efficient regeneration through somatic embryogenesis. *Hypoxis hemerocallidea* has shown antidiabetic properties by reducing hyperglycemia and hyperglycemia-induced oxidative stress (Oguntibeju et al., 2016; Mahomed and Ojewole, 2003). The findings suggest that the use of 2,4-Din combination with, N6-benzyladenine (BA) is essential for embryogenic callus formation. Whereas, the germination relies on the strength of the medium which determines the osmotic potential. In this study, the use of a half-strength culture medium gave a higher frequency of germination compared to a full-strength culture medium. In addition to this, gibberellic acid showed a higher frequency of germination and augmented the conversion of somatic embryos to plantlets (Kumar et al., 2017).

Another important biotechnology-based method to enhance secondary metabolite production is transgenesis. This topic is discussed in detail in the later titles.

11.4 TRANSGENESIS AS A TOOL TO ENHANCE SECONDARY METABOLITE PRODUCTION

Transgenesis is a technique that allows the incorporation of foreign genes into any plant system through genetic engineering, thereby giving rise to a transgenic plant. Secondary metabolite production in medicinal plants can be enhanced using transgenic approaches. This technique has also benefited agriculturally important crops by modifying them to be pest-resistant or drought tolerant. The main objective of transgenesis is to produce plants with ideal traits, high quality, and yield. The success of this technique is highly dependent on factors such as gene construct, vectors, and transformation technique (Figure 11.1).

A basic gene construct includes regions of promoter and terminator sometimes enhancer, silencer with or without reporter genes. For successful transgenesis, it is important to include the gene of interest and appropriate selectable markers. In plant genetic modifications, appropriate promoters and enhancers are used depending on gene expression levels required. In case a mild toxicity is observed due to the transgene, the expression levels are kept at a basal level. Whereas a higher level of expression is required when there is a need to detect an under expressed transgene. Reporter genes are usually linked to the gene of interest. These genes help to detect the quantity of gene expression and localize the protein products. The protein products of reporter genes can be detected using appropriate methods. Conversely, properties like antibiotic-resistant, resistance to herbicides or anti-metabolic properties that provide resistance to specific chemicals, preventing the growth of non-transgenic plants are usually generated by selectable marker genes. For plant systems, the following are some of the commonly used reporter genes namely galactosidase (LacZ), β-glucuronidase (GUS), luciferase (Luc), and presently popular green fluorescent protein (GFP) (Low et al., 2018).

Vectors act as carriers of the transgene to the target plant cell. A vector has three main components – origin of replication, multiple cloning site or polylinker and a selectable marker. The origin of replication (AT-rich) binds to a particular protein and begins replication. The act of binding to the protein helps to unwind the vector and the polymerases facilitate the replication. The multiple cloning site contains a set of restriction sites also known as polylinker. Restriction enzyme cuts at a unique site, and allows for the integration of the transgene which can be detected by incorporation of selectable marker in the vector (Low et al., 2018).

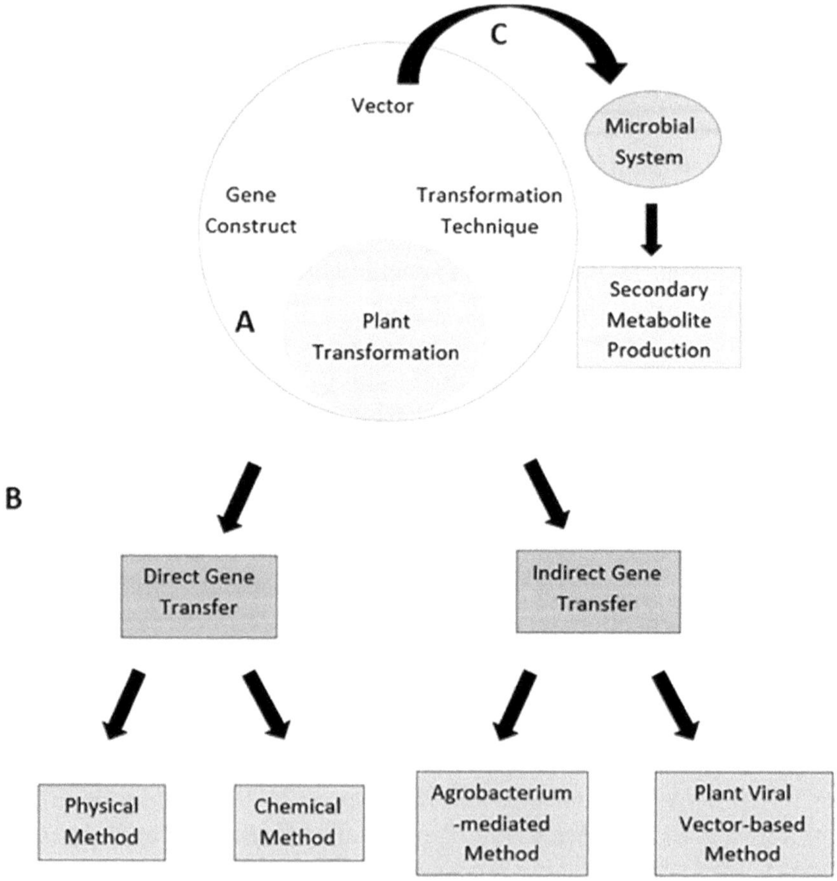

FIGURE 11.1 (A) Factors affecting plant transformation; (B) plant transformation methods; (C) heterologous biosynthesis using microbial systems.

Source: Original flowchart; Designed by: Lekha Bhagtaney.

Successful plant transformation can be obtained by two transformation techniques – direct or indirect gene transfer (Figure 11.2). Indirect gene transfer utilizes a biological vector for the introduction of a transgene into the plant. Whereas, in the method of direct gene transfer the transgene is introduced into the plant directly by physical or chemical methods (Low et al., 2018).

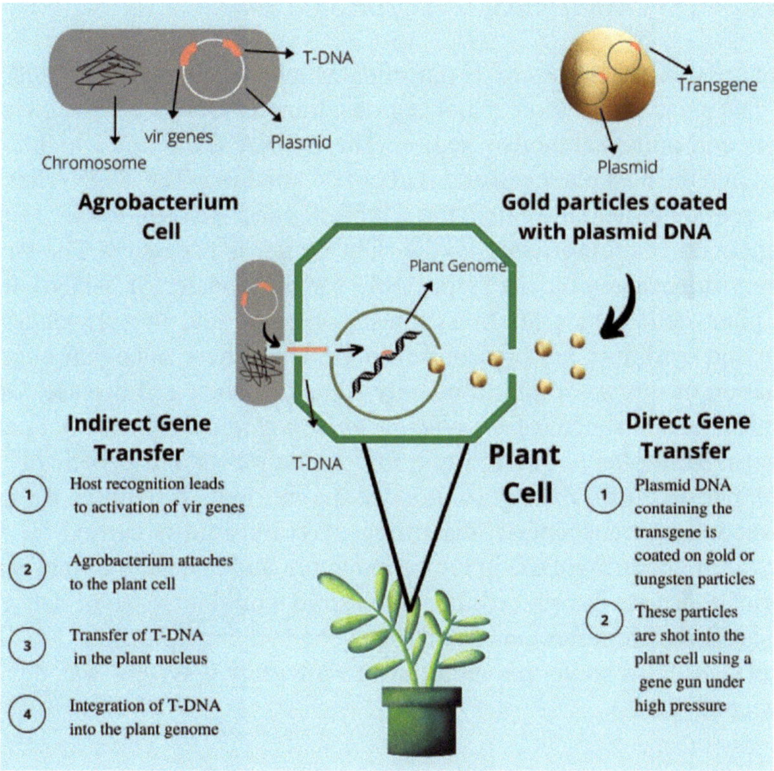

FIGURE 11.2 Direct and indirect gene transfer techniques in plant transformation.
Source: Original illustration; Designed by: Lekha Bhagtaney.

11.4.1 INDIRECT GENE TRANSFER

Gene transfer mediated by *Agrobacteria* is a frequently utilized method for indirect gene transfer. The pathogenic, gram-negative *Agrobacteria* can induce diseases like crown gall or hairy root in plants. *Agrobacterium tumefaciens* and *Agrobacterium rhizogenes* are commonly used species used for plant transformation. This is because these species show presence of tumor-inducing (Ti) plasmid and root inducing (Ri) plasmid. Ti plasmid induces crown gall disease whereas hairy root disease is caused by Ri plasmid in plants.

11.4.1.1 TI PLASMID-BASED VECTOR SYSTEM

There are various classes of Ti plasmids which vary in size from 200 to 800 kbp. The plasmid has three major regions: transfer DNA (T-DNA), virulence region, and opine catabolism region. The T-DNA is the one which is transferred into the host plant cell and is about 24 kbp long. The T-DNA has repeat sequences on either side called the left border and the right border sequence. For induction of tumorigenesis, the right border is necessary. The *vir* genes in the virulence region are responsible for the transfer of T-DNA into the host plant cell. Genes for biosynthesis of opine and the phytohormones-auxin and cytokinin are present in the T-DNA. These oncogenes cause the formation of tumors which ultimately leads to crown gall disease. Opine is the main carbon source of *Agrobacterium tumefaciens* but it is not naturally synthesized in plants. Therefore, it infects the plants and genetically modifies the host cell machinery to produce the nutrients it requires for growth. For successful transgenesis, disarming of Ti plasmid is carried out by the removal of genes involved in tumor induction and instead the transgene and reporter genes are introduced. The Ti plasmid would increase in size with the introduction of reporter genes and transgene and large plasmids are difficult to work with. To solve this issue, the co-integrative system was developed (Low et al., 2018).

11.4.1.2 CO-INTEGRATED VECTOR

A vector designed by homologous recombination (HR) of the disarmed Ti plasmid and intermediate vector are called co-integrative vectors. A plasmid from *Escherichia coli* bearing the desired gene is normally used as the intermediate vector. HR is possible because of the presence of similar sequences between these plasmids. The co-integrative vector thus formed is a large vector that is then introduced in the host plant cell for transformation. However, the large size of the co-integrative vector makes its manipulation laborious. Therefore, after the occurrence of binary vectors, the use of a co-integrative vector system was discontinued (Low et al., 2018).

11.4.1.3 BINARY VECTOR

The binary vector system is a two-plasmid system that was developed after the realization that the T-DNA can function autonomously without being attached to the Ti plasmid. Helper vector and mini vector comprise of the two-plasmid system. The small-sized vector is the mini vector that contains the T-DNA as well as the origin of replication of both *E. coli* and *A. tumefaciens*. Thereby allowing for the cloning of the plasmid in both these organisms. Whereas, the helper vector contains all the genes of the Ti plasmid barring the T-DNA region.

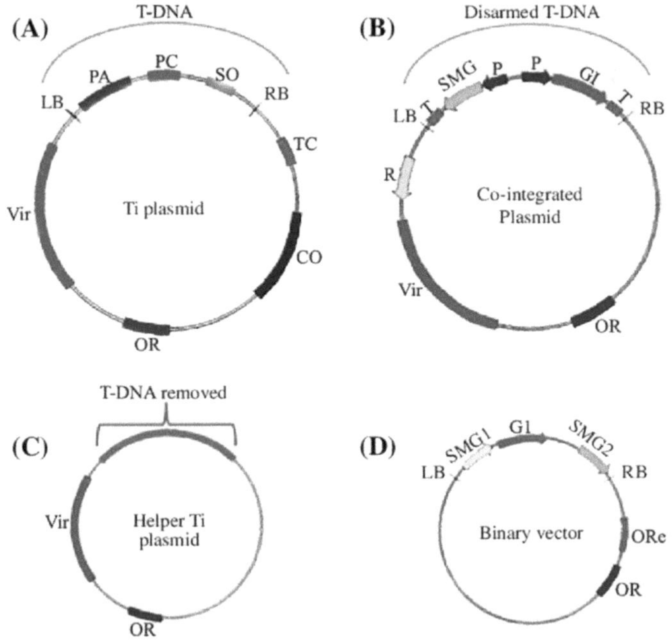

FIGURE 11.3 Structure of Ti plasmid with different modifications used in plant transformation: (A) Structure of Ti plasmid (LB: left border; RB: right border; PA: gene for auxin biosynthesis; PC: gene for cytokinin biosynthesis; SO: gene for opine synthesis; TC: gene for conjugative transfer; CO: genes involved in catabolism of opine; OR: origin of replication of *A. tumefaciens*; Vir: virulence region); (B) structure of co-integrative plasmid, when the Ti plasmid is disarmed the PA, PC, SO, TC, and CO genes are removed (P: promoter; T: terminator; SMG: selective marker gene; GI: gene of interest); (C) structure of helper Ti plasmid-genes for transfer and integration of desired gene; (D) structure of binary/mini vector, along with gene of interest, it consists of two selective marker genes for *A. tumefaciens* (SMG2) and plants (SMG1), and both *E. coli* origin of replication (ORe) as well as for *A. tumefaciens* (OR).

Source: Reprinted with permission from: Zerbini et al. (2014). © 2014 Elsevier.

The reason it is named as the helper vector is because its codes for the genes responsible for the transfer and integration of genes. These plasmid vectors are introduced in *A. tumefaciens*, which is then used for plant transformation (Low et al., 2018). The structure of Ti plasmid with its different modifications has been given above (Figure 11.3).

11.4.1.4 RI PLASMID-BASED VECTOR SYSTEM

Infection of plant tissue by *Agrobacterium rhizogenes* gives rise to hairy root cultures. *Agrobacterium rhizogenes* has a root inducing Ri plasmid which causes unlimited growth of plant tissue. The plasmid DNA is incorporated into the plant genome with the help of transfer genes namely *rol* A, *rol* B, *rol* C. The advantages of these cultures are rapid growth, they do not need plant growth regulators, high genetic stability and they are possible to grow on a large scale. These cultures are unaffected by the natural environment and able to produce important metabolites in high amounts. These cultures can grow on simple media which makes them economically attractive. There is also no risk of human or plant pathogen transmission (Kowalczyk et al., 2020).

Agrobacterium-mediated gene transfer involves two processes known as agroinoculation and agroinfiltration. Agroinoculation involves the plant tissue's surface to be inoculated with *Agrobacterium* containing the transgenic plasmid (either Ti or Ri). Whereas, the agrofiltration procedure requires a syringe or vacuum. Syringe infiltration method includes injection of transformed *Agrobacterium* onto the underside of the leaf and counter pressure being applied on the opposite side of the leaf. This technique, though very challenging is applicable only for gene expression on a small scale. Conversely, a more efficient and faster process is vacuum infiltration, where the entire explant is submerged in an environment containing recombinant *Agrobacterium*. This takes place under conditions of vacuum that facilitates the recombinant *Agrobacterium* to infiltrate into the explant. This technique is suitable for large-scale production. It has been reported that before plant transformation if the plant tissues undergo heat and hydrolysis treatment, the efficiency of transformation is increased (Low et al., 2018).

Another example of indirect gene transfer is the utilization of plant vectors. Plant vectors are viruses that have been altered to use as substitute sources for transgenesis. Tobacco mosaic virus (TMV), Cauliflower mosaic virus (CaMV), Alfalfa mosaic virus (AMV), Cowpea mosaic virus (CPMV) and Potato Virus X (PVX) are frequently used as vectors for transgenesis. To increase efficiency through both ways, the plant viruses have been altered to enable their utilization with *Agrobacteria* as well as plant cells. Designing a virus with similarity to its wild type which has the ability of infecting the host plant is one of the ways. Another way would be to create a 'deconstruct' virus that replaces all the undesirable viral genes with reporter genes or selectable markers that help with the screening process of transgenesis (Low et al., 2018).

11.4.1.5 PLANT VIRAL VECTORS

One of the most frequently used expression vectors is CaMV it is easy to manipulate *in vitro*. There are seven genes in the CaMV genome, namely gene I to gene VII (Figure 11.4) and the CaMV35S promoter (P35S) and terminator has been used to express novel genes in plants since the 1980s. P35S is known to be a strong promoter which is constitutive and induces elevated levels of gene expression in plants (Podevin and du Jardin, 2012) (Figure 11.4). The drawback of this vector is that there is a limited capacity for insertion of transgene, thereby preventing the insertion of larger genes. Potato Virus X (PVX) is another expression vector that is widely used. The PVX genome contains 5 open reading frames (ORFs), of which the first ORF is required for the synthesis and replication of RNA and is an RNA-dependent RNA polymerase (RdRp). There is a presence of three overlapping ORFs called triple gene block present in the central genome coding for p25, p12, and p8 proteins that mediate the transport of viral particles from one cell to another. The coat protein is encoded by the fifth ORF and is also used for cell-to-cell transport as well as long-distance transport. The PVX genome can accommodate larger heterologous sequence inserts as compared to many other viral expression vectors (Manske and Schiemann, 2005). Tobacco Mosaic Virus (TMV) is one of the best-characterized plant viral vectors. TMV is an RNA virus that produces high amounts of coat protein from a viral subgenomic promoter. TMV genome contains three ORFs from which encodes for four proteins. Two genes under separate subgenomic promoters produces the capsid protein and the protein for viral movement. In TMV either the viral gene or the coat protein gene is replaced by the gene of interest or the viral genome inserted with an additional subgenomic promoter will drive its expression. TMV vectors can be inoculated in the target plant using agroinfection (Lindbo, 2007).

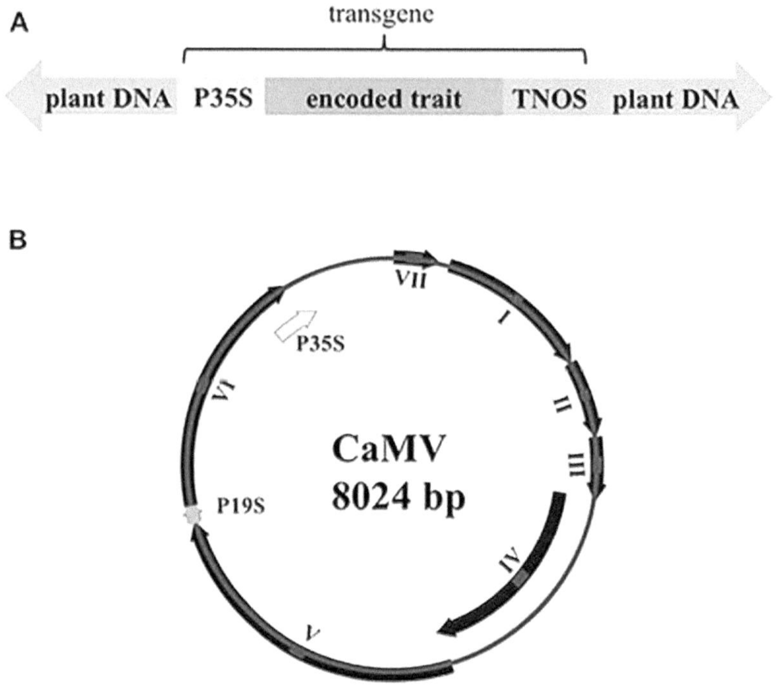

FIGURE 11.4 Structure of cauliflower mosaic virus (CaMV): (A) representation of a model plant construct used in genetic engineering; P35S: CaMV promoter; TNOS: Nos terminator; (B) structure of CaMV genome, I-VI: seven CaMV genes, P35S, P19S: promoters.

Source: Reprinted with permission from: Bak and Emerson (2020), Open access. https://creativecommons.org/licenses/by/4.0/.

T1DM occurs due to autoimmune destruction of pancreatic cells. Human glutamic acid decarboxylase has a 65 kDa isoform (hGAD65) which is a major T1DM autoantigen. Several studies have shown its therapeutic potential by inducing immune tolerance. There have been reports of production of hGAD65 in plants as the production is less expensive and has a lesser risk of contamination as compared to the other platforms. Preclinical studies of hGAD65 show prevention and delay of onset of T1DM. However, there is a need for a source of production that can produce hGAD65 in higher quantities. A study was conducted showing that an edible plant system can be used for the production of this autoantigen which will not require extensive purification, thereby cutting down on downstream processing costs. For this study, the edible plants used were red beet (*Beta vulgaris* cv Moulin Rouge) and spinach (*Spinacia oleracea* cv Industra). A deconstructed TMV-based system known as magnICON was used for transient expression in these edible plants. The magnICON system produces a higher amount of recombinant protein in a shorter period of time. The vectors were inserted in *Agrobacterium tumefaciens* strain GV3101. The gene for GFP was used as a reporter to check the transient expression. Red beet showed higher levels of GFP when vacuum agroinfiltration method was used. The results of this study suggest that red beet can be used as a candidate for production of oral vaccine based on hGAD65 (Bertini et al., 2018).

11.4.2 DIRECT GENE TRANSFER

In direct gene transfer, the transgene is introduced into the plant cell by disrupting its outer membrane. Direct gene transfer can be done via physical or chemical methods.

11.4.2.1 GENE TRANSFER BY PHYSICAL METHOD

In the physical method, the outer cell wall and cell membrane are mechanically disrupted. The most common method in this type is particle bombardment biolistic wherein gold or tungsten particles are coated with DNA and discharged into the plant cell of interest under high pressure with a gene gun. The high velocity of the particles enables the coated DNA to permeate straight through the outer cell wall, guiding it towards the nucleus of the target plant cell. The biggest drawbacks in this method are the accessibility to specific equipment and the efficiency of the delivery of DNA into the nucleus rather than other cell organelles (Low et al., 2018).

Electroporation is another physical method that depends on electrical impulses to introduce the desired gene into plant cells. The cells of the plant to be transformed are incubated in a buffer having the DNA of interest, which is then subjected to electrical impulses resulting in the formation of transient pores which are temporary on the plant cell membrane facilitating the entry of the DNA containing the desired gene into plant cells. This method is relatively easy; however, it can only be applied to protoplasts, i.e., cell without a cell wall (Low et al., 2018).

In a study, *Bidens pilosa* L., a medicinal plant with antidiabetic properties was transformed via the *Agrobacterium*-mediated method. The transformation was carried out successfully using the vector construct carrying the cDNA of *Petunia* chalcone synthase (CHS), 35S promoter of CaMV, neomycin phosphotransferase (nptII) as selectable marker and nopaline synthase terminator (nos). A strain of *A. tumefaciens* LBA4404 was transformed with the vector construct by electroporation (Wang et al., 2012).

11.4.2.2 GENE TRANSFER BY CHEMICAL METHOD

Another method for transfer of desired genes into plant cells involves the use of chemicals and is called the gene transfer by chemical method. This method is not commonly used for plant transformation since it can be only carried out with plant protoplasts. In this method, the chemical that is commonly used is polyethylene glycol (PEG) along with a divalent cation that enhances the permeability of the cell membrane by destabilizing it. Although not well understood, it is believed that the osmotic pressure of the cell membrane is increased by PEG causing the protoplast to shrink. This allows the endocytosis of the DNA into the protoplast. The protoplast can also be transformed using liposomes which act as transporters of a foreign gene of interest to the protoplast. Because of its lipophilic property, it can easily access the protoplast of the transforming plant (Low et al., 2018).

11.4.3 INTEGRATION OF TRANSGENES

There exist two kinds of transgene integration: transient or stable. Stable transformation involves introducing the nucleus of the plant with the transgene, which later goes on to integrate with the plant genome. This way transgenes replicate along with the plant genome and are inherited by the following generation. Whereas in transient transformation, the transgene remains in the nucleus and does not integrate with the plant genome. Consequently, no replication takes place and the copy number remains the same. In transient transformation, the expression occurs for a brief time. These transgenes will be lost after a certain period of time in cell division.

In *Agrobacterium*-mediated gene transfer, the T-DNA integrates with the plant genome giving rise to a stable transformant. In the biolistic or the electroporation method of direct gene transfer, the stability of the transformant is dependent on the gene construct. Virus-mediated vectors are usually non-integrative vectors giving rise to transient transformants.

Inheritance of the transgene, depends on two factors, namely the site at which it gets integrated as well as the copy number. The methods to analyze inheritance and transmission of the transgene are its molecular characterization and phenotypic expression patterns.

Transgene integration can be confirmed depending on the gene construct as well as the selectable marker or reporter gene used during plant transformation. The plants carrying genes having anti herbicidal or antibiotic properties can be tested by adding herbicides or antibiotics. This can be an important distinguishing characteristic between transformed and non-transformed plants. However, the herbicides and antibiotics usually have high costs causing this method to be disadvantageous. Another disadvantage is also the possibility of horizontal gene transfer from various other bacteria. Therefore, more sensitive methods like screening through expression of reporter gene and polymerase chain reaction (PCR) are used commonly. The expression of reporter genes such as Luc, GUS, and GFP are either colorimetric or fluorimetric which can easily be observed under the microscope. Quantification of gene expression is possible by using a spectrophotometer. GUS expression is also helpful in providing information on localization of transgenes using histochemical assays. Other techniques which are also used as screening methods are Southern blotting. These molecular techniques give information on transgene integrity and the number of transgenes. PCR is a simple but powerful tool in these regards which uses primers that are specific to the desired gene and confirmation of transgene integration is possible if successful amplification takes place. It can be further verified by the use of DNA sequencing methods (Low et al., 2018).

11.4.4 HETEROLOGOUS BIOSYNTHESIS

Heterologous biosynthesis is a method of increasing secondary metabolite production by reconstruction of the desired compound's biosynthetic pathway in a host cell. The original producer can be difficult to cultivate or sometimes is not amenable. Mostly, microorganisms are used as host producers but more compliant plant species also serve as good alternatives like tobacco or *Arabidopsis* sp. (Atanasov et al., 2015).

A study was conducted on the plant *Catharanthus roseus* (L.) G. Don which is known to produce valuable alkaloids having anti-inflammatory properties and hence can be used for the treatment of diabetes. In this study, the incomplete biosynthetic pathway of secoiridoid in *C. roseus* (L.) G. Don was described and the complete pathway expression was observed using *Nicotiana benthamiana* Domin as host, thereby achieving sustainable production of the alkaloids (Tiong et al., 2013).

11.4.5 EXAMPLES OF STUDIES INVOLVING PLANT TRANSGENESIS AND HETEROLOGOUS BIOSYNTHESIS OF METABOLITES

The transformation of medicinal plants is highly dependent on factors such as gene construct, vector, and transformation techniques as discussed above. Some examples of successful transformation of medicinal plants with anti-diabetic properties based on these factors are as follows.

In certain plants, overexpression of a particular gene can improve the yield of valuable secondary metabolites. This displays the importance of a good gene construct. A study was done on transgenic ginseng roots to overexpress squalene epoxidase gene (PgSE1) and understand their effects. It was found that the overexpression of the gene enhanced the production of secondary metabolites. There was a higher accumulation of the secondary metabolites – ginsenosides and phytosterols when PgSE1 gene is overexpressed in ginseng urged on by the 35S promoter (Han et al., 2020).

The *Agrobacterium rhizogenes* strains TR105, A4, R-1000, 15834, R-1601, MAFF03-01724 and LBA9402 are commonly utilized for induction of hairy roots in medicinal plants. However, the efficiency of a strain is host-dependent. This is reported in a study performed on the leaf discs of *Linum flavum* where the strain LBA9402 caused the most efficient transformation. When tested with the highly virulent A4 strain, it was not effective on *Linum flavum* leaf discs (Faraz et al., 2020). *Linum* sp. show the presence of lignan-type secondary metabolites (Alejandre-García et al., 2015). Lignans display antioxidant properties which have shown effective control in BGL (Pan et al., 2007).

A study was conducted on developing an efficient method of transformation for *Citrus maxima* using *A. tumefaciens*. The results suggested that an increased number of shoots were observed after incubating the seedlings for two weeks in the dark accompanied by parafilm covering around the seedling wounds. Various binary vectors in *A. tumefaciens* were used namely pCAMBIA2301G-hp20 (constituting of CTV p20 gene, neomycin phosphotransferase as the selective marker gene and GUS marker gene) pCAMBIA1301-DlWUS (constituting of WUS gene of DIWUS (Dimocarpus longan Lour), hygromycin selective gene and GIS marker gene) or pFGC5941 (with Bar selective gene). The confirmation of transgenic integration suggested that the highest efficiency of transformation was observed when the selective marker gene used was hygromycin. Therefore, the efficiency of transformation of *C. maxima* is increased after a 2-week dark incubation and when the selective marker gene was hygromycin (Zhang et al., 2017).

There was also a report on establishing an efficient plant transformation via *A. rhizogenes*-mediated method for *Achyranthes aspera* L. Studies on *A. aspera* have previously reported strong anti-hyperglycemic activities when tested on diabetic mice induced with alloxan. *A. aspera* extracts show strong radical scavenging activity because of the occurrence of many phenolic compounds and flavonoids (Talukder et al., 2012). A variation was observed in the virulence pattern of four different strains of *A. rhizogenes* – MTCC 2364, A4, NCIM 5140 and ATCC 15834, that were assessed for efficient induction of hairy roots. Maximum transformation efficiency was observed using the strain *A. rhizogenes* was ATCC 15834 when tested on young leaf explants. There have been previous studies that have reported that acetosyringone can improve the T-DNA transfer and thus the hairy root transformation. This phenolic compound is secreted by wounded cells of a plant. In this study, a concentration of 100 μM acetosyringone was added externally and it resulted in an enhanced transformation frequency. However, increasing the concentration of acetosyringone (300 μM) had a negative impact on the transformation efficiency and reduced the induction of hairy roots. The confirmation of transgene integration was done using the DNA from hairy root clones and amplifying it using PCR. HPLC and mass spectra results show the presence of increased levels (~ 4 times) of 20-hydroxyecdysone in transgenic *A. aspera* plants as compared to wild *A. aspera* plants. 20-hydroxyecdysone is an important secondary metabolite with many important pharmacological activities. Thus, these transgenic lines can be used as a source of production of 20-hydroxyecdysone (John et al., 2017).

Scoparia dulcis shows the presence of compounds like diterpenes, triterpenes, and flavonoids which have contributed to its antidiabetic properties. These compounds show strong antioxidant activities which are important to combat hyperglycemia-induced oxidative stress and inflammatory reactions observed in diabetes (Pamunuwa et al., 2016). A study was carried out using axenic leaf explants of *S. dulcis* was transformed using the LBA4404 strain of *A. tumefaciens* bearing pCAMBIA1301 as the binary vector. The pCAMBIA1301 T-DNA consists of GUS gene which is intron interrupted and hygromycin phosphotransferase selectable marker genes having CaMV 35S promoter. The integration of transgene was validated by PCR amplifying the GUS and hygromycin phosphotransferase genes in the DNA separated from the leaves of the transformed as well as from the non-transformed plants. Cocultivation is an important step in *A. tumefaciens*-mediated transformation as it increases transformation efficiency. In this study, the cocultivation medium consisted of optimum concentrations of acetosyringone and certain phytohormones which significantly increased the transformation efficiency (54.6%) in *S. dulcis*. This sort of an optimized protocol can allow for the production of higher levels of important secondary metabolites in the future using metabolic engineering (Aileni et al., 2011).

A recent study reported artemisinin to be an important therapeutic agent for DM (Atanasov et al., 2015). Artemisinin can promote β-cells proliferation, enhanced insulin secretion and improved glucose homeostasis (Yuan et al., 2017). Artemisinin can be produced in *S. cerevisiae* strain genetically engineered to produce high yields of artemisinic acid. Thereafter, by a procedure involving simple chemical conversion, artemisinic acid is converted to artemisinin. Artemisinin can directly be produced in tobacco plants that have been transformed. In a study, a construct of mega vector was developed which included yeast-derived and five plant genes associated with the biosynthetic pathway of mevalonate and artemisinin. This mega vector construct was transferred into tobacco plants. However, around 7 μg/g biomass dry weight was aggregated and that is lower in comparison to the other production alternates (Atanasov et al., 2015).

11.5 APPLICATION OF TRANSGENESIS TO PLANT SECONDARY METABOLITES WITH POTENTIAL ANTIDIABETIC PROPERTIES

Some of the most effective secondary metabolites of plants that have been used in many antidiabetic studies have been classified into phenolic compounds, terpenoids, alkaloids, chalcones, polysaccharides, and others (Hussein and El-Anssary, 2019).

11.5.1 VARIOUS PHENOLIC COMPOUNDS IN PLANTS

There are various phenolic compounds which have favorable impacts on human health as compared to other groups of secondary metabolites. There have been many epidemiological study reports that propose a positive correlation between ingestion of a diet rich in phenols and a reduction in the prevalence of diabetes. Phenolic compounds constitute one or more phenol groups and range from simple structures with one aromatic ring to complex structures involving polymerization (Hussein and El-Anssary, 2019). Phenolic compounds are valued as effective radical scavenging antioxidants and their mechanism of action involves inhibition of enzymes participating in radical production (Zhao et al., 2019). These compounds can be classified based on structure into flavonoids, resveratrol, xanthones, tannins, and others.

11.5.1.1 FLAVONOIDS

Flavonoids are one of the most plentiful secondary metabolites present in various plant parts like fruits, seeds, stems, flowers, etc. Because of this abundance, flavonoids have been one of the most well studied bioactive compounds, and thus isolation techniques have helped in the characterization of many unique flavonoids. These are low molecular weight compounds that have important therapeutic effects including cell synthesis (Hasan et al., 2018). Flavonoids come across as acylglycosides or glycosides, however, sulfated, methylated, and acylated forms of flavonoids are present in lower concentrations and are not very common (Gothai et al., 2016). Many studies have displayed their potential in the treatment of diabetes due to its extensive mechanisms of action (Hasan et al., 2018).

It has been reported that polyphenols help in inhibiting pancreatic enzymes like α-amylase or α-glucosidases. This contributes to their hypoglycemic activity. Flavonoids include several polyphenolic compounds whose bioactivity is attributed to important differences in structures. They have shown important effects in the regulation of postprandial hyperglycemia. For instance, apigenin raises serum insulin levels and regulates hyperglycemia, whereas rutin increases glucose uptake. In diabetes, the flavonoid quercetin stimulates the secretion of insulin and inhibits the production of aldolase reductase. Thus, flavonoids have displayed many antidiabetic properties along with their well-recognized antioxidant properties (Zhao et al., 2019).

Apigenin is also known to reduce the production of cytokines like TNF-α, IL-1 β and IL-6, which are pro-inflammatory, by regulating the numerous intracellular signaling pathways in cells, thereby reducing hyperglycemia and improving antioxidants via oxidative stress based signaling pathways (Gothai et al., 2016).

Diosmin is another naturally occurring flavonoid that is known to produce important antioxidants. These antioxidants cause the deactivation of proinflammatory chemokines and cytokines that are responsible for diabetes-induced ROS, thereby subduing diabetes (Gothai et al., 2016).

11.5.1.2 RESVERATROL

Resveratrol is a natural phenol compound and is known to be a powerful scavenger of ROS (Gothai et al., 2016). It has been reported in some studies that in diabetic rats and patients' administration of resveratrol is known to increase insulin sensitivity. Many studies have observed that activation of histone deacetylase Sirtuin1 (Sirt1) is induced by resveratrol which can improve the sensitivity to insulin as well as provide protection from the harm caused by the intake of a high-fat diet. Deacetylase activity of Sirt1 suppresses the transactivation of pyruvate dehydrogenase lipoamide kinase 4 (PDK4) induced by Foxo1. PDK4 acts as a negative regulator for the enzyme pyruvate dehydrogenase which is a glycolytic enzyme (Ota and Ulrih, 2017). Resveratrol as therapy in diabetes can improve regulation of BGL, insulin sensitivity and other related metabolic problems (Zhu et al., 2017). Along with this, resveratrol decreases the release of proinflammatory cytokines caused due to activation of immune cells through the inhibition of NF-kappaB, leading to downregulation in the expression of COX-2 gene in diabetic animal models (Gothai et al., 2016).

11.5.1.3 XANTHONES

Xanthones have similarities to flavonoids concerning their structure and chromatographic behavior. However, xanthones have a limited presence in nature compared to flavonoids (Negi et al., 2013). Many reports have shown xanthones to have antidiabetic properties. The glycoside element of mangiferin, the most common xanthanoid, decreases insulin resistance and can inhibit the activity of many carbohydrate metabolizers (Zhao et al., 2019). Studies performed on bellidifolin and swerchirin have reported strong hypoglycemic activity. The compound, 1,8-dihydroxy3,5-dimethoxyxanthone (swerchirin) isolated from *Swertia chirayita* has shown a considerable decrease in levels of blood glucose when tested on albino rats *in vivo* (Negi et al., 2013).

11.5.1.4 TANNINS

Tannins are antioxidant polyphenols found in several plants. Tannins act as radical scavengers and activate antioxidant enzymes (Kumari and Jain, 2012). Tannins are capable of triggering non-specific inhibition of α-glucosidase.

Inhibition of α-amylase is seen due to condensed tannins; there is an increase in inhibition corresponding to the increase in the degree of polymerization (Zhao et al., 2019). A natural tannin called epicatechin can induce regeneration of β-cells leading to decrease in hyperglycemia (Kumari and Jain, 2012). *Atriplex halimus* L. extract containing tannins shows a considerable decrease in levels of blood glucose at a dosage of 20 mg/kg in diabetic rats induced with streptozotocin (Gupta et al., 2018).

11.5.2 TERPENOIDS

Terpenes are one of the biggest classes of plant secondary metabolites consisting of five-carbon isoprene units which are assembled in various ways. Altered terpenes called terpenoids have different functional groups attached at various positions (Perveen, 2018). Terpenoids act by reducing hyperglycemia, enhancing glycogenesis, and reducing glycogenolysis, also by reducing the enzyme aldose reductase activity. For instance, curcumin studies have shown to inhibit glucose transporter activity. Intestinal epithelial cells show the presence of GLUT proteins. The inhibition of these GLUT proteins can decrease the absorption of glucose thereby creating a hypoglycemic effect (Zhao et al., 2019). *Cassia auriculata* extracts containing terpenoids have been shown to reduce hyperglycemia in diabetic rats induced with streptozotocin. Extracts of *Tinospora crispa* injected intravenously increased the plasma insulin levels in diabetic rats (Gupta et al., 2018).

11.5.3 VARIOUS ALKALOIDS IN PLANTS

Alkaloids are high molecular mass, complex compounds containing nitrogen and are attached to a minimum of two carbon atoms (Teoh and Das, 2018). Alkaloids have two mechanisms to combat diabetes, stimulating proliferation of β-cells to increase insulin secretion and decrease resistance or decrease the glucogenic enzyme levels (Zhao et al., 2019). Total alkaloid extract of *Catharanthus roseus* (L.) G. Don. resulted in a prominent reduction in fasting BGL and also increased the levels of insulin. Protein Tyrosine Phosphatase 1B (PTP1B) causes the negative regulation of insulin and leptin signaling pathway and thus has a significant role in T2DM. In the *in vitro* studies using myoblast C2C12 or pancreatic β-TC6 cell lines with alkaloids extracted from *C. roseus* (L.) G. Don. namely vindolinine, vindolicine, vindolidine, and vindoline was found to increase the uptake of glucose and increased inhibition of PTP1B. Berberine is an alkaloid extracted from *Coptis chinensis* (CC), which has shown a considerable reduction in HbA1c levels along with triglycerides and total cholesterol and also raised insulin levels as seen in diabetic rats. It also shows a decrease in fasting and postprandial hyperglycemia in diabetes patients (Teoh and Das, 2018).

11.5.4 CHALCONES

Chalcones are known to be precursors of flavonoids and isoflavonoids. Licochalcones A isolated from *Glycyrrhiza inflata* is known to inhibit PTP1B (Mahapatra et al., 2015). A type of chalcone namely 2-hydroxy-40-methoxychalcone are known to exhibit antidiabetic properties by exhibiting anti-atherosclerotic activity and stimulating the expression of PPAR-γ in aortic smooth muscle cells. Thus, it is seen that chalcones increases insulin sensitivity of cells by activating PPAR-γ, mimicking the action of thiazolidinediones which are pharmacological agonists for these nuclear transcription factors (Zhao et al., 2019).

11.5.5 POLYSACCHARIDES

Polysaccharides have also shown properties that prevent T2DM. These polysaccharides have desirable effects on the regulation of glycemia and decrease insulin resistance. A polysaccharide isolated from *Morus alba* L. show prominent antihyperglycemic activity in diabetic rats induced with streptozotocin. The study reported a reduction in fasting blood glucose (FBG), glycated serum protein and improved insulin levels after seven weeks of treatment (Ganesan and Xu, 2019). Ganoderma Lucidum Polysaccharides inhibit the levels of high blood glucose by modulating the expression of major enzymes involved in the glucose metabolism pathway (Ma et al., 2015). Plasma insulin levels were reduced and reverse in high-fat diet-induced insulin resistance was observed because of Ganoderma Lucidum Polysaccharides (GLP). GLP brings about these effects by inducing lipolysis in adipose tissues, improving low-grade chronic inflammation, and reducing plasma triglyceride levels. GLP regulates the gut microbiota composition which influences obesity and development of T2DM (Zhao et al., 2019).

A few other bioactive compounds include thiosugar derivatives which have shown high inhibitory activity against the enzyme α-glucosidase

(Zhao et al., 2019). A study performed on morin, a compound isolated from medicinal plants like almond and guava, reported that it reduced the level of inflammatory cytokines produced during diabetes. Therapy with morin regulated BGL, enzymes involved in glucose metabolism and enhanced levels of insulin in diabetic models (Gothai et al., 2016). Curcumin is an important bioactive compound found in turmeric. It is a polyphenol compound that demonstrates antioxidant, anti-inflammatory, anticancer, and antimicrobial properties. Curcumin causes a reduction of ROS production and ICAM-1 expression, which leads to a reduction in vascular inflammation, commonly observed in diabetes (Gothai et al., 2016). Colchicine is a compound derived from the plant *Colchicum autumnale*, which is commonly used in the management of inflammatory diseases like pericarditis and gout. One of the related conditions of diabetes is diabetic neuropathy, where there is a high level of inflammatory cells which infiltrate the renal tubulointerstitium. Treatment with colchicine alleviates inflammatory cell infiltration observed in diabetic neuropathy through inhibition of MCP-1 and ICAM-1 expression (Gothai et al., 2016). Transgenic hairy root cultures of Cucurbits have shown an increase in the production of the secondary metabolite charantin. Charantin is a steroidal glycoside that has shown significant hypoglycemic effects when studied on diabetic rabbits (Rekha and Thiruvengadam et al., 2017). Extracts of safflower and Japanese kelp show the presence of high amounts of important bioactive compounds. These compounds demonstrated significant suppression of the activity of α-glucosidase, thus controlling the uptake of glucose in the gut. It has also been reported that regulation of Glucagon-like peptide-1 (GLP-1) homeostasis by the soluble fiber, Inulin. Plant extracts such as polysaccharides of Dioscorea, anthocyanins of blueberries, cinnamon, and fenugreek seeds show improvement of insulin resistance. Along with this, many bioactive compounds found in various plants such as chili peppers, tea extracts, bitter melon, turmeric, and ginseng demonstrate a combinatorial effect (Gothai et al., 2016).

11.5.6 TESTS PERFORMED TO CHECK ANTIDIABETIC POTENTIAL OF SECONDARY METABOLITES

Experiments performed *in vivo* and *in vitro* are essential as they provide a good understanding of the pathology and pathogenesis, thereby aiding in finding new and better treatment options (Vijayaraj et al., 2019).

11.5.6.1 IN VITRO ANTIDIABETIC ASSAYS

In vitro antidiabetic assays screen for primary and secondary antidiabetic biological activities (Haddad et al., 2012). Some of the common *in vitro* experiments involve enzyme inhibition assays, assays on uptake of glucose and experiments on cell lines isolated from pancreatic islet cells (Karthikeyan et al., 2016).

The core idea of enzyme inhibition assays is that enzymes that are involved in inhibition of carbohydrate metabolism like α-amylase and α-glucosidase that delay the release of glucose from digested food, thereby acting on post-prandial hyperglycemia (Lankatillake et al., 2019). The function of α-amylase is to breakdown starch to oligosaccharides in the regions they are found such as the mouth and small intestine. Due to the enzyme's inhibition, there is a delay in carbohydrate metabolism leading to slower glucose absorption in the small intestine, ultimately reducing postprandial hyperglycemia. Oligosaccharides and sucrose are broken down into glucose in the small intestine by α-glucosidase enzyme. The inhibition of this enzyme causes a delay in carbohydrate metabolism leading to slower glucose absorption in the small intestine, thereby reducing postprandial hyperglycemia (Lankatillake et al., 2019). Dipeptidyl peptidase-4 (DPP-4) is an enzyme of importance as it regulates the glucose levels in blood through the catabolism of incretin peptides. The inhibition of this enzyme provides the potential of reducing postprandial hyperglycemia (Dilshad et al., 2020).

A study was performed to check the antidiabetic potential of *A. rhizogenes* transformed *Artemisia carvifolia* with high artemisinin and high flavonoid content. Both α-amylase and α-glucosidase inhibition assays were performed. The results suggested that to *rol* genes transformant extracts show a reduction in the IC50 value as compared to their wild type counterparts. Along with this, DPP-4 inhibition assay was also performed, the results of which suggested that the plant extracts had an inhibitory effect on the enzyme. The observed results are significant as it displays the potential of the *A. carvifolia* plant extract as well as better antidiabetic activity due to transformation of the plant (Dilshad et al., 2020). The inhibitory activity of two bioactive compounds, phloretin, and eupatilin, were studied against the α-glucosidase and α-amylase enzyme activity. Phloretin is a flavonoid found in many fruits and vegetables, while eupatilin is a flavonoid molecule with potential antioxidant properties found in the Artemisia plant. The study results suggest a good inhibitory activity (Gulçin et al., 2018).

Glucose can be transported in and out of the cell by two mechanisms, active transport or facilitated diffusion. Mammalian cell membranes show the presence of glucose transporters (GLUT) that allow for the movement of glucose per the chemical gradient. The regulation of BGL that is followed by glucose uptake is influenced by the level of GLUT expression on the cell membrane. Stimulation of glucose uptake in muscle and adipose tissue by insulin is necessary to attain normoglycemia. One of the most important sites for insulin-stimulated glucose utilization is the muscle, and therefore, the basal expression levels of GLUT4 in the muscle are generally correlated to the insulin-stimulated glucose utilization of the entire body (Yamamoto et al., 2011). In T2DM, it was found that there was a decrease in the uptake of glucose-stimulated by insulin in muscle and adipose tissue due to insulin resistance. In muscle tissues, this could be due to defective stimulation which prevents the translocation of GLUT receptors to the cell surface. Whereas in adipose tissue, this occurs due to defective stimulation as well as a decrease in expression levels of GLUT receptors. Reduced GLUT expression levels have been observed in human obesity and T2DM as well as rat models exhibiting resistance to insulin (Wang et al., 2009). In yeast cells, the uptake of glucose is via facilitated diffusion and the glucose carriers transport glucose down the concentration gradient. This action emphasizes the importance of the clearance of intracellular glucose. Only if the intracellular glucose is cleared, there is effective transport of glucose (Pitchaipillai and Ponniah, 2016). Similarly, if there are increased glucose uptake and glucose metabolism observed in yeast cells treated with medicinal plant extracts, these extracts must be tested *in vivo* as well (Yamamoto et al., 2011). Glucose uptake assay using yeast cells is a famous *in vitro* screening method to study the hypoglycemic effect of medicinal plants (Sairam and Urooj, 2013).

A study was performed using the extract of *Cassia nemophila* pod where the glucose uptake assay using yeast cells was used to check the antidiabetic potential. It was observed that in the presence of the pod extracts there was improved uptake of glucose and increase in the concentration of extract resulted in an increase in uptake of glucose by the yeast cells. An inverse relationship was observed with molar concentrations of glucose which implies that a lower glucose concentration in the solution leads to a higher uptake by yeast cells (Rehman et al., 2018). Berberine, an alkaloid isolated from medicinal plants such as *Coptidis rhizoma* and *Cortex phellodendri* demonstrated antidiabetic activity when tested on 3T3-L1 adipocytes. Improved basal glucose uptake was reported in both normal and insulin-resistant states. Another bioactive compound, 4-Hydroxypipecolic acid (4-HPA) isolated from the seed of *Peganum harmala* was studied for its glucose uptake activity on L6 skeletal muscle cell lines. The results suggested an improved glucose uptake as well as an increase in GLUT translocation to cell surface depending on the concentration of 4-HPA (Ga and Mahalingam, 2016).

Certain antidiabetic studies are also performed using isolated pancreatic islet cells. In these studies, extracts of plants or important isolated bioactive compounds are tested to understand their impact on glucose-induced secretion of insulin by β-cells of the pancreas. *Morus nigra* extracts have been reported to increase insulin secretion when tested on isolated mouse islet cells. In a similar study performed using the extract of *Orthosiphon aristatus* on INS-1 β-cell line of the pancreas, increased secretion of insulin by the cells was observed. (Lankatillake et al., 2019).

11.5.6.2 IN VIVO STUDIES

Positive results obtained from *in vitro* antidiabetic studies of medicinal plant extracts could be used to demonstrate *in vivo* also the regulation of levels of glucose in blood, hence *in vivo* studies using animal models are necessary (Haddad et al., 2012; Hasan et al., 2018). To test this potentiality, several obesity and diabetes animal models exist. These models are obtained from genetic defects that predispose these animals to diabetes and obesity like Zucker diabetic fatty (ZDF) rats and Ob/ob mice or through induction of certain chemicals like streptozotocin and alloxan. There also exist other methods that rely on metabolic disturbances caused by dietary intervention where the models demonstrate a period from insulin resistance to T2DM (Haddad et al., 2012). The hypoglycemic activity is understood by elucidating the blood sugar levels in animal models or by hypoglycemic seizures in mice. Alternatively, there are studies wherein the insulin-like activity is tested on cells and tissues utilizing methods like the epididymal fat pad, 3T3-L1 adipocytes, collagenase digested isolated fat cells, assays involving transport of glucose, etc. (Hasan et al., 2018).

11.6 GENOMICS OF MEDICINAL PLANTS WITH ANTIDIABETIC PROPERTIES

Progress in the development of various sequencing technologies has allowed for rapid completion of genome sequencing projects of several organisms. In the past few years, a lot of knowledge has emerged due to next-generation sequencing (NGS) technology. There are many ongoing genome projects of medicinal plants. These projects can prove to be extremely useful in terms of understanding and analysis of the different biochemical pathways of important secondary metabolites, genome annotation and data mining. These studies can help in the development of treatments and also help in discovering genes that can improve the yield of important secondary metabolites (Singh et al., 2019). Since there is an abundance of secondary metabolites in medicinal plants, their genomic/transcriptome studies must be explored. Consequently, based on the information sequences and expression data, characterization of genes and biochemical pathways associated with the synthesis of these secondary metabolites can be carried out (Chaudhary and Sharma, 2016).

One of the first genomic sequence drafts of a medicinal plant was of *Azadirachta indica. A. indica* has a less complex genome compared to other medicinal plants, consisting of 20,000 genes representing 364 Mb genome. It has many important therapeutic properties such as antidiabetic, antifungal, antimicrobial, antiviral, and others. An NGS study was conducted on *A. indica* to understand its terpenoid biosynthesis pathway. Another study was conducted on *Juglans regia*, known for its antioxidant, anti-cancer properties and is also used for the treatment of symptoms of diabetes to obtain information on expression of genes involved in the biosynthesis of non-structural polyphenols. The significance of nuclear genomic studies based on NGS technology, of important antidiabetic medicinal plants is demonstrated through these examples. In recent years such NGS-based studies were used to discover several genes involved in encoding of unique characteristics in medical plants. Along with this, a deep understanding has been provided of the expression of genes and the process associated with the biosynthesis of therapeutically important compounds (Chaudhary and Sharma, 2016).

The main objective of transcriptomic studies based on NGS technology is to determine the genes associated with the biosynthesis of bioactive compounds. The main focus of transcriptome analysis of *A. indica* was to characterize the metabolic pathways, perform comparative evolutionary studies and study all the metabolic pathways connected to azadirachtin synthesis. Azadirachtin is a bioactive compound found in *A. indica* responsible for its medicinal properties (Chaudhary and Sharma, 2016). Studies have reported the significant antihyperglycemic as well as islet regeneration/insulin production effects of *A. indica* extracts containing azadirachtin on diabetic rats induced with streptozotocin (Bhat et al., 2011; Chandra et al., 2007). One such study found their results of antihyperglycemic effects to be similar to that of insulin (Chandra et al., 2007). NGS-based transcriptome analysis of *Catharanthus roseus* was performed to develop a metabolic pathway database known as "CathaCyc." *C. roseus* is known for its antidiabetic, antimicrobial, antiviral, anticancer properties among others. "CathaCyc" provides information on different genes, metabolic pathways, enzymes, and proteins utilized in the synthesis of various secondary metabolites. This database provides significant information in order to study the different primary and secondary metabolic pathways of medicinal plants including *C. roseus* (Chaudhary and Sharma, 2016). Another transcriptome study was conducted on *Lotus corniculatus* which has shown potential as a hypoglycemic agent. This study had the aim of identifying genes of key enzymes and transcription factors that are involved in the biosynthesis pathway of flavonoids (Chaudhary and Sharma, 2016). *Rehmannia glutinosa* is used to treat many conditions such as diabetes, anemia, fever, osteoporosis, and allergies. The main focus of its transcriptome analysis was to identify genes related to the iridoid biosynthesis (Chaudhary and Sharma, 2016). Iridoid compounds have demonstrated antihyperglycemic activities. For instance, an iridoid glycoside known as catalpol which is present in *R. glutinosa* extracts has shown a decrease in hyperglycemia when given as treatment to diabetic rats induced with streptozotocin and alloxan. Catalpol helped in improving the structural impairment of the pancreas along with maintaining the balance between oxidative and antioxidative enzymes present in diabetic rats induced with streptozotocin fed on high fat and high sugar diet (Zhu et al., 2016).

In a study of genome-transcriptome data annotation of *Lignosus rhinocerus* to PlantCyc enzymes database some transcripts exhibiting antidiabetic properties were identified. It was observed that these genes also code for carotenoid and abscisic acid biosynthesis along with superoxide dismutases, glyoxalase I and catalase-peroxidases. The study indicates that in *L. rhinocerus,* the presence of these genes can be associated with the anti-glycation properties of the secondary metabolites. There is an increasing focus on anti-glycation compounds found naturally in medicinal plants. These compounds show the potential to reduce or prevent the likelihood of chronic complications in diabetes patients. anti-glycation activities were demonstrated in the fractions of medium molecular weight (MMW) from the extracts of *L. rhinocerus*. The fractions of MMW mainly inhibited by about 500-fold the formation of compounds responsible for glycation end product structures when tested on a human serum albumin-glucose system. This kind of activity of curbing of protein glycation can be attributed to its potent radical scavenging activity (Yap et al., 2018).

The plant metabolic pathways are very diverse and only a small part has been discovered till now. To discover the unknown metabolic pathways and enzymes, genomic studies on medicinal plants with the help of large-scale sequencing tools are essential (Chakraborty, 2018). With the help of NGS-based technology, genome sequencing has become more cost and labor effective and increases opportunities to carry out in-depth study of important medicinal plants at genomic and transcriptomic levels. The data obtained from these analyzes is a valuable source from which the selection of candidate genes coding for important bioactive compounds can be done. These genes can be further investigated for the development of transgenic medicinal plants that can improve the potential of therapeutic compounds (Chaudhary and Sharma, 2016).

Various medicinal plants producing secondary metabolites having antidiabetic properties have been reported. Biotechnology-based transgenic methods can be used to further enhance the levels of secondary metabolites with antidiabetic properties produced by these plants. Discovery of the gene/biosynthetic pathway can further improve the production of the secondary metabolite. A summary of transgenic approaches on medicinal plants showing antidiabetic properties along with their secondary metabolite and gene/biosynthetic pathway information has given in Table 11.2.

11.7 METABOLITE PROFILING OF COMPOUNDS WITH ANTIDIABETIC PROPERTIES FROM MEDICINAL PLANTS

Primary and secondary metabolites constitute the highly complex metabolome of plants. To obtain a better understanding of biochemical pathways and the underlying genes, plant metabolites must be identified and quantified. The highly complex metabolome of plants profiling of metabolites, a targeted analysis of metabolites and fingerprinting of metabolites (Kumar et al., 2018). There has been development in the field of metabolomics in the last few years allowing for detailed and precise metabolite profiling of several organisms. The various metabolites present in the single plant extract can be identified using metabolomics techniques thereby helping in rapid and accurate analysis of metabolites. Exploration of metabolites present in mutants and transgenic lines enable the understanding of metabolic networks and help determine the candidate genes. Metabolomics helps in understanding the function of genes, including the regulation of transcription factors and the link between interconnected pathways which is otherwise not possible in conventional assays like microarrays (Kumar et al., 2017).

TABLE 11.2 Summary of Transgenic Plants with Antidiabetic Properties

Plant	Vector/ Technique	Secondary Metabolite	Gene/Biosynthetic Pathway	Antidiabetic Properties	References
Panax ginseng	*A. tumefaciens*-mediated gene transfer	Phytosterol and Ginsenoside	PgSE1	Hypoglycemic activity is exhibited by ginsenosides in diabetic rats induced with streptozotocin.	Han et al. (2020); Srivastav et al. (2020)
Citrus maxima	*A. tumefaciens*-mediated gene transfer	Naringin	Phenylpropanoid pathway	Naringin exhibits anti-hyperglycemic activity when given as treatment to diabetic rats induced with streptozotocin.	Zhang et al. (2017); Reshmi (2018); Sharma et al. (2019)
Achyranthes aspera L.	*A. rhizogenes*-mediated gene transfer	20-hydroxy-ecdysone	Shares most steps of the biosynthetic pathway with cholesterol as cholesterol is its direct precursor. Conversion of cholesterol to 7-dehydrocholesterol with further modifications gives rise to 20-hydroxyecdysone.	20-hydroxyecdysone can reduces the risk of obesity, resistance to insulin and hyperglycemia.	Kizelsztein et al. (2009); John et al. (2017); Talukder et al. (2012)
Scoparia dulcis	*A. tumefaciens* strain LBA4404 bearing the binary vector pCAMBIA1301	Diterpenes, Triterpenes, and Flavonoids	Diterpene biosynthesis mainly involves two genes SdCPS2 and SdKSL1. Triterpenes are synthesized from farnesyl pyrophosphate. Farnesyl phosphate is further converted to 2,3-oxidosqualene which ultimately gives rise to various triterpenes. Phenylpropanoid pathway is involved in the biosynthesis of flavonoids.	These compounds show strong antioxidant activities which are important to combat hyperglycemia-induced oxidative stress and inflammatory reactions observed in diabetes.	Pamunuwa et al. (2016); Aileni et al. (2011); Yamamura et al. (2017); Singh and Sharma. (2015)

TABLE 11.2 *(Continued)*

Plant	Vector/ Technique	Secondary Metabolite	Gene/Biosynthetic Pathway	Antidiabetic Properties	References
Swertia chirayita	*A. rhizogenes*-mediated gene transfer	Swechirin, Swertiamerin	Swerchirin and swertiamerin belong to xanthone class of secondary metabolites. Xanthones are derived from a mixture of shikimate and acetate.	Swechirin has shown a considerable decrease in the levels of blood glucose when tested on albino rats. Swertiamerin can cause an increase in levels of insulin with decrease in hyperglycemia in diabetic rats induced with streptozotocin.	Negi et al. (2013); Leong et al. (2016); Samaddar et al. (2019).
Catharanthus roseus (L.) G. Don	Heterologous biosynthesis using host plant *Nicotiana benthamiana* Domin.	Vindoline, Vindolidine, Vindolicine, and Vindolinine	Secoiridoid biosynthesis pathway	Various alkaloid extracts from the plant namely vindolinine, vindolicine, vindolidine, and vindoline caused higher uptake of glucose and increased the inhibition of PTP1B as observed in *in vitro* studies.	Atanasov et al. (2015); Teoh and Das. (2018)
Curcubits	*A. rhizogenes*-mediated gene transfer	Charantin	Charantin is a triterpenoid molecule and hence follows a pathway similar to triterpene where there is further conversion of 2,3-oxidosqualene to cycloartenol which gives rise to charantin.	Charantin has shown significant hypoglycemic effects when studied on diabetic rabbits.	Cuong et al. (2017); Rekha and Thiruvengadam (2017)

TABLE 11.2 *(Continued)*

Plant	Vector/ Technique	Secondary Metabolite	Gene/Biosynthetic Pathway	Antidiabetic Properties	References
Artemisia carvifolia	*A. rhizogenes*-mediated gene transfer	Artemisinin, Flavonoid	The genes involved in artemisinin biosynthesis include aldehyde dehydrogenase 1, CYP71AV1and amorpha-4, 11 diene synthase.	Artemisinin has shown significant results in *in vitro* studies. It is a potent inhibitor of α-amylase, α-glucosidase, and DPP-4 enzyme.	Dilshad et al. (2015, 2020)
Azadirachta indica	*Agrobacterium*-mediated gene transfer	Azadirachtin	Terpenoid biosynthesis pathway	Studies have reported the significant antihyperglycemic as well as islet regeneration/insulin production effects of *A. indica* extracts containing azadirachtin on diabetic rats induced with streptozotocin.	Chaudhary and Sharma. (2016); Bhat et al. (2011); Chandra et al. (2007)
Rehmannia glutinosa	*Agrobacterium*-mediated gene transfer	Catalpol	Iridoid biosynthesis	Catalpol has shown a decrease in hyperglycemia when given as treatment to diabetic rats induced with streptozotocin and alloxan. It also helped in improving the structural impairment of the pancreas along with maintaining the balance between oxidative and antioxidative enzymes present in diabetic rats induced with streptozotocin fed on high fat and high sugar.	Chaudhary and Sharma. (2016); Zhu et al. (2016)

The main objective of metabolite profiling is the determination of the relative quantities of a maximum number of metabolites among a predefined group that is structurally related but it is not concerned with the determination of precise concentrations (Kumar et al., 2018). The two most commonly tools used for metabolite profiling are Nuclear Magnetic Resonance (NMR) and Mass Spectrometry (MS) (Kumar et al., 2017).

All the current metabolomics platforms utilize MS and NMR as tools for generating metabolome data. Metabolite profiling based on NMR utilizes the magnetic properties possessed by the nuclei of atoms which is controlled by a magnetic field for detection. NMR is widely used for the detection of small molecular weight metabolites (<50 kDa) as it is a non-destructive method of detection. The drawback of this technique is that it has low sensitivity and is unable to identify biomarkers with low abundance. This is why there is more extensive use of MS which demonstrates higher sensitivity which allows for greater coverage of the metabolome. MS aided in identifying novel biomarkers and molecules that can further help in the reconstruction of metabolic pathways. Progress in ionization techniques like atmospheric pressure MALDI-TOF, electrospray ionization (ESI) and chemical ionization (APCI) has increased the efficiency of MS. To improve the output, MS is generally clubbed with chromatographic technologies like liquid chromatography (LC), gas chromatography (GC), capillary electrophoresis (CE), field asymmetric waveform ion mobility spectrometry (FAIMS) and Fourier transform ion cyclotron resonance (FT-ICR). However, NMR has important properties such as its potential to determine the ligands physical properties, protein binding sites, determination of arrangement of protein-ligand complexes, and its direct interaction with the target protein, which is why it is still used and sometimes preferred over MS (Kumar et al., 2017).

In metabolomics GC-MS is the "gold standard" due to its ability to identify, separate, and quantify metabolites (Kumar et al., 2018). It is a platform extensively used for non-targeted analysis. This technique relies on compounds being converted to volatile form by derivatization (Kumar et al., 2017). For metabolite profiling as well as targeted analysis GC-MS is a widely used approach and many protocols for a given machine setup are available with data mining and interpretation (Kumar et al., 2018). GC x GC-TOF-MS has brought about an improvement in separation of peaks which co-elute as well as for higher sample throughput. The LC-MS platform with ESI and APCI is commonly used for both targeted and non-targeted studies which allow for the detection of high mass metabolites (<1,500 Da) (Kumar et al., 2017).

The data sets which are generated from the various high throughput tools that are mentioned above are processed by using different platforms like MET-XAlign, Met-Align, MET-COFEA, etc. The process of identification of metabolites includes correction of baseline, normalization, alignment, co-eluting peaks being separated, etc. Other metabolome databases used for the identification of metabolites include GOLM, NIST, METLIN, etc. Afterward, the data of the identified metabolite is subjected to statistical analysis by the use of web-based tools and software like MetaboAnalyst, Cytoscape, Statistical analysis tool to name a few (Kumar et al., 2017).

A study was conducted on *Ficus benjamina* extracts for their enzyme inhibitory activities. In addition to this, the metabolite profiling of the plant was also performed using UHPLC-QTOF-MS/MS and 1H NMR. The biologically active components present in the ethanolic extracts of *F. benjamina* were determined using a combination of techniques such as 1H-NMR and UHPLC-QTOF-MS/MS. The metabolites present in the extract were identified by the peaks of 1H-NMR that were compared with database Chenomx. Whereas the metabolites when identified using UHPLC-QTOF-MS/MS were compared with the ACD/Labs library and other relevant references based on the patterns of fragmentation, peaks of the molecular ions and accurate masses. Some of the bioactive compounds identified in the extract were quercetin, rutin, catechin, epicatechin, kaempferol, morin, apigenin, among others. Several studies have reported these compounds to have inhibitory activity towards enzymes such as α-glucosidase and α-amylase, thereby proving their potential options for therapy and management of diabetes. From the study, it was demonstrated that the extracts of *F. benjamina* contain bioactive compounds which demonstrated good antioxidant and inhibitory activities towards the enzymes α-amylase and α-glucosidase (Mumtaz et al., 2018).

Another metabolite profiling study was conducted on transgenic *Artemisia carvifolia*. The transgenic *Artemisia carvifolia* was generated by *rol* genes and had overexpression of genes coding for enzymes of isoprenoid with high artemisinin content and phenylpropanoid pathway with high flavonoids content. The objective of the study was to obtain information on any changes in the other secondary metabolite profile due to the overexpression of these genes. Metabolite profiling was performed using ESI/MS-TOF. The results suggested that there was a higher level of vital amino acids like proline and arginine in *rol* B transgenic lines and a higher level of glutamine in the *rol* C transgenic lines. Arginine, as reported in other studies, has displayed antidiabetic potential where it can increase insulin sensitivity. Whereas glutamine is an amino acid that has an important effect in plant metabolism since it provides nitrogen for biosynthesis of nucleic acids and various amino acids as well as a significant amino group donor in transamination reactions. There was also an increased presence of notable phenylpropanoids such as shikimic acid, caffeic acid, cinnamic acid in transgenic lines expressing *rol* B. These compounds which are phenolic in nature act as intermediary metabolites of the phenylpropanoid pathway and show important antioxidant properties which are important in the therapy of diabetes. When the transgenic lines were subjected to metabolite fingerprinting, the results show an increased presence of metabolites that are precursors to the secondary metabolites. This increase can be due to the increased artemisinin and flavonoids content in the transgenic plants. This is why there is also an enhanced antidiabetic activity observed in the transgenic plants (Dilshad et al., 2020).

Montbretin A (MbA) was recently identified as a very potent and precise inhibitor of enzyme α-amylase and thus is a potential drug candidate for T2DM therapy. MbA is an acylated flavonol glycoside which can be isolated from Montbretia corms. Depending on the cultivar of montbretia, the amounts of MbA range between ~0.3 to 3 mg per g corm. This quantity is enough to conduct pre-clinical studies and initial clinical trials, however, they are too low to be used for commercial production. High amounts of MbA may be achieved through cultivating improved montbretia cultivars or by heterologous synthesis of MbA using metabolic engineering for which the knowledge of the biosynthetic pathway of MbA as well as the genes involved are essential. In order to understand the biosynthetic pathway of MbA methods such as metabolite and enzyme activity profiling, transcriptome-based analysis were used, based on which the enzymes that catalyze the initial three steps in MbA biosynthesis was determined. The flavonol myricetin is the first compound in the biosynthetic pathway of MbA and its accumulation is observed in the early stages of corm development. Mini-MbA which is myricetin 3-O-(6'-O-caffeoyl)-glucosyl rhamnoside, was found to be a precursor of MbA and it displayed inhibitory activity against α-amylase enzyme. Elucidation of the biosynthetic pathway can be useful to develop a strategy for the production of mini-MbA or MbA by transgenic methods. From a thorough investigation of metabolite and gene expression profiles, the researchers found four genes involved in the building of the bioactive core structure of MbA and mini-MbA. Two of these four enzymes are UDP-dependent glycosyltransferases and the other two are acyltransferases. Mini-MbA could be produced in transgenic *Nicotiana benthamiana* which expressed these enzymes and hence proved it as a potential heterologous plant system for metabolic engineering (Irmisch et al., 2018).

11.8 CONCLUSION

For a disease as prevalent as DM there is extensive research for better treatment options. Current treatment for DM is expensive and has many side effects. Investigation of medicinal plants have shown positive effects in treatment of the disease and its complications. Since the main therapeutic activity of these plants is due to their secondary metabolites, several studies focus on biotechnological methods in order to improve the production of these metabolites. Transgenic medicinal plants are an invaluable tool for the production of various secondary metabolites which are of therapeutic importance and have antidiabetic properties. Hence, transgenic approach can be explored for the discovery and production of new bioactive compounds in medicinal plants. There are ongoing genome projects for several medicinal plants, however, more of these plants need to be studied. Advances in genomics allow for extensive studies on metabolic and biosynthetic pathways. This can be used for discovering new pathways for the synthesis of secondary metabolites, which can be further applied to transgenic studies. Genomics and metabolomics can not only help in the discovery of novel metabolites but also improve the quality and yield of the medicinal plants.

KEYWORDS

- **antidiabetic property**
- **binary vectors**
- **biotechnology approach**
- **metabolite profiling**
- **secondary metabolites**
- **transgenic plants**

REFERENCES

Aileni, M., Abbagani, S., & Zhang, P., (2011). Highly efficient production of transgenic *Scoparia Dulcis* L. mediated by *Agrobacterium* tumefaciens: Plant regeneration via shoot organogenesis. *Plant Biotechnol. Rep., 5*(2), 147–156.

Alejandre-García, I., Álvarez, L., Cardoso-Taketa, A., González-Maya, L., Antúnez, M., Salas-Vidal, E., Díaz, J. F., et al., (2015). Cytotoxic activity and chemical composition of the root extract from the Mexican species *Linum scabrellum*: Mechanism of action of the active compound 6-methoxypodophyllotoxin. *Evid. Based. Complement. Alternat. Med., 2015*, 298463.

Ali, H., Khan, M. A., Kayani, W. K., Khan, T., Mashwani, Z. U. R., Nazif -Ullah, & Khan, R. S., (2018). Thidiazuron regulated growth, secondary metabolism and essential oil profiles in shoot cultures of *Ajuga bracteosa*. *Ind. Crops Prod., 121*, 418–427.

Atanasov, A. G., Waltenberger, B., Pferschy-Wenzig, E. M., Linder, T., Wawrosch, C., Uhrin, P., Temml, V., et al., (2015). Discovery and resupply of pharmacologically active plant-derived natural products: A review. *Biotechnol. Adv., 33*(8), 1582–1614.

Bajaj, S., & Khan, A., (2012). Antioxidants and diabetes. *Indian J. Endocrinol. Metab., 16*(Suppl 2), S267–271.

Bak, A., & Emerson, J. B., (2020). Cauliflower mosaic virus (CaMV) biology, management, and relevance to GM plant detection for sustainable organic agriculture. *Front. Sustain. Food Syst, 4*.

Bertini, E., Merlin, M., Gecchele, E., Puggia, A., Brozzetti, A., Commisso, M., Falorni, A., et al., (2018). Design of a type-1 diabetes vaccine candidate using edible plants expressing a major autoantigen. *Front. Plant Sci., 9*.

Bhat, M., Kothiwale, S. K., Tirmale, A. R., Bhargava, S. Y., & Joshi, B. N., (2011). Antidiabetic properties of *Azardiracta indica* and bougainvillea spectabilis: *In vivo* studies in murine diabetes model. *Evid. Based. Complement. Alternat. Med., 2011*, 561625.

Chakraborty, P., (2018). Development of new antidiabetic drug from medicinal plant-genomic research. *Journal of Diabetology, 2*(1).

Chakraborty, P., (2018). Herbal genomics as tools for dissecting new metabolic pathways of unexplored medicinal plants and drug discovery. *Biochim. Open, 6*, 9–16.

Chandra, A., Mahdi, A. A., Ahmad, S., & Singh, R. K., (2007). Indian herbs result in hypoglycemic responses in streptozotocin-induced diabetic rats. *Nutr. Res., 27*(3), 161–168.

Chaudhary, S., & Sharma, P. C., (2016). Next generation sequencing-based exploration of genomes and transcriptomes of medicinal plants. *Indian J. Plant Physiol., 21*(4), 489–503.

Cuong, D. M., Jeon, J., Morgan, A. M. A., Kim, C., Kim, J. K., Lee, S. Y., & Park, S. U., (2017). Accumulation of charantin and expression of triterpenoid biosynthesis genes in bitter melon (*Momordica charantia*). *J. Agric. Food Chem., 65*(33), 7240–7249.

Dilshad, E., Cusido, R. M., Ramirez, E. K., Bonfill, M., & Mirza, B., (2015). Genetic transformation of *Artemisia carvifolia* Buch with rol genes enhances artemisinin accumulation. *PLoS One, 10*(10), e0140266.

Dilshad, E., Ismail, H., Khan, M. A., Cusido, R. M., & Mirza, B., (2020). Metabolite profiling of *Artemisia carvifolia* Buch transgenic plants and estimation of their anticancer and antidiabetic potential. *Biocatal. Agric. Biotechnol., 24*(101539), 101539.

Ebrahimi, M., Mokhtari, A., & Amirian, R., (2018). A highly efficient method for somatic embryogenesis of *Kelussia odorotissima* mozaff., an endangered medicinal plant. *Plant Cell Tissue Organ Cult., 132*(1), 99–110.

Faraz, R., Gokhale, M., & Gothalwa, R., (2020). *Hairy Root Culture Through Agrobacterium Rhizogenes for Enhancement of Secondary Metabolites Production in Medicinal PLANTS: A Review*. ResearchGate. [Online] *2020* https://www.researchgate.net/publication/342878314 (accessed on 18 December 2021).

Ga, G., & Mahalingam, G., (2016). Review on enhancement of glucose uptake and up-regulation of glucose transporters by antidiabetic medicinal plants. *Asian J. Pharm. Clin. Res.*, 34–39.

Ganesan, K., & Xu, B., (2019). Antidiabetic effects and mechanisms of dietary polysaccharides. *Molecules, 24*(14), 2556.

Gothai, S., Ganesan, P., Park, S. Y., Fakurazi, S., Choi, D. K., & Arulselvan, P., (2016). Natural phyto-bioactive compounds for the treatment of type 2 diabetes: Inflammation as a target. *Nutrients, 8*(8), 461.

Gulçin, İ., Taslimi, P., Aygün, A., Sadeghian, N., Bastem, E., Kufrevioglu, O. I., Turkan, F., & Şen, F., (2018). Antidiabetic and antiparasitic potentials: Inhibition effects of some natural antioxidant compounds on α-glycosidase, α-amylase and human glutathione S-transferase enzymes. *Int. J. Biol. Macromol., 119*, 741–746.

Gupta, N., Gudipati, T., & Prasad, G. B. K. S., (2018). Plant secondary metabolites of pharmacological significance in reference to diabetes mellitus: An update. *Int. J. Curr. Microbiol. Appl. Sci., 7*(05), 3409–3448.

Haddad, P. S., Musallam, L., Martineau, L. C., Harris, C., Lavoie, L., Arnason, J. T., Foster, B., et al., (2012). Comprehensive evidence-based assessment and prioritization of potential antidiabetic medicinal plants: A case study from Canadian eastern James Bay Cree traditional medicine. *Evid. Based. Complement. Alternat. Med., 2012*, 893426.

Han, J. Y., Jo, H. J., & Choi, Y. E., (2020). Overexpression of the squalene epoxidase gene (PgSE1) resulted in enhanced production of ginsenosides and phytosterols in transgenic ginseng. *Plant Biotechnol. Rep.*

Haque, S. M., Chakraborty, A., & Ghosh, B., (2018). Callus mediated shoot organogenesis and regeneration of cytologically stable plants of *Ledebouria revoluta*: An ethnomedicinal plant with promising antimicrobial potency. *J. Genet. Eng. Biotechnol., 16*(2), 645–651.

Hasan, M. M., Ahmed, Q. U., Mat, S. S. Z., & Tunna, T. S., (2018). Animal models and natural products to investigate *in vivo* and *in vitro* antidiabetic activity. *Biomed. Pharmacother., 101*, 833–841.

Hussein, R. A., & El-Anssary, A., (2019). A. plants secondary metabolites: The key drivers of the pharmacological actions of medicinal plants. In: *Herbal Medicine*. IntechOpen.

Irmisch, S., Jo, S., Roach, C. R., Jancsik, S., Man, S. Y. M., Madilao, L. L., O'Neil-Johnson, M., et al., (2018). Discovery of UDP-glycosyltransferases and BAHD-acyltransferases involved in the biosynthesis of the antidiabetic plant metabolite montbretin A. *Plant Cell, 30*(8), 1864–1886.

John, R., Shajitha, P. P., Devassy, A., & Mathew, L., (2017). Hairy-root cultures of *Achyranthes aspera* Linn. as a novel route for the production of 20-hydroxyecdysone. *Acta Physiol. Plant, 39*(12).

Karthikeyan, M., Balasubramanian, T., & Pawan, K., (2016). In-vivo animal models and in-vitro techniques for screening antidiabetic activity. *J. Dev. Drugs, 5*(02).

Khan, M. Y., Aliabbas, S., Kumar, V., & Rajkumar, S., (2009). Recent advances in medicinal plant biotechnology. *Indian J. Biotechnol.*

Kitisripanya, T., Laoburee, M., Puengsiricharoen, L., Pratoomtong, P., Daodee, S., Wangboonskul, J., & Putalun, W., (2020). Production of carbazole alkaloids through callus and suspension cultures in *Clausena harmandiana*. *Nat. Prod. Res., 34*(3), 434–440.

Kizelsztein, P., Govorko, D., Komarnytsky, S., Evans, A., Wang, Z., Cefalu, W. T., & Raskin, I., (2009). 20-hydroxyecdysone decreases weight and hyperglycemia in a diet-induced obesity mice model. *Am. J. Physiol. Endocrinol. Metab., 296*(3), E433–439.

Kowalczyk, T., Wieczfinska, J., Skała, E., Śliwiński, T., & Sitarek, P., (2020). Transgenesis as a tool for the efficient production of selected secondary metabolites from *in vitro* plant cultures. *Plants, 9*(2), 132.

Kumar, A., Mosa, K. A., Ji, L., Kage, U., Dhokane, D., Karre, S., Madalageri, D., & Pathania, N., (2018). Metabolomics-assisted biotechnological interventions for developing plant-based functional foods and nutraceuticals. *Crit. Rev. Food Sci. Nutr., 58*(11), 1791–1807.

Kumar, R., Bohra, A., Pandey, A. K., Pandey, M. K., & Kumar, A., (2017). Metabolomics for plant improvement: Status and prospects. *Front. Plant Sci., 8*, 1302.

Kumar, V., Moyo, M., & Van, S., J., (2017). Somatic embryogenesis in *Hypoxis hemerocallidea*: An important African medicinal plant. *S. Afr. J. Bot., 108*, 331–336.

Kumari, M., & Jain, S., (2012). *Tannin: An Antinutrient with Positive Effect to Manage Diabetes*. ResearchGate. [Online] https://www.researchgate.net/publication/236143118 (accessed on 18 December 2021).

Lankatillake, C., Huynh, T., & Dias, D. A., (2019). Understanding glycaemic control and current approaches for screening antidiabetic natural products from evidence-based medicinal Plants. *Plant Methods, 15,* 105.

Leong, X. Y., Thanikachalam, P. V., Pandey, M., & Ramamurthy, S., (2016). A systematic review of the protective role of swertiamarin in cardiac and metabolic diseases. *Biomed. Pharmacother., 84*, 1051–1060.

Lindbo, J. A., (2007). TRBO: A high-efficiency tobacco mosaic virus RNA-based overexpression vector. *Plant Physiol., 145*(4), 1232–1240.

Low, L. Y., Yang, S. K., Kok, D. X. A., Ong-Abdullah, J., Tan, N. P., & Lai, K. S., (2018). Transgenic plants: Gene constructs, vector and transformation method. In: *New Visions in Plant Science*. InTech.

Ma, H. T., Hsieh, J. F., & Chen, S. T., (2015). Antidiabetic effects of *Ganoderma lucidum. Phytochemistry, 114*, 109–113.

Mahapatra, D. K., Asati, V., & Bharti, S. K., (2015). Cheminform abstract: Chalcones and their therapeutic targets for the management of diabetes: Structural and pharmacological perspectives. *ChemInform, 46*(15).

Mahomed, I. M., & Ojewole, J. A. O., (2003). Hypoglycemic effect of *Hypoxis hemerocallidea* corm (African potato) aqueous extract in rats. *Methods Find. Exp. Clin. Pharmacol., 25*(8), 617–623.

Manske, U., & Schiemann, J., (2005). Development and assessment of a potato virus x-based expression system with improved biosafety. *Environ. Biosafety Res., 4*(1), 45–57.

Mumtaz, M. W., Al-Zuaidy, M. H., Abdul, H. A., Danish, M., Akhtar, M. T., & Mukhtar, H., (2018). Metabolite profiling and inhibitory properties of leaf extracts of *Ficus benjamina* towards α-glucosidase and α-amylase. *Int. J. Food Prop., 21*(1), 1560–1574.

Negi, J. S., Bisht, V. K., Singh, P., Rawat, M. S. M., & Joshi, G. P., (2013). naturally occurring xanthones: Chemistry and biology. *J. Appl. Chem., 2013*, 1–9.

Noipha, K., Thongthoom, T., Songsiang, U., Boonyarat, C., & Yenjai, C., (2010). Carbazoles and coumarins from *Clausena harmandiana* stimulate glucose uptake in L6 myotubes. *Diabetes Res. Clin. Pract., 90*(3), e67–71.

Oguntibeju, O. O., Meyer, S., Aboua, Y. G., & Goboza, M., (2016). Hypoxis hemerocallidea significantly reduced hyperglycaemia and hyperglycaemic-induced oxidative stress in the liver and kidney tissues of streptozotocin-induced diabetic male Wistar rats. *Evid. Based. Complement. Alternat. Med., 2016*, 8934362.

Ota, A., & Ulrih, N. P., (2017). An overview of herbal products and secondary metabolites used for management of type two diabetes. *Front. Pharmacol., 8,* 436.

Pamunuwa, G., Karunaratne, D. N., & Waisundara, V. Y., (2016). Antidiabetic properties, bioactive constituents, and other therapeutic effects of scoparia dulcis. *Evid. Based. Complement. Alternat. Med., 2016*, 8243215.

Pan, A., Sun, J., Chen, Y., Ye, X., Li, H., Yu, Z., Wang, Y., et al., (2007). Effects of a flaxseed-derived lignan supplement in type 2 diabetic patients: A randomized, double-blind, cross-over trial. *PLoS One, 2*(11), e1148.

Perveen, S., (2018). Introductory chapter: Terpenes and terpenoids. In *Terpenes and Terpenoids*. IntechOpen.

Pitchaipillai, R., & Ponniah, T., (2016). *In vitro* antidiabetic activity of ethanolic leaf extract of *Bruguiera cylindrica* L. – glucose uptake by yeast cells method. *International Biological and Biomedical Journal, 2*(4), 171–175.

Podevin, N., & Du Jardin, P., (2012). Possible consequences of the overlap between the CaMV 35S promoter regions in plant transformation vectors used and the viral gene VI in transgenic plants. *GM Crops Food, 3*(4), 296–300.

Rehman, G., Hamayun, M., Iqbal, A., Ul Islam, S., Arshad, S., Zaman, K., Ahmad, A., et al., (2018). *In vitro* antidiabetic effects and antioxidant potential of cassia nemophila pods. *Biomed. Res. Int., 2018*, 1–6.

Rekha, K., & Thiruvengadam, M., (2017). Secondary metabolite production in transgenic hairy root cultures of cucurbits. In: *Reference Series in Phytochemistry* (pp. 267–293). Springer International Publishing: Cham.

Reshmi, S. K., (2018). Food Formulations for Diabetic Population Based on the Bioactives of *Citrus Maxima* (Burm.) Merr. (Pomelo) Fruit. Doctoral thesis, University of Mysore. Chapter 2: Isolation and characterization of antidiabetic bioactive compound from pomelo fruit. Section 2.3.5: Efficacy of bioactive compounds against antidiabetic activity by in vitro assays, pp. 41–42. Repository link: http://ir.cftri.res.in/id/eprint/14076 (accessed on 8 March 2022).

Sairam, S., & Urooj, A., (2013). *Artocarpus altilis* – mode of anti-hyperglycemic activity: Elucidation by suitable in-vitro and ex-vivo techniques. *International Journal of Pharmaceutical Sciences and Research*. [Online] https://ijpsr.com/bft-article/artocarpus-altilis-mode-of-anti-hyperglycemic-activity-elucidation-by-suitable-in-vitro-and-ex-vivo-techniques/?view=fulltext (accessed on 18 December 2021).

Salehi, B., Ata, A., Anil, K. N. V., Sharopov, F., Ramírez-Alarcón, K., Ruiz-Ortega, A., Abdulmajid, A. S., et al., (2019). Antidiabetic potential of medicinal plants and their active components. *Biomolecules, 9*(10), 551.

Samaddar, T., Sarkar, S., & Jha, S., (2019). *Agrobacterium rhizogenes* mediated transformation of the critically endangered species, *Swertia chirayita*. *Plant Tissue Cult. Biotechnol., 29*(2), 231–244.

Satake, H., Koyama, T., Bahabadi, S. E., Matsumoto, E., Ono, E., & Murata, J., (2015). Essences in metabolic engineering of lignan biosynthesis. *Metabolites, 5*(2), 270–290.

Sharma, P., Kumar, V., & Guleria, P., (2019). Naringin: Biosynthesis and pharmaceutical applications. *IJPS, 81*(6).

Singh, B., & Sharma, R. A., (2015). Plant terpenes: Defense responses, phylogenetic analysis, regulation and clinical applications. *3 Biotech, 5*(2), 129–151.

Singh, S., Singh, D. B., Singh, S., Shukla, R., Ramteke, P. W., & Misra, K., (2019). Exploring medicinal plant legacy for drug discovery in post-genomic era. *Proc. Natl. Acad. Sci. India Sect. B Biol. Sci., 89*(4), 1141–1151.

Srivastav, V. K., Egbuna, C., & Tiwari, M., (2020). Plant secondary metabolites as lead compounds for the production of potent drugs. In: *Phytochemicals as Lead Compounds for New Drug Discovery* (pp. 3–14). Elsevier.

Talukder, F. Z., Khan, K. A., Uddin, R., Jahan, N., & Alam, M. A., (2012). *In vitro* free radical scavenging and anti-hyperglycemic activities of achyranthes aspera extract in alloxan-induced diabetic mice. *Drug Discov. Ther., 6*(6), 298–305.

Teoh, S. L., & Das, S., (2018). Phytochemicals and their effective role in the treatment of diabetes mellitus: A short review. *Phytochem. Rev., 17*(5), 1111–1128.

Tiong, S. H., Looi, C. Y., Hazni, H., Arya, A., Paydar, M., Wong, W. F., Cheah, S. C., et al., (2013). Antidiabetic and antioxidant properties of alkaloids from *Catharanthus roseus* (L.) G. don. *Molecules, 18*(8), 9770–9784.

Tiwari, P., (2015). Recent trends in therapeutic approaches for diabetes management: A comprehensive update. *J. Diabetes Res., 2015*, 340838.

Tripathi, L., & Tripathi, J. N., (2005). Role of biotechnology in medicinal plants. *Trop. J. Pharm. Res., 2*(2).

Vijayaraj, R., Sri Kumaran, & Thamada, S., (2019). *In Vivo and In Vitro Models for Biological Screening of Antidiabetic Drugs*. ResearchGate. [Online] https://www.researchgate.net/publication/332979614 (accessed on 18 December 2021).

Wang, C. K., Hsu, S. Y., Chen, P. Y., & To, K. Y., (2012). Transformation and characterization of transgenic *Bidens pilosa* L. *Plant Cell Tissue Organ Cult., 109*(3), 457–464.

Wang, Y., Nishina, P. M., & Naggert, J. K., (2009). Degradation of IRS1 leads to impaired glucose uptake in adipose tissue of the type 2 diabetes mouse model TALLYHO/Jng. *J. Endocrinol., 203*(1), 65–74.

Who.int. Diabetes. https://www.who.int/health-topics/diabetes (accessed on 18 December 2021).

Yamamoto, N., Ueda, M., Sato, T., Kawasaki, K., Sawada, K., Kawabata, K., & Ashida, H., (2011). Measurement of glucose uptake in cultured cells. *Curr. Protoc. Pharmacol., 1*, Unit 12.14.1–22. Chapter 12.

Yamamura, Y., Kurosaki, F., & Lee, J. B., (2017). Elucidation of terpenoid metabolism in scoparia dulcis by RNA-seq analysis. *Sci. Rep., 7,* 43311.

Yap, H. Y. Y., Tan, N. H., Ng, S. T., Tan, C. S., & Fung, S. Y., (2018). Inhibition of protein glycation by tiger milk mushroom [*Lignosus rhinocerus* (Cooke) ryvarden] and search for potential antidiabetic activity-related metabolic pathways by genomic and transcriptomic data mining. *Front. Pharmacol., 9.*

Yuan, D. S., Chen, Y. P., Tan, L. L., Huang, S. Q., Li, C. Q., Wang, Q., & Zeng, Q. P., (2017). Artemisinin: A panacea eligible for unrestrictive use? *Front. Pharmacol., 8,* 737.

Zerbini, F. M., Silva, F. N. D., Urquiza, G. P. C., & Basso, M. F., (2014). Transgenic plants. In: *Biotechnology and Plant Breeding* (pp. 179–199). Elsevier.

Zhang, Y. Y., Zhang, D. M., Zhong, Y., Chang, X. J., Hu, M. L., & Cheng, C. Z., (2017). A simple and efficient in planta transformation method for pommelo (*Citrus maxima*) using *Agrobacterium* tumefaciens. *Sci. Hortic. (Amsterdam), 214*, 174–179.

Zhao, C., Yang, C., Wai, S. T. C., Zhang, Y., Portillo, M. P., Paoli, P., Wu, Y., et al., (2019). Regulation of glucose metabolism by bioactive phytochemicals for the management of type 2 diabetes mellitus. *Crit. Rev. Food Sci. Nutr., 59*(6), 830–847.

Zhu, H., Wang, Y., Liu, Z., Wang, J., Wan, D., Feng, S., Yang, X., & Wang, T., (2016). Antidiabetic and antioxidant effects of catalpol extracted from *Rehmannia glutinosa* (Di Huang) on Rat diabetes induced by streptozotocin and high-fat, high-sugar feed. *Chin. Med., 11*(1), 25.

Zhu, X., Wu, C., Qiu, S., Yuan, X., & Li, L., (2017). Effects of resveratrol on glucose control and insulin sensitivity in subjects with type 2 diabetes: Systematic review and meta-analysis. *Nutr. Metab. (Lond.), 14*(1).

CHAPTER 12

Transcriptomic Studies of Medicinal Plants Elucidating Their Role in Diabetes

ROHIT KUMAR,[1] LATA SHARMA,[2] and NEELESH KAPOOR[3]

[1]*Advanced Centre for Treatment, Research, and Education in Cancer, Kharghar, Navi Mumbai – 410210, Maharashtra, India, E-mail: rohitbiomed@hotmail.com*

[2]*Molecular Diagnosis Laboratory, Post-Graduate Institute of Veterinary Education and Research, Jaipur – 302031, Rajasthan, India*

[3]*Department of Fingerprinting, College of Biotechnology, Sardar Vallabhbhai Patel University of Agriculture and Technology, Meerut – 250110, Uttar Pradesh, India, E-mail: drneeleshbiotech@gmail.com*

ABSTRACT

Diabetes is a multifaceted metabolic disorder that results in hyperglycemia due to either insufficient secretion of insulin or non-functional insulin production by the pancreas. The patient requires lifestyle changes, but often, insulin therapy is an important part of diabetes treatment. The disease's chronic persistence can lead to potential complications, including heart disease, stroke, and kidney and nerve damage. There are also several complications associated with drugs used for diabetes treatment, including drug resistance, side effects, and even toxicity; therefore, initial remedy in diabetes sometimes recommends medicinal plants. The plant contains alkaloids, carotenoids, flavonoids, terpenoids, and glycosides that successfully decrease the free radicals and reactive oxygen species from systematic circulation and potentially have antidiabetic effects. Medicinal plants also act as antidiabetic

Antidiabetic Potential of Plants in the Era of Omics. Deepu Pandita, PhD, Anu Pandita, MSc, & Chander Bhanu, MBBS (Eds.)

by decreasing the glucose uptake from the intestine or directly improve the pancreatic tissue performance by increasing insulin secretion. Therefore, the antidiabetic mechanism of any phytochemical or plant product varies with its mechanism of action. Herbal extracts of a medicinal plant acts on one or several druggable targets of diabetes and therefore a comprehensive evaluation of medicinal property and antidiabetic effects needs to be evaluated at molecular levels. These systematic molecular studies will not only help to decipher the molecular mechanisms of medicinal plant but can help to identify potent and more specific therapeutic targets that will be immensely beneficial to diabetic patients. Studies of these molecular mechanisms have been made possible with the advent of next-generation transcriptome sequencing technology (RNA-Seq).

Transcriptome profiling (or transcriptomics) uses deep-sequencing technologies to study the genome's functional elements and reveal the major constituents of cells and tissues working during a biological process. The major advantage of RNA-Seq technology is its efficiency in detecting novel molecules and does not require any prior information about the genes. This makes transcriptome sequencing (RNA-Seq) applicable, particularly for new plant species whose genomic sequences are yet not available. This approach can be exploited to understand the variation in gene expression level in the sample treated with plant extract by comparing the expression level of the same gene in control. It is also useful in finding the novel genes linked to the diseased condition by tracking the entire transcriptome. The advancement of bioinformatics approaches aids RNA-Seq technology and helps determine the putative functional regulators' ultimate pathways. RNA-Seq technique provides an opportunity to study the antidiabetic effect of plant extract. The preventive measures formulated using these specific studies are surely more applicable instead of adopting them from comparative studies. This chapter aims to introduce the transcriptome studies on antidiabetic medicinal plants and their associated mechanisms that help reduce the glucose levels and increase insulin secretion and ultimately help in diabetes maintenance.

12.1 HISTORY AND ADVANCEMENT OF TRANSCRIPTOMIC TECHNOLOGY

Transcriptomics typically meant studying the complete set of RNA transcripts or the genome's coding region using high-throughput methods, such as microarray analysis and next-generation sequencing. Transcriptomics is used to reveal the molecular constituents of cells and tissues to interpret the genome's functional elements (Wang et al., 2009a). The technology is generally implemented to decode or compare the molecular changes during exposure to specific physical or biochemical changes. The comparative analysis of transcriptomes helps in identification of genes that are modulated in a distinct cell population in response to environmental, biochemical, or biological changes. In the past three decades, plant transcriptomics has evolved rapidly and is being studied using multiple techniques, including hybridization-based approaches, sequence-based approaches, and RNA sequencing (Tan et al., 2009; Wang et al., 2009a). The hybridization-based methods involve the synthesis of high-density oligo microarrays followed by incubation of fluorescently-labeled cDNA to screen the binding on microarray chips. The binding pattern and signal generated by these hybridizations can be further converted into gene expression profiling (Kiyama and Zhu, 2014). With the help of probes spanning adjacent exon-exon junctions, they are also used to identify and quantify different spiced isoform variants (Lee and Roy, 2004). However, hybridization-based approaches require prior knowledge about genome sequences and have a limited dynamic range of detection due to the saturation of signals and background noise (Wang et al., 2009b; Zhao et al., 2014).

On the contrary, sequence-based approaches such as Sanger sequencing can directly determine the cDNA sequences but have low throughput, low quantification, and higher cost. These limitations were resolved with the advent of tag-based approaches such as massively parallel signature sequencing (MPSS) (Brenner et al., 2000), serial analysis of gene expression (SAGE) (Hu and Polyak, 2006) and cap analysis of gene expression (CAGE) (Kodzius et al., 2006). Although tag-based methods provide precise quantification of gene expression in high-throughput, only a small portion of transcripts can be analyzed in one experiment. The isoforms are not distinguishable from each other (Brenner et al., 2000). Sequence subtractive hybridization (SSH) relies on removing dsDNA annealed by hybridization of common cDNA between a test sample and control followed by removal of these hybridized cDNAs and retaining of the differentially expressed for expression analysis. However, sometimes low-abundance cDNAs are not enriched equally and creates a "false positive" expression in the library that remain a challenging weakness of the SSH technique.

RNA sequencing (RNA-Seq) using next-generation sequencing technologies is the most recent and advanced approach used in plant transcriptomic studies. The technology does not rely on prior information of genome or any reference genome; therefore, it can be used for any non-model plant species (Strickler et al., 2012). There are several advantages associated with RNA-Seq: Firstly, the major advantage of studying transcriptome is identifying novel genes or genetic interactions for complex genetic traits as sometimes the genotype and phenotypes may not be linked directly to each other. Secondly, it is possible to study a non-model organism and identify novel gene expression responses to an environmental variable. Thirdly transcriptome sequencing can predict the precise position of transcription boundaries and least prone to background signal or false results. There is also no upper limit of transcript quantification.

Further, it is more feasible to identify the acute environmental effect on gene transcription using transcriptomics compared to phenotypic variation or phenotypic responses. Most of the transcriptomic studies require a control design where environmental impact is either incorporated in study design, or the transcript expression is corelated to the environmental variable. There are also a few limitations associated with transcriptomic studies. The transcription of genes may differ from one cell to another cell type to coordinate the response against the environmental variable. Thus, transcriptome sequencing might be misleading to track these cell-specific responses (Cloonan et al., 2008; Marioni et al., 2008; Nagalakshmi et al., 2008). The genetic mechanisms may also not be mediated through changes in transcription or protein levels and could be mediated through epigenetic variations that are difficult to identify using transcriptome sequencing (Madlung et al., 2002; Slotkin and Martienssen, 2007; Heo et al., 2013).

12.2 CHALLENGES IN TRANSCRIPTOMIC SEQUENCING OF NON-MODEL PLANTS

The transcriptome sequencing has advanced from a few million-base sequencings to terabyte base sequencing in a single run. However, few challenges slightly restrict the implementation of RNA-seq in transcriptomics of non-model plant species. The RNA-seq majorly use the short-gun sequencing method. Therefore, the transcript's construction relies on two assembly methods; a) reference assembly where the genome sequence of species of interest is available, and b) *de novo* assembly, where assembly exploits the specific algorithm to construct nearly accurate transcripts. Genome-guided assembly is less efficient in the detection of sequencing error rates and polymorphisms. Also, reference-based assemblies are more prone to lose paralogous loci information than *de novo* (Vijay et al., 2013).

The genome size is also a challenge in plant genomic studies. Plant genome size varies by four orders of magnitude ranging from 60 Mb of *Genliseaaurea* to 152 Gb of *Paris japonica* (Pellicer and Leitch, 2020). These large genome sizes are majorly attributed to invading DNA/RNA, transposable elements, whole-genome duplication, and polyploidy (Michael, 2014). In comparison to their genome sizes, the plant transcriptome also varies to a great extent. The transcriptome size varies based on cell size, ploidy level, growth condition, and any associated disease (Coate and Doyle, 2015). The estimates of relative expression sometimes falsely represented by the unrelated transcript abundance of the duplicated genome. Therefore, the gene expression normalization is carried out on implicit yet rarely validated assumptions size of transcriptome is parallel among the analyzed samples (Coate and Doyle, 2010; Loven et al., 2012; Aanes et al., 2014; Coate and Doyle, 2015).

Several approaches have been adapted to improve the *de novo* transcriptome assemblies. One of the transcriptome assembly metrics is the evaluation of the completeness of single-copy orthologs characterized across higher taxonomic groups. Benchmarking universal single-copy orthologs (BUSCO) analysis compares the assembled contigs with orthologous genes' dataset and provides a statistical measure of the completeness of transcriptome assemblies (Simao et al., 2015; Seppey et al., 2019). A well-refined *de novo* transcriptome assemblies of *Secale cereale* L. (Mahmood et al., 2020), *Camellia sinensis* (Qiao et al., 2019), *Croton tiglium* (Haak et al., 2018), *Persea americana* (Vergara-Pulgar et al., 2019), etc., are a few of the example of usage of validated tools for transcriptome assemblies. Although such refined transcriptomes of plant species that comprise antidiabetic properties are still lacking in literature, these studies will definitely overlay the foundations for more studies in the future.

12.3 ROLE OF TRANSCRIPTOMICS IN DISEASE MANAGEMENT

Genetic and environmental risk factors play a combinative role in the pathophysiology of diabetes. Diabetes is mostly caused by polygenetic traits; however, there is a rare form of inherent diabetes reported at very low frequency *viz.* maturity-onset diabetes in the young (MODY) and mitochondrial DNA mutation induced diabetes. Therefore, genomic resources assist in understanding the disease physiology and mechanism of disease progression to identify the biomarkers and develop the targeted treatment approaches.

Type 1 diabetes (T1D) is majorly the result of destruction of insulin producing beta cells of the pancreas by self-immune system. There are several reasons attributed to autoimmune response activation against the pancreatic beta cells. Bacterial and viral infections have been linked to the disease with strong evidence. Recent transcriptomic studies, along with classical pathological studies, attributed the enterovirus infection to induction and of T1D (Krogvold et al., 2015; Richardson and Morgan, 2018; Poma et al., 2020). Abdellatif et al. (2019) isolated the human islets cells and experimentally exposed them to three different bacterial species (Davis-Richardson et al., 2014; Kostic et al., 2015; Vatanen et al., 2016; Stewart et al., 2018) earlier linked to T1D and finally analyzed the transcriptomic response of islet cells. The study suggested a complex anti-bacterial immune response elicited by islets through the release of leukocyte recruitment factors and oxidative damage and that may initiate the damage and subsequent autoimmune reaction to beta-cell. Another predisposing factor of T1D could be attributed to leakage of microbial antigens into the pancreas that may elicit the immune response (Korsgren et al., 2012). Transcriptome data also helped to identify the expression pattern of candidate genes induced by pro-inflammatory cytokines, i.e., interleukin 6 (IL-6), interleukin-1β (IL-1β) and interferon-γ (IFN-γ). The upregulation of chemokines and cytokines after exposure of islets cells to cytokine strengthens the idea of crosstalk between the immune system and the islet cells in T1D (Eizirik et al., 2012).

Type 2 diabetes (T2D) is a chronic metabolic disease majorly associated with a sedentary lifestyle accompanied with cigarette smoking and/or generous alcohol consumption (Hu et al., 2001; Centers for Disease and Prevention, 2004). However, there is strong evidence of an inheritable genetic connection in T2D (Camastra et al., 1999; Walley et al., 2006; McCarthy, 2010). Therefore, transcriptome sequencing could be a better approach to understand the linkage and co-contribution of genetic and environmental factors in T2D induction and progression. Christodoulou et al. (2019) used RNA-seq methodology followed by real-time qPCR validation to investigate the expression profile of T2D-associated genes (n=24), and their respective transcript variants (tv) in T2D patients and healthy individuals. The data revealed the deregulation of CAPN10 (transcript variant 3) and linked it to T2D as an independent predictor for disease prognosis. Transcriptome profiling of Asian Indian T2D patients revealed the role of peripheral adipose tissue pathology with disease prognosis. The study also indicated that pathologic changes such as adipocyte hypertrophy and up-regulation of genes associated with inflammatory pathways in peripheral subcutaneous adipose tissue is linked to T2D (Saxena et al., 2019). The transcriptome signature of T2D and non-diabetic patients is expected to differ from one another; however, single-cell transcriptome data analysis indicates different molecular signatures within the diabetic as well as non-diabetic alpha, beta, and delta cells. The study identified vital cell-population-specific signatures of different pancreatic cells thus provided a comprehensive overview of islet biology and disease pathogenesis (Lawlor et al., 2017).

Studies comparing transcriptome responses of different types of diabetes revealed distinct epidemiological, clinical, immunological, and genetic signatures at the molecular level. A comprehensive analysis of microarray data from T1D and T2D diabetic patients showed that *PNP* and *CCR1* genes may be used for T1D and T2D diagnosis, whereas *MGAM* and *NAMPT* can be used as a biomarker of T2D prognosis (Gao et al., 2019). An integrative analysis of T1D, T2D, and gestational diabetes mellitus (DM) transcriptome profiles revealed clustering of GDM and T1D patients together, whereas the inflammatory response of T2D was significantly distinct (Evangelista et al., 2014). Various biological conditions can also induce hyperglycemia. The increase in blood glucose level has been reported during gestation. Transcriptome studies revealed that gestational diabetes shared more similarities with T1D than T2D (Collares et al., 2013). T1D and T2D are also a major risk factors associated with cardiovascular disease (CVD), and both shared common gene signatures and pathways with myocardial infarction (MI). The genes majorly shared by T1D and T2D with MI belonging to the metabolic process of cellular proteins and downregulation of genes associated with endopeptidase activity pathway (Song et al., 2019). A study also linked the upregulation of HMGCS2 with T1D induced cardiac dysfunction (Shukla et al., 2017). Therefore, the transcriptome sequencing studies have aided the advancement in the molecular level understanding of diabetes and the identification of risk factors and biomarkers of disease progression.

12.4 TRANSCRIPTOMIC STUDIES OF ANTIDIABETIC PLANTS

Gymnema sylvestre is a perennial woody vine majorly grow in the tropical forests of Southern India, Africa, and Australia. The plant extract contains Gymnemic acid with anti-obesity and antidiabetic properties and helps decrease body weight and inhibits glucose absorption from the intestine (Pothuraju et al., 2014). Kalariya et al. (2018) sequenced the transcriptome of leaf, flower, and fruits of *G. sylvestre* to categorize the putative genes associated with gymnemic acid biosynthesis pathway. Different *in vivo* studies carried out on diabetic models of rat (Gupta, 1963; Srivastava et al., 1985) and rabbit (Shanmugasundaram et al., 1983, 1988; Porchezhian and Dobriyal, 2003) have specified a substantial decrease in blood glucose level as well as enhancement in glycogen synthesis, glycolysis, and gluconeogenesis when treated with different extracts of *Gymnemasylvestre*.

Ginkgo biloba also known as maidenhair tree is one of the oldest living plant species. The plant extract has been used in traditional medicines as an adjuvant and improves metformin treatment outcomes in patients with hysterical T2D (Aziz et al., 2018). Metformin is generally used with other antidiabetic medications to control hyperglycemia with a proper diet and exercise program. Besides, the bioactive compound Ginkgolide A of *G. biloba* also decreases low-grade vascular inflammation caused by hyperglycemia through the STAT3-mediated pathway (Zhao et al., 2015). He et al. (2015) carried out the transcriptome sequencing of *G. biloba* and found upregulation of genes associated with biosynthesis of terpenoid backbone, ginkgolide, and bilobalide. Chang et al. (2013) also provide evidence through *in vivo* studies on the rat model that *G. biloba* controls hyperglycemia and hyperlipidemia and exerts antioxidant activities in streptozotocin-induced chronic diabetic rats.

Allium sativum L. (Garlic) is also an antidiabetic plant that contains various bioactive compounds, such as allicin, alliin, diallyl mono-di and tri-sulfide, ajoene, and S-allyl-cysteine that possess antidiabetic properties (Shang et al., 2019). Ashraf et al. (2011) carried out a study on T2D patients (n=60) diagnosed with fasting blood sugar (FBS) level >126 mg/dl and evaluated the outcome of garlic (300 mg trice daily) with standard antidiabetic therapy (metformin) on the blood glucose level. Surprisingly the combination treatment of garlic with metformin indicated improved glycemic control compared to standard treatment combined with placebo (Ashraf et al., 2011). A comprehensive metanalysis of all the randomized controlled trials carried out till 2017 using garlic supplementation on diabetic control (DC) stipulate that garlic supplement exert a positive impact in controlling blood glucose and total cholesterol level in T2D (Wang et al., 2017). The alliin treatment for eight weeks increased glucose homeostasis, sensitivity to insulin as well as ameliorate the lipid profile in the diet-induced obese mice (Zhai et al., 2018). Sun et al. (2012) and Mehra et al. (2020) carried out the transcriptome sequencing and characterized the organic sulfur biosynthesis associated genes of *A. sativum*

Plants belonging to the Berberis group, e.g., *Berberis aquifolium*, *B. vulgaris*, *B. aristata*, and few other plant species are a major source of berberine. Berberine is a potent antidiabetic compound with significant anti-inflammatory and antioxidant activity (Cicero and Baggioni, 2016). Although the mechanism of berberine action is complex, there is growing evidence that it activates both insulin-mediated and exercise-induced uptake of glucose (Kim et al., 2007) whereas inhibit the active absorption of glucose from the intestinal (Pan et al., 2003; Liu et al., 2008). Berberine activates the Glucose transporters (GLUT) or AMPK pathway (Jing and Ismail-Beigi, 2007; Jing et al., 2008; Hang et al., 2018; Zhang et al., 2018a), thus helping in glucose uptake in T2D (Meyer et al., 2008). *In vivo* mice model (Zhang et al., 2008a; Chueh and Lin, 2011; Ni et al., 2015) as well as randomized preclinical studies (Yin et al., 2008; Zhang et al., 2008b) have indicated a positive effect of berberine in diabetes. In an *in vitro* insulin resistant cell line model, a transcriptome study showed that berberine treatment ameliorates insulin resistance via the ROS-dependent pathway (Lin and Mao, 2020).

Momordica charantia is another important medicinal plant traditionally used for T2D mellitus treatment (Habicht et al., 2014). A metanalysis of 10 different clinical studies of T2D patients (n=1045) revealed that *M. charantia* herbal formulation significantly decreases FPG, PPG, and HBA_{1c} levels compared to placebo control (Peter et al., 2019). A recent clinical trial revealed that *M. charantia* administration also lowers the hyperglycemia and increases total insulin secretion (Cortez-Navarrete et al., 2018). Studies on *in vivo* rat models provided concrete evidence that *M. charantia* treatment activates pancreatic beta cells and ameliorate insulin resistance as well as protects liver tissue in T2D rats (Ma et al., 2017; Malekshahi et al., 2019). The active polysaccharides found *M. charantia* have hypoglycemic effect and improve glucose metabolism. RNA sequencing analysis of liver tissue of T2D rat model identified major differentially expressed genes that might be responsible for modulation of T2D (Bai et al., 2018b). Another transcriptomic study on mice model suggested that aqueous extract of *M. charantia* regulates glucose metabolism through the insulin signaling pathway (Lo et al., 2013).

Stevia rebaudiana is a potent source of Steviol glycoside, which is also called a bio-sweetener (Goyal et al., 2010). Refined steviol glycosides are usually safe for consumption without any prescription (Samuel et al., 2018). The aqueous extract of stevia have indicated antidiabetic (Ahmad and

Ahmad, 2018) as well as α-amylase inhibitory activity (Singla et al., 2019) in different *in vivo* studies. There are several *in vitro* and clinical studies as well as comprehensive reviews that support the promising nutraceutical usage of *S. rebaudiana* for controlling diabetes and its associated problems (Kujur et al., 2010; Mohd-Radzman et al., 2013; Ritu and Nandini, 2016; Momtazi-Borojeni et al., 2017; Gholizadeh et al., 2019). A transcriptomic and metabolomic study carried out by Holvoet *et al.* (2015) assessed the outcome of stevioside, rebaudioside A, and steviol on a mouse model of obesity and insulin resistance. Enhanced glucose and bile acid metabolism and catabolism of fat demonstrated that all the compounds attenuated hepatic steatosis (Holvoet et al., 2015).

Panax ginseng (ginseng) also comprises antidiabetic properties. Different studies supported by *in-vitro* and *in-vivo* models reported to ameliorate hyperglycemia and insulin resistance in T2D. Ginseng activates promotes fatty acid oxidation through activation of AMPK (adenosine monophosphate-activated protein kinase) pathway. AMPK also enhances mitochondrial biogenesis and improve glucose metabolism by activating various mitochondrial and antihyperglycemic response genes (Lee et al., 2009). Ginseng also improves lipid metabolism and inhibits obesity in diabetic mice models (Shin and Yoon, 2018). Ginsenosides are the major bioactive compounds of ginseng that promotes hypoglycemia in streptozotocin-induced diabetic mice (Liu et al., 2009; Liu et al., 2013). Integrated transcriptomic analysis revealed that squalene epoxidase, dammar enediol synthase, and cytochrome P450 (CYP) gene superfamily majorly contribute to ginsenoside biosynthesis (Lee et al., 2017; Zhao et al., 2019). Also, the transcriptome and metabolome data of five different *Panax ginseng* cultivars identified the pathways and genes involved in ginsenoside biosynthesis. A clinical study indicated that ginseng improves glucose metabolism and insulin regulation in T2D patients (Vuksan et al., 2008). Jayakodi *et al.* (2018) have created an open-access genomic database of *Panax ginseng*. Zou *et al.*(2016) also reviewed the transcriptome studies carried out on ginseng species with major emphasis on identifying novel genes and elucidation of their metabolic regulation.

Catharanthus roseus, is a traditionally used medicinal plant that accumulates various antidiabetic terpenoid indole alkaloids (Verma et al., 2014). Vindoline and bis-indole alkaloids are the main bioactive compounds of C. roseus that induces insulin secretion from islet cells and thus help in T2D (Goboza et al., 2019; Lunagariya et al., 2019). Transcriptome analysis carried out by Verma *et al.* (2014) that antidiabetic vindoline biosynthesis genes are majorly express by aerial tissues, whereas other bioactive compounds such as bisindole are synthesized by roots and leaves (Verma et al., 2014). There is also a functional genomic database 'croFGD' created from 53 RNA-seq datasets and comprises gene expression profiles from global, tissue, and treatment specific experiments (She et al., 2019).

Trigonella foenum-graecum L. (Fenugreek) is also used traditionally in diabetes due to their hypoglycemic activity. Plant extract comprises steroids, alkaloids, saponins, polyphenols, and flavonoids as an active component showing potent antidiabetic, antihyperlipidemic, anti-obesity activities along with other pharmacological effects (Nagulapalli Venkata et al., 2017). Preclinical and clinical studies have outlined the hypoglycemic and antihyperlipidemic properties of fenugreek (Baquer et al., 2011; Sankar et al., 2012; Geberemeskel et al., 2019). Along with chronic hyperglycemia T2D also shows oxidative stress. Fenugreek seed extract decreases inflammation and oxidative stress and increase the exocrine function in diabetic rats (Joshi et al., 2015). Fenugreek also contains diosgenin regulating glycolipid metabolism, apoptosis, inflammation, and oxidative stress associated pathways to improve diabetes and related complications (Gan et al., 2020). Transcriptome studies also revealed that indicated that diosgenin biosynthesis is induced by methyl jasmonate (MeJA) (Ciura et al., 2017; Zhou et al., 2019a).

Costus pictus is recognized as the "Insulin Plant" in Southern India. The leaves of the plant are used to increase the insulin level in systematic circulation. Although, the precise antidiabetic mechanism of *C. pictus* is unknown; however, animal studies have indicated a rapid increase in insulin secretion post-treatment of plant extract (Al-Romaiyan et al., 2010). The leaf transcriptome sequenced by Annadurai *et al.*(2012) attributed *C. pictus* leaves' antidiabetic properties to genes associated with bixin, abscisic acid, geranial, and geraniol biosynthesis pathways. Diabetic rats fed orally with an aqueous extract of *C. pictus* showed significant hypoglycemia as well as a substantial increase in plasma insulin level after 14 days of treatment (Gireesh et al., 2009).

Besides, there are genomic resources available for other medicinal plants that comprise antidiabetic properties (Grover et al., 2002). The antidiabetic properties of aerial roots and bark extracts of *Ficus bengalensis* tested in rat models revealed a substantial decrease in blood glucose level and restoration of serum electrolytes and glycolytic enzymes (Gayathri and Kannabiran, 2008; Singh et al., 2009). *Hibiscus rosa-sinensis* flowers also comprise antidiabetic and antioxidant properties (Pethe et al., 2017). Alcohol extracts of *Mucuna pruriens* significantly decrease plasma glucose level of chronic alloxanized rats at chronic treatment (Rathi et al., 2002; Majekodunmi et al., 2011). The seeds of *M. pruriens* may directly or indirectly induce insulin-like action and thus exerts a hypoglycemic effect in alloxan-diabetic rabbits. Systematic reviews accentuated a positive impact of *Nigella sativa* on T2D (Heshmati and Namazi, 2015; Hamdan et al., 2019). Ethanol extract of *N. sativa* induces insulin-like effects on adiposities and skeletal muscle cells by working as an agonist of the PPAR-gamma gene that plays an important function in glucose homeostasis (Benhaddou-Andaloussi et al., 2010). *Helianthus tuberosus* is used to extract fructans (inulin) that shows anti-diabetes and anti-obesity properties. Inulin is used as mild sweeteners due to its indigestible properties. Transcriptome study carried out by Jung *et al.*(2014) supplemented the information genes responsible for inulin biosynthesis in Jerusalem artichoke.

12.5 ANTIDIABETIC PROPERTIES OF PLANTS DECIPHERED USING TRANSCRIPTOMICS

There are very limited studies available describing the antidiabetic properties of plants using transcriptomic approaches. Most of the transcriptomic analyzes either carried out to describe the transcriptome of different parts (e.g., root, leaf, seeds, etc.), of plants or describe the biosynthesis pathways of compounds comprising antidiabetic properties in respective plant species (Table 12.1). Few of the studies mentioned below have uncovered the antidiabetic properties of medicinal plants and their bioactive compounds using next-generation sequencing technologies.

Flavonoids and glycosides present in plants are the major bioactive compounds comprising various health benefits. Zhu et al. (2019) compared the molecular signatures of liver transcriptome and proteome using RNA-seq and iTRAQ analysis of the control and avicularin treated mice. The study revealed that genes regulating the glycolytic gluconeogenetic and arachidonic acid pathways, as well as the gene responsible for controlling the oxidative phosphorylation pathway, majorly play a role in the hypoglycemic effect of avicularin.

Dietary habitats also increase the risk of diabetes, and intake of saturated fat could be a key factor related to non-alcoholic fatty liver as well as insulin resistance in diabetes. A clinical trial comprising 14 healthy individuals ascertained that a fat-rich diet increases lipid storage and energy metabolism of kidney cells while decreasing their insulin sensitivity (Hernandez et al., 2017). Therefore, a balanced diet with a low-calorie diet devoid of saturated fat could be useful to lower the risk of diabetes (Schoenaker et al., 2016; Islam et al., 2019).

TABLE 12.1 Antidiabetic Plant, Their Bioactive Compounds and Transcriptome Resources Available for Respective Plant Species

SL. No.	Plant Species	Bioactive Compound	Antidiabetic Property	References	
				Antidiabetic Effects	Transcriptome Resources
1.	Gymnema Sylvestre	Gymnemic acid	Gymnemic acid decreases the glucose absorption from the intestine	Leach (2007); Pothuraju et al. (2014); Tiwari et al. (2017)	Kalariya et al. (2018)
2.	Ginkgo biloba	Ginkgolide A	Adjuvant to metformin, antihyperglycemic, antioxidant, and antihyperlipidemic activities	He et al. (2015); Aziz et al. (2018); Shi et al. (2019)	Gu et al. (2009); He et al. (2015)
3.	Allium sativum	Allicin, alliin,	Allicin, alliin are antioxidants	Islam and Choi (2008); Ashraf et al. (2011); Padiya and Banerjee (2013)	Sun et al. (2012); Kamenetsky et al. (2015)
4.	Berberis vulgaris, Berberis aristata	Berberine	Glucose transporters (GLUT) or AMPK mediated uptake of glucose	Jing and Ismail-Beigi (2007); Jing et al. (2008); Di Pierro et al. (2013)	–
5.	Momordica charantia	Polysaccharides, Charantin, charine, cryptoxanthin, cucurbitins	Insulin receptor binding activity, insulin-releasing activity, and reduction of fructosamine levels	Fuangchan et al. (2011); Lo et al. (2013); Bai et al. (2018b)	Lo et al. (2013); Shukla et al. (2015); Bai et al. (2018b)
6.	Nigella sativa	Thymoquinone	Insulin and AMP-activated protein kinase (AMPK) pathways	Hamdan et al. (2019)	Ismail et al. (2016)
7.	Stevia rebaudiana	Steviol glycosides	Biological sweetener	Samuel et al. (2018)	Chen et al. (2014); Modi et al. (2016); Dhandapani et al. (2020); Samsulrizal et al. (2020)

TABLE 12.1 *(Continued)*

SL. No.	Plant Species	Bioactive Compound	Antidiabetic Property	References	
				Antidiabetic Effects	Transcriptome Resources
8.	Panax ginseng	Ginsenosides	Insulinotropic action, reduces gluconeogenesis and promotes glucose transport	Bai et al. (2018a); Zhou et al. (2019b)	Cao et al. (2015); Lee et al. (2017); Xu et al. (2017); Zhao et al. (2019; Jayakodi et al. (2018)[#]
9.	Trigonella foenum-graecum	Steroids, alkaloids, flavonoids, polyphenols, saponins	Hypoglycemic and anti-oxidative activity	Baquer et al. (2011); Sankar et al. (2012); Geberemeskel et al. (2019)	Ciura et al. (2017); Zhou et al. (2019a)
10.	Catharanthus roseus	Vindoline and Bis-indole	Induce the insulin secretion from pancreatic beta cells	Goboza et al. (2019); Lunagariya et al. (2019)	Shukla et al. (2006; Van Moerkercke et al. (2013); She et al. (2019)[#]
11.	Helianthus tuberosus	Fructans (inulin)	Due to their indigestible properties fructans are used as mild sweeteners	Ahn et al. (2018); Zhang et al. (2018b); Shao et al. (2020)	Jung et al. (2014)

[#]Comprehensive transcriptomic/genomic databases are available.

Transcriptome studies are also focused on diabetes-associated complications. The diabetic patient also has high reactive oxygen species and inflammatory reactions. Studies have indicated that various bioactive compounds found in medicinal plants comprising antioxidant and anti-inflammation properties are helpful in overcoming diabetes-associated complications. The molecular mechanism of moringa isothiocyanate a bioactive constituent of *Moringa oleifera* comprising potent anti-inflammatory and anti-oxidative properties, was studied using next-generation sequencing technology in mice models (Li et al., 2019). Kidney mesangial cells of mice were initially challenged with high glucose, followed by treatment with moringa isothiocyanate. RNA-Seq analysis revealed that more than 20 canonical pathways showed inverse correlations between high glucose challenged vs. moringa isothiocyanate treated kidney mesangial cells (Li et al., 2019). Another RNA-seq study indicated that moringa isothiocyanate activates Nrf2-ARE signaling and induces TGFβ1 signaling mediated anti-inflammatory anti-oxidative responses proximal tubule HK-2 (Human kidney) cells (Cheng et al., 2019). Therefore, RNA-seq analysis can help to identify potential molecular targets and therapeutic strategies for diabetes. Arctigeninis isolated from *Fructus arctii* and used to reduce the proteinuria in diabetic patients. Arctigenin attenuates proteinuria and podocyte injury and effectively modulates the cell adhesion and inflammation pathways of diabetic glomeruli of mice as indicated by transcriptome analysis (Zhong et al., 2019).

12.6 FUTURE PROSPECTIVE OF TRANSCRIPTOMIC STUDIES ON ANTIDIABETIC PLANTS

In the last two decades, molecular research has extensively focused on diabetes and associated complications. However, these studies are fragmented, and the complexity of the disease mechanism has increased rather than decreased. Researchers have either focused their transcriptomic studies either on pathways associated to biosynthesis of active compounds or evaluating the effect of plant extract in mice or rat models. A comprehensive analysis of different bioactive components on molecular, physiological, and metabolic levels are needed to be addressed. Also, despite our knowledge of hypoglycemic or insulin like properties of bioactive components, there are very few studies evaluating the molecular mechanism of their action mediating the ultimate function of controlling diabetes. There is also a requirement of a uniform methodology framework to analyze the transcriptome data. As mentioned in earlier sections, plant genomes vary by four order magnitude due to invading DNA/RNA, transposable elements, whole-genome duplication, and polyploidy. Therefore, the gene expression normalization is based on implicit yet rarely validated assumptions of a similar transcriptome size among analyzed samples. Although some of the quality control measures have already been implemented to overcome these biases, their uniform usage in transcriptome analysis is still not adapted. Therefore, the RNA-Seq studies require the intervention of a comprehensive understanding of researchers working on diabetes and their treatment options. Protocols for experimental design, sample processing (e.g., RNA processing for sequencing), and associated parameters of diabetes improvement indicators need to be defined and studied properly. Uniform guidelines for analyzing, validating, interpreting, and finally translating the massive data generated from next-generation sequencing studies must be prepared and adopted as an integrated exertion towards diabetes remedy. Most of the transcriptomic data is generated and interpreted as an independent experiment rather than integrating the complementary information produced through different NGS approaches for one uniform goal. There is an urgent need to design a holistic approach to get a significant biological insight into the disease to discover better therapeutic methods rather than restricting to interpret a single transcriptome dataset.

12.7 CONCLUSION

Diabetes is a chronic disease with a complex pathophysiology and requires a combined intervention of treatment and dietary control. Medicinal plants possessing antidiabetic properties have traditionally been used to treat the disease and ease diabetes complications. Different high throughput approaches, including transcriptomic, proteomic, epigenetic accompanied with clinical trials, have been performed to verify medicinal plants' treatment efficacy. The technology has advanced from tissue-specific to single-cell-omics approaches and has opened the new horizon to discover broader cellular level mechanisms modulated by bioactive components. However, these approaches were either used independently or in a single experiment to provide insight into the hypoglycemic or anti-oxidative properties of plant extract or bioactive compounds. We strongly believe that these high throughputs next-generation molecular techniques complement each other. Thus, a comprehensive approach needs to be established to interlink the disease physiology with molecular mechanisms modulated by bioactive medicinal plant components and discover a more potent and effective remedy for the disease.

KEYWORDS

- **analysis of gene expression**
- **bioactive compounds**
- **diabetes**
- **medicinal plants**
- **transcriptome**

REFERENCES

Aanes, H., Winata, C., Moen, L. F., Ostrup, O., Mathavan, S., Collas, P., Rognes, T., & Alestrom, P., (2014). Normalization of RNA-sequencing data from samples with varying mRNA levels. *PLoS One, 9*(2), e89158.

Abdellatif, A. M., Jensen, S. H., Harms, R. Z., & Sarvetnick, N. E., (2019). Human islet response to selected type 1 diabetes-associated bacteria: A transcriptome-based study. *Front. Immunol., 10*, 2623.

Ahmad, U., & Ahmad, R. S., (2018). Antidiabetic property of aqueous extract of *Stevia rebaudiana* bertoni leaves in Streptozotocin-induced diabetes in albino rats. *BMC Complement Altern. Med., 18*(1), 179.

Ahn, H. Y., Kim, M., Seo, C. R., Yoo, H. J., Lee, S. H., & Lee, J. H., (2018). The effects of *Jerusalem artichoke* and fermented soybean powder mixture supplementation on blood glucose and oxidative stress in subjects with prediabetes or newly diagnosed type 2 diabetes. *Nutr. Diabetes, 8*(1), 42.

Al-Romaiyan, A., Jayasri, M. A., Mathew, T. L., Huang, G. C., Amiel, S., Jones, P. M., & Persaud, S. J., (2010). *Costus pictus* extracts stimulate insulin secretion from mouse and human islets of Langerhans *in vitro*. *Cell Physiol. Biochem., 26*(6), 1051–1058.

Annadurai, R. S., Jayakumar, V., Mugasimangalam, R. C., Katta, M. A., Anand, S., Gopinathan, S., Sarma, S. P., et al., (2012). Next-generation sequencing and de novo transcriptome analysis of *Costus pictus* D. don, a non-model plant with potent antidiabetic properties. *BMC Genomics, 13*, 663.

Ashraf, R., Khan, R. A., & Ashraf, I., (2011). Garlic (*Allium sativum*) supplementation with standard antidiabetic agent provides better diabetic control in type 2 diabetes patients. *Pak. J. Pharm. Sci., 24*(4), 565–570.

Aziz, T. A., Hussain, S. A., Mahwi, T. O., Ahmed, Z. A., Rahman, H. S., & Rasedee, A., (2018). The efficacy and safety of *Ginkgo biloba* extract as an adjuvant in type 2 diabetes mellitus patients ineffectively managed with metformin: A double-blind, randomized, placebo-controlled trial. *Drug Des. Devel. Ther., 12*, 735–742.

Bai, L., Gao, J., Wei, F., Zhao, J., Wang, D., & Wei, J., (2018a). Therapeutic potential of Ginsenosides as an adjuvant treatment for diabetes. *Front. Pharmacol., 9,* 423.

Bai, W., Zhang, C., & Chen, H., (2018b). Transcriptomic analysis of *Momordica charantia* polysaccharide on streptozotocin-induced diabetic rats. *Gene, 675*, 208–216.

Baquer, N. Z., Kumar, P., Taha, A., Kale, R. K., Cowsik, S. M., & McLean, P., (2011). Metabolic and molecular action of *Trigonella foenum-graecum* (fenugreek) and trace metals in experimental diabetic tissues. *J. Biosci., 36*(2), 383–396.

Benhaddou-Andaloussi, A., Martineau, L. C., Vallerand, D., Haddad, Y., Afshar, A., Settaf, A., & Haddad, P. S., (2010). Multiple molecular targets underlie the antidiabetic effect of *Nigella sativa* seed extract in skeletal muscle, adipocyte and liver cells. *Diabetes Obes. Metab., 12*(2), 148–157.

Brenner, S., Johnson, M., Bridgham, J., Golda, G., Lloyd, D. H., Johnson, D., Luo, S., et al., (2000). Gene expression analysis by massively parallel signature sequencing (MPSS) on microbead arrays. *Nat. Biotechnol., 18*(6), 630–634.

Camastra, S., Bonora, E., Del, P. S., Rett, K., Weck, M., & Ferrannini, E., (1999). Effect of obesity and insulin resistance on resting and glucose-induced thermogenesis in man. EGIR (European Group for the Study of Insulin Resistance). *Int. J. Obes Rela.t Metab. Disord., 23*(12), 1307–1313.

Cao, H., Nuruzzaman, M., Xiu, H., Huang, J., Wu, K., Chen, X., Li, J., et al., (2015). Transcriptome analysis of methyl jasmonate-elicited *Panax ginseng* adventitious roots to discover putative ginsenoside biosynthesis and transport genes. *Int. J. Mol. Sci., 16*(2), 3035–3057.

Centers for Disease, C., & Prevention. (2004). Prevalence of overweight and obesity among adults with diagnosed diabetes--United States, 1988–1994 and 1999–2002. *MMWR Morb. Mortal Wkly. Rep., 53*(45), 1066–1068.

Chen, J., Hou, K., Qin, P., Liu, H., Yi, B., Yang, W., & Wu, W., (2014). RNA-Seq for gene identification and transcript profiling of three *Stevia rebaudiana* genotypes. *BMC Genomics, 15,* 571.

Cheng, D., Gao, L., Su, S., Sargsyan, D., Wu, R., Raskin, I., & Kong, A. N., (2019). Moringa isothiocyanate activates Nrf2: Potential role in diabetic nephropathy. *AAPS J, 21*(2), 31.

Cheng, D., Liang, B., & Li, Y., (2013). Antihyperglycemic effect of *Ginkgo biloba* extract in streptozotocin-induced diabetes in rats. *Biomed. Res. Int., 2013*, 162724.

Christodoulou, M. I., Avgeris, M., Kokkinopoulou, I., Maratou, E., Mitrou, P., Kontos, C. K., Pappas, E., et al., (2019). Blood-based analysis of type-2 diabetes mellitus susceptibility genes identifies specific transcript variants with deregulated expression and association with disease risk. *Sci. Rep., 9*(1), 1512.

Chueh, W. H., & Lin, J. Y., (2011). Berberine, an isoquinoline alkaloid in herbal plants, protects pancreatic islets and serum lipids in nonobese diabetic mice. *J. Agric. Food Chem., 59*(14), 8021–8027.

Cicero, A. F., & Baggioni, A., (2016). Berberine and its role in chronic disease. *Adv. Exp. Med. Biol., 928*, 27–45.

Ciura, J., Szeliga, M., Grzesik, M., & Tyrka, M., (2017). Next-generation sequencing of representational difference analysis products for identification of genes involved in diosgenin biosynthesis in fenugreek (*Trigonella foenum-graecum*). *Planta, 245*(5), 977–991.

Cloonan, N., Forrest, A. R., Kolle, G., Gardiner, B. B., Faulkner, G. J., Brown, M. K., Taylor, D. F., et al., (2008). Stem cell transcriptome profiling via massive-scale mRNA sequencing. *Nat. Methods, 5*(7), 613–619.

Coate, J. E., & Doyle, J. J., (2010). Quantifying whole transcriptome size, a prerequisite for understanding transcriptome evolution across species: An example from a plant allopolyploid. *Genome Biol. Evol., 2*, 534–546.

Coate, J. E., & Doyle, J. J., (2015). Variation in transcriptome size: Are we getting the message? *Chromosoma, 124*(1), 27–43.

Collares, C. V., Evangelista, A. F., Xavier, D. J., Takahashi, P., Almeida, R., Macedo, C., Manoel-Caetano, F., et al., (2013). Transcriptome meta-analysis of peripheral lymphomononuclear cells indicates that gestational diabetes is closer to type 1 diabetes than to type 2 diabetes mellitus. *Mol. Biol. Rep., 40*(9), 5351–5358.

Cortez-Navarrete, M., Martinez-Abundis, E., Perez-Rubio, K. G., Gonzalez-Ortiz, M., & Mendez-Del Villar, M., (2018). *Momordica charantia* administration improves insulin secretion in type 2 diabetes mellitus. *J. Med. Food, 21*(7), 672–677.

Davis-Richardson, A. G., Ardissone, A. N., Dias, R., Simell, V., Leonard, M. T., Kemppainen, K. M., Drew, J. C., et al., (2014). *Bacteroides dorei* dominates gut microbiome prior to autoimmunity in Finnish children at high risk for type 1 diabetes. *Front Microbiol., 5*, 678.

Dhandapani, S., Kim, M. J., Chin, H. J., Leong, S. H., & Jang, I. C., (2020). Identification and functional characterization of tissue-specific terpene synthases in *Stevia rebaudiana*. *Int. J. Mol. Sci., 21*(22).

Di Pierro, F., Putignano, P., Villanova, N., Montesi, L., Moscatiello, S., & Marchesini, G., (2013). Preliminary study about the possible glycemic clinical advantage in using a fixed combination of *Berberis aristata* and *Silybum marianum* standardized extracts versus only *Berberis aristata* in patients with type 2 diabetes. *Clin. Pharmacol., 5*, 167–174.

Eizirik, D. L., Sammeth, M., Bouckenooghe, T., Bottu, G., Sisino, G., Igoillo-Esteve, M., Ortis, F., et al., (2012). The human pancreatic islet transcriptome: Expression of candidate genes for type 1 diabetes and the impact of pro-inflammatory cytokines. *PLoS Genet., 8*(3), e1002552.

Evangelista, A. F., Collares, C. V., Xavier, D. J., Macedo, C., Manoel-Caetano, F. S., Rassi, D. M., Foss-Freitas, M. C., et al., (2014). Integrative analysis of the transcriptome profiles observed in type 1, type 2 and gestational diabetes mellitus reveals the role of inflammation. *BMC Med. Genomics, 7*, 28.

Fuangchan, A., Sonthisombat, P., Seubnukarn, T., Chanouan, R., Chotchaisuwat, P., Sirigulsatien, V., Ingkaninan, K., et al., (2011). Hypoglycemic effect of bitter melon compared with metformin in newly diagnosed type 2 diabetes patients. *J. Ethnopharmacol., 134*(2), 422–428.

Gan, Q., Wang, J., Hu, J., Lou, G., Xiong, H., Peng, C., Zheng, S., & Huang, Q., (2020). The role of diosgenin in diabetes and diabetic complications. *J. Steroid Biochem. Mol. Biol., 198*, 105575.

Gao, L., Sun, N., Xu, Q., Jiang, Z., & Li, C., (2019). Comparative analysis of mRNA expression profiles in Type 1 and Type 2 diabetes mellitus. *Epigenomics, 11*(6), 685–699.

Gayathri, M., & Kannabiran, K., (2008). Antidiabetic and ameliorative potential of *Ficus bengalensis* bark extract in streptozotocin-induced diabetic rats. *Indian J. Clin. Biochem., 23*(4), 394–400.

Geberemeskel, G. A., Debebe, Y. G., & Nguse, N. A., (2019). Antidiabetic effect of fenugreek seed powder solution (*Trigonella foenum-graecum* L.) on hyperlipidemia in diabetic patients. *J. Diabetes Res., 2019*, 8507453.

Gholizadeh, F., Dastghaib, S., Koohpeyma, F., Bayat, E., & Mokarram, P., (2019). The protective effect of *Stevia rebaudiana* bertoni on serum hormone levels, key steroidogenesis enzymes, and testicular damage in testes of diabetic rats. *Acta Histochem., 121*(7), 833–840.

Gireesh, G., Thomas, S. K., Joseph, B., & Paulose, C. S., (2009). Antihyperglycemic and insulin secretory activity of *Costus pictus* leaf extract in streptozotocin-induced diabetic rats and in *in vitro* pancreatic islet culture. *J. Ethnopharmacol., 123*(3), 470–474.

Goboza, M., Aboua, Y. G., Chegou, N., & Oguntibeju, O. O., (2019). Vindoline effectively ameliorated diabetes-induced hepatotoxicity by docking oxidative stress, inflammation and hypertriglyceridemia in type 2 diabetes-induced male Wistar rats. *Biomed. Pharmacother., 112,* 108638.

Goyal, S. K., Samsher, & Goyal, R. K., (2010). Stevia (*Stevia rebaudiana*) a bio-sweetener: A review. *Int. J. Food Sci. Nutr., 61*(1), 1–10.

Grover, J. K., Yadav, S., & Vats, V., (2002). Medicinal plants of India with antidiabetic potential. *J. Ethnopharmacol., 81*(1), 81–100.

Gu, X., Xie, Z., Wang, Q., Liu, G., Qu, Y., Zhang, L., Pan, J., et al., (2009). Transcriptome profiling analysis reveals multiple modulatory effects of *Ginkgo biloba* extract in the liver of rats on a high-fat diet. *FEBS J., 276*(5), 1450–1458.

Gupta, S. S., (1963). Experimental studies on pituitary diabetes. III. Effect of indigenous antidiabetic drugs against the acute hyperglycaemic response of anterior pituitary extract in glucose fed albino rats. *Indian J. Med. Res., 51*, 716–724.

Haak, M., Vinke, S., Keller, W., Droste, J., Ruckert, C., Kalinowski, J., & Pucker, B., (2018). High quality de novo transcriptome assembly of *Croton tiglium*. *Front Mol. Biosci., 5*, 62.

Habicht, S. D., Ludwig, C., Yang, R. Y., & Krawinkel, M. B., (2014). *Momordica charantia* and type 2 diabetes: From *in vitro* to human studies. *Curr. Diabetes Rev., 10*(1), 48–60.

Hamdan, A., Haji, I. R., & Mokhtar, M. H., (2019). Effects of *Nigella sativa* on type-2 diabetes mellitus: A systematic review. *Int. J. Environ. Res. Public Health, 16*(24).

Hang, W., He, B., Chen, J., Xia, L., Wen, B., Liang, T., Wang, X., et al., (2018). Berberine ameliorates high glucose-induced cardiomyocyte injury via AMPK signaling activation to stimulate mitochondrial biogenesis and restore autophagic flux. *Front Pharmacol., 9*, 1121.

He, B., Gu, Y., Xu, M., Wang, J., Cao, F., & Xu, L. A., (2015). Transcriptome analysis of *Ginkgo biloba* kernels. *Front Plant Sci., 6,* 819.

Heo, J. B., Lee, Y. S., & Sung, S., (2013). Epigenetic regulation by long noncoding RNAs in plants. *Chromosome Res., 21*(6, 7), 685–693.

Hernandez, E. A., Kahl, S., Seelig, A., Begovatz, P., Irmler, M., Kupriyanova, Y., Nowotny, B., et al., (2017). Acute dietary fat intake initiates alterations in energy metabolism and insulin resistance. *J. Clin. Invest., 127*(2), 695–708.

Heshmati, J., & Namazi, N., (2015). Effects of black seed (*Nigella sativa*) on metabolic parameters in diabetes mellitus: A systematic review. *Complement Ther. Med., 23*(2), 275–282.

Holvoet, P., Rull, A., Garcia-Heredia, A., Lopez-Sanroma, S., Geeraert, B., Joven, J., & Camps, J., (2015). Stevia-derived compounds attenuate the toxic effects of ectopic lipid accumulation in the liver of obese mice: A transcriptomic and metabolomic study. *Food Chem. Toxicol., 77*, 22–33.

Hu, F. B., Manson, J. E., Stampfer, M. J., Colditz, G., Liu, S., Solomon, C. G., & Willett, W. C., (2001). Diet, lifestyle, and the risk of type 2 diabetes mellitus in women. *N. Engl. J. Med., 345*(11), 790–797.

Hu, M., & Polyak, K., (2006). Serial analysis of gene expression. *Nat. Protoc., 1*(4), 1743–1760.

Islam, M. A., Amin, M. N., Siddiqui, S. A., Hossain, M. P., Sultana, F., & Kabir, M. R., (2019). Trans fatty acids and lipid profile: A serious risk factor to cardiovascular disease, cancer and diabetes. *Diabetes Metab. Syndr., 13*(2), 1643–1647.

Islam, M. S., & Choi, H., (2008). Comparative effects of dietary ginger (*Zingiber officinale*) and garlic (*Allium sativum*) investigated in a type 2 diabetes model of rats. *J. Med. Food, 11*(1), 152–159.

Ismail, N., Ismail, M., Azmi, N. H., Abu, B. M. F., Basri, H., & Abdullah, M. A., (2016). Modulation of hydrogen peroxide-induced oxidative stress in human neuronal cells by thymoquinone-rich fraction and thymoquinone via transcriptomic regulation of antioxidant and apoptotic signaling genes. *Oxid. Med. Cell Longev., 2016*, 2528935.

Jayakodi, M., Choi, B. S., Lee, S. C., Kim, N. H., Park, J. Y., Jang, W., Lakshmanan, M., et al., (2018). Ginseng genome database: An open-access platform for genomics of *Panax ginseng*. *BMC Plant Biol., 18*(1), 62.

Jing, M., & Ismail-Beigi, F., (2007). Critical role of 5'-AMP-activated protein kinase in the stimulation of glucose transport in response to inhibition of oxidative phosphorylation. *Am. J. Physiol. Cell Physiol., 292*(1), C477–487.

Jing, M., Cheruvu, V. K., & Ismail-Beigi, F., (2008). Stimulation of glucose transport in response to activation of distinct AMPK signaling pathways. *Am J. Physiol. Cell Physiol., 295*(5), C1071–1082.

Joshi, D. V., Patil, R. R., & Naik, S. R., (2015). Hydro-alcohol extract of *Trigonella foenum-graecum* seed attenuates markers of inflammation and oxidative stress while improving exocrine function in diabetic rats. *Pharm. Biol., 53*(2), 201–211.

Jung, W. Y., Lee, S. S., Kim, C. W., Kim, H. S., Min, S. R., Moon, J. S., Kwon, S. Y., et al., (2014). RNA-seq analysis and *de novo* transcriptome assembly of Jerusalem artichoke (*Helianthus tuberosus* Linne). *PLoS One, 9*(11), e111982.

Kalariya, K. A., Minipara, D. B., & Manivel, P., (2018). *De novo* transcriptome analysis deciphered polyoxypregnane glycoside biosynthesis pathway in *Gymnema sylvestre*. *3 Biotech, 8*(9), 381.

Kamenetsky, R., Faigenboim, A., Shemesh, M. E., Ben, M. T., Gershberg, C., Kimhi, S., Esquira, I., et al., (2015). Integrated transcriptome catalogue and organ-specific profiling of gene expression in fertile garlic (*Allium sativum* L.). *BMC Genomics, 16*, 12.

Kim, S. H., Shin, E. J., Kim, E. D., Bayaraa, T., Frost, S. C., & Hyun, C. K., (2007). Berberine activates GLUT1-mediated glucose uptake in 3T3-L1 adipocytes. *Biol. Pharm. Bull, 30*(11), 2120–2125.

Kiyama, R., & Zhu, Y., (2014). DNA microarray-based gene expression profiling of estrogenic chemicals. *Cell Mol. Life Sci., 71*(11), 2065–2082.

Kodzius, R., Kojima, M., Nishiyori, H., Nakamura, M., Fukuda, S., Tagami, M., Sasaki, D., et al., (2006). CAGE: Cap analysis of gene expression. *Nat. Methods, 3*(3), 211–222.

Korsgren, S., Molin, Y., Salmela, K., Lundgren, T., Melhus, A., & Korsgren, O., (2012). On the etiology of type 1 diabetes: A new animal model signifying a decisive role for bacteria eliciting an adverse innate immunity response. *Am. J. Pathol., 181*(5), 1735–1748.

Kostic, A. D., Gevers, D., Siljander, H., Vatanen, T., Hyotylainen, T., Hamalainen, A. M., Peet, A., et al., (2015). The dynamics of the human infant gut microbiome in development and in progression toward type 1 diabetes. *Cell Host. Microbe, 17*(2), 260–273.

Krogvold, L., Edwin, B., Buanes, T., Frisk, G., Skog, O., Anagandula, M., Korsgren, Oet al., (2015). Detection of a low-grade enteroviral infection in the islets of Langerhans of living patients newly diagnosed with type 1 diabetes. *Diabetes, 64*(5), 1682–1687.

Kujur, R. S., Singh, V., Ram, M., Yadava, H. N., Singh, K. K., Kumari, S., & Roy, B. K., (2010). Antidiabetic activity and phytochemical screening of crude extract of *Stevia rebaudiana* in alloxan-induced diabetic rats. *Pharmacognosy Res., 2*(4), 258–263.

Lawlor, N., George, J., Bolisetty, M., Kursawe, R., Sun, L., Sivakamasundari, V., Kycia, I., et al., (2017). Single-cell transcriptomes identify human islet cell signatures and reveal cell-type-specific expression changes in type 2 diabetes. *Genome Res., 27*(2), 208–222.

Leach, M. J., (2007). *Gymnema sylvestre* for diabetes mellitus: A systematic review. *J. Altern. Complement Med., 13*(9), 977–983.

Lee, C., & Roy, M., (2004). Analysis of alternative splicing with microarrays: Successes and challenges. *Genome Biol., 5*(7), 231.

Lee, H. J., Lee, Y. H., Park, S. K., Kang, E. S., Kim, H. J., Lee, Y. C., Choi, C. S., et al., (2009). Korean red ginseng (*Panax ginseng*) improves insulin sensitivity and attenuates the development of diabetes in Otsuka Long-Evans Tokushima fatty rats. *Metabolism, 58*(8), 1170–1177.

Lee, Y. S., Park, H. S., Lee, D. K., Jayakodi, M., Kim, N. H., Koo, H. J., Lee, S. C., et al., (2017). Integrated transcriptomic and metabolomic analysis of five *Panax ginseng* cultivars reveals the dynamics of ginsenoside biosynthesis. *Front Plant Sci., 8*, 1048.

Li, S., Li, W., Wu, R., Yin, R., Sargsyan, D., Raskin, I., & Kong, A. N., (2019). Epigenome and transcriptome study of moringa isothiocyanate in mouse kidney mesangial cells induced by high glucose, a potential model for diabetic-induced nephropathy. *AAPS J., 22*(1), 8.

Lin, M., & Mao, Z. J., (2020). lncRNA-mRNA competing endogenous RNA network in IR-hepG2 cells ameliorated by APBBR decreasing ROS levels: A systematic analysis. *PeerJ., 8*, e8604.

Liu, L., Deng, Y., Yu, S., Lu, S., Xie, L., & Liu, X., (2008). Berberine attenuates intestinal disaccharidases in streptozotocin-induced diabetic rats. *Pharmazie, 63*(5), 384–388.

Liu, Z., Li, W., Li, X., Zhang, M., Chen, L., Zheng, Y. N., Sun, G. Z., & Ruan, C. C., (2013). Antidiabetic effects of malonyl ginsenosides from *Panax ginseng* on type 2 diabetic rats induced by high-fat diet and streptozotocin. *J. Ethnopharmacol., 145*(1), 233–240.

Liu, Z., Wang, L. J., Li, X., Hu, J. N., Chen, Y., Ruan, C. C., & Sun, G. Z., (2009). Hypoglycemic effects of malonyl-ginsenosides extracted from roots of *Panax ginseng* on streptozotocin-induced diabetic mice. *Phytother. Res., 23*(10), 1426–1430.

Lo, H. Y., Ho, T. Y., Lin, C., Li, C. C., & Hsiang, C. Y., (2013). *Momordica charantia* and its novel polypeptide regulate glucose homeostasis in mice via binding to insulin receptor. *J. Agric. Food Chem., 61*(10), 2461–2468.

Loven, J., Orlando, D. A., Sigova, A. A., Lin, C. Y., Rahl, P. B., Burge, C. B., Levens, D. L., et al., (2012). Revisiting global gene expression analysis. *Cell, 151*(3), 476–482.

Lunagariya, J., Bhadja, P., Zhong, S., Vekariya, R., & Xu, S., (2019). Marine natural product Bis-indole alkaloid caulerpin: Chemistry and biology. *Mini Rev. Med. Chem., 19*(9), 751–761.

Ma, C., Yu, H., Xiao, Y., & Wang, H., (2017). *Momordica charantia* extracts ameliorate insulin resistance by regulating the expression of SOCS-3 and JNK in type 2 diabetes mellitus rats. *Pharm Biol., 55*(1), 2170–2177.

Madlung, A., Masuelli, R. W., Watson, B., Reynolds, S. H., Davison, J., & Comai, L., (2002). Remodeling of DNA methylation and phenotypic and transcriptional changes in synthetic Arabidopsis allotetraploids. *Plant Physiol., 129*(2), 733–746.

Mahmood, K., Orabi, J., Kristensen, P. S., Sarup, P., Jorgensen, L. N., & Jahoor, A., (2020). *De novo* transcriptome assembly, functional annotation, and expression profiling of rye (*Secale cereale* L.) hybrids inoculated with ergot (*Claviceps purpurea*). *Sci. Rep., 10*(1), 13475.

Majekodunmi, S. O., Oyagbemi, A. A., Umukoro, S., & Odeku, O. A., (2011). Evaluation of the antidiabetic properties of *Mucuna pruriens* seed extract. *Asian Pac. J. Trop. Med., 4*(8), 632–636.

Malekshahi, H., Bahrami, G., Miraghaee, S., Ahmadi, S. A., Sajadimajd, S., Hatami, R., Mohammadi, B., & Keshavarzi, S., (2019). *Momordica charantia* reverses type II diabetes in rat. *J. Food Biochem., 43*(11), e13021.

Marioni, J. C., Mason, C. E., Mane, S. M., Stephens, M., & Gilad, Y., (2008). RNA-seq: An assessment of technical reproducibility and comparison with gene expression arrays. *Genome Res., 18*(9), 1509–1517.

McCarthy, M. I., (2010). Genomics, type 2 diabetes, and obesity. *N. Engl. J. Med., 363*(24), 2339–2350.

Mehra, R., Jasrotia, R. S., Mahajan, A., Sharma, D., Iquebal, M. A., Kaul, S., & Dhar, M. K., (2020). Transcriptome analysis of snow mountain garlic for unraveling the organosulfur metabolic pathway. *Genomics, 112*(1), 99–107.

Meyer, J. A., Froelich, J. M., Reid, G. E., Karunarathne, W. K., & Spence, D. M., (2008). Metal-activated C-peptide facilitates glucose clearance and the release of a nitric oxide stimulus via the GLUT1 transporter. *Diabetologia, 51*(1), 175–182.

Michael, T. P., (2014). Plant genome size variation: Bloating and purging DNA. *Brief Funct. Genomics, 13*(4), 308–317.

Modi, A., Kumar, N., & Narayanan, S., (2016). Transcript quantification of genes involved in steviol glycoside biosynthesis in *Stevia rebaudiana* bertoni by real-time polymerase chain reaction (RT-PCR). *Methods Mol. Biol., 1391*, 289–301.

Mohd-Radzman, N. H., Ismail, W. I., Adam, Z., Jaapar, S. S., & Adam, A., (2013). Potential roles of *Stevia rebaudiana* bertoni in abrogating insulin resistance and diabetes: A review. *Evid Based Complement Alternat Med., 2013*, 718049.

Momtazi-Borojeni, A. A., Esmaeili, S. A., Abdollahi, E., & Sahebkar, A., (2017). A review on the pharmacology and toxicology of steviol glycosides extracted from *Stevia rebaudiana*. *Curr. Pharm Des., 23*(11), 1616–1622.

Nagalakshmi, U., Wang, Z., Waern, K., Shou, C., Raha, D., Gerstein, M., & Snyder, M., (2008). The transcriptional landscape of the yeast genome defined by RNA sequencing. *Science, 320*(5881), 1344–1349.

Nagulapalli, V. K. C., Swaroop, A., Bagchi, D., & Bishayee, A., (2017). A small plant with big benefits: Fenugreek (*Trigonella foenum-graecum* Linn.) for disease prevention and health promotion. *Mol. Nutr. Food Res., 61*(6).

Ni, W. J., Ding, H. H., & Tang, L. Q., (2015). Berberine as a promising antidiabetic nephropathy drug: An analysis of its effects and mechanisms. *Eur. J. Pharmacol., 760*, 103–112.

Padiya, R., & Banerjee, S. K., (2013). Garlic as an antidiabetic agent: Recent progress and patent reviews. *Recent Pat. Food Nutr. Agric., 5*(2), 105–127.

Pan, G. Y., Huang, Z. J., Wang, G. J., Fawcett, J. P., Liu, X. D., Zhao, X. C., Sun, J. G., & Xie, Y. Y., (2003). The antihyperglycaemic activity of berberine arises from a decrease of glucose absorption. *Planta Med., 69*(7), 632–636.

Pellicer, J., & Leitch, I. J., (2020). The plant DNA c-values database (release 7.1): An updated online repository of plant genome size data for comparative studies. *New Phytol., 226*(2), 301–305.

Peter, E. L., Kasali, F. M., Deyno, S., Mtewa, A., Nagendrappa, P. B., Tolo, C. U., Ogwang, P. E., & Sesaazi, D., (2019). *Momordica charantia* L. lowers elevated glycaemia in type 2 diabetes mellitus patients: Systematic review and meta-analysis. *J. Ethnopharmacol., 231*, 311–324.

Pethe, M., Yelwatkar, S., Manchalwar, S., & Gujar, V., (2017). Evaluation of biological effects of hydroalcoholic extract of *Hibiscus rosa sinensis* flowers on alloxan induced diabetes in rats. *Drug Res. (Stuttg), 67*(8), 485–492.

Poma, A. M., Genoni, A., Broccolo, F., Denaro, M., Pugliese, A., Basolo, F., & Toniolo, A., (2020). Immune transcriptome of cells infected with enterovirus strains obtained from cases of type 1 diabetes. *Microorganisms, 8*(7).

Porchezhian, E., & Dobriyal, R. M., (2003). An overview on the advances of *Gymnema sylvestre*: chemistry, pharmacology and patents. *Pharmazie, 58*(1), 5–12.

Pothuraju, R., Sharma, R. K., Chagalamarri, J., Jangra, S., & Kumar, K. P., (2014). A systematic review of *Gymnema sylvestre* in obesity and diabetes management. *J. Sci. Food Agric., 94*(5), 834–840.

Qiao, D., Yang, C., Chen, J., Guo, Y., Li, Y., Niu, S., Cao, K., & Chen, Z., (2019). Comprehensive identification of the full-length transcripts and alternative splicing related to the secondary metabolism pathways in the tea plant (*Camellia sinensis*). *Sci. Rep., 9*(1), 2709.

Rathi, S. S., Grover, J. K., & Vats, V., (2002). The effect of *Momordica charantia* and *Mucuna pruriens* in experimental diabetes and their effect on key metabolic enzymes involved in carbohydrate metabolism. *Phytother. Res., 16*(3), 236–243.

Richardson, S. J., & Morgan, N. G., (2018). Enteroviral infections in the pathogenesis of type 1 diabetes: New insights for therapeutic intervention. *Curr. Opin. Pharmacol., 43*, 11–19.

Ritu, M., & Nandini, J., (2016). Nutritional composition of *Stevia rebaudiana*, a sweet herb, and its hypoglycaemic and hypolipidaemic effect on patients with non-insulin-dependent diabetes mellitus. *J. Sci. Food Agric., 96*(12), 4231–4234.

Samsulrizal, N. H., Khadzran, K. S., Shaarani, S. H., Noh, A. L., Sundram, T. C., Naim, M. A., & Zainuddin, Z., (2020). *De novo* transcriptome dataset of *Stevia rebaudiana* accession MS007. *Data Brief, 28*, 104811.

Samuel, P., Ayoob, K. T., Magnuson, B. A., Wolwer-Rieck, U., Jeppesen, P. B., Rogers, P. J., Rowland, I., & Mathews, R., (2018). Stevia leaf to stevia sweetener: Exploring its science, benefits, and future potential. *J. Nutr., 148*(7), 1186S–1205S.

Sankar, P., Subhashree, S., & Sudharani, S., (2012). Effect of *Trigonella foenum-graecum* seed powder on the antioxidant levels of high fat diet and low dose streptozotocin-induced type II diabetic rats. *Eur. Rev. Med. Pharmacol. Sci., 16*(Suppl 3), 10–17.

Saxena, A., Tiwari, P., Wahi, N., Soni, A., Bansiwal, R. C., Kumar, A., Sharma, B., et al., (2019). Transcriptome profiling reveals association of peripheral adipose tissue pathology with type-2 diabetes in Asian Indians. *Adipocyte, 8*(1), 125–136.

Schoenaker, D. A., Mishra, G. D., Callaway, L. K., & Soedamah-Muthu, S. S., (2016). The role of energy, nutrients, foods, and dietary patterns in the development of gestational diabetes mellitus: A systematic review of observational studies. *Diabetes Care, 39*(1), 16–23.

Seppey, M., Manni, M., & Zdobnov, E. M., (2019). BUSCO: Assessing genome assembly and annotation completeness. *Methods Mol. Biol., 1962*, 227–245.

Shang, A., Cao, S. Y., Xu, X. Y., Gan, R. Y., Tang, G. Y., Corke, H., Mavumengwana, V., & Li, H. B., (2019). Bioactive compounds and biological functions of garlic (*Allium sativum* L.). *Foods, 8*(7).

Shanmugasundaram, E. R., Venkatasubrahmanyam, M., Vijendran, N., & Shanmugasundaram, K. R., (1988). Effect of an isolate from *Gymnema sylvestre*, R. Br. In the control of diabetes mellitus and the associated pathological changes. *Anc. Sci. Life, 7*(3, 4), 183–194.

Shanmugasundaram, K. R., Panneerselvam, C., Samudram, P., & Shanmugasundaram, E. R., (1983). Enzyme changes and glucose utilization in diabetic rabbits: The effect of *Gymnema sylvestre*, R.Br. *J. Ethnopharmacol., 7*(2), 205–234.

Shao, T., Yu, Q., Zhu, T., Liu, A., Gao, X., Long, X., & Liu, Z., (2020). Inulin from Jerusalem artichoke tubers alleviates hyperglycaemia in high-fat-diet-induced diabetes mice through the intestinal microflora improvement. *Br. J. Nutr., 123*(3), 308–318.

She, J., Yan, H., Yang, J., Xu, W., & Su, Z., (2019). croFGD: *Catharanthus roseus* functional genomics database. *Front. Genet., 10,* 238.

Shi, R., Wang, Y., An, X., Ma, J., Wu, T., Yu, X., Liu, S., et al., (2019). Efficacy of co-administration of liuwei dihuang pills and *Ginkgo biloba* tablets on albuminuria in type 2 diabetes: A 24-month, multicenter, double-blind, placebo-controlled, randomized clinical trial. *Front Endocrinol (Lausanne), 10*, 100.

Shin, S. S., & Yoon, M., (2018). Korean red ginseng (*Panax ginseng*) inhibits obesity and improves lipid metabolism in high fat diet-fed castrated mice. *J. Ethnopharmacol., 210*, 80–87.

Shukla, A. K., Shasany, A. K., Gupta, M. M., & Khanuja, S. P., (2006). Transcriptome analysis in *Catharanthus roseus* leaves and roots for comparative terpenoid indole alkaloid profiles. *J. Exp. Bot., 57*(14), 3921–3932.

Shukla, A., Singh, V. K., Bharadwaj, D. R., Kumar, R., Rai, A., Rai, A. K., Mugasimangalam, R., et al., (2015). *De novo* assembly of bitter gourd transcriptomes: Gene expression and sequence variations in gynoecious and monoecious lines. *PLoS One, 10*(6), e0128331.

Shukla, S. K., Liu, W., Sikder, K., Addya, S., Sarkar, A., Wei, Y., & Rafiq, K., (2017). HMGCS2 is a key ketogenic enzyme potentially involved in type 1 diabetes with high cardiovascular risk. *Sci. Rep., 7*(1), 4590.

Simao, F. A., Waterhouse, R. M., Ioannidis, P., Kriventseva, E. V., & Zdobnov, E. M., (2015). BUSCO: Assessing genome assembly and annotation completeness with single-copy orthologs. *Bioinformatics, 31*(19), 3210–3212.

Singh, R. K., Mehta, S., Jaiswal, D., Rai, P. K., & Watal, G., (2009). Antidiabetic effect of *Ficus bengalensis* aerial roots in experimental animals. *J. Ethnopharmacol., 123*(1), 110–114.

Singla, R., Singla, N., & Jaitak, V., (2019). *Stevia rebaudiana* targeting alpha-amylase: An *in-vitro* and *in-silico* mechanistic study. *Nat. Prod. Res., 33*(4), 548–552.

Slotkin, R. K., & Martienssen, R., (2007). Transposable elements and the epigenetic regulation of the genome. *Nat. Rev. Genet., 8*(4), 272–285.

Song, L., You, W., Wang, P., Li, F., & Liu, H., (2019). Transcriptomic analysis of the association between diabetes mellitus and myocardial infarction. *Exp. Clin. Endocrinol. Diabetes, 127*(9), 603–614.

Srivastava, Y., Nigam, S. K., Bhatt, H. V., Verma, Y., & Prem, A. S., (1985). Hypoglycemic and life-prolonging properties of *Gymnema sylvestre* leaf extract in diabetic rats. *ISR J. Med. Sci., 21*(6), 540–542.

Stewart, C. J., Ajami, N. J., O'Brien, J. L., Hutchinson, D. S., Smith, D. P., Wong, M. C., Ross, M. C., et al., (2018). Temporal development of the gut microbiome in early childhood from the TEDDY study. *Nature, 562*(7728), 583–588.

Strickler, S. R., Bombarely, A., & Mueller, L. A., (2012). Designing a transcriptome next-generation sequencing project for a nonmodel plant species. *Am. J. Bot., 99*(2), 257–266.

Sun, X., Zhou, S., Meng, F., & Liu, S., (2012). *De novo* assembly and characterization of the garlic (*Allium sativum*) bud transcriptome by Illumina sequencing. *Plant Cell Rep., 31*(10), 1823–1828.

Tan, K. C., Ipcho, S. V., Trengove, R. D., Oliver, R. P., & Solomon, P. S., (2009). Assessing the impact of transcriptomics, proteomics and metabolomics on fungal phytopathology. *Mol. Plant Pathol., 10*(5), 703–715.

Tiwari, P., Ahmad, K., & Baig, M. H., (2017). *Gymnema sylvestre* for diabetes: From traditional herb to future's therapeutic. *Curr. Pharm. Des., 23*(11), 1667–1676.

Van, M. A., Fabris, M., Pollier, J., Baart, G. J., Rombauts, S., Hasnain, G., Rischer, H., et al., (2013). CathaCyc, a metabolic pathway database built from *Catharanthus roseus* RNA-seq data. *Plant Cell Physiol., 54*(5), 673–685.

Vatanen, T., Kostic, A. D., D'Hennezel, E., Siljander, H., Franzosa, E. A., Yassour, M., Kolde, R., et al., (2016). Variation in microbiome LPS immunogenicity contributes to autoimmunity in humans. *Cell, 165*(4), 842–853.

Vergara-Pulgar, C., Rothkegel, K., Gonzalez-Aguero, M., Pedreschi, R., Campos-Vargas, R., Defilippi, B. G., & Meneses, C., (2019). *De novo* assembly of *Persea americana* cv. 'Hass' transcriptome during fruit development. *BMC Genomics, 20*(1), 108.

Verma, M., Ghangal, R., Sharma, R., Sinha, A. K., & Jain, M., (2014). Transcriptome analysis of *Catharanthus roseus* for gene discovery and expression profiling. *PLoS One, 9*(7), e103583.

Vijay, N., Poelstra, J. W., Kunstner, A., & Wolf, J. B., (2013). Challenges and strategies in transcriptome assembly and differential gene expression quantification. A comprehensive in silico assessment of RNA-seq experiments. *Mol. Ecol., 22*(3), 620–634.

Vuksan, V., Sung, M. K., Sievenpiper, J. L., Stavro, P. M., Jenkins, A. L., Di Buono, M., Lee, K. S., et al., (2008). Korean red ginseng (*Panax ginseng*) improves glucose and insulin regulation in well-controlled, type 2 diabetes: Results of a randomized, double-blind, placebo-controlled study of efficacy and safety. *Nutr. Metab. Cardiovasc Dis., 18*(1), 46–56.

Walley, A. J., Blakemore, A. I., & Froguel, P., (2006). Genetics of obesity and the prediction of risk for health. *Hum. Mol. Genet., 15*(Spec. No 2), R124–130.

Wang, J., Zhang, X., Lan, H., & Wang, W., (2017). Effect of garlic supplement in the management of type 2 diabetes mellitus (T2DM): A meta-analysis of randomized controlled trials. *Food Nutr. Res., 61*(1), 1377571.

Wang, X., Elling, A. A., Li, X., Li, N., Peng, Z., He, G., Sun, H., et al., (2009a). Genome-wide and organ-specific landscapes of epigenetic modifications and their relationships to mRNA and small RNA transcriptomes in maize. *Plant Cell, 21*(4), 1053–1069.

Wang, Z., Gerstein, M., & Snyder, M., (2009b). RNA-Seq: A revolutionary tool for transcriptomics. *Nat. Rev. Genet., 10*(1), 57–63.
Xu, J., Chu, Y., Liao, B., Xiao, S., Yin, Q., Bai, R., Su, H., et al., (2017). *Panax ginseng* genome examination for ginsenoside biosynthesis. *Gigascience, 6*(11), 1–15.
Yin, J., Xing, H., & Ye, J., (2008). Efficacy of berberine in patients with type 2 diabetes mellitus. *Metabolism, 57*(5), 712–717.
Zhai, B., Zhang, C., Sheng, Y., Zhao, C., He, X., Xu, W., Huang, K., & Luo, Y., (2018). Hypoglycemic and hypolipidemic effect of S-allyl-cysteine sulfoxide (alliin) in DIO mice. *Sci. Rep., 8*(1), 3527.
Zhang, B., Pan, Y., Xu, L., Tang, D., Dorfman, R. G., Zhou, Q., Yin, Y., et al., (2018a). Berberine promotes glucose uptake and inhibits gluconeogenesis by inhibiting deacetylase SIRT3. *Endocrine, 62*(3), 576–587.
Zhang, Q., Yu, H., Xiao, X., Hu, L., Xin, F., & Yu, X., (2018b). Inulin-type fructan improves diabetic phenotype and gut microbiota profiles in rats. *PeerJ, 6*, e4446.
Zhang, W., Xu, Y. C., Guo, F. J., Meng, Y., & Li, M. L., (2008a). Antidiabetic effects of cinnamaldehyde and berberine and their impacts on retinol-binding protein 4 expression in rats with type 2 diabetes mellitus. *Chin. Med. J. (Engl), 121*(21), 2124–2128.
Zhang, Y., Li, X., Zou, D., Liu, W., Yang, J., Zhu, N., Huo, L., et al., (2008b). Treatment of type 2 diabetes and dyslipidemia with the natural plant alkaloid berberine. *J. Clin. Endocrinol. Metab., 93*(7), 2559–2565.
Zhao, M., Lin, Y., Wang, Y., Li, X., Han, Y., Wang, K., Sun, C., et al., (2019). Transcriptome analysis identifies strong candidate genes for ginsenoside biosynthesis and reveals its underlying molecular mechanism in *Panax ginseng* C.A. Meyer. *Sci. Rep., 9*(1), 615.
Zhao, Q., Gao, C., & Cui, Z., (2015). Ginkgolide A reduces inflammatory response in high-glucose-stimulated human umbilical vein endothelial cells through STAT3-mediated pathway. *Int. Immunopharmacol., 25*(2), 242–248.
Zhao, S., Fung-Leung, W. P., Bittner, A., Ngo, K., & Liu, X., (2014). Comparison of RNA-Seq and microarray in transcriptome profiling of activated T cells. *PLoS One, 9*(1), e78644.
Zhong, Y., Lee, K., Deng, Y., Ma, Y., Chen, Y., Li, X., Wei, C., et al., (2019). Arctigenin attenuates diabetic kidney disease through the activation of PP2A in podocytes. *Nat. Commun., 10*(1), 4523.
Zhou, C., Li, X., Zhou, Z., Li, C., & Zhang, Y., (2019a). Comparative transcriptome analysis identifies genes involved in diosgenin biosynthesis in *Trigonella foenum-graecum* L. *Molecules, 24*(1).
Zhou, P., Xie, W., He, S., Sun, Y., Meng, X., Sun, G., & Sun, X., (2019b). Ginsenoside Rb1 as an antidiabetic agent and its underlying mechanism analysis. *Cells, 8*(3).
Zhu, X., Qiu, Z., Ouyang, W., Miao, J., Xiong, P., Mao, D., Feng, K., et al., (2019). Hepatic transcriptome and proteome analyses provide new insights into the regulator mechanism of dietary avicularin in diabetic mice. *Food. Res. Int., 125,* 108570.
Zou, L. Q., Kuang, X. J., Li, Y., & Sun, C., (2016). Advance in transcriptomic studies of ginseng species. *Zhongguo Zhong Yao Za Zhi, 41*(22), 4138–4143.

CHAPTER 13

How Metabolomics Shapes the Future of Drugs in Antidiabetic Plants

JAVID AHMAD MALIK,[1] SHAZA HUSSINY ALY,[2] YOUNIS AHMAD HAJAM,[3] and SAMEER QUAZI[4]

[1]*Department of Zoology Guru Ghasidas Vishwavidyalaya, Bilaspur, Chhattisgarh – 495001, India, Phone: 7006970581; 7051625284, E-mail: javidmalikr@gmail.com*

[2]*Department of Pharmacognosy, Badr University, Cairo, Egypt*

[3]*Department of Life Sciences and Allied Health Sciences, Sant Baba Bagh Singh University, Khalia Padhiana, Jalandhar, Punjab–144030, India*

[4]*Department of Genetics, Indian Academy Degree College, Bangalore, Karnataka, India*

ABSTRACT

Diabetes mellitus (DM) is the main common endocrine infection on the planet which is probably going to be the significant pandemic in history. In recent years, numerous latest medicines against diabetes have been delivered and brought into the markets, nonetheless, long lasting treatment utilizing manufactured medications is restricted. Therapeutic plants assume an incredible part in diabetes treatment. Numerous medicinal herbs and their related medicines for treating diabetes are utilized all through the world addressing promising choices for the administration of diabetes treatment. Metabolomics investigations relating to diabetes have added to numerous aspects of investigating biomarkers and to understand the pathophysiology of the

Antidiabetic Potential of Plants in the Era of Omics. Deepu Pandita, PhD, Anu Pandita, MSc, & Chander Bhanu, MBBS (Eds.)

disease at metabolism levels. Moreover, in the most recent decade, various metabolomic approaches have stressed on examining the action mechanism of different herbal medications. So, the aim of this chapter is to focus on metabolomic studies carried out to investigate the role of phytomedicines on diabetes, uncovering biomarkers.

13.1 INTRODUCTION

Diabetes mellitus (DM) is a complex chronic condition in which insulin is not released enough by the pancreas (type 1) or the human body is not able to utilize insulin well (type 2). The main clinical characteristic is hyperglycemia resulting from glucose intolerance. Besides, alterations in lipid and protein metabolism in the long run. Due to the adverse effects of synthetic drugs the use of natural products to control diabetes is of great demands especially that herbal medicine has been shown to provide an effective treatment and prevention of diabetes by disturbance of metabolic pathways and biomarkers by different mechanisms. Here in highlights and investigation of the application of different metabolomics studies on the roles of medicinal plants in the treatment of diabetes and unraveling various biomarkers in the *in vivo* studies or herbal medicine effects on different enzymes *in vitro*.

Diabetes is a complex disorder brought about by a complete absence of insulin or comparative absence of insulin because of scarcity in insulin secretion, activity, or a mix of both. Glucose intolerance bringing about hyperglycemia is characterized as the quality of diabetes pathology. In the long run, alterations in protein and lipid metabolism lead to the development of different perturbations, for example, microvascular (nephropathy, neuropathy, and retinopathy) and macrovascular (coronary illness, atherosclerosis-related vascular illness, cerebrovascular infection, and fringe vascular sickness) problems (Zimmet et al., 2001). The disease can be categorized into two kinds; Type 1 diabetes mellitus (T1DM) and Type 2 diabetes mellitus (T2DM). Between these two types, about 5–10% of the diabetic persons are determined to have T1DM, which is related with complete or most of the insulin inadequacy linked with auto-immune destruction of pancreatic β-cells. However, the episode of T2DM has now been a scourge that is recognized by β-cell dysfunction, fluctuating levels of insulin resistance, and elevated production of glucose (ADA, 2015). In late years, an elevation in T2DM prevalence has happened in both youngsters and adults.

An unhealthful diet, way of life, obesity, and genetic elements are the danger factors for T2DM and the occurrence of diabetes increments significantly with age and diminished active work levels (Hu and Satija, 2015). Worldwide, the commonness of diabetes is assessed at 2.8% in 2000, which is expected to increment to 4.4% by 2030 (Wild et al., 2004). Treatment of diabetes utilizing engineered drugs is commonly followed by unfavorable results including gastrointestinal responses, an inclination in body weight, variance in blood glucose (Bischoff, 1995; Thulé and Umpierrez, 2004), threat of cardiovascular illness (Ahmadian et al., 2013; Gerstein et al., 2008). By restraining intestinal adsorption (acarbose) (Bischoff, 1995) and hepatic gluconeogenesis (metformin) (Hundal et al., 2000), exciting insulin receptors (metformin and rosiglitazone) (Ahmadian et al., 2013), and stimulation of pancreatic insulin release (sitagliptin and glipizide) (Thulé and Umpierrez, 2004) most of the drugs against diabetes work on the regulation of glycemia. Numerous antidiabetic medicines have been created and put available lately to treat diabetic patients, yet the majority of them have their side effects (Palanisamy et al., 2018). Moreover, numerous conventional plant medicines for diabetes are utilized all through the world and address promising choices for the administration of T2DM management (Bailey and Day, 1989) Given the decreased expense and less difficult openness of medicinal plants contrasted with professionally prescribed medications, natural treatment has now been generally used to treat T2DM (Karimi et al., 2015).

Most plants possess bioactive components, including flavonoids, terpenoids, carotenoids, alkaloids, glycosides, and can regularly have antidiabetic effects (Afrisham et al., 2015). Obviously, information on the bioactive substances of medicinal plants for carrying out a clinical preliminary investigation for the appropriate medical utilization of natural medication is necessary. Metabolomics helps to have a thorough investigation of bio-sample metabolites to untangle metabolite profile changes. Metabolomics approaches have given experiences into novel techniques throughout the most recent decade to clarify the conduct of a medicine or wholesome intercession by characterizing the altered metabolites among control and treatment groups (Kaddurah-Daouk et al., 2015; Kaddurah-Daouk et al., 2014). The metabolites are treated as biomarkers and can thus be utilized to perceive targets and pathways related with efficacy, results, and in case toxicity. Metabolomics can investigate the pharmacological impacts and the activity mechanism of various spices also (Shi et al., 2016).

A few metabolomics examines have assessed the relationship between various metabolites, T2DM and insulin resistance, utilizing in vivo and in vitro studies. Albeit numerous kinds of progressed technical approaches could be used in metabolomics studies, however two procedures including mass spectrometry (MS) combined with gas-or fluid stage chromatography (GCeMS and LCeMS separately) and proton atomic magnetic resonance (NMR) (Nicholson and Lindon, 2008) are majorly used. Some studies have shown the role of metabolomics research in T2DM and other metabolic disorders (Dunn et al., 2011; Lu et al., 2013; Ng et al., 2012). It would surely direct researchers to find the best medicinal plant or new drug details for potential developments in diabetes treatment. Consequently, this chapter means to feature and survey the recent metabolomics studies, which have shown expected biomarkers of natural examples and herbal medication in vivo tries.

13.2 *IN VIVO* STUDIES

The aqueous extract of the dry stem bark of *Berberis vernae* subjected to *in vivo* assay by high-fat food and streptozotocin (STZ) incited type 2 diabetic rats. The outcomes uncovered that B. vernae extract significantly reduced ($P < 0.01$) serum fasting blood glucose (FBG), insulin, glycated serum protein, IL-1β, TNF-α, and IL-6 levels, while as elevated serum insulin sensitivity index ($P < 0.01$) in diabetic rats, compared with the control group. By ^{1}H NMR based metabolomics, an aggregate of 28 endogenous metabolites were distinguished where 9 metabolites that were altered by diabetic condition were essentially switched by B. vernae extract and rat serum levels were significantly recovered towards normal by the extract.

Out of which, six metabolites including low density lipoprotein/very low density lipoprotein, valine, isoleucine, N-acetyl glycoproteins (NAG), aceto-acetate, and lipids were elevated, whereas three other betaine, trimethylamine N-oxide (TMAO), and glucose were decreased after *B. vernae* treatment for 30 days. Additionally, Representative HPLC chromatogram revealed that *B. vernae* extract mainly comprise six compounds (bufotenidine, palmatine, magnoflorine, ferulic acid 4-O-β-D-glucopyranoside, jatrorrhizine, and berberine).

The antidiabetic action of extract *B. vernae* could be related to its regulation of multiple pathways because the butanoate metabolism and the disease-related signal pathways as adipocytokine signaling pathway (Li et al., 2020).

Leaves of *Lycium barbarum* L. (LLB) (Solanaceae) was widely used in China and Southeast Asia for centuries due to its hypoglycemic effect. In this study it is investigated *in vivo* by high fat diet and STZ-induced T2DM rat model. The contents of chlorogenic acid, neochlorogenic acid, caffeic acid, and rutin in LLB extract were 32.41 mg/g, 13.47 mg/g, 3.67 mg/g, and 11.02 mg/g, respectively. The biochemical indexes were evaluated, and the results revealed that the FBG levels were reduced by 30.5%, 19.5%, and 19.9%, respectively, after the administration of metformin hydrochloride and high and low dose of LLB extract. Also, total cholesterol (T-CHO), triglyceride (TG), low density lipoprotein cholesterin (LDL-C), free fatty acid (FFA), and oxidized low density lipoprotein (OX-LDL) tended to be normal significantly different after treatment. The levels of ALT, AST, and proteinuria were reduced after administration of LLB, indicating that LLB could improve liver and kidney injury from T2DM. By UPLC-Q-TOF/MS serum and urine metabolites were identified, where among the total 22 metabolites, the levels of nine metabolites, including neriantogenin, prostaglandin H2, lysophosphatidylcholine (lysoPC) (18:0), xanthosine, histidinal, pantetheine 4′-phosphate, L-allothreonine, homovanillin, and niacinamide, significantly elevated in the model group as compared to the control group, whereas other 13 metabolite levels dramatically decreased. By using one-way analysis of variance (ANOVA), the variations between the model group and other groups were compared. The contents of 12 biomarkers were reversed ($P < 0.05$) after administration of LLB extract. The administration of LLB extract could effectively regulate the abnormal changes in the potential biomarkers as presence of neriantogenin (M2) and prostaglandin H2 (M3) in serum, and presence of xanthosine (M8), malonic acid (M9), histidinal (M10), homovanillin (M14), niacinamide (M15), 5-amino-1-(5-phospho-D-ribosyl) imidazole-4-carboxylate (M16), hippuric acid (M17), 2-phenylacetamide (M18), L-allothreonine (M19), and trans-S-(1-propenyl)-L-cysteine (M20) in urine.

The changes of biomarkers in serum and urine induced by T2DM were regained after the LLB administration that correlated mainly with the pathways like arachidonic acid metabolism, nicotinate, and nicotinamide metabolism, and purine metabolism. Moreover, sequencing of the 16S rDNA taken from fecal samples witnessed that the LLB extract could undo the gut microbiota dysbiosis induced by T2DM. Therefore, here it was concluded that T2DM, which is responsible for altering the metabolic profiles and also gut microbiota, could be effectively recovered by LLB extract (Zhao et al., 2020).

At 250 mg/kg dose, the antihyperglycemic effect of various *Ficusdeltoidea* (Moraceae) varieties was assessed on streptozotocin-nicotinamide-induced diabetic rat models and their urinary metabolites were further evaluated using ^{1}H-NMR. The 30 day treatment therapy with extracts of *F. deltoidea var. trengganuensis, var. intermedia* and *var. kunstleri.* hyperglycemic blood levels returned to normal glycemic status. In diabetic rats, the extracts have substantially regulated the biochemical parameters. The findings of Metabolomics showed that these extracts were able to handle the changed diabetic rat metabolites by transferring a few metabolites back to their usual state (Noor et al., 2020).

Another plant of family Moracea, bark of *Morus multicaulis* was subjected to explore its mechanism on HFD and STZ-induced T2DM in rats and the metabolite alterations were monitored using GC/MS analysis. The metabolic ailments in model group comprise improvement of the pathway for Sorbitol and galactose metabolite activation, improved gluconeogenesis activity, proteins catabolism, fatty acid oxidation and induced activities of PPP (pentose phosphate pathway), glycolysis, and oxidation pathways, cholesterol synthesis, inositol production.

Treatment by *M. multicaulis* ameliorate these abnormal changes as showed by differential metabolites compared with that of model mice, such as reducing ketone body accumulation, improving NADPH biosynthesis, partly reversing oxidative stress, and disrupting energy metabolism. The extract aerial parts of *Melastoma dodecandrum* (MDE) (Melastomataceae) in a dose of 600 mg/kg were investigated for their antidiabetic activity using HFD and STZ-induced T2DM in rats.

Where the hypoglycemic effect of MDE was seen and a significant low blood glucose concentration was seen when compared to the diabetic group. Also, the serum insulin levels, triglyceride (TG), total cholesterol (TC), malondialdehyde (MDA), LDL, alanine aminotransferase, aspartate aminotransferase, and urea nitrogen in the MDE-treated group were significantly reduced towards normal. The metabolomic profiling by UHPLC/QTOF-MS explored 17 identified biomarkers mainly involved in protein, lipid, taurine arachidonic acid and nicotinic acid metabolism (Weng et al., 2019).

Lycium barbarum polysaccharide (LBP) (Solanaceae) was assessed for its metabolic impacts of LBP on the liver and urine metabolomics on a HFD and STZ-prompted diabetic rat model. Metabolomics revealed that LBP positively regulated FBP, hemoglobin-A1c, SOD levels, glycogen, and homeostasis model evaluation for insulin resistance significantly when contrasted with the diabetic control (DC) group. By GC/MS investigation, liver metabolomics demonstrated more significant levels of myo-inositol and reduced levels of fumaric acid, D-arabitol, L-malic acid, O-phosphorylethanolamine,

L-allothreonine 1, xylitol, 5-methoxytryptamine 2, ribitol, and digitoxose 2 in the LBP-High dose group versus the DC group, which shows that LBP may manage the citrate cycle, alanine, glutamate, and aspartate digestion, glyoxylate, and dicarboxylate digestion. The urine metabolomics indicated elevated degrees of D-galacturonic acid 2, creatinine, 2,3-dihydroxybutyric acid and citrus acid, and diminished degrees of methylmalonic acid, benzoic acid and xylitol between the LBP-High portion and DC gatherings.

As per the statistical standard utilized for assessing metabolic profiles in liver, 16 metabolites and 7 metabolites were recognized in urine. The LBP elevated degrees of D galacturonic acid 2, creatinine, 2,3-dihydroxybutyric acid and citrus acid, while diminished degrees of methylmalonic acid, xylitol, and benzoic acid fundamentally in urine, contrasted with the DC group. However, no able pathway was found with KEGG pathway online database (Xia et al., 2019).

Flavonoids of Trigonella foenum-graecum Linn. (Fabaceae) applied a solid antidiabetic impact described by huge decrease of FBP ($P < 0.01$), increment in serum insulin level ($P < 0.01$) and HEPATIC glycogen content ($P < 0.01$), weakening of weight reduction, and improvement of pancreatic islet and kidney conditions. Luteolin and Apigenin were the fundamental aglycones and xylose, glucose, galactose, arabinose, and rhamnose represented as the principle glycosides associated with aglycones generally through C-glycoside linkage. Furthermore, metabolomics studies were also carried out to study the antidiabetic impact of fenugreek flavonoids. Serum tests of healthy and diabetic rat models with and without application of fenugreek flavonoids were assessed by UPLC-Q-TOF-MS, trailed by PC and OPLS-DA. The PCA model uncovered ample contrasts among different animal groups, while as OPLS-DA distinguished fenugreek flavonoids-instigated changes of 11 potential biomarkers engaged with lipid digestion, amino corrosive digestion, and kidney function related digestion. Flavonoids apply their restorative impacts by various ways, including decreasing insulin resistance, enhancing gluconeogenesis, and shielding kidneys and islet cells from harm (Jiang et al., 2018)

A combination of *Scutellariae Radix* (SR) (Lamiacea) and *Coptis chinensis* (CC) (Ranunculaceae) extracts have been investigated for their antidiabetic activity using T2DM rats caused by HFD and a low dose of STZ. The levels of FBG, fasting insulin (FINS), TC, TG, LDL-C, and FFA levels in T2DM rodents were diminished from 51.1% to 85.4% after administration of the combined extract. UPLC-Q-TOF-MS/MS was utilized to produce metabolic profiles of urine and plasma from control and T2DM rodents before and after oral injection of the combined extracts. The metabolic profiles of urine and plasma were significantly improved in T2DM rodents by the combined extract. Around 14 biomarkers (five in urine and nine in plasma) were recognized. The consolidated concentrate viably alter metabolic pathway remembering purine digestion for T2DM rodents (Cui et al., 2018)

Introduction of Rehmannia glutinosa (RG) (Orobanchaceae), CC (Ranunculaceae) alone and their blend on high-fat-diet-actuated diabetes in mice could diminish the FBG level, improve the glucose tolerance, and recuperate the degrees of TG, TC, HDL-C, and LDL-C diabetic mice. The results of UPLC-Q-TOF-MS/MS and OPLS-DA on serum metabolomes uncovered that 79 ESI+ and 76 ESI−metabolites were altered by DM contrasted with the typical control. Heatmaps on these diabetes-related metabolites revealed that both CC and RG/CC were grouped nearer with the typical control, showing that they had the better antidiabetic impacts at the metabolite level. 15 of the differential metabolites in DM serum were clarified and their connected metabolic pathways were lipid digestion. This information proposed that RG and CC alone and in mix treatment had the antidiabetic movement in bringing down glycemia and improving lipid digestion (Qin et al., 2018).

In STZ-administered diabetic rats, extract of Moringa oleifera leaves creates a most extreme decrease to 47.86% in acute impact while, 44.5% in chronic effect when contrasted with control. The FBG, lipid profile, liver catalyst levels were altogether ($p < 0.05$) reestablished in both experimental models.

Orthosiphon stamineus (OS) (Lamiaceae)500 mg/kg bw is exposed to a ^{1}H-NMR based pee metabolomics device to distinguish the metabolic defensive system of OS in DM utilizing STZ-initiated T2DM in rodents. The sinensetin and rosmarinic acid levels in OS extract, as identified by HPLC evaluation technique, were 0.19% and 17.86%, separately. An aggregate of 15 urinary metabolites, whose levels changed fundamentally upon treatment were distinguished as the biomarkers of OS watery extract in diabetes recognized by ^{1}H-NMR including allantoin, choline, betaine, citrate, acetoacetate, creatine, creatinine, glutamate, hippurate, glucose, hydroxybutyrate, leucine, pyruvate, N-PAG, and taurine. PLS-DA on the ^{1}H-NMR information of the control group, diabetic, and OSE treated gatherings' rodent pee tests uncovered that the different OS watery extract has critical antidiabetic action tantamount to glibenclamide. A systematic metabolic pathways investigation distinguished that OS added to the antidiabetic action basically through controlling the Krebs cycle, glycolysis/gluconeogenesis, fat, and amino acid digestion (Azam et al., 2017).

The progressions of metabolites due organization of ethanol extract of Ipomoea aquatica (Convolvulaceae) in large diabetic-incited STZ rodents were recognized. The outcomes recommended 19 biomarkers with variable significance projections (VIP) above 0.5, which comprise creatine/creatinine, creatinine, citrate, glucose, carnitine, 2-oxoglutarate, hippurate, leucine, succinate, 1-methylnicotinamice (MNA), taurine, tryptophan, 3-hydroxybutyrate (3-HB), lysin. Ipomoea aquatica ethanolic remove increased metabolite levels, and diminishes some metabolite levels, including glucose and carboxyl corrosive (TCA) intermediates. This infers abilities of I. aquatica ethanolic extract in glycolysis, gut fauna and nicotinate metabolism, improving the glomerular filtration rate and lessening the β-oxidation rate. Nonetheless, the introduction of I. aquatica has a few downsides, for example, in lessening branched-chain amino acid (BCAA) formation pathways and lipid digestion. e, xallantoin, formiate, acetoacetate (AcAc) and dimethylamine (Sajak et al., 2017)

Wan et al. explored the preventive impacts of Vine tea *Ampelopsis grossedentata* (Vitaceae) on the irregularities of lipid and carbohydrate digestion in HFD-instigated diabetic rodents to elaborate the potential ways included. Their outcomes indicated that the introduction of vine tea decreased serum glucose levels and cholesterol. The outcome showed that plant tea induction increased xanthine proline, histidine, and threonine levels, yet diminished the degree of uric acid, hypoxanthine, allantoin, and inosine. Hence, it could turn around HFD-instigated purine debasement (Wan et al., 2017).

Antidiabetic potential of combination of four herbal extracts: Pueraria lobata (Fabaceae) *Scutellaria baicalensis* (Lamiaceae), CC (Ranunculaceae), and Glycyrrhiza uralensis (Fabaceae) was assessed in vivo utilizing streptozotocin-prompted diabetic rodents' model in a 4.95 g crude dose. Plasma tests were investigated utilizing UHPLC-Q-TOF-MS which uncovered various possible biomarkers. Two PCs addressed >70% of the all the variance in positive ion model and negative-ion model of Mass spectrometry.

Great partition among the three gatherings was gotten utilizing the metabonomics strategy. Pantothenic acid is one of the identified biomarkers where, it is urgent for the metabolism and formation of carbohydrates, amino acids, and lipids since it is needed for coenzyme-A (CoA) arrangement. Expanded levels of CoA in DM hearts in vivo may result from diminished degrees of insulin and the initiation of a synthetic pathway. Insulin has been discovered to be a solid inhibitor of CoA biosynthesis in confined perfused hearts. These biochemical changes are related to the metabolism of sphingolipids, acetyl coenzyme-A and fats, which could be valuable to additionally explore the attributes of STZ-initiated diabetes mellitus (DM) and the remedial system of activity of GGQLD. This metabonomic investigation could give a valuable beginning stage to elucidate the restorative impacts and instrument of activity of this decoction in DM (Zhang et al., 2016).

^{1}H-NMR-based metabolomics of the serum of STZ-actuated diabetic rodents after administration of ethanol concentrate of airborne pieces of Phyllanthus niruri (Phyllanthaceae) 250 and 500 mg/kg uncovered improvement of diabetic rodents towards the control and the concentrate diminished serum glucose and improved the fat profile in obese diabetic rats. From ^{1}H-NMR spectroscopic information the extract of P. niruri contained different bioactive metabolites, including ellagic corrosive, epicatechin, catechin, quercetin rhamnoside, quercetin, chlorogenic acid, phyllanthin, and hypophyllanthin.

13.3 *IN VITRO* STUDIES

Umar et al. used Fourier-transform infrared (FTIR) spectra and UHPLC-Q-Orbitrap-HRMS profiling to discriminate the plant parts, geographical origin, and metabolic profile of *Curculigo orchioides* and *C. latifolia* (Hypoxidaceae) that contribute to their α-glucosidase inhibition activity *in vitro*. The best α-glucosidase antagonostic activity is presented by *C. latifolia* from a high altitude region with IC_{50} value of 155.25 ± 0.68 mg/mL extract. Partial least squares (PLS) examination has recognized a few useful gatherings, for example, O–H, C=O, C–C, C–H, C–O, and substance compounds, unknown-85 and unknown-185, that are destined to add to the cell reinforcement and α-glucosidase inhibitory exercises (Umar et al., 2021). The metabolomic studies examined were led by ^{1}H-NMR and UHPLC-MS to recognize and evaluate the metabolites in *Phyllanthus acidus* leaves (Phyllanthaceae) responsible for the critical α-glucosidase antagonistic activity with IC_{50} estimation of 6.29, 1.53 μg/mL. Around 38 metabolites were probably distinguished by ^{1}H-NMR and 39 by UHPLC-MS. The incomplete least-square investigation (PLS) uncovered the metabolites that added to α-glucosidase inhibitory movement, including choline, arginine, phyllanthusol B, valine, phyllanthin, quercetin, myricetin, kaempferol, lupeol acetic acid derivation, malic, gallic, citrus, 4-aminobutyric, quinic, cinnamic, ellagic, and chlorogenic acids.(Ghafar et al., 2020).

The hydromethanolic concentrates of leaves of *Tetracera scandens* (Dilleniaceae) were assessed for its α-glucosidase inhibitory action and to describe the metabolites answerable for such movement through GC/MS metabolomics. The methanol extricates showed strong α-glucosidase inhibitory potentials In addition, the set up OPLS multivariate model distinguished

16 metabolites to be answerable for the α-glucosidase inhibitory action of T. scandens that have a place with have a place with various substance classes,

Sorbus alnifolia (Siebold and Zucc.) K. Koch (Rosaceae) was tested for its α-amylase and α-glucosidas inhibition and it showed significant inhibition with IC_{50} values of 4.6 ± 0.4 and 0.9 ± 0.02. The variation in the metabolites in the extract was investigated using ^{1}H-NMR metabolomics approach and PCA (Broholm et al., 2019).

The secondary metabolites profile of three traditionally used *Ziziphus; Z. jujube, Z. spinachristi* and *Z. mauritiana* leaf specimens (Rhamnaceae) was investigated *via* UHPLC/PDA/ESI-MS. Results of PCA and HCA revealed identical metabolite semblance among *Z. spinachristi* and *Z. mauritiana* leaf samples found rich in saponins and different from those of *Z. jujuba* where quercetin-3-O-(2-pentosyl)-rhamnoside is most abundantly found. Further, *in vitro* antidiabetic assays by α-glucosidase inhibition discovered for *Z. spinachristi* and *Z. mauritiana* high effects as paralleled to *Z. jujuba* (27.47 ± 2.30, 24.65 ± 1.97 and 5.99 ± 1.62 mg ACAE/g extract, respectively) and in correlation with their metabolites range (Sakna et al., 2019).

Different fractions of *Clinacanthus nutans* Lindau (Acanthaceae) leaves evaluated for the α-glucosidase antagonostic activity with identification of the metabolites responsible for the bioactivity using GC/MS analysis. The highest activity was seen for hexane fraction, which was comparable or even better to that obtained for quercetin, the positive control and the methanol extract exhibited the lowest activity. Moreover GC/MS data generated a suitable partial least square (PLS) model resulting in 11 bioactive compounds, namely phytol, hexadecanoic acid (methyl ester), palmitic acid, 1-monopalmitin, pentadecanoic acid, stigmast-5-ene, heptadecanoic acid, glycerol monostearate, 1-linolenoylglycerol, alpha-tocospiro B, and stigmasterol.

Aqueous extract of *Moringa oleifera* leaf remarkably repressed α-glucosidase and α-amylase activities besides an improved glucose uptake and glucose tolerance rate in yeast cell. The IC_{50} values against α-glucosidase and α-amylase were 33.4 and 52.5 μg/mL, respectively. (Khan et al., 2017).

The PLS analysis helped to assess the association between metabolites of germinated rice extract (GRE) and its biological activity. Among the metabolites which could be accountable for the higher α-glucosidase inhibitory activity are γ-oryzanol, α-linolenic acid, α-tocopherol, GABA, glutamic acid, α-aminobutyric acid, leucine, 3-hydroxybutyric acid, hydroxyl-L-proline, 2, 3-butanediol, fatty acids, fumaric acid, vanillic acid, valine, and phenylalanine. Thereby, it might be suggested that the metabolites play a vital role towards associated with these GRE (Pramai et al., 2017).

TABLE 13.1 Summary of the Metabolomics Studies of Antidiabetic Plants

Plant (Family)	Extract	Analytical Method	Type of Study	Dose/IC_{50}	Statistical Method	References
Phyllanthus acidus (Phyllanthaceae)	Ethanol extract	^{1}H-NMR UHPLC-MS	*In vitro* α-glucosidase inhibition	IC_{50} = 6.29, 1.53 μg/mL	PLS	Ghafar et al. (2020)
Tetracera scandens (Dilleniaceae)	100%, 80%, 60% and 40% methanol extracts	GC/MS	*In vitro* α-glucosidase inhibition	IC_{50} values ranged from 19.54 ± 2.5 μg/mL to 23.46 ± 3.91 μg/mL	PCA OPLS-DA	Nokhala et al. (2020)
Paederia foetida (Rubiaceae)	Chloroform extract	GC/MS	*In vitro* α-amylase and α-glucosidase inhibition	IC_{50} = 600.287 ± 0.06 and 1349.01 ± 0.01 μg/mL	PCA OPLS-DA	Tan et al. (2019)
Moringa oleifera (Moringaceae)	Aqueous extract	GC/MS	*In vitro* α-amylase and α-glucosidase inhibition	IC_{50} values against α-amylase and α-glucosidase were 52.5 and 33.4 μg/mL, respectively	PCA OPLS-DA	Khan et al. (2017)
Melicope lunu-ankenda (Rutaceae)	Ethanol extract	^{1}H-NMR	*In vitro* α-glucosidase inhibition	IC_{50} = 37 μg/mL	OPLS-DA	Al-Zuaidy et al. (2017)
Sorbus alnifolia (Rosaceae)	Aqueous extract	^{1}H-NMR	*In vitro* α-amylase and α-glucosidase inhibition	IC_{50} = 4.6 ± 0.4 and 0.9 ± 0.02 μg/mL	PCA	Broholm et al. (2019)

TABLE 13.1 *(Continued)*

Plant (Family)	Extract	Analytical Method	Type of Study	Dose/IC_{50}	Statistical Method	References
Z. spinachristi, *Z. mauritiana* and *Z. jujube* (Rhamnaceae)	Methanol extract	UHPLC/PDA/ESI-MS	*In vitro* α-glucosidas inhibition	27.47 ± 2.30, 24.65 ± 1.97 and 5.99 ± 1.62 mg ACAE/g extract	PCA HCA	Sakna et al. (2019)
Curculigo orchioides and *C. latifolia* (Hypoxidaceae)	Ethanol extract	FTIR spectra and UHPLC-Q-Orbitrap-HRMS	*In vitro* α-glucosidas inhibition	IC_{50} = 155.25 ± 0.68 mg/mL extract	PCA	Umar et al. (2021)
Clinacanthus nutans (Acanthaceae)	Hexane extract	GC/MS	*In vitro* α-glucosidas inhibition	IC_{50} = 3.05 μg/mL	PLS-DA	Murugesu et al. (2018)
Berberis vernae (Berberidaceae)	Aqueous decoction	^{1}H NMR	*In vivo* High fat diet and STZ-induced T2DM	0.84 g/kg b.w.	PCA PLS-DA OPLS-DA	Li et al. (2020)
Lycium barbarum (Solanaceae)	Aqueous extract	UPLC-Q-TOF/MS	*In vivo* High fat diet and STZ-induced T2DM	1.04 g/kg 2.08 g/kg	PCA OPLS-DA	Zhao et al. (2020)

TABLE 13.1 *(Continued)*

Plant (Family)	Extract	Analytical Method	Type of Study	Dose/IC_{50}	Statistical Method	References
Orthosiphon stamineus(Lamiaceae)	Aqueous extract	^{1}H-NMR	*In vivo* STZ-induced T2DM	500 mg/kg bw	PLS-DA	Azam et al. (2017)
Scutellariae Radix (Lamiacea) and *Coptis chinensis* (Ranunculaceae)	Aqueous extract	UPLC-Q-TOF-MS/MS	*In vivo* High fat diet and STZ-induced T2DM	6.3 g/kg 12.6 g/kg	PCA OPLS-DA	Cui et al. (2018)
Rehmannia glutinosa (Orobanchaceae), *Coptis chinensis* (Ranunculaceae)	Aqueous extract	UPLC-Q-TOF-MS/MS	*In vivo* High fat diet-induced T2DM	10 g/kg/day	PCA OPLS-DA	Qin et al. (2018)
Moringa oleifera (Moringaceae)	Aqueous extract	GC/MS	*In vivo* High fat diet and STZ-induced T2DM	100 mg/kg	PCA OPLS-DA	Khan et al. (2017)
Melicope lunu-ankenda (Rutaceae)	Ethanol extract	^{1}H-NMR	*In vivo* High fat diet and STZ-induced T2DM	200 and 400 mg/kg BW	PCA	Al-Zuaidy et al. (2017)
Melastoma Dodecandrum (Melastomataceae)	Ethanol extract	UHPLC/QTOF-MS	*In vivo* High fat diet and streptozotocin-induced T2DM	600 mg/kg	PCA OPLS-DA	Weng et al. (2019)

TABLE 13.1 *(Continued)*

Plant (Family)	Extract	Analytical Method	Type of Study	Dose/IC_{50}	Statistical Method	References
Trigonella foenum-graecum Linn. (Fabaceae)	Flavonoids	UHPLC/ QTOF-MS	*In vivo* STZ-induced T2DM	0.5 g fenugreek flavonoids in 10 mL of water intragastrically (10 mL/ kg/day)	PCA OPLS-DA	Jiang et al. (2018)
Lycium barbarum (Solanaceae)	Polysaccharides	GC/MS	*In vivo* High fat diet and STZ-induced T2DM	50, 100 and 200 mg/kg/day	PCA OPLS-DA	Xia et al. (2019)
Pueraria lobata (Fabaceae) *Scutellariabaicalensis* (Lamiaceae), *CoptisChinensis* (Ranunculaceae), and *Glycyrrhiza uralensis* (Fabaceae)	Decoction	UHPLC/ PDA/ ESI-MS	*In vivo* STZ-induced T2DM	4.95 g crude drug/kg body weight	PCA	Zhang et al. (2016)
Phyllanthus niruri (Phyllanthaceae)	Ethanol extract	^{1}H-NMR	*In vivo* STZ-induced T2DM	250 and 500 mg/kg bw	PCA HCA PLS-DA	Mediani et al. (2016)
Ipomoea aquatica (Convolvulaceae)	Ethanol extract	^{1}H-NMR	*In vivo* High fat diet and STZ-induced T2DM	150, 250, and 500 mg/ kg bw	PCA PLS-DA VIP	Sajak et al. (2017)

TABLE 13.1 *(Continued)*

Plant (Family)	Extract	Analytical Method	Type of Study	Dose/IC_{50}	Statistical Method	References
Ampelopsis grossedentata (Vitaceae)	Aqueous extract	LC-MS	*In vivo* High fat diet-induced T2DM	500 mg/L or 2,000 mg/L	PLS-DA	Wan et al. (2017)
Morus multicaulis (Moraceae)	The bark powder were weighed and mixed with standard diet.	GC/MS	*In vivo* High fat diet and STZ-induced T2DM	5, 10, 20%	PCA	Qiu et al. (2019)
Ficus deltoidea (Moraceae)	Ethanol extract	^{1}H-NMR	*In vivo* STZ-Nicotinamide-induced T2DM	250 mg/kg	PCA PLS-DA	Noor et al. (2020)

Melicope lunu-ankenda (ML) (Rutaceae) leaf extract was investigated *in vitro* for its inhibition of α-glucosidase enzyme and the results revealed significant inhibition by the ethanolic extract of ML with IC_{50} value of 37 μg/mL and evaluated *in vivo* with high doses of 400 mg/Kg of the plant extract and it witnessed a great decrement by about 62.75% in FBG. The metabolites responsible have been assessed effectively utilizing ^{1}H-NMR-based metabolomics approaches.

The prominent bioactive compounds characterized which attribute towards the characterization included: skimmianine, isorhamnetin, isorhamnetin, scopoletin, and melicarpinone. Another report by Al-Zuaidy et al. revealed *in vivo* antidiabetic activity in rats that the ML leaf extract at 400 mg/kg significantly elevated both insulin level as well as insulin sensitivity of diabetic rats, paralleled with declination in glucose and insulin resistance.

Significant decline in TG, cholesterol, and LDL have been observed post treatment. Fascinatingly, a significant increment in HDL was observed of the treated rats in addition to a declination in of liver enzymes activities. Furthermore, metabolomic studies also demonstrated that the plant extract significantly revoked the disturbances associated with carbohydrate metabolism, TCA cycle, fat, and amino acid metabolism (Table 13.1) (Al-Zuaidy et al., 2017).

KEYWORDS

- **biomarkers**
- **diabetes mellitus**
- **herbal medicine**
- **metabolomics**
- **streptozotocin**

REFERENCES

A.D. Association, (2015). *Standards of Medical Care in Diabetes—2015: Summary of Revisions*. Am Diabetes Assoc.

Abd, G. S. Z., Mediani, A., Maulidiani, M., Rudiyanto, R., Ghazali, H. M., Ramli, N. S., & Abas, F., (2020). Complementary NMR-and MS-based metabolomics approaches reveal the correlations of phytochemicals and biological activities in *Phyllanthus acidus* leaf extracts. *Food Research International, 136,* 109312.

Afrisham, R., Aberomand, M., Ghaffari, M. A., Siahpoosh, A., & Jamalan, M., (2015). Inhibitory effect of *Heracleum persicum* and Ziziphus jujuba on activity of alpha-amylase. *J. Bot., 2015.*

Ahmadian, M., Suh, J. M., Hah, N., Liddle, C., Atkins, A. R., Downes, M., & Evans, R. M., (2013). PPARγ signaling and metabolism: The good, the bad and the future, *Nat. Med., 19*, 557–566.

Al-Zuaidy, M. H., Mumtaz, M. W., Hamid, A. A., Ismail, A., Mohamed, S., & Razis, A. F. A., (2017). Biochemical characterization and 1 H NMR based metabolomics revealed *Melicope lunu-ankenda* leaf extract a potent antidiabetic agent in rats. *BMC Complementary and Alternative Medicine, 17*(1), 1–17.

Azam, A. A., Pariyani, R., Ismail, I. S., Ismail, A., Khatib, A., Abas, F., & Shaari, K., (2017). Urinary metabolomics study on the protective role of *Orthosiphon stamineus* in Streptozotocin-induced diabetes mellitus in rats via 1 H NMR spectroscopy. *BMC Complementary and Alternative Medicine, 17*(1), 1–13.

Bailey, C. J., & Day, C., (1989). Traditional plant medicines as treatments for diabetes. *Diabetes Care, 12*, 553–564.

Bischoff, H., (1995). The mechanism of alpha-glucosidase inhibition in the management of diabetes. *Clin. Invest. Med., 18*, 303–311.

Broholm, S. L., Gramsbergen, S. M., Nyberg, N. T., Jäger, A. K., & Staerk, D., (2019). Potential of Sorbus berry extracts for management of type 2 diabetes: Metabolomics investigation of 1H NMR spectra, α-amylase and α-glucosidase inhibitory activities, and *in vivo* anti-hyperglycemic activity of *S. norvegica*. *Journal of Ethnopharmacology, 242*, 112061.

Cui, X., Qian, D. W., Jiang, S., Shang, E. X., Zhu, Z. H., & Duan, J. A., (2018). Scutellariae radix and *Coptidis rhizoma* improve glucose and lipid metabolism in T2DM rats via regulation of the metabolic profiling and MAPK/PI3K/Akt signaling pathway. *International Journal of Molecular Sciences, 19*(11), 3634.

Dunn, W. B., Goodacre, R., Neyses, L., & Mamas, M., (2011). Integration of metabolomics in heart disease and diabetes research: Current achievements and future outlook. *Bioanalysis, 3*, 2205–2222.

Gerstein, H. C., Miller, M. E., Byington, R. P., Goff, D. C., Bigger, J. T., Buse, J. B., Cushman, W. C., et al., (2008). Effects of intensive glucose lowering in type 2 diabetes. *N. Engl. J. Med., 358*, 2545–2559.

Hu, F. B., Satija, A., & Manson, J. E., (2015). Curbing the diabetes pandemic: The need for global policy solutions. *JAMA, 313*, 2319–2320.

Hundal, R. S., Krssak, M., Dufour, S., Laurent, D., Lebon, V., Chandramouli, V., Inzucchi, S. E., et al., (2000). Mechanism by which metformin reduces glucose production in type 2 diabetes. *Diabetes, 49*, 2063–2069.

Jiang, W., Si, L., Li, P., Bai, B., Qu, J., Hou, B., Zou, H., et al., (2018). Serum metabonomics study on antidiabetic effects of fenugreek flavonoids in streptozotocin-induced rats. *Journal of Chromatography B, 1092*, 466–472.

Kaddurah-Daouk, R., Weinshilboum, R. M., & P. R. Network, (2014). Pharmacometabolomics: Implications for clinical pharmacology and systems pharmacology. *Clin. Pharmacol. Ther., 95*, 154–167.

Kaddurah-Daouk, R., Weinshilboum, R., & P. R. Network, (2015). Metabolomic signatures for drug response phenotypes: Pharmacometabolomics enables precision medicine. *Clin. Pharmacol. Ther., 98*, 71–75.

Khan, W., Parveen, R., Chester, K., Parveen, S., & Ahmad, S., (2017). Hypoglycemic potential of aqueous extract of *Moringa oleifera* leaf and *in vivo* GC-MS metabolomics. *Frontiers in Pharmacology, 8,* 577.

Li, Q., Zhao, C., Zhang, Y., Du, H., Xu, T., Xu, X., Zhang, J., et al., (2020). 1H NMR-based metabolomics coupled with molecular docking reveal the antidiabetic effects and potential active components of *Berberis vernae* on type 2 diabetic rats. *Frontiers in Pharmacology, 11,* 932.

Lu, J., Xie, G., Jia, W., & Jia, W. (2013). Metabolomics in human type 2 diabetes research. *Front. Med. (Lausanne), 7,* 4–13.

Mediani, A., Abas, F., Maulidiani, M., Khatib, A., Tan, C. P., Ismail, I. S., Shaari, K., et al., (2016). Metabolic and biochemical changes in streptozotocin-induced obese-diabetic rats treated with *Phyllanthus niruri* extract. *Journal of Pharmaceutical and Biomedical Analysis, 128,* 302–312.

Murugesu, S., Ibrahim, Z., Ahmed, Q. U., Nik, Y. N. I., Uzir, B. F., Perumal, V., Abas, F., et al., (2018). Characterization of α-glucosidase inhibitors from *Clinacanthus nutans* Lindau leaves by gas chromatography-mass spectrometry-based metabolomics and molecular docking simulation. *Molecules, 23*(9), 2402.

Nicholson, J. K., & Lindon, J. C., (2008). Metabonomics. *Nature, 455,* 1054–1056.

Nokhala, A., Siddiqui, M. J., Ahmed, Q. U., & Ahamad, B. M. S., (2020). Investigation of α-glucosidase inhibitory metabolites from *Tetracera scandens* leaves by GC–MS metabolite profiling and docking studies. *Biomolecules, 10*(2), 287.

Noor, H. S. M., Ismail, N. H., Kasim, N., Mediani, A., Zohdi, R. M., Ali, A. M., Mat, N., & Al-Mekhlafi, N. A., (2020). Urinary metabolomics and biochemical analysis of antihyperglycemic effect of *Ficus deltoidea* jack varieties in streptozotocin-nicotinamide–induced diabetic rats. *Applied Biochemistry and Biotechnology*, 1–21.

Ota, & Ulrih, N. P., (2017). An overview of herbal products and secondary metabolites used for management of type two diabetes. *Front. Pharmacol., 8,* 436.

Palanisamy, S., Yien, E. L. H., Shi, L. W., Si, L. Y., Qi, S. H., Ling, L. S. C., Lun, T. W., & Chen, Y. N., (2018). Systematic review of efficacy and safety of newer antidiabetic drugs approved from 2013 to 2017 in controlling HbA1c in diabetes patients. *Pharmacy, 6,* 57.

Pramai, P., Hamid, N. A. A., Mediani, A., Maulidiani, M., Abas, F., & Jiamyangyuen, S., (2017). Metabolite profiling, antioxidant, and a-glucosidase inhibitory activities of germinated rice: Nuclear-magnetic-resonance-based metabolomics study. *Journal of Food and Drug Analysis, 30*(1), e1–1.

Qin, Z., Wang, W., Liao, D., Wu, X., & Li, X. E., (2018). UPLC-Q/TOF-MS-based serum metabolomics reveals hypoglycemic effects of *Rehmannia glutinosa*, *Coptis chinensis* and their combination on high-fat-diet-induced diabetes in KK-Ay mice. *International Journal of Molecular Sciences, 19*(12), 3984.

Qiu, F., & Zhang, Y. Q., (2019). Metabolic effects of mulberry branch bark powder on diabetic mice based on GC-MS metabolomics approach. *Nutrition & Metabolism, 16*(1), 1–16.

Sajak, A. A. B., Mediani, A., Dom, N. S. M., Machap, C., Hamid, M., Ismail, A., Khatib, A., & Abas, F., (2017). Effect of Ipomoea aquatica ethanolic extract in streptozotocin (STZ) induced diabetic rats via 1H NMR-based metabolomics approach. *Phytomedicine, 36,* 201–209.

Sakna, S. T., Mocan, A., Sultani, H. N., El-Fiky, N. M., Wessjohann, L. A., & Farag, M. A., (2019). Metabolites profiling of Ziziphus leaf taxa via UHPLC/PDA/ESI-MS in relation to their biological activities. *Food Chemistry, 293,* 233–246.

Shi, J., Cao, B., Wang, X. W., Aa, J. Y., Duan, J. A., Zhu, X. X., Wang, G. J., & Liu, C. X., (2016). Metabolomics and its application to the evaluation of the efficacy and toxicity of traditional Chinese herb medicines. *J. Chromatogr. B, 1026*, 204–216.

Tan, D. C., Kassim, N. K., Ismail, I. S., Hamid, M., & Ahamad, B. M. S., (2019). Identification of antidiabetic metabolites from *Paederia foetida* L. twigs by gas chromatography-mass spectrometry-based metabolomics and molecular docking study. *BioMed Research International, 2019.*

Thulé, P. M., & Umpierrez, G., (2014). Sulfonylureas: A new look at old therapy. *Curr. Diab. Rep., 14*, 473.

Umar, A. H., Ratnadewi, D., Rafi, M., & Sulistyaningsih, Y. C., (2021). Untargeted metabolomics analysis using FTIR and UHPLC-Q-orbitrap HRMS of two curculigo species and evaluation of their antioxidant and α-glucosidase inhibitory activities. *Metabolites, 11*(1), 42.

Wan, W., Jiang, B., Sun, L., Xu, L., & Xiao, P., (2017). Metabolomics reveals that vine tea (*Ampelopsis grossedentata*) prevents high-fat-diet-induced metabolism disorder by improving glucose homeostasis in rats. *PloS One, 12*(8), e0182830.

Weng, J., Zhou, J., Liang, L., & Li, L., (2019). UHPLC/QTOF-MS-based metabolomics reveal the effect of *Melastoma dodecandrum* extract in type 2 diabetic rats. *Pharmaceutical Biology, 57*(1), 807–816.

Wild, S., Roglic, G., Green, A., Sicree, R., & King, H., (2004). Global prevalence of diabetes: Estimates for the year 2000 and projections for 2030. *Diabetes Care, 27*, 1047–1053.

Xia, H., Tang, H., Wang, F., Yang, X., Wang, Z., Liu, H., Pan, D., et al., (2019). An untargeted metabolomics approach reveals further insights of *Lycium barbarum* polysaccharides in high fat diet and streptozotocin-induced diabetic rats. *Food Research International, 116*, 20–29.

Zhang, Q., Xu, G., Li, J., Guo, X., Wang, H., Li, B., Tu, J., & Zhang, H., (2016). Metabonomic study on the plasma of streptozotocin-induced diabetic rats treated with Ge gen qin Lian decoction by ultra-high performance liquid chromatography-mass spectrometry. *Journal of Pharmaceutical and Biomedical Analysis, 120*, 175–180.

Zhao, X. Q., Guo, S., Lu, Y. Y., Hua, Y., Zhang, F., Yan, H., Shang, E. X., ET AL., (2020). *Lycium barbarum* L. leaves ameliorate type 2 diabetes in rats by modulating metabolic profiles and gut microbiota composition. *Biomedicine & Pharmacotherapy, 121,* 109559.

Zimmet, P., Alberti, K., & Shaw, J., (2001). Global and societal implications of the diabetes epidemic. *Nature, 414*, 782–787.

Zimmet, P., Alberti, K., & Shaw, J., (2001). Global and societal implications of the diabetes epidemic. *Nature, 414*(6865), 782–787; (b) Hasanpour, M., Iranshahy, M., & Iranshahi, M., (2020). The application of metabolomics in investigating antidiabetic activity of medicinal plants. *Biomedicine & Pharmacotherapy, 128,* 110263.

CHAPTER 14

Zinc Finger Nuclease Based Genome Editing in Enhancement of Antidiabetic Activity of Plants: Prospective Applications

PARAMA MAJUMDAR,[1] GEETANJALI HARALE,[2] SUMAN GANGER,[3] and SHEETAL PARDESHI[4]

[1]*Department of Botany, Hislop College, Nagpur, Maharashtra, India, E-mail: parama.majumdar@gmail.com*

[2]*Department of Biotechnology, Vivekanand Education Society's College of Arts, Science, and Commerce, Mumbai, Maharashtra, India, E-mail: haralegeetanjali@gmail.com*

[3]*Department of Microbiology, Vivekanand Education Society's College of Arts, Science, and Commerce, Mumbai, Maharashtra, India, E-mail: sumanganger@gmail.com*

[4]*Department of Microbiology, PES Modern College of Arts, Science, and Commerce (Autonomous), Shivajinagar, Pune, Maharashtra, India, E-mail: sheetalppardeshi@gmail.com*

ABSTRACT

Zinc finger nucleases (ZFN) are synthetic, chimeric, endonuclease systems that can be programmed to target specific DNA sequences to induce mutagenesis or replacement of genes. They are two-component systems of which the first component is a non-specific cleavage unit belonging to *Fok I* restriction endonuclease and the second component is a repeating unit of Zn

Antidiabetic Potential of Plants in the Era of Omics. Deepu Pandita, PhD, Anu Pandita, MSc, & Chander Bhanu, MBBS (Eds.)

finger domains. Each Zn finger can identify a unique tandem sequence of 3 nucleotides on the target and thus the modular system of Zn-fingers can be tailor-made for different targets across the genome. They introduce double-stranded breaks (DSB) in the DNA, which significantly improve specific transformation rate. Such customizable nucleases expand the horizons of gene editing particularly aimed at quality improvement and metabolome manipulation. Antidiabetic potential of plants is a property that is innate to its secondary metabolite composition and enhancement of this activity by a high probability gene-editing method like ZFN, is an area continuously being explored. In this chapter, the authors intend to delve into the prospects of ZFN-based genome editing for enhancement of antidiabetic property of plants.

14.1 INTRODUCTION

Plants host an imposing chemical diversity, that is primarily attributed by class of compounds called secondary metabolites. The role played by these compounds is crucial to the life of plants as they instill characteristics like fragrance of flowers, peculiarity of taste, pathogen resistance, anti-herbivory, amongst many others. It is also incredible to note that plants from different evolutionary lineages produce compounds meeting the same functional purpose in the plant, but may belong to an entirely different chemical class. Throughout the taxonomic hierarchy, plants produce indigenous blends of these chemicals, which can be provide taxonomic identification at family, class or even genus level. (Pichersky and Gang, 2000).

By 1988, NAPRALERT had already reported more than 88,000 secondary metabolites also called the phytochemicals, with an approximate increment of 4,000 new metabolites every year. Out of the 13,207 species of monocots and dicots, 239 species of gymnosperms and 506 species in all of pteridophytes, bryophytes, and lichens that were studied 37,528 secondary metabolites were reported belonging to a single type of biological active group like alkaloids. (Verpoorte, 1998). These secondary metabolites have also been of tremendous importance to mankind as the observed herbal capacities have all been linked to these bioactive compounds. Ancient medicinal systems around the world have heavily relied on medicines of plant origin (Wink, 2015). Contagious and non-contagious diseases have been treated with herbal medicines all around the world. Diabetes mellitus (DM) is one such disease which has become the most common non-communicable disease with an approximate 25% of the world population affected by it. Diabetes poses as a grim predicament of 21st century. The ethnobotanical records report a large number of plants with antidiabetic potential. Several Indian ethnobotanical records as well as Ayurveda, mention the use of plants in the treatment of various ailments (Arumugam et al., 2013) A large number of prescriptions record the use of plants for infectious as well as systemic diseases including diabetes and the medicinal attributes have now been credited to the secondary metabolites. (Salehi et al., 2019)

Plants have been extensively studied for their properties ranging from antidiabetic to α amylase inhibitory, to antihyperglycemic to hypoglycemic. Some phytochemicals have also been found to interact with insulin and stimulate the complex metabolic pathways associated with glucose metabolism. In a diabetic mouse model, phytochemicals like Morin found in *Morus alba, Maclura pomifera* and *M. tinctoria*, *Prunus dulcis* and *Chlorophora tinctoria* have been found to activating the Akt/eNOS pathway thereby rescuing endothelial dysfunction. (Taguchi et al., 2016). Extracts of plants like Accacianilotica, Allium cepa and Aloe vera have also been proven to be antidiabetic, antihyperglycemic, and Alpha-amylase inhibitors respectively (Salehi et al., 2019). Lupanine from *Lupinus* spp., enhances the secretion of insulin and boosts glucose homeostasis by impelling the K+ATP channels. It has also been found to interact with the insulin-related genes (García López et al., 2004). Kaempferol which is a low molecular weight favonol found in many medicinally acclaimed plants like *Equisetum* spp., *Sophora japonica*, and *Ginkgo biloba,* promote tissue sensitivity to insulin and conserves pancreatic β-cell mass (Alkhalidy et al., 2015)

With thousands of such bioactive compounds, still under research, the extrapolation of these numbers for all the plant species known opens an enormous and promising cue to the exploration of new bioactive compounds and also to the upcoming drug development programs in the field of diabetology.

Enhancement of antidiabetic activity by high probability molecular methods, is an area continuously being explored. Many of the specific biochemical pathways that lead to the production of secondary metabolites are yet to be elucidated. However, it is clearly understood that their synthesis is a part of a complex and branched, multi-enzyme pathway involving different enzymes for different plants and different classes of compounds (Pichersky and Gang, 2000). Further, by studies on *Arabidopsis*, it has been predicted that approximately 10% of the total genes are assigned to the cellular role of secondary metabolism. (Bevan et al., 1998). One of the strategies for both analysis and production of antidiabetic phytochemicals is crop improvement through approaches of genetic and metabolic intervention. Endeavors to use molecular-assisted approach in order to improve the quality traits and modify the phytochemical profile are rapidly evolving from the classical plant breeding. (Borrelli and Trono, 2016b)

The approaches to augment the production of desirable phytochemicals in plants can be broadly categorized into two groups based on the nature of intervention. The enhancement of production can be realized by the elicitation using suitable chemicals. This is the conventional method wherein suspension cultures are used to invoke the elicitor response triggering secondary metabolite production in the suspension. The second method to achieve overproduction of phytochemicals is by genetic intervention enhancing the basal level of phytochemical production. The latter is the point of consideration of this chapter. This chapter particularly deals with the molecular approach to enhance antidiabetic activity, with special reference to editing tool of Zinc Finger Nuclease.

14.2 GENETIC TOOLS FOR ENHANCING BIO-ACTIVE COMPONENTS IN PLANTS

The conventional method to harness the phytochemical potential of the medicinal plants involves suspension culture of the regarded plant part wherein the production of bioactive components has been triggered by broad range elicitors for secondary metabolism.

Developments in the field of plant tissue culture, along with the advancement of genetic engineering, have resulted in immense development in technologies producing phyto-nutraceuticals. Recent advances in the field of fermentation technology, enzymology, and molecular biology have the widened scope of manipulation and provided a viable source of important secondary metabolites. Genome manipulation via engineered viruses have yielded larger quantities of desired compounds, whereas transgenic plants that maintain high basal levels of desirable proteins without further external intervention have also been obtained. Such suspension cultures ensure largescale production of secondary metabolites as compared to traditional methods of plantation as it offers controlled supply of biochemical independent of plant availability (Hansen and Wright, 1999; Hussain et al., 2012).

The study of tools of genetic manipulation, to elevate the production of important phytochemicals, has received a global upthrust because of the efforts in investigation of the metabolic pathways responsible for the accumulation of secondary metabolites. Enormous research has been invested to establish a link between the secondary metabolite levels and the enzymes that regulate key steps in the biosynthetic pathways. Studies are also abreast to understand the genes encoding these enzymes. The identified genes can be used as markers and quality indicators in order to select improved genotypes achieved through various molecular breeding and biotechnological techniques. A lot of effort has been put in to clone these genes and identify the transcription factors and signal transduction mechanism which functions post elicitation in order to program the plant cells for enhanced production of required secondary metabolites.

Arabidopsis has been studied intently to understand the flavonoid synthesis pathway, its regulation and subcellular distribution. Studies in *Medicago* have led to the identification of *BANYLUS* (*BAN*) gene which encodes anthocyanidin reductase (ANR) that catalyzes the conversion of anthocyanidin to flavanol, which in turn is a precursor for tannin production (Xie et al., 2003) Analysis of expressed sequence tag (EST) has ushered the elucidation of biosynthetic pathways involved in the production of tannin in *Camelia*. The PCR-based cDNA subtraction compares isolated cDNAs with the mRNA populations obtained from two groups of plants. This technique has been used to analyze genes that are either overexpressed or solely expressed in one group of plants as compared to others. (Park et al., 2004).

Multiple public repositories serve to provide functional information for identification of candidate genes. Accessibility to next-generation sequencing technologies, public databases based on transcriptome analysis, whole transcriptome and genome sequencing methods have aided the exploration of genes expressed in plants in a tissue or organ-specific manner at a particular stage of growth or environmental conditions. Databases containing metabolite information like in PubChem, KEGG compound, etc.; information about the biochemical reaction, like the sources KEGG enzyme/KEGG reaction; and information about metabolic pathways exclusive to plants, viz. KEGG pathway are available. A number of genomes sequencing projects aiming to analyze cultivated plant genomes or their wild relatives are also underway. (Borrelli and Trono, 2016a)

Agrobacterium tumefaciens is a natural plant-pathogen thriving in soil and it uses a type IV secretion system (T4SS) to naturally transfer the Ti(or Ri) plasmid-borne T-DNA into host plant cells at the wounded sites and causes crown gall disease. Comprehensive knowledge on the mechanism of infection of *A. tumefaciens* has made it one of the most used plant transformation tools. The gene of interest is easily introduced to the host by inserting it in place of the oncogenes that located in the region of T-DNA of the plasmid. Thus, various types of binary vectors can be accomplished using *A. tumefaciens,* for genetic transformation in plants. *A. tumefaciens* has been used in *Arabidopsis*, to bring about stable and transient transformation and in spite of inconsistent efficiencies of transformation among the various accessions of *Arabidopsis*, the system continues to be the most prevalent technique (Hwang et al., 2017).

Agrobacterium was first studied in order to understand the pathogenic agent of diseases like crown gall, hairy root, and cane gall in plants. The Crown gall disease is attributed to the tumor-inducing (Ti) plasmid of *Agrobacterium tumefaciens* that enables transfer of some oncogenes to the host DNA. The oncogenes on the transferred-DNA (T-DNA) express to prompt neoplastic tumor-like cell growth. *Agrobacterium rhizogenes*, which is also a gram-negative pathogenic bacteria causes abnormal root formation in the host plants. This is also a manifestation of the oncogenes present on the of root inducing plasmid (Ri) of the bacteria (Anami et al., 2013; Hwang et al., 2017).

Manoeuvring the phytochemical profile of the plant by enhancing the production of desirable secondary metabolites while suppressing the undesirable ones, by metabolic engineering of endogenous pathways, altering the native genes, incorporation of genetic traits (genes) of new or non-native metabolic pathways in the plant has been one of the major service provided *Agrobacterium*-mediated transformation systems (Sinha et al., 2019).

Medicinal plant transformation is also aided by *A. rhizogenes* which causes profuse hairy root development by transformation with Ri (Root inducing) plasmid. This method of transformation is vital in hyperproduction of phytochemicals based in the roots of the plants (Muranaka and Saito, 2010).

In spite of the success of *A. tumefaciens* in dicot transformation, their efficiency in monocots, is an unrelenting lag. One of the attributing reasons for this poor efficiency is the absence of a natural phenolic compound acetosyringone in monocot defense mechanism. This compound is an inducer of *vir* genes that are required to successfully establish *Agrobacterium* infection. A future direction of study can be understanding this defense signaling mechanism in monocots. The use of efficient *A. tumefaciens* mediated transformation technique along with CRISPR technique is a growing trend which empowers genome editing to remove undesirable gene fragments like as antibiotic markers and dispensable foreign genes within the plant genome. This can further diminish the concerns of possible toxicity of the GM crops by entire T-DNA integration (Hwang et al., 2017).

14.3 HISTORICAL BACKGROUND OF GENOME EDITING TOOLS

Genome editing has been known to mankind since the 1980s. The pioneering work of Capecchi and Smithies on homologous recombination (HR) was amongst the first hopes towards synthetic biology with the development of controlled genome editing. The HR can be brought about artificially by a sequence of enzymatic reactions involving DNA topoisomerase Spo11 to catalyze double-stranded breaks (DSB) working along with Mre11 which is DSB repair protein, a DNA repair protein Rad50 and meiotic recombination protein NBS1 (Mohanta et al., 2017). Another method reported in early years of genome editing was Non-homologous end joining (NHEJ). However, these suffered drawbacks of uncertainty, less control over recombining segments and low frequency of recombination (Adli, 2018). Meganucleases, also called homing nucleases, are a family of enzymes found in every domain of life and are usually encoded in intron regions. Formation of hybrid nucleases using DNA binding and DNA cleavage domains from meganucleases and other proteins are well reported in genome editing since the last two decades (Silva et al., 2011). Construction of artificial Zinc finger nucleases (ZFNs) was reported in 1996 where the domain of zinc finger proteins (ZFPs) that binds to DNA was fused with DNA cleaving domain of FokI endonucleases, a type IIs restriction endonuclease from *Flavobacterium okeanokoites,* resulting in hybrid molecules that could bind to specific target sequence and thus could cleave in site-specific manner (Kim, Cha, and Chandrasegaran, 1996). ZFNs will be examined in detail in the ensuing segments of this chapter.

In 2008, Engler et al. introduced a type IIs restriction enzymes based 'Golden gate' technology for the creation of recombinant plasmids devoid of unwanted sequences (Engler, Kandzia, and Marillonnet, 2008). Transcription activator-like effector nucleases (TALENs) are another well-established technology for genome editing. The TALE proteins, discovered in *Xanthomonas,* were found to activate transcription of specific plant genes increasing susceptibility of plants to *Xanthomonas.* TALE proteins were found to bind to specific DNA sequences by recognizing individual bases and were then used for making chimeric nucleases by fusing with DNA binding domains with endonucleases (Nemudryi et al., 2014). The target sequences for action by TALENs are limited due to the requirement of thymidine residue at the first position (Doyle et al., 2012) and may fail *in vivo* signifying validation using first *in vitro* and then *in vivo* experimentation prior to use (Hwang et al., 2013). A detailed account of TALENs is discussed in a separate chapter of this book.

CRISPR (clustered regularly interspaced short palindromic repeat)/ Cas9 (CRISPR-associated protein 9) is the most recent technology in genome editing which can facilitate tailor-made constructs of genes through site-specific editing. The system was elucidated in *Streptococcus pyogenes*, where it acts as a part of adaptive immunity against viruses and plasmids, and has won Nobel prize in Chemistry in the year 2020 (Bortesi and Fischer, 2015; Jinek et al., 2012). In 2015, Zetsche et al. described a new Cas protein, Cpf1, with the ability to produce 5' 5 nucleotides overhang upon cleavage (Kanchiswamy et al., 2016) that could expand the CRISPR based genome editing (Zetsche et al., 2015). The use of the CRISPR system in genome editing in plants is well documented in literature (Liu et al., 2017; Mao et al., 2013). In a study on human pluripotent stem cell genome editing, CRISPR/Cas9 was reported to perform superior to TALENs (Ding et al., 2013). Nevertheless, the function of both TALENs and CRISPR/Cas9, are significantly affected by high order chromatin structure and thus the epigenetic context of the target regions (Chen et al., 2016).

Apart from the above well documented systems of genome editing, other tools such as cisgenesis, intragenesis, RNA interference (RNAi) and pentatricopeptide repeat proteins (PPRs) have been reported to be used in plants. Site as well as oligonucleotide-directed mutagenesis can be used for single nucleotide changes. Adenine base editors (ABEs), as the name indicates, can be used to edit A-T base pair to G-C base pair with the help of deoxyadeninedeaminase enzyme (Mohanta et al., 2017). Recently, prokaryotic Argonaute proteins are being tapped for genome editing potential (Hegge, Swarts, and van der Oost, 2018).

14.4 COMPONENTS AND MECHANISM OF ACTION OF ZFN

Zinc finger nucleases are hybrid artificial fusion proteins with DNA binding domain from zinc finger proteins (ZFP) and a DNA cleaving domain of FokI, (Bortesi and Fischer, 2015; Kim, Cha, and Chandrasegaran, 1996), a methylase domain (M), transcription activator domain (A) and transcription repressor domain (R) to form ZFN (Mohanta et al., 2017). Each zinc finger motif can bind one specific codon, and many such motifs can be linked together to give ZFP oligomer (Figure 14.1).

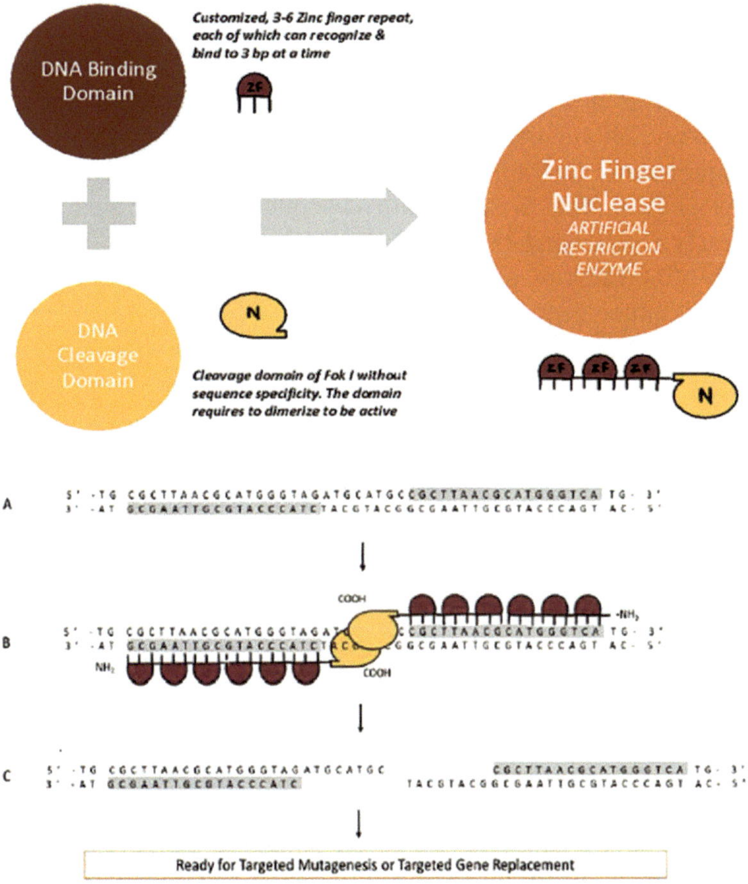

FIGURE 14.1 The programmable nuclease molecule is composed of the DNA binding domain, i.e., zinc finger (ZF; represented as maroon tri-pronged structure) and the DNA cleavage domain, i.e., Fok-I nuclease molecule (N; represented as the yellow lariat). Each ZF unit binds to a sequence of 3 bp on the DNA with the N-terminal oriented towards the 3' end of the strand. (A) To identify a certain sequence of interest a customized sequence of ZF linked to Fok-I Nuclease is used. (B) 2 ZFN units attach to the target site on opposite strands as dictated by their orientation. The Fok-I on the C-terminal of the two units dimerize and this dimerization triggers asymmetric cleavage of the DNA. (C) The double-stranded break thus generated is a suitable site for site-directed mutagenesis.

The DNA binding domain of ZFP contains approximately 30 amino acids with tandem repeats of Cys2-His2 that recognize and bind approximately 18 to 36 base pairs of DNA per cleavage site. The Cys2His2 zinc finger proteins contain tandem array of sequences (Tyr, Phe)-Xaa-Cys-Xaa2-4-Cys-Xaa3-Phe-Xaa5-Leu-Xaa2-His-Xaa3-5-His. This region of the protein binds zinc (II) ions forming a finger-like structural domain and hence the name. ZFPs associate with DNA by inserting the alpha helix region to the major groove of the DNA helix. The amino acids at positions –1 to +6 of this alpha-helix are crucial in sequence-specific interactions. Although the sequence specificity of ZNPs is complex phenomenon, it's shown to be modulated by changing some key amino acids in the sequence and it is possible to construct ZFPs binding to each of the 64 codons, however, such a designing is extremely challenging (Kim, Cha, and Chandrasegaran, 1996). Thus, it is the DNA binding domain of ZFN that may be customized to recognize the precise target sequence (Figure 14.1) on genome followed by the cleavage by nuclease leading to target specific gene editing (Cai and Yang, 2014; Kanchiswamy et al., 2016). The intervening linker sequences may also be modified to improve sequence specificity (Maggio and Gonçalves, 2015). The recognition site of FokI is a pentadeoxyribonucleotide sequence 5'-GGATG-3':5'-CATCC-3' and it cleaves approximately 9 to 13 nucleotides downstream to this site. The activity of FokI requires dimerization (Guha, Wai, and Hausner, 2017) and hence ZFNs are constructed as dimers whose DNA binding domain dictates the specificity of the reaction. ZFNs produce 5' overhangs of 4 to 5 nucleotides upon introducing DSB which makes it possible to insert foreign sequence which was treated using the same nuclease producing cohesive ends (Cristea et al., 2013; Kanchiswamy et al., 2016). The cleavage carried out by FokI nuclease domain happens in a context determined manner, i.e., it depends on the sequence composition of flanking regions and thus designing ZFNs is very challenging and time consuming. (Bortesi and Fischer, 2015; Sander et al., 2011; Maggio and Gonçalves, 2015). Upon introduction of DSB, a gene can be targeted for insertion at DSB site using HR or NHEJ making both rapid insertion or disruption of sequence possible. The hetero-dimerization of two variants of ZFNs created by site-directed mutagenesis of the overlapping dimerization regions of FokI cleavage domain was found to be an effective tool in controlling the genotoxicity of ZFNs as they displayed higher accuracy in binding target sequence (Miller et al., 2007; Maggio and Gonçalves, 2015). The other mode that has been reported to enhance the applicability of ZFN is by using base skipping linker sequences (Figure 14.2).

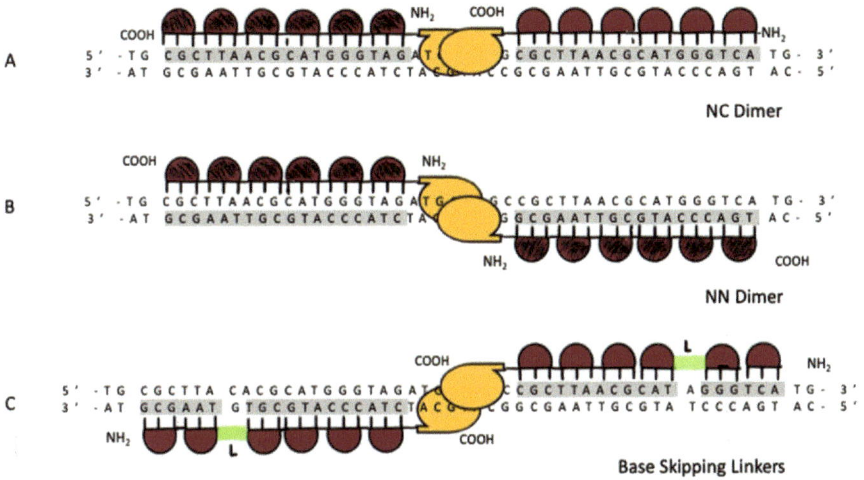

FIGURE 14.2 To enhance the target precision, diverse forms of ZFNs have been implemented using linkers that attach the Fok I to N-terminal of the zinc finger chain (represented as the shaded version of ZFN). (A) NC dimers: Use of one unit of ZFN with N-terminal Fok-I along with a canonical ZFN with C-terminal Fok-I allows the selection of a single strand of DNA as the target. (B) NN dimers: Use of 2 units of ZFN with N-terminal Fok-I leads to dimers with target selection that is a mirror image of the typical C-C dimer ZFN units. In order to further attune the attachment to the target sequence, linkers are added between the Zinc Finger chain. (C) Base skipping linkers: By using base skipping linker (*L*), usually between the zinc finger 2 and 3 or 4 and 5, the versatility of target binding is enhanced.

The linker molecules are the stretches of sequences that are inserted between adjacent individual zinc fingers. Use of these linker sequences allows for skipping such regions of target sequence for which high affinity zinc finger motifs are not available, thus, increasing the number of possible targets. Bioinformatics tools are also available to calculate the success of linkers *in silico* before actual experimentation. These linker sequences have also been used to change the dimer architecture of ZFN such as shown in. Use of appropriate linkers allows NC dimer (amino-terminal of one FokI domain interacting with carboxyl-terminal of other FokI domain) where left and right zinc finger motifs bind the same DNA strand of target dsDNA; and NN dimer (amino terminals of both FokI domains interact with each other) which results in inverse binding of zinc finger motifs as compared to classical design of CC dimer where carboxyl terminals of both FokI domains interact with each other (Figure 14.2). Such design rearrangements and reversals have resulted in increasing number of target sequences that can be processed using ZFNs (Anand et al., 2013; Paschon et al., 2019).

One major concern with ZFNs and other genome editing technologies is their delivery inside the targeted cell/tissue. The delivery systems developed so far can be broadly categorized into two groups, that is, viral vector-based and non-viral ones. Viral vector-based delivery includes the use of adeno-associated virus (AAV), integrase-defective lentivirus (IdLV) and adenoviruses. The non-viral tools include electroporation, nanomaterial-based delivery, use of cell-penetrating peptides (CPP), microfluidic technologies (Yin, Kauffman, and Anderson, 2017) and lipid-based delivery (Zhang, Zhang, and Yin, 2019). In the case of plants, the use of *Agrobacterium* for delivery or transformations of protoplasts have been the methods of choice (Yin, Gao, and Qiu, 2017). While using lentiviruses, a cotransduction by two lentiviral vectors containing monomers of functional heterodimer is followed (Lombardo et al., 2007; Pelascini et al., 2013). Delivery of ZFNs via receptor-mediated endocytosis was reported by conjugating ZFN with transferrin (Chen et al., 2013) and electroporation is also documented to be useful (Sahin, Karikó, and Türeci, 2014).

As mentioned earlier, the efficiency of DNA binding and cleavage by ZFNs is target contextual, that is, the upstream and downstream sequences to the target site as well as interaction of adjacent zinc fingers with DNA sequence making it very difficult to design and more difficult to predict the efficiency. To aid in these designing problems, *in-silico* methods such as oligomerized pool engineering (OPEN) and context-dependent assembly (CoDA) are available in open platforms to design sequence-specific DNA binding domains of ZFNs. The CoDA method evaluates context-dependent effects of the adjacent zinc fingers to give better prediction of specificity and homologous sequence binding thus giving closer *in silico* estimate of functional efficiency (Sander et al., 2011). Certain *in-silico* prediction tools such as PROGNOS have been developed to predict the possible off-target editing events by ZFNs.

Although a significant amount of literature indicates use of CRISPR/Cas9 in the editing of plant genome; there are reports of successful implementation of ZFNs in *Arabidopsis*, *Petunia*, tomato, maize, *Nicotiana*, rice, rapeseed, soybean, fig, and apple (Zhang et al., 2018). Artificial zinc-finger protein (AZP)-staphylococcal nuclease (SNase), (AZP-SNase), was found effective in antiviral therapies (Guha, Wai, and Hausner, 2017). Use of ZFNs in single nucleotide mutants as well as codon replacements in some systems (Melanie et al., 2012), antiviral therapy (Didigu et al., 2014; Tebas et al.,

2014), genome editing in nematodes (Wood et al., 2011) is known. This chapter will focus on the applications of ZFNs in enhancement of antidiabetic activity of plants.

14.5 ZFN BASED STRATEGIES OF PLANT IMPROVEMENT WITH RESPECT TO ANTIDIABETIC ACTIVITY

For medicinal or nutraceutical applications of phytochemicals, the biosynthetic pathways are steered by targeting explicit compounds involved in a biosynthetic pathway, which is achieved by up/down-regulation or silencing of definite pathway genes, genes of competing or cooperative pathways, or transcription factors, as well as the introduction of heterologous genes to enable the development of non-native compounds (Wilson and Roberts, 2014).

However, the implementation of specific gene targeting in higher plants is yet to reach its desired efficacy due to the high frequency of unsolicited recombination of the DNA at non-homologous sites. The high number of these recombinations give rise to a practical impediment in the selection of the individuals that have undergone HR. (Wright et al., 2005). An effective method of enhancing the probability of HRs resulting in successful transformation is to create a break at a specific target site within the chromosome. Custom designed Zinc finger nucleases are an effective strategy for achieving this.

The antidiabetic property of the plants is an interplay of multiple genes that drive the complex secondary metabolic pathway of the plants. The involvement of ZFN to manipulate the activity of precise genes that are associated with a specific phytochemical in the desired plants is the key to enhancing their antidiabetic activity. If a plant has an established genome for the reference of ZFN production, phenotypes of high phytochemical production can be identified through quantitative trait loci (QTL) followed by genome-wide association studies (GWAS). Yet, the characteristic of secondary metabolism is its species-specific nature and thus, for plants that are not model and the genome information is not available for reference, application of ZFN has to rely on the identification of specific genes by transcriptome data and construction of metabolic pathway databases. Therefore, in spite of the unavailability of genomes sequences of medicinal plants, large EST or RNA-Seq compendiums are available which can aid the understanding of gene function in specific secondary metabolite synthesis. (Wilson and Roberts, 2014)

Using the RNA-seq data, a database of metabolic pathways for *Catharanthus roseus* called CathaCyc, has been generated. *C. roseus* is a potent medicinal plant with established anti-cancer and antidiabetic properties. (Salehi et al., 2019) This database allows the comprehension of transcriptomic data, especially that of secondary metabolism and the plant hormones involved in the regulation of it. Biosynthetic pathway for terpene indole alkaloid (TIA) and biosynthetic pathway for the phytohormone jasmonic acid were studied using the annotated data. The PhytoMetaSyn Project has been gathering transcriptomic data for several significant medicinal plant species that produce alkaloids, terpenoids, etc. The data generated by these methods is the key to formulating effective gene editing and metabolic engineering strategies. (Van Moerkercke et al., 2013). Some of the strategies implemented to maneuver the profile of plants have been explored in subsections.

14.5.1 ALTERING THE ACTIVITY OF RATE LIMITING ENZYMES FOR ENHANCED TERPENE PRODUCTION

The mevalonate (MVA) and methylerythritol phosphate (MEP) pathways are key for the production of various groups of terpenes which are one of the most prevalent groups of natural products that are immensely significant as medicine. The triterpenes and sesquiterpenes are synthesized in the cytoplasm of eukaryotes via the MVA pathway. The MEP pathway chiefly occurs in eubacteria and also in the chloroplasts of photosynthetic eukaryotes. They are responsible for the synthesis of mono and diterpenes along with certain hormone precursors. (Rohdich et al., 2001; Zhang et al., 2020)

1-deoxy-D-xylulose 5-phosphate synthase (DXS) catalyzes the rate-determining first step of the MEP pathway, which catalyzes the reaction between pyruvate and glyceraldehyde-3-phosphate to form 1-deoxy-D-xylulose-5-phosphate. (Querol et al., 2002; Takahashi et al., 1998). This enzyme DSX has been found to be coded by multicopy genes that have been identified to be belonging to 3 distinct clades. (Rodríguez-Concepción and Boronat, 2002; Zhang et al., 2020). Characterization of three *DXS* genes from *Morus notabilis* viz. *MnDXS1*, *MnDXS2A*, and *MnDXS2B* led to the assortment of MnDXS1 to the clade 1, which included the already characterized DXS1 proteins acknowledged in *A. thaliana, Z. mays*, and other plant species. Clade 2 included MnDXS2A and MnDXS2B, that are also represented in plants with antidiabetic potential like *Medicago truncatula* and *Catharanthus roseus*. The *MnDXS* genes allocated to clade 2 are thought to be responsible for the synthesis of terpenoids (Zhang et al., 2020)

DXR gene expression has been positively correlated to production of isoprenoids in *Amomum villosum* (Yang et al., 2012), *A. thaliana* (Carretero-Paulet et al., 2002), *Catharanthus roseus* (Veau et al., 2000), *Salvia miltiorrhiza* (Dongfeng Yang et al., 2012)

Another potential site for manipulation has been identified as the rate-limiting flavodoxin-dependent 4-hydroxy-3-methylbut-2-enyl-diphosphate synthase (IspG), which catalyzes the conversion of (2*E*)-4-hydroxy-3-methylbut-2-enyl diphosphate to 2-*C*-methyl-D-erythritol 2,4-cyclic diphosphate in the MEP pathway. (https://www.uniprot.org/uniprot/F4K0E8). Overexpression of a heterologous ispG gene in *Escherichia coli* has led to a 30% increase in isoprene production (Gao et al., 2016). *ispF* sequences from *A. thaliana* and *Catharanthus roseus* have been found to be orthologous, carrying 88% similarity with these genes and putative plastid-targeting sequences. (Rohdich et al., 2001)

14.5.2 ALTERING FLAVONOID PRODUCTION

Flavonoids are synthesized through the pervasive phenylpropanoid pathway which is among the central pathways of secondary metabolism. During this pathway phenylalanine ammonia-lyase (PAL) catalyzes the formation of cinnamic acid from phenylalanine. This is followed trans-cinnamate 4-monooxygenase (C4H) catalyzed oxidation of cinnamic acid to 4-coumaric acid (https://www.uniprot.org/uniprot/Q94IP1) followed by formation of (*E*)-4-coumaroyl-CoA by the action of 4-coumaroyl CoA ligase (4CL) (https://www.uniprot.org/uniprot/Q42524). For the formation of flavonoid, the first commitment step involves condensation of one 4-coumaroyl-CoA molecule with three malonyl-CoA molecules to form 2,'4,4,'6'-tetrahydroxychalcone also known as naringenin chalcone. This step is catalyzed by chalcone synthase (CHS) (https://www.uniprot.org/uniprot/P13114). Tyrosine ammonia-lyase (TAL) found in some plants and bacteria, can bypass these reactions, and directly convert tyrosine to 4-coumaric acid. (https://www.uniprot.org/uniprot/Q3IWB0)

Using *E. coli* as the model organism, CHS, and TAL, 4CL and CHI were introduced, after codon-optimization, to 2 strains obtaining remarkably high production of naringenin. The production was further elevated by the inhibition of the fatty acid metabolic pathway that steals away malonyl-CoA, thus presenting a competition. This competition, however, presents itself as a restriction in genetic manipulation of flavonoid biosynthesis in *E. coli.* (Santos et al., 2011; Wang et al., 2011). Naringenin has been clinically proven to slow down the diabetic nephropathy because of the anti-fibrotic and anti-inflammatory properties. This naturally occurring flavanone also decreased the expression of interleukin (IL)-1β, IL-6, type IV collagen, fibronectin, and transforming growth factor β. (Tsai et al., 2012). The peel color of tomato is usually associated with the quantity of naringenin chalcone (NarCh). The lack of NarCh in recessive yellow (y) mutants makes tomatoes discolored and results in a pink-colored peel. The gene corresponding to this deformation was identified as SlMYB12, which encodes an R2R3-MYB transcription factor. (Ballester et al., 2010). An experiment to carry out targeted disruption of SlMYB12 led to the development of pink-fruited tomatoes, which are more popular than red-fruited tomatoes in Asian countries. (Ku and Ha, 2020).

Transformation studies on *Saussurea involucrata* callus using 2 transcription factors *PAP1* (Production of Anthocyanin Pigment 1; gene encoding a MYB transcription factor from *Arabidopsis*) and *Lc* (Leaf color; gene encoding a bHLH-type transcription factor from maize) were performed and the overall transcriptome analysis along with the blastX analysis using the KEGG protein database was carried out on the assembled unigenes to identify the genes that were induced by the factors. It was observed that the genes of the phenylpropanoid pathway that were among the 5 major pathways that were triggered. Overexpression of phenylpropanoid pathway genes including CHS and C4H led to elevated levels of individual phenylpropanoids, total flavonoids, total phenolics and the overall antioxidant capacity of the callus extracts of transgenic *S. involucrata*. The major compounds to be enhanced were syringin, chlorogenic acid (5-caffeoyl quinic acid), cyanrine (1,5-caffeoylquinic acid) and rutin. (Qiu et al., 2013) Of these rutin is a flavonoid glycoside which has reported antidiabetic activity as it alters the activities of glycolytic and gluconeogenic enzymes, thereby enhancing glucose homeostasis. It also interacts with insulin receptor kinases and insulin-dependent glucose transporters (GLUT). Other flavonoids like fisetin, catechins, luteolin, morin, genistein, naringenin, quercetin, kaempferol have also been established as phytochemicals with antidiabetic effects (Salehi et al., 2019).

The production of isoflavone genistein, has also been achieved in *Allium cepa* L. by introducing *Glycine max* Isoflavone synthase (GmIFS) gene. (Malla et al., 2021)

14.5.3 METABOLIC ENGINEERING OF MICROORGANISMS FOR BIOSYNTHESIS OF SECONDARY METABOLITES

The introduction of heterologous secondary metabolic pathways in microbial hosts provides an alternative approach to the production of phytochemicals while avoiding the regulatory mechanisms of the native plant cell. Introduction of mevalonate pathway of plant cell origin, in E. coli has been found to enhance isoprene synthesis. Mevalonate (MVA) pathway was exogenously expressed by engineered plasmids to supply the precursors necessary for the synthesis of an assortment of isoprenoid alcohols. This, along with the overexpression of NudB (dihydroneopterin triphosphate diphosphohydrolase) and IspA (FPP synthase) effectively hydrolyzed IPP and DMAPP to elevate the levels of isoprene in the titre. (Zada et al., 2018) Recombinant expression of MVA in *R. capsulatus* leading to the formation of sesquiterpenes under phototrophic growth conditions has been observed (Troost et al., 2019).

In *Pseudomonas putida* enhanced isoprenoid synthesis was achieved by enhancing the supply of IPP and DMAPP using lycopene as read-out (Hernandez-Arranz et al., 2019). Optimized metabolic flux has also been attained by enhancing the precursor pool to achieve increased accumulation of natural products in *E. coli.* Overexpression of acetyl-CoA synthase and acetyl-CoA carboxylase which is a rate-controlling enzyme and simultaneous knockdown of pabA (encoding p-aminobenzoate synthetase) have ensued higher levels of malonyl-CoA leading to enhanced production of naringenin, resveratrol, 6-methylsalicylic acid, aloesone, etc. Knockdown of genes like fabB, fabI, and fabF involved in fatty acid biosynthesis have also been put to practice to elevate intracellular concentration of malonyl-CoA. Genome editing tools have been profusely employed to achieve this feet (Wu et al., 2017; Yang et al., 2018, 2020).

14.5.4 EXECUTION OF STRATEGIES USING ZFN

Zinc finger nucleases (ZFNs) have been used to stimulate HR and gene targeting events by introducing targeted double-stranded breaks (DSBs) at specific genomic loci. Depending on the strategy selected for enhancing the titre concentration of antidiabetic compounds, one of the following commonly used methods may be proposed.

14.5.4.1 SITE-DIRECTED MUTAGENESIS TO ELEVATE GENE EXPRESSION

In order to mitigate the plant cells from low fidelity of NHEJ, the ZFN induced site-directed mutagenesis has been employed heavily in the plants. The delivery of ZFN expression vector into the *Arabidopsis* genome using *Agrobacterium*-mediated transformation along with suitable heat shock protein HSP18.2 gene promoter induced specific mutations in *ABA-INSENSITIVE4* (*ABI4*) gene. Upon induction by heat shock, ZFN expression triggered mutations were analyzed using nuclease. A PCR using flanking regions of ZFN along with subsequent mismatch-specific Surveyor endonuclease assay led to the detection of mutations in two lines out of nine. (Osakabe et al., 2010; Wright et al., 2005a)

It is therefore proposed that the ZFN based genetic tool may be implemented to bring about site-directed mutagenesis in order to alleviate the repression of key rate-determining enzymes of terpenoid synthesis like 1-deoxy-D-xylulose 5-phosphate synthase (DSX) and 4-hydroxy-3-methylbut-2-enyl-diphosphate synthase (IspG). It may also be used to transform suitable host cells at a much higher rate, with genes like CHS or Tyrosine ammonia-lyase (TAL) to enhance the flavonoid content of the titre. Promoter engineering using ZFN can also come in handy.

14.5.4.2 GENE KNOCKOUT TO ESTABLISH GENE FUNCTION

Lack of knowledge on the gene function continues to be one of the major impediments in metabolic engineering of natural products. This lack of understanding is a manifestation of the extremely diverse nature of secondary metabolites across the plant kingdom. A projected use of ZFN can therefore be a site-specific disarmament of a probable gene, and the knock out can be related to the altered metabolite composition.

Tomato has been subjected to similar gene disruption by the use of ZFN to establish the relation between genes and the phytochemical profile of tomato. Disruption of the heterotrimeric nuclear transcription factor Y (NF-Y) transcription factor (TF) gene LEAFY-COTYLEDON1-LIKE4 (L1L4, NF-YB6), using ZFN led to the observation that L1L4, a member of the L1L gene family, was expressed in green developing fruits and flowers (Hilioti et al., 2014) and the L1L4 (or NF-YB6), encodes transcription factor specific to the embryo, involved in plant development and regulation of levels of seed storage proteins and fatty acid (Mu et al., 2008). The L1L4 TF in tomato is also related to the overall fruit quality in terms of the metabolites generated. (Gago et al., 2017)

14.5.4.3 DIRECT DELIVERY OF ZFN PROTEINS

Recent advances in designed endonucleases have allowed precision in the genome editing techniques in plants. In a novel approach to ZFN delivery method, the plant cells were transfected with ZFN protein. This eliminated the requirement of stable nuclease genomic integration and allowed the generation of non-transgenic edited, cells or tissues in wheat. ZFNs were used to disrupt *INOSITOL PENTAKISPHOSPHATE KINASE* 1 (*IPK1*) gene, meant for the enzyme that catalyzes the final steps of phytic acid production, which is an anti-nutritional compound. Wheat microspores or embryos were directly transfected with active ZFN protein purified from *E. coli* BL21 (DE3) (Novagene) cultures, which was complexed with CPP. *In vitro* cleavage assay assessed the activity of purified ZFN proteins and the CPP mediated delivery of ZFN proteins was examined with Alexa Fluor 647 labeled ZFN R monomer. Successful plant microspore genome editing was established using this protocol. (Bilichak et al., 2020)

14.6 ADVANTAGES AND LIMITATIONS OF THE ZFN BASED TECHNOLOGY

ZFNs are shown to confer an advantage of less frequency of off-target editing events. Off-target editing can happen with measurable but low rate at some defined genomic sites which are exceedingly homologous to the target site (Pattanayak et al., 2011; Gabriel et al., 2011; Cai and Yang, 2014). Generation of cohesive overhangs after double-strand break is one of the advantages associated with using ZFNs, which makes insertion of genes at the site of cleavage through NHEJ or HR easier. However, this advantage is also associated with TALENs and CRISPR/Cpf1 proteins. ZFNs can be used for gene disruption, addition, or correction as well as tag ligation (Cai and Yang, 2014). The small size of the ZFN cassette makes it possible to introduce it in various virus vectors making delivery in multiple systems easier (Guha, Wai, and Hausner, 2017). Some challenges in ZFN technology include construction of target-specific ZFNs being expensive as compared to TALENs or CRISPR/Cas9 systems. Moreover, ZFNs preferably require guanine rich repeats in target sequence which limits their application (Kanchiswamy et al., 2016). Ramirez et al. in 2008 reported very high failure rates (around 76%) of modular ZNFs with an interesting observation that if the target site has two or less GXX codon (where X is any base) at the binding sites for ZFPs, then the efficiency of target-specific genome editing decreases considerably as compared to the sites with 3 or more GXX sites (Ramirez et al., 2008). This requirement and the difficulty in targeting sequences with poor guanosine content led to exclusion of many target sites. Considering the difficulty involved in designing, multiplexing, and the cost of making working ZFNs, high-throughput technology seems to be farfetched (Reyon et al., 2012).

14.7 ETHICAL CONCERNS ASSOCIATED WITH GENOME EDITING

Genome editing is a technology of future opening unfathomable avenues in synthetic biology; however, it has associated concerns and conflicts. These concerns have been discussed widely in the case of human and plant genome modifications with CRISPR/Cas9 editing at the center. Hong et al. have given an excellent review about various issues that need to be tackled to make commercialization of GM crops. Technical issues such as selection of appropriate genes for mutation and minimizing off-target editing events, regulatory issues in terms of having some uniformity in laws globally, intellectual property concerns, ethical issues in terms of maintaining 'naturalness' of the product to make it consumer acceptable are very important to be addressed urgently (Hong et al., 2020). It must be clearly understood that there is a difference in 'genetically modified' (GMOs) and 'genome edited' (GEOs) products, as in genetic modification; usually a foreign gene is introduced in the genome of the recipient. On the contrary, in 'GEOs' products, one may expect to have controlled and precise editing of natural genome without introduction of foreign genes in it making the product near-natural without traces of genetic engineering (Carroll, 2015; Hong et al., 2020). Although many scientific groups have advocated GMO/edited crop plants in relation to green revolution, nutritive properties and sustainability, the approach still remains controversial, tightly regulated and less popular among consumers (Bartkowski, 2018). However, there are some instances where genome-edited plants were approved for commercial production, such as, CRISPR-Cas9 modified *Agaricusbisporus,* white button mushroom, created by knock-out of polyphenol oxidase gene responsible for browning was sanctioned by US Department of Agriculture Animal and Plant Health Inspection Service (USDA-APHIS) (Brinegar et al., 2017.; Bartkowski, 2018). The decision was based on the fact that the resulting product did not have any traces of genetic engineering and was treated as a mutation that may happen in natural conditions. The USDA also declared eight cases of ZFN and TALEN mutations as beyond their regulatory jurisdiction allowing commercialization of these products (Brinegar et al., 2017).

Though the technology has many advantages, the safety concerns cannot be ignored. The deleterious or unwanted effects of mutation, the transfer of mutation to plant progenies, long term consumption effects, sustainability in or evasion of natural ecosystems can be some major concerns which require long term study affecting the investments in the area. In a recent review on the similar lines, Kawall et al. have suggested revision in current EU risk assessment guidance for GMOs to encompass all possible genomic irregularities such as plasmid integration or large off-target deletions/editing (Zhao and Wolt, 2017) that may arise due to the advancing technology and to have additional risk assessment tools for the same (Kawall, Cotter, and Then, 2020). In an interesting communication by Meyer (2020), it has been stated by author that considering general public to be a passive group, they must be informed and educated about the benefits of gene editing, therefore, increasing the 'public/consumer acceptance,' for which, scientific community needs to go beyond the academic publications to reach masses and study the 'fabric of the public' (Meyer, 2020). It must be considered that this technology has potential to meet sustainable development goals and may help in dealing with the food crisis projected in future with increasing population (Schiemann et al., 2020). Nevertheless, the unpredictable future outcomes and challenges with respect to genome-edited plants are always debated and limit their acceptability.

14.8 FUTURE PROSPECTS OF ZFN BASED TECHNOLOGY AND GENOME EDITING IN PLANTS

When it comes to genome editing with rapidly advancing techniques and technologies, the future course of the GEOs plants/crops seems to be very promising. Increased production of specific secondary metabolites to treat various metabolic disorders and other ailments, creation of disease and drought-resistant varieties can be near future targets for plant genome editing.

KEYWORDS

- **antidiabetic activity of plants**
- **antidiabetic phytochemicals**
- **antihyperglycemic**
- **endonucleases**
- **genome editing**
- **hypoglycemic**
- **metabolic engineering**
- **secondary metabolites**
- **zinc finger nucleases**
- **α-amylase inhibitors**
- **α-glucosidase inhibitors**

REFERENCES

Adli, M., (2018). The CRISPR tool kit for genome editing and beyond. *Nat. Commun., 9*(1), 1911.

Allina, S. M., Pri-Hadash, A., Theilmann, D. A., Ellis, B. E., & Douglas, C. J., (1998). 4-coumarate: Coenzyme a ligase in hybrid poplar: Properties of native enzymes, CDNA cloning, and analysis of recombinant enzymes. *Plant Physiol., 116*(2), 743–754.

Anami, S., Njuguna, E., Coussens, G., Aesaert, S., & Van, L. M., (2013). Higher plant transformation: Principles and molecular tools. *Int. J. Dev. Biol., 57*(6, 7, 8), 483–494.

Anand, P., Schug, A., & Wenzel, W., (2013). Structure-based design of protein linkers for zinc finger nuclease. *FEBS Lett., 587*, 3231–3235.

Arumugam, G., Manjula, P., & Paari, N., (2013). A review: Antidiabetic medicinal plants used for diabetes mellitus. *Journal of Acute Disease, 2*(3), 196–200.

Baenas, N., García-Viguera, C., & Moreno, D., (2014). Elicitation: A tool for enriching the bioactive composition of foods. *Molecules, 19*(9), 13541–13563.

Bahramnejad, B., Naji, M., Bose, R., & Jha, S., (2019). A critical review on the use of *Agrobacterium rhizogenes* and their associated binary vectors for plant transformation. *Biotechnology Advances, 37*(7), 107405.

Ballester, A. R., Molthoff, J., De Vos, R., Hekkert, B. Te, L., Orzaez, D., Fernández-Moreno, J. P., Tripodi, P., et al., (2010). biochemical and molecular analysis of pink tomatoes: Deregulated expression of the gene encoding transcription factor SlMYB12 leads to pink tomato fruit color. *Plant Physiol., 152*(1), 71–84.

Bartkowski, B., (2018). *Snipping Around for Food: Economic, Ethical and Policy Implications of CRISPR/Cas Genome Editing, 9*, 172–180.

Baset, M., Ali, T., Elshamy, H., El Sadek, A., Sami, D., Badawy, M., Abou-Zekry, S., et al., (2020). Antidiabetic effects of fenugreek (*Trigonella foenum-graecum*): A comparison between oral and intraperitoneal administration – an animal study. *Int. J. Funct. Nutr.*

Bevan, M., Bancroft, I., Bent, E., Love, K., Goodman, H., Dean, C., Bergkamp, R., et al., (1998). The EU *Arabidopsis* genome project. Analysis of 1.9 Mb of contiguous sequence from chromosome 4 of *Arabidopsis thaliana*. *Nature, 391*(6666), 485–488.

Bhutkar, M. A., & Bhise, S. B., (2012). *In vitro Assay of Alpha-Amylase Inhibitory Activity of Some Indigenous Plants, 6.*

Bhutkar, M. A., Bhinge, S. D., Randive, D. S., & Wadkar, G. H., (2017). Hypoglycemic effects of *Berberis aristata* and *Tamarindus indica* Extracts *in vitro*. *Bulletin of Faculty of Pharmacy, Cairo University, 55*(1), 91–94.

Borrelli, G., & Trono, D., (2016). Molecular approaches to genetically improve the accumulation of health-promoting secondary metabolites in staple crops—A case study: The lipoxygenase-B1 genes and regulation of the carotenoid content in pasta products. *IJMS, 17*(7), 1177.

Bortesi, L., & Fischer, R., (2015). The CRISPR/Cas9 system for plant genome editing and beyond. *Biotechnol. Adv., 33*(1), 41–52.

Brinegar, K., Yetisen, A. K., Choi, S., Vallillo, E., Ruiz-Esparza, G. U., Prabhakar, A. M., Khademhosseini, A., & Yun, S. H., (2017). The commercialization of genome-editing technologies. *Crit. Rev. Biotechnol., 37*(7), 924–932.

Cai, M., & Yang, Y., (2014). Targeted genome editing tools for disease modeling and gene therapy. *Curr. Gene Ther., 14*(1), 2–9.

Carretero-Paulet, L., Ahumada, I., Cunillera, N., Rodríguez-Concepción, M., Ferrer, A., Boronat, A., & Campos, N., (2002). Expression and molecular analysis of the *Arabidopsis dxr* gene encoding 1-deoxy-D-xylulose 5-phosphate reductoisomerase, the first committed enzyme of the 2- *C* -methyl-D-erythritol 4-phosphate pathway. *Plant Physiol., 129*(4), 1581–1591.

Carroll, D., & Charo, R. A., (2015). The Societal Opportunities and Challenges of Genome Editing, *Genome Biol., 16,* 242.

Chen, X., Rinsma, M., Janssen, J. M., Liu, J., Maggio, I., & Gonçalves, M. A. F. V., (2016). Probing the impact of chromatin conformation on genome editing tools. *Nucleic Acids Res., 44*(13), 6482–6492.

Chen, Z., Jaafar, L., Agyekum, D. G., Xiao, H., Wade, M. F., Kumaran, R. I., Spector, D. L., et al., (2013). Receptor-mediated delivery of engineered nucleases for genome modification. *Nucleic Acids Res., 41*(19), e182–e182.

Cheng, Z., Wang, R., Shen, J., Yang, Z., & Chen, Y., (2019). Cloning, characterization, and expression of 1-deoxy-d-xylulose-5-phosphate reductoisomerase gene from *Dioscorea zingiberensis*. *Biologia plant., 63*(1), 262–267.

Cristea, S., Freyvert, Y., Santiago, Y., Holmes, M. C., Urnov, F. D., Gregory, P. D., & Cost, G. J., (2013). *In vivo* cleavage of transgene donors promotes nuclease-mediated targeted integration. *Biotechnol. Bioeng., 110*(3), 871–880.

Dey, A., (2021). CRISPR/Cas genome editing to optimize pharmacologically active plant natural products. *Pharmacological Research, 164,* 105359.

Didigu, C. A., Wilen, C. B., Wang, J., Duong, J., Secreto, A. J., Danet-Desnoyers, G. A., Riley, J. L., et al., (2014). Simultaneous zinc-finger nuclease editing of the HIV coreceptors Ccr5 and Cxcr4 protects CD4+ T cells from HIV-1 infection. *Blood, 123*(1), 61–69.

Ding, Q., Regan, S. N., Xia, Y., Oostrom, L. A., Cowan, C. A., & Musunuru, K., (2013). Enhanced efficiency of human pluripotent stem cell genome editing through replacing TALENs with CRISPRs. *Cell Stem Cell, 12*(4), 393–394.

Doyle, E. L., Booher, N. J., Standage, D. S., Voytas, D. F., Brendel, V. P., VanDyk, J. K., & Bogdanove, A. J., (2012). TAL effector-nucleotide targeter (TALE-NT) 2.0: Tools for TAL effector design and target prediction. *Nucleic Acids Res., 40*(Web Server issue):W117–22.

Engels, B., Dahm, P., & Jennewein, S., (2008). Metabolic engineering of taxadiene biosynthesis in yeast as a first step towards Taxol (paclitaxel) production. *Metabolic Engineering, 10*(3, 4), 201–206.

Engler, C., Kandzia, R., & Marillonnet, S., (2008). A one pot, one step, precision cloning method with high throughput capability. *PLoS One, 3*(11), e3647.

Farag, M. A., Rasheed, D. M., Kropf, M., & Heiss, A. G., (2016). Metabolite profiling in *Trigonella* seeds via UPLC-MS and GC-MS analyzed using multivariate data analyses. *Anal. Bioanal. Chem., 408*(28), 8065–8078.

Ferguson, L. A., Docherty, H. M., MacKenzie, A. E., & Docherty, K., (2009). An engineered zinc finger protein reveals a role for the insulin VNTR in the regulation of the insulin and adjacent *IGF2* genes. *FEBS Letters, 583*(19), 3181–3186.

Gabriel, R., Lombardo, A., Arens, A., Miller, J. C., Genovese, P., Kaeppel, C., Nowrouzi, A., et al., (2011). An unbiased genome-wide analysis of zinc-finger nuclease specificity. *Nat. Biotechnol., 29*(9), 816–823.

Gago, C., Drosou, V., Paschalidis, K., Guerreiro, A., Miguel, G., Antunes, D., & Hilioti, Z., (2017). Targeted gene disruption coupled with metabolic screen approach to uncover the LEAFY COTYLEDON1-LIKE4 (L1L4) function in tomato fruit metabolism. *Plant Cell Rep., 36*(7), 1065–1082.

Gao, X., Gao, F., Liu, D., Zhang, H., Nie, X., & Yang, C., (2016). Engineering the methylerythritol phosphate pathway in cyanobacteria for photosynthetic isoprene production from CO_2. *Energy Environ. Sci., 9*(4), 1400–1411.

Gómez-Galera, S., Pelacho, A. M., Gené, A., Capell, T., & Christou, P., (2007). The genetic manipulation of medicinal and aromatic plants. *Plant Cell Rep, 26*(10), 1689–1715.

Guha, T. K., Wai, A., & Hausner, G., (2017). Programmable genome editing tools and their regulation for efficient genome engineering. *Comput. Struct. Biotechnol. J., 15*, 146–160.

Gutensohn, M., Nguyen, T. T. H., McMahon, R. D., Kaplan, I., Pichersky, E., & Dudareva, N., (2014). Metabolic engineering of monoterpene biosynthesis in tomato fruits via introduction of the non-canonical substrate neryl diphosphate. *Metabolic Engineering, 24*, 107–116.

Hasanpour, M., Iranshahy, M., & Iranshahi, M., (2020). The application of metabolomics in investigating antidiabetic activity of medicinal plants. *Biomedicine & Pharmacotherapy, 128,* 110263.

Havinga, R. M., Hartl, A., Putscher, J., Prehsler, S., Buchmann, C., & Vogl, C. R., (2010). *Tamarindus indica* L. (Fabaceae): Patterns of use in traditional African medicine. *Journal of Ethnopharmacology, 127*(3), 573–588.

He, X., Zhao, X., Gao, L., Shi, X., Dai, X., Liu, Y., Xia, T., & Wang, Y., (2018). Isolation and characterization of key genes that promote flavonoid accumulation in purple-leaf tea (*Camellia sinensis* L.). *Sci. Rep., 8*(1), 130.

Hegge, J. W., Swarts, D. C., & Van, D. O. J., (2018). Prokaryotic argonaute proteins: Novel genome-editing tools? *Nat. Rev. Microbiol., 16*(1), 5–11.

Hernandez-Arranz, S., Perez-Gil, J., Marshall-Sabey, D., & Rodriguez-Concepcion, M., (2019). Engineering *Pseudomonas putida* for isoprenoid production by manipulating endogenous and shunt pathways supplying precursors. *Microb Cell Fact, 18*(1), 152.

Hong, J., Shi, Q., Biswas, S., Jiang, S. C., & Shi, J., (2020). Moving genome-edited crops forward from the laboratory bench to the kitchen table. *Food Control.*

Hussain Md, S., Rahman Md, A., Fareed, S., Ansari, S., Ahmad, I., & Mohd, S., (2012). Current approaches toward production of secondary plant metabolites. *J. Pharm. Bioall. Sci, 4*(1), 10.

Hwang, H. H., Yu, M., & Lai, E. M., (2017). *Agrobacterium*-mediated plant transformation: Biology and applications. *The Arabidopsis Book, 15,* e0186.

Hwang, W. Y., Fu, Y., Reyon, D., Maeder, M. L., Tsai, S. Q., Sander, J. D., Peterson, R. T., et al., (2013). Efficient genome editing in zebrafish using a CRISPR-Cas system. *Nat. Biotechnol., 31*(3), 227–229.

Iqbal, Z., Iqbal, M. S., Ahmad, A., Memon, A. G., & Ansari, M. I., (2020). New prospects on the horizon: Genome editing to engineer plants for desirable traits. *Current Plant Biology, 24,* 100171.

Irmisch, S., Jancsik, S., Man, S. Y. M., Madilao, L. L., & Bohlmann, J., (2020). Complete biosynthesis of the antidiabetic plant metabolite montbretin A. *Plant Physiol., 184*(1), 97–109.

Jain, A., Bhadoriya, S., Ganeshpurkar, A., Narwaria, J., & Rai, G., (2011). *Tamarindus indica*: Extent of explored potential. *Phcog. Rev., 5*(9), 73.

Jinek, M., Chylinski, K., Fonfara, I., Hauer, M., Doudna, J. A., & Charpentier, E., (2012). A Programmable dual-RNA-guided DNA endonuclease in adaptive bacterial immunity. *Science, 337*(6096), 816–821.

Kabera, J. N., Semana, Mussa, A., & He, X., (2014). Plant secondary metabolites: Biosynthesis, classification, function and pharmacological properties. *Journal of Pharmacy and Pharmacology, 2*, 377–392.

Kanchiswamy, C. N., Maffei, M., Malnoy, M., Velasco, R., & Kim, J. S., (2016). Fine-tuning next-generation genome editing tools. *Trends Biotechnol., 34*(7), 562–574.

Kawall, K., Cotter, J., & Then, C., (2020). Broadening the GMO risk assessment in the EU for genome editing technologies in agriculture. *Environ. Sci. Eur., 32*(1), 106.

Kim, Y. G., Cha, J., & Chandrasegaran, S., (1996). Hybrid restriction enzymes: Zinc finger fusions to fok I cleavage domain. *Proc. Natl. Acad. Sci., 93*(3), 1156–1160.

Kong, F., Yamaoka, Y., Ohama, T., Lee, Y., & Li-Beisson, Y., (2019). Molecular genetic tools and emerging synthetic biology strategies to increase cellular oil content in *Chlamydomonas reinhardtii*. *Plant and Cell Physiology, 60*(6), 1184–1196.

Ku, H. K., & Ha, S. H., (2020). Improving nutritional and functional quality by genome editing of crops: Status and perspectives. *Front. Plant Sci., 11,* 577313.

Lee, J., Chung, J. H., Kim, H. M., Kim, D. W., & Kim, H., (2016). Designed nucleases for targeted genome editing. *Plant Biotechnol. J., 14*(2), 448–462.

Liu, H., Ding, Y., Zhou, Y., Jin, W., Xie, K., & Chen, L. L., (2017). CRISPR-P 2.0: An improved CRISPR-Cas9 tool for genome editing in plants. *Mol. Plant, 10*(3), 530–532.

Liu, X., Ding, W., & Jiang, H., (2017). Engineering microbial cell factories for the production of plant natural products: From design principles to industrial-scale production. *Microb. Cell Fact., 16*(1), 125.

Lombardo, A., Genovese, P., Beausejour, C., Colleoni, S., Lee, Y. L., Kim, K., Ando, D., et al., (2007). Gene editing in human stem cells using zinc finger nucleases and integrase-defective lentiviral vector delivery. *Nat. Biotechnol., 25*, 1298–1306.

Luan, H., Liao, W., Song, Y., Niu, H., Hu, T., & Zhi, H., (2020). Transgenic plant generated by RNAi-mediated knocking down of soybean Vma12 and soybean mosaic virus resistance evaluation. *AMB Expr, 10*(1), 62.

Ludwig-Müller, J., (2014). Metabolic engineering of selected secondary metabolites. In: Paek, K. Y., Murthy, H. N., & Zhong, J. J., (eds.), *Production of Biomass and Bioactive Compounds Using Bioreactor Technology* (pp. 509–536). Springer Netherlands: Dordrecht.

Maggio, I., & Gonçalves, M. A. F. V., (2015). Genome editing at the crossroads of delivery, specificity, and fidelity. *Trends Biotechnol., 33*(5), 280–291.

Maiti, R., Jana, D., Das, U. K., & Ghosh, D., (2004). Antidiabetic effect of aqueous extract of seed of *Tamarindus indica* in streptozotocin-induced diabetic rats. *Journal of Ethnopharmacology, 92*(1), 85–91.

Mao, Y., Zhang, H., Xu, N., Zhang, B., Gou, F., & Zhu, J. K., (2013). Application of the CRISPR–cas system for efficient genome engineering in plants. *Mol. Plant, 6*(6), 2008–2011.

Meyer, M., (2020). *The Fabric of the Public in Debates About Gene Editing*, 6.

Meyer, M., Ortiz, O., Angelis De, M. H., Wurst, W., & Kühn, R., (2012). Modeling disease mutations by gene targeting in one-cell mouse embryos. *Proc. Natl. Acad. Sci., 109*(24), 9354–9359.

Miller, J. C., Holmes, M. C., Wang, J., Guschin, D. Y., Lee, Y. L., Rupniewski, I., Beausejour, C. M., et al., (2007). An improved zinc-finger nuclease architecture for highly specific genome editing. *Nat. Biotechnol., 25*(7), 778–785.

Mohanta, T., Bashir, T., Hashem, A., Abd_Allah, E., & Bae, H., (2017). Genome editing tools in plants. *Genes, 8*(12), 399.

Moreno, P. R. H., Van, D. H. R., & Verpoorte, R., (1993). Effect of terpenoid precursor feeding and elicitation on formation of indole alkaloids in cell suspension cultures of *Catharanthus roseus*. *Plant Cell Reports, 12*(12), 702–705.

Mouradov, A., & Spangenberg, G., (2014). Flavonoids: A metabolic network mediating plants adaptation to their real estate. *Front. Plant Sci., 5.*

Muranaka, T., & Saito, K., (2010). Production of pharmaceuticals by plant tissue cultures. In: *Comprehensive Natural Products II* (pp. 615–628). Elsevier.

Nemudryi, A. A., Valetdinova, K. R., Medvedev, S. P., & Zakian, S. M., (2014). TALEN and CRISPR/Cas genome editing systems: Tools of discovery. *Acta Naturae, 6*(3), 19–40.

Osakabe, K., Osakabe, Y., & Toki, S., (2010). Site-directed mutagenesis in *Arabidopsis* using custom-designed zinc finger nucleases. *Proceedings of the National Academy of Sciences, 107*(26), 12034–12039.

Pandey, M., Debnath, M., Gupta, S., & Chikara, S. K., (2011). Phytomedicine: An Ancient Approach Turning into Future Potential Source of Therapeutics, *Journal of Pharmacognosy and Phytotherapy, 3*(2), 27–37.

Park, J. S., Kim, J. B., Hahn, B. S., Kim, K. H., Ha, S. H., Kim, J. B., & Kim, Y. H., (2004). EST analysis of genes involved in secondary metabolism in *Camellia sinensis* (Tea), using suppression subtractive hybridization. *Plant Science, 166*(4), 953–961.

Paschon, D. E., Lussier, S., Wangzor, T., Xia, D. F., Li, P. W., Hinkley, S. J., Scarlott, N. A., et al., (2019). Diversifying the structure of zinc finger nucleases for high-precision genome editing. *Nat. Commun., 10*(1), 1133.

Pattanayak, V., Ramirez, C. L., Joung, J. K., & Liu, D. R., (2011). Revealing off-target cleavage specificities of zinc-finger nucleases by *in vitro* selection. *Nat. Methods, 8*(9), 765–770.

Pelascini, L. P. L., Maggio, I., Liu, J., Holkers, M., Cathomen, T., & Gonçalves, M. A. F. V., (2013). Histone deacetylase inhibition rescues gene knockout levels achieved with integrase-defective lentiviral vectors encoding zinc-finger nucleases. *Hum. Gene Ther. Methods, 24*(6), 399–411.

Pelletier, M. K., Burbulis, I. E., & Shirley, B. W., (1999). Disruption of specific flavonoid genes enhances the accumulation of flavonoid enzymes and end products in *Arabidopsis* seedlings. *Plant Molecular Biology, 40*(1), 45–54.

Pichersky, E., & Gang, D. R., (2000). Genetics and biochemistry of secondary metabolites in plants: An evolutionary perspective. *Trends in Plant Science, 5*(10), 439–445.

Pouvreau, B., Vanhercke, T., & Singh, S., (2018). From plant metabolic engineering to plant synthetic biology: The evolution of the design/build/test/learn cycle. *Plant Science, 273*, 3–12.

Qiu, J., Gao, F., Shen, G., Li, C., Han, X., Zhao, Q., Zhao, D., Hua, X., & Pang, Y., (2013). Metabolic engineering of the phenylpropanoid pathway enhances the antioxidant capacity of *Saussurea involucrata*. *PLoS One, 8*(8), e70665.

Querol, J., Campos, N., Imperial, S., Boronat, A., & Rodríguez-Concepción, M., (2002). Functional analysis of the *Arabidopsis thaliana* GCPE protein involved in plastid isoprenoid biosynthesis. *FEBS Letters, 514*(2, 3), 343–346.

Reyon, D., Tsai, S. Q., Khayter, C., Foden, J. A., Sander, J. D., & Joung, J. K., (2012). FLASH assembly of TALENs for high-throughput genome editing. *Nat. Biotechnol., 30*(5), 460–465.

Rodriguez-Concepcion, M., (2002). Elucidation of the methylerythritol phosphate pathway for isoprenoid biosynthesis in bacteria and plastids. a metabolic milestone achieved through genomics. *Plant Physiology, 130*(3), 1079–1089.

Rohdich, F., Kis, K., Bacher, A., & Eisenreich, W., (2001). The non-mevalonate pathway of isoprenoids: Genes, enzymes and intermediates. *Current Opinion in Chemical Biology, 5*(5), 535–540.

Sahin, U., Karikó, K., & Türeci, Ö., (2014). MRNA-based therapeutics — Developing a new class of drugs. *Nat. Rev. Drug Discov., 13*(10), 759–780.

Saini, D. K., Chakdar, H., Pabbi, S., & Shukla, P., (2020). Enhancing production of microalgal biopigments through metabolic and genetic engineering. *Critical Reviews in Food Science and Nutrition, 60*(3), 391–405.

Salehi, Ata, Anil, K. V., Sharopov, Ramírez-Alarcón, Ruiz-Ortega, Abdulmajid, A., et al., (2019). Antidiabetic potential of medicinal plants and their active components. *Biomolecules, 9*(10), 551.

Sander, J. D., Dahlborg, E. J., Goodwin, M. J., Cade, L., Zhang, F., Cifuentes, D., Curtin, S. J., et al., (2011). Selection-free zinc-finger-nuclease engineering by context-dependent assembly (CoDA). *Nat. Methods, 8*(1), 67–69.

Schiemann, J., Robienski, J., Schleissing, S., Spök, A., Sprink, T., & Wilhelm, R. A., (2020). Editorial: Plant genome editing – policies and governance. *Front. Plant Sci., 11*, 284.

Siahsar, B., Rahimi, M., Tavassoli, A., & Raissi, A., (2011). Application of biotechnology in production of medicinal plants. *Environ. Sci.,* 6.

Silva, G., Poirot, L., Galetto, R., Smith, J., Montoya, G., Duchateau, P., & Pâques, F., (2011). Meganucleases and other tools for targeted genome engineering: Perspectives and challenges for gene therapy. *Curr. Gene Ther., 11*(1), 11–27.

Sinha, S., Sandhu, K., Bisht, N., Naliwal, T., Saini, I., & Kaushik, P., (2019). Ascertaining the paradigm of secondary metabolism enhancement through gene-level modification in therapeutic plants. *JYP, 11*(4), 337–343.

Takahashi, S., Kuzuyama, T., Watanabe, H., & Seto, H., (1998). A 1-deoxy-D-xylulose 5-phosphate reductoisomerase catalyzing the formation of 2-C-methyl-D-erythritol 4-phosphate in an alternative nonmevalonate pathway for terpenoid biosynthesis. *Proceedings of the National Academy of Sciences, 95*(17), 9879–9884.

Tebas, P., Stein, D., Tang, W. W., Frank, I., Wang, S. Q., Lee, G., Spratt, S. K., et al., (2014). Gene editing of CCR5 in autologous CD4 T cells of persons infected with HIV. *N. Engl. J. Med., 370*(10), 901–910.

Troost, K., Loeschcke, A., Hilgers, F., Özgür, A. Y., Weber, T. M., Santiago-Schübel, B., Svensson, V., et al., (2019). Engineered *Rhodobacter capsulatus* as a phototrophic platform organism for the synthesis of plant sesquiterpenoids. *Front. Microbiol., 10,* 1998.

Tundis, R., Loizzo, M. R., & Menichini, F. (2010). Natural products as alpha-amylase and alpha-glucosidase inhibitors and their hypoglycaemic potential in the treatment of diabetes: an update. *Mini Rev Med Chem. 10*(4), 315–31.

Uchenna, U. E., Shori, A. B., & Baba, A. S., (2018). *Tamarindus indica* seeds improve carbohydrate and lipid metabolism: An *in vivo* study. *Journal of Ayurveda and Integrative Medicine, 9*(4), 258–265.

Van, M. A., Fabris, M., Pollier, J., Baart, G. J. E., Rombauts, S., Hasnain, G., Rischer, H., et al., (2013). CathaCyc, a metabolic pathway database built from *Catharanthus roseus* RNA-seq data. *Plant and Cell Physiology, 54*(5), 673–685.

Verpoorte, R., (1998). Exploration of nature's chemodiversity: The role of secondary metabolites as leads in drug development. *Drug Discovery Today, 3*(5), 232–238.

Voytas, D. F., (2013). Plant genome engineering with sequence-specific nucleases. *Annu. Rev. Plant Biol., 64*(1), 327–350.

Wang, R., Xu, S., Wang, N., Xia, B., Jiang, Y., & Wang, R., (2017). transcriptome analysis of secondary metabolism pathway, transcription factors, and transporters in response to methyl jasmonate in *Lycoris aurea. Front. Plant Sci., 7.*

Wang, Y., Chen, S., & Yu, O., (2011). Metabolic engineering of flavonoids in plants and microorganisms. *Appl. Microbiol. Biotechnol., 91*(4), 949–956.

Wilson, S. A., & Roberts, S. C., (2014). Metabolic engineering approaches for production of biochemicals in food and medicinal plants. *Current Opinion in Biotechnology, 26*, 174–182.

Wink, M., (2015). Modes of action of herbal medicines and plant secondary metabolites. *Medicines, 2*(3), 251–286.

Wood, A. J., Lo, T. W., Zeitler, B., Pickle, C. S., Ralston, E. J., Lee, A. H., Amora, R., et al., (2011). Targeted genome editing across species using ZFNs and TALENs. *Science, 333*(6040), 307–307.

Wright, D. A., Thibodeau-Beganny, S., Sander, J. D., Winfrey, R. J., Hirsh, A. S., Eichtinger, M., Fu, F., et al., (2006). Standardized reagents and protocols for engineering zinc finger nucleases by modular assembly. *Nat Protoc, 1*(3), 1637–1652.

Wright, D. A., Townsend, J. A., Winfrey, R. J., Irwin, P. A., Rajagopal, J., Lonosky, P. M., Hall, B. D., et al., (2005). High-frequency homologous recombination in plants mediated by zinc-finger nucleases: Recombination and zinc-finger nucleases. *The Plant Journal, 44*(4), 693–705.

Xie, D. Y., (2003). Role of anthocyanidin reductase, encoded by BANYULS in plant flavonoid biosynthesis. *Science, 299*(5605), 396–399.

Xu, P., Marsafari, M., Zha, J., & Koffas, M., (2020). Microbial coculture for flavonoid synthesis. *Trends in Biotechnology, 38*(7), 686–688.

Yang, D., Du, X., Liang, X., Han, R., Liang, Z., Liu, Y., Liu, F., & Zhao, J., (2012). Different roles of the mevalonate and methylerythritol phosphate pathways in cell growth and tanshinone production of *Salvia miltiorrhiza* hairy roots. *PLoS One*, *7*(11), e46797.

Yang, D., Kim, W. J., Yoo, S. M., Choi, J. H., Ha, S. H., Lee, M. H., & Lee, S. Y., (2018). Repurposing type III polyketide synthase as a malonyl-CoA biosensor for metabolic engineering in bacteria. *Proc. Natl. Acad. Sci. USA, 115*(40), 9835–9844.

Yang, D., Park, S. Y., Park, Y. S., Eun, H., & Lee, S. Y., (2020). Metabolic engineering of *Escherichia coli* for natural product biosynthesis. *Trends in Biotechnology, 38*(7), 745–765.

Yin, H., Kauffman, K. J., & Anderson, D. G., (2017). Delivery technologies for genome editing. *Nat. Rev. Drug Discov., 16*(6), 387–399.

Yin, K., Gao, C., & Qiu, J. L., (2017). Progress and prospects in plant genome editing. *Nat. Plants, 3*(8), 17107.

Zada, B., Wang, C., Park, J. B., Jeong, S. H., Park, J. E., Singh, H. B., & Kim, S. W., (2018). Metabolic engineering of *Escherichia coli* for production of mixed isoprenoid alcohols and their derivatives. *Biotechnol Biofuels, 11*(1), 210.

Zetsche, B., Gootenberg, J. S., Abudayyeh, O. O., Slaymaker, I. M., Makarova, K. S., Essletzbichler, P., Volz, S. E., et al., (2015). Cpf1 Is a single RNA-guided endonuclease of a class 2 CRISPR-Cas system. *Cell, 163*(3), 759–771.

Zhang, H. X., Zhang, Y., & Yin, H., (2019). Genome editing with MRNA encoding ZFN, TALEN, and Cas9. *Mol. Ther., 27*(4), 735–746.

Zhang, S., Ding, G., He, W., Liu, K., Luo, Y., Tang, J., & He, N., (2020). Functional characterization of the 1-deoxy-D-xylulose 5-phosphate synthase genes in *Morus notabilis*. *Front. Plant Sci., 11*, 1142.

Zhang, Y., Massel, K., Godwin, I. D., & Gao, C., (2018). Applications and potential of genome editing in crop improvement. *Genome Biol., 19*(1), 210.

Zhao, H., & Wolt, J. D., (2017). Risk Associated with off-target plant genome editing and methods for its limitation. *Emerg. Top. Life Sci., 1*(2), 231–240.

Zhao, J., Zhu, W. H., & Hu, Q., (2001). Enhanced catharanthine production in *Catharanthus roseus* cell cultures by combined elicitor treatment in shake flasks and bioreactors. *Enzyme and Microbial Technology, 28*(7, 8), 673–681.

Zu, Y., Prather, K. L., & Stephanopoulos, G., (2020). Metabolic engineering strategies to overcome precursor limitations in isoprenoid biosynthesis. *Current Opinion in Biotechnology, 66*, 171–178.

CHAPTER 15

Transcription Activator-Like Effector Nucleases (TALENs): Genome Editing Tool to Explore Enhanced Activity of Antidiabetic Plants

GEETANJALI HARALE,[1] SHEETAL PARDESHI,[2] PARAMA MAJUMDAR,[3] and SUMAN GANGER[4]

[1]*Department of Biotechnology, Vivekanand Education Society's College of Arts, Science, and Commerce, Mumbai, Maharashtra, India, E-mail: haralegeetanjali@gmail.com*

[2]*Department of Microbiology, PES Modern College of Arts, Science, and Commerce (Autonomous), Shivajinagar, Pune, Maharashtra, India, E-mail: sheetalppardeshi@gmail.com*

[3]*Department of Botany, Hislop College, Nagpur, Maharashtra, India, E-mail: parama.majumdar@gmail.com*

[4]*Department of Microbiology, Vivekanand Education Society's College of Arts, Science, and Commerce, Mumbai, Maharashtra, India, E-mail: sumanganger@gmail.com*

ABSTRACT

Development of genome editing tools based on chimeric nucleases have led us to target and modify specific genome sequences with high precision. These sequence specific nuclease technologies are gaining popularity in the field of medical science for detection, prevention, and treatment of diseases. A major prospect of this gene-editing tool is its application in plant improvement with

Antidiabetic Potential of Plants in the Era of Omics. Deepu Pandita, PhD, Anu Pandita, MSc, & Chander Bhanu, MBBS (Eds.)

special reference to the introduction of novel genes and modulating gene expression. The primary impediment to transformation, however, is the high rate of illegitimate recombination. Genome engineering tools that induce DNA double-strand breaks (DSBs) at desired genomic locations stimulate homology-directed repair (HDR) or error-prone nonhomologous end joining (NHEJ), thus improving transformation rate significantly. Transcription Activator-Like Effector Nucleases (TALENs) is one such tool which is an engineered endonuclease composed of programmable, sequence-specific DNA-binding domain attached to a nonspecific DNA cleavage domain. The competency to quickly and efficiently alter genes using TALENs finds a profound influence on biological research and yields potential therapeutic approaches for chronic illness like diabetes from phytomedicine stand point. Production of plant metabolites that supplement or replace the traditional antidiabetic drug therapy can be manipulated at the molecular level using the genome editing tools. Use of TALENs for pathway elucidation, enhanced secondary metabolite production and creation of knockout mutants for desired traits in medicinal plants have been contemplated in this chapter, and it also intends on giving detailed information on TALENs technology.

15.1 INTRODUCTION

Herbal medicine (herbalism or herbology) is a traditional approach to explore plant kingdom for medicinal purposes. Plants are looked as treasure for disease treatments even before the prehistoric period. According to WHO (World Health Organization), 'around 80% of the world population depend on herbal medicines for managing or treating part of their primary health care needs. It is estimated that 21,000 plant species have the potential to be tapped as medicinal plants.' AYUSH (Ayurveda, Yoga, and Naturopathy, Unani, Siddha, and Homeopathy) systems of India have classified about 8,000 herbal remedies (NHP Admin, 2021). Even with such a rich repository of traditional medicine, the research data on these medicinal plants is in its infancy. For the last three decades, governments around the world, through their National Policies, have given a lot of impetus to study and explore these traditional medicines, which is also perceived as a booming global commercial enterprise (Sissi and Iris, 2011).

The number of known chemical structures in plants, called phyto-chemicals or phyto-compounds, is projected to be nearly four-fold higher than that of the microbial kingdom (Ramachandra Rao and Ravishankar,

2002). Most of these phytochemicals are secondary metabolites, products of supplementary pathways in the plants . (Hussein and El-Anssary, 2019). Secondary Metabolites (SM) play a lesser significant role in plant development or growth but are required for the plant's interaction with the environment (Pagare et al., 2015). In general, they facilitate the plant to protect itself from the biotic and abiotic stresses and give peculiar characteristics of fragrance, color, etc. These low molecular weight compounds belong to steroids, alkaloids, phenolic, lignin, carbohydrates, and glycosides. They not only play an important role in adapting plants to its environment but also act as an important source of pharmaceuticals, with proven activities of antiallergic, anticancer, antimicrobial, anti-inflammatory, antidiabetic, and antioxidants (Tran et al., 2020). They also have wide applications in agrochemicals, agents for flavors, fragrances, colors, biopesticides, and food additives (Thirumurugan et al., 2018).

Evidence suggests that SM interact with important targets in cells like protein, membranes, and nucleic acid and exert a broad spectrum of bioactivities. Their mode of action could be a simple specific interaction with the cell receptors or less specific by forming hydrogen, hydrophobic, and ionic bonds with proteins resulting in 3D conformational changes (Wink, 2015). With current advancement in understanding of biochemical pathways, molecular biology, genome analysis and bioinformatics, a humongous amount of data is generated which needs compilation and exploration. Table 15.1 lists some public repositories for medicinal plants and their derived phytochemicals, assembled through literature mining and then by manual curation from scientific text and books. Assembling and accessibility to all-inclusive information on medicinal plants and their natural chemical compounds via database may facilitate direct use of plant extracted natural compounds for computer-simulated drug discovery (Kumar et al., 2018; Shanak et al., 2019).

Diabetes mellitus (DM) is a group of metabolic diseases distinguished by chronic hyperglycemia (high blood sugar) with altered metabolism of carbohydrates, proteins, and fats due to defects in insulin action, insulin secretion, or both (Richard et al., 2000). It is a global health problem distressing human society in all walks of life (Reitsma and Terpstra, 1981). In 2021, as per WHO 422 million people world over have diabetes, predominantly in low-and-middle-income countries. Modern antidiabetic drug acquisition is beyond the reach of people from such countries and the in-general side effects associated with long-term drug use is significant. Hence to manage and/or treat diabetes, plant based therapy is sought for. Right in 1980, the

WHO expert committee on DM recommended the investigation of traditional methods of use of plants with medicinal properties for diabetes management.

TABLE 15.1 Few of the Medicinal Plant Databases and Reference Papers on Them

Databases	URL	References
Indian Medicinal Plants, Phytochemistry And Therapeutics (IMPPAT), India	https://cb.imsc.res.in/imppat/home	Mohanraj et al. (2018)
Prelude Medicinal Plants Database, Africa	https://www.africamuseum.be/research/collections_libraries/biology/prelude	
Medicinal Plants Database for Drug Designing (MPD3), USA	https://www.bioinformation.info/	Mumtaz et al. (2017)
Dr. Duke's Phytochemical and Ethnobotanical Databases, USA	https://phytochem.nal.usda.gov/phytochem/search	–
Taiwan indigenous plants database-3D (TIPdb-3D)	https://cwtung.kmu.edu.tw/tipdb/	Tung et al. (2014)
Traditional Chinese Medicine Integrated Database (TCMID), Taiwan	http://tcm.cmu.edu.tw/	Chen (2011)

For centuries, plants as remedies have been used for the treatment of diabetes, but few have been scientifically evaluated (Ozougwu and Eyo, 2011). Literature cites and/or recommends many plants for diabetes management (Grover et al., 2002; Kayarohanam, 2015; Sidhu, 2013; Shanak et al., 2019). More than 1,200 plant species find their names in ethnomedicine for remedying diabetes (Behzad et al., 2018; Lankatillake, Huynh, and Dias, 2019) but additional pharmacological and chemical research to decipher the exact mechanism of their antidiabetic property is warranted. Research suggests that different parts of the medicinal plants are used. Their phytochemical analysis shows the presence of polyphenols, terpenoids, and flavonoids. Numerous mechanisms of action are proposed for the plant extracts but there is inadequate mention of the exact antidiabetic mechanism of the bioactive compounds (Aggarwal and Shishu, 2011; Kayarohanam, 2015; Sidhu, 2013; Patel et al., 2012). Many authors suggest that the hypoglycemic effects of the plant extract concurs with standard drugs and is due to its antioxidant property (Kayarohanam, 2015). Type 2 diabetes is more prevalent than Type 1 diabetes. Around 95% of diabetic cases in the USA belong to type 2 as per the Centre of Disease Control and Prevention's

2020 National Diabetes Statistics report titled Estimates of Diabetes and Its Burden in the United States. Type 2 Diabetes is a more complex disorder which involves genetics, epigenetics, and environmental factors (Prasad and Groop, 2015). Medicinal plants with antidiabetic properties have shown to effect this multiple levels of interaction, suggesting the plant extracts to be of complex origin having manifold targets to act upon (Wink, 2015). Anti-diabetic plants have been classified based on parts of plant used (bark, bulb, flower, fruit, seed, stem, tuber, and whole plant) and their mode of reduction in blood glucose (insulinomimetic or insulin secretagogues activity). Studies on animal models suggest the intake of plant extracts or concoction increases interaction between insulin and its cell receptors, i.e., reducing insulin-resistance. In some beta-cell rejuvenation occurs to release more insulin and increase in plasma insulin concentration. There is a boost in the metabolism of glucose in tissues and cells and lowering of glycogen hydrolysis in the liver by controlling the action of hormones as seen in some studies (Ansari et al., 2021; Patel et al., 2012; Shanak et al., 2019). Cataloguing of such active constituents is done using the phytochemical analysis, biochemical pathway studies and bioinformatics. Understanding the genetic makeup of such plants, metabolic steps involved in the production of these secondary metabolites and their regulation, co-metabolites, and their toxicity will facilitate the plant genetics and molecular biologist to unleash the power of genome editing tools to create plants with modified traits.

Considering the complexity of the plant genome, its manipulation using conventional genetic engineering techniques are tedious, time consuming and unpredictable. Restriction enzymes play an important part in gene manipulation, but their application is limited to only those genes, which are near the restriction site. Precise genome editing by engineered nucleases such as homing meganucleases (Stoddard, 2005), Zinc-finger nuclease (ZFNs), transcription activator-like effector nucleases (TALENs) (Wood et al., 2011) and Clustered Regularly Interspaced Short Palindromic Repeats/ CRISPR-associated 9 (CRISPR/Cas9) (Cong et al., 2013) have reformed the possibility to alter the target genomic site with high accuracy. These engineered endonucleases are bound to modular DNA recognition domain, and recognize any target sequence. They give the advantage of selected possible outcomes such as mutations, insertions, deletions, replacements, and rearrangements of chromosome structure like inversion and translocation (Liang et al., 2014). They facilitate the generation of site-specific double-stranded DNA breaks (DSBs), which are vital for effective gene editing (Bi and Yang, 2017). In eukaryotes, this triggers inherent repair mechanisms viz.

error-prone Non-Homologous end-joining (NHEJ) pathway or via the error-free homology-directed repair (HDR) pathway (Shrivastav et al., 2008) which attempts to repair the error created by TALENs and invariably produces genes with expected specific modified sequences resulting in desired characteristics.

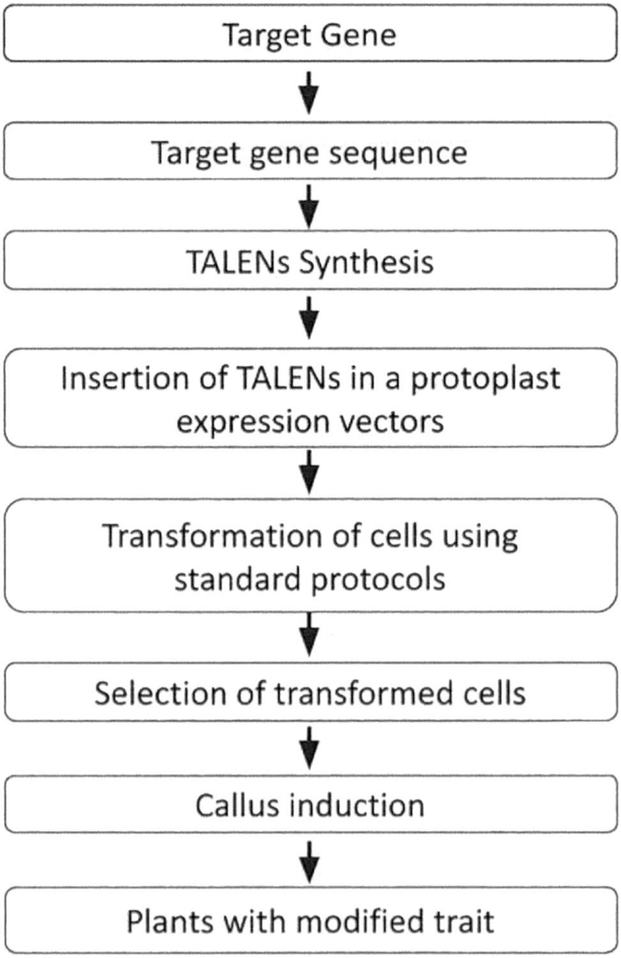

FIGURE 15.1 Basic steps to create plants with modified traits using TALENs technology.

TALENs is one such promising system available (Bacman et al., 2014) in this era of Genome Editing. Figure 15.1 outlines the basic steps to create plants with modified traits using TALENs technology. The current chapter presents an overview on the technological aspect of TALENs and suggests specific approaches with respect to enhancing the antidiabetic potential of plants. History of Genome editing has been covered in this book elsewhere.

15.2 TALENS TECHNOLOGY

TALENs is a chimeric or fusion protein comprising of two parts, a specific DNA sequence binding domain (TALES) and a non-specific nuclease enzyme domain (*FokI*)

15.2.1 TALES

The DNA binding domain is developed from the Transcription Activator Like (TAL) effector. The TAL effector protein is naturally found in *Xanthomonas*, a plant pathogen (Beurdeley et al., 2013; Christian et al., 2012; Takasu et al., 2013). During infection, *Xanthomonas* delivers TAL effectors into the plant cells and activates transcription of certain genes which is linked either to resistance or susceptibility to the infection (Mak et al., 2013). This mechanism enables the pathogen to rapidly synthesize a new DNA specific activator, thereby leading to colonization of the pathogen (Christian et al., 2012; White et al., 2009).

15.2.1.1 STRUCTURE OF TALE

TAL effector protein is composed of a DNA binding repeat region sandwiched between a non-repeat region at N-terminal and C-terminal. The number of amino acid repeats can vary from five to over 30. On an average 17 tandem repeats are present (Mak et al., 2013). Each repeat consists of a 34 amino acid motif (Christian et al., 2012; Joung and Sander, 2013; Mak et al., 2013). This amino acid sequence is constant except at the 12th and 13th position (refer to Figure 15.2). The amino acids at these positions are variable in different types of repeats and are responsible for specific nucleotide recognition. These residues are called repeat variable di-residues (RVDs). Structurally, each repeat forms two alpha-helices joined by a short loop that contains the RVD. Out of the two, amino acid at the 12th position interacts with one of the alpha helices to stabilize the loop, and the 13th residue contacts a specific base within the major groove of DNA (Christian et al., 2012). RCSB's Protein DataBase (https://www.rcsb.org/structure/3UGM) can be referred to understand TALE-DNA interaction. Around two dozen RVDs are identified in the known TAL effector molecules. Seven most common RVDs are HD, NG, HG, NN, NS, NI, and 'N*'; where N* corresponds to a 33 residue repeat with a missing residue within the RVD loop (Mak et al., 2013). Table 15.2 shows the affinity of RVDs to specific nucleotides. Within them HD, NN, NI, and NG, account for 75% of total RVDs (Chen and Gao, 2013). At the carboxyl end TAL effector protein has a nuclear localization signal (NLS), which directs TALE towards the nucleus (Boch et al., 2009).

TABLE 15.2 Affinity of RDVs to Specific Nucleotide

SL. No.	RVDs	Preferred Nucleotide
1.	HD	Cytosine
2.	NG	Thymine
3.	NI	Adenine
4.	NN	Guanine or adenine
5.	N*	Cytosine
6.	IG	Thymine
7.	NS	Adenine, cytosine, guanine, or thymine

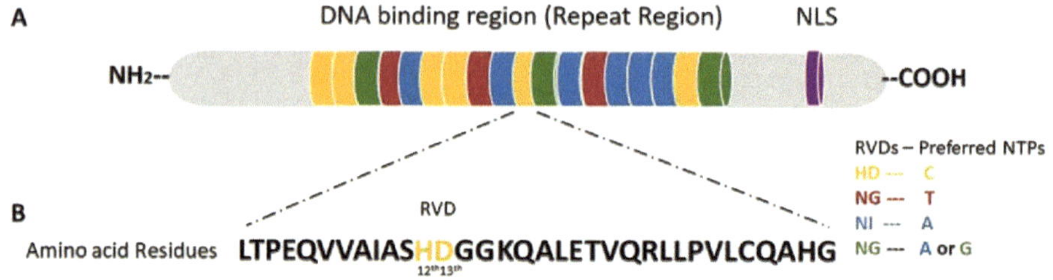

FIGURE 15.2 TALE protein. (A) General structure: It consists of a DNA binding region, which has a highly repetitive, specific nucleotide-binding domain, sandwiched between non-repetitive sequences. At the carboxyl end, a nuclear localization signal (NLS) sequence is present which assists TALE towards the target sequence. (B) Each nucleotide-binding domain consists of a 34 amino acid motif. This amino acid sequence is constant for all motifs except at the 12th and 13th position. These are referred to as repeat variable di-residues (RVDs). RDVs are responsible for the specificity of the domains. The most common RVDs are HD, NG, NI, and NS with cytosine, thymine, adenine, and adenine or guanine as target nucleotides, respectively.

15.2.2 FOKI ENDONUCLEASES

FokI is a type II restriction endonuclease which is isolated from *Flavobacterium okeanokoites. FokI* recognizes asymmetrical sequences and makes a double-strand break outside the recognition sequence. It consists of 587 amino acids with a molecular weight of 65.4 kDa. Two separate *FokI* domains, one for DNA recognition and other for cleavage were identified post proteolytic studies. It was observed that the N-terminal proteolytic fragment (41 kDa) recognizes the DNA and C-terminal proteolytic fragment (25 kDa) nonspecifically cleaves DNA in the presence of Mg^{2+} ions. Studies have proved that *FokI* catalytic domain can be fused with other DNA, recognizing domain (protein) to construct an actively fused or chimeric protein with unique recognition specificity (Bitinaite et al., 1998). Bitinaite et al. postulated that 'dimerization of two *FokI* molecules in the presence of specific DNA and metal ions are required for DSBs. Also, that dimerization occurs at the interface of two catalytic domains of *FokI.*'

15.2.3 TALEN ASSEMBLY METHODS

The DNA binding domains of TALEs are highly repetitive in nature, therefore, constructing engineered TALENs by regular PCR and cloning methods is very challenging (Hommelsheim et al., 2014). Recent knowledge of the composition of TALENs has opened doors for the invention of many TALEs assembly methods, which are briefly mentioned in Table 15.3.

TABLE 15.3 TALEs Assembly Method

TALEs Assembly Method	Description	References
Standard cloning	This method employs a classic restriction digestion approach. First single vector carrying one of the four nucleotides recognizing TALE sequences is constructed. Then using restriction enzymes, any two of vectors are specifically cleaved following ligation. These are called double unit vectors. Further, double unit vectors are digested and ligated to obtain specific sequences of TALEs. Using this approach, three methods, namely Unit Assembly, REAL, and REAL-FAST Assembly have been developed.	Chen and Gao (2013); Dey (2021)

TABLE 15.3 *(Continued)*

TALEs Assembly Method	Description	References
Golden gate	It is one of the most used types of assembly method. Golden Gate uses Type IIS restriction endonuclease (BsaI, Esp3I). Type II restriction endonucleases cleave DNA outside their recognition sequence by creating a four base overhanging cuts. These overhanging bases are unique for a particular DNA fragment as it depends on the DNA construct and not on the restriction site. In Golden Gate, a series of vectors containing one RDV each, are designed in such a way that once digested, vectors have compatible overhangs. This will help the digests to assemble in a specific order according to the target sequence. It is a single step reaction in which digestion and ligation is performed. At a time 9 TALES can be aligned. Different such modules are created and then ligated together to get desired TALE sequence.	Chen and Gao (2013); Engler et al. (2009); Weber et al. (2011)
Solid phase	Four assembly methods are developed using solid phase, which are high throughput. They are FLASH (fast ligation-based automatable solid-phase high-throughput), ICA (iterative capped assembly), chip, and LIC (ligation independent cloning). In the solid phase, the assembly of TALEs is carried out on a solid bead which is magnetic. Depending on the method, different approaches are employed for the restriction and ligation reaction to extend the chain	Chen and Gao (2013)
STAR (simple TALE assembly reaction)	It is also called the Gibson Assembly which uses an isothermal that is rapid, accessible, scalable, also labor, and cost-effective. In this method, a new sequence for the 21 bp ends of TALEs is developed with 16 unique base sequences for each TALE repeat domain. This helps in the correct sequential assembly in the solution. A group of four TALE sequences are ligated with the vector called the 4mer. In a separate reaction four 4mers are fused for a complete TALE expression vector.	Gogolok et al. (2016)

As shown in Figure 15.3, for a single target site, two TALE sequences spanning approximately 16 to 20 nucleotides are constructed (one for each forward and reverse strand). These TALE sequences are then linked to a monomeric *FokI* cleavage module. As explained previously, dimerization of *the FokI* cleavage module is required for DBS. This dimerization when combined with the specificity of two TALE DNA binding domains facilitates high sequence-specific cleavage in complex genomes. The effectiveness of TALE-*FokI* is seen in a number of plant species (De Lange et al., 2014).

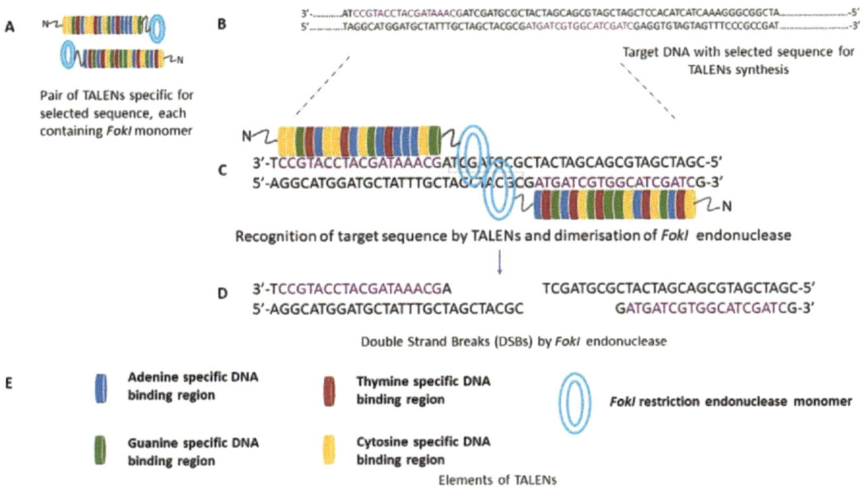

FIGURE 15.3 Diagrammatic representation of target DNA recognition and cleavage by TALENs. (A) Generation of a pair of TALENs, one for forward strand and other for reverse strand, specific for selected sequence. Each TALEN contains a *FokI* monomer. (B) Target DNA with TALENs recognition sequence (purple colored). (C) Once TALENs recognizes the sequence, it binds and leads to dimerization of *FokI* endonucleases. Dimerised, *FokI* makes a double-strand break in the target DNA. (D) Cleaved target DNA with staggered cuts. (E) Different elements present in TALENs.

15.3 DELIVERY SYSTEMS AND GENERATION OF MODIFIED PLANT

Delivery of TALENs is an area of challenge more so as compared to its design. The success of genome editing tools largely depend upon the genetic transformation and plant regeneration, both considered as a setback in the process (Altpeter et al., 2016).

Depending on the objective of the study, specific plasmids can be constructed with genes encoding the TALE proteins. Study on improving soybean oil quality using TALENs involved individual TALENs cloned into protoplast expression vectors containing nopaline synthase (NOS) promoter and terminator. The N-terminus of the TALENs was modified by fusing nuclear localization signal sequence (NLS) of SV40, to direct it to the plant cell nucleus (Demorest et al., 2016). Recombinant plasmids containing the designer TALENs genes can then be introduced using a variety of plant transformation methods.

Protoplasts of *Nicotiana benthamiana* were transformed with direct delivery of purified two TALEN monomers ALS2T1L and ALS2T1R as against nucleic acid. The proteins translocated into the nucleus and cleaved the host DNA at a prearranged endogenous site. Though the desired mutation in gene *acetolactate synthase 2* (NbALS2) was established the frequency of mutation was marginal at 1.4% as compared to plasmid-mediated transformation which gave 18.6%. Nonetheless, this attempt suggests the possibility of delivering large TALEN protein directly in to the plant cell (Luo et al., 2015).

Methods tapped successfully to carry out transformation in plants implementing TALENs technology include PEG protoplast transformation in *Solanum tuberosum* (Clasen et al., 2016); *Nicotiana tabacum* (Yong Zhang et al., 2013); *Agrobacterium*-mediated system in *Zea mays* (Kelliher et al., 2017; Tung et al., 2014) *and* particle bombardment in case of *Solanum tuberosum* (Clasen et al., 2016; Nicolia et al., 2015). From medicinal plants point of view which include shrubs and trees, transformation using the above mentioned methods have been successfully carried out in *Veratrum dahuricum* (Ma et al., 2020); *Catharanthus roseus* (Niazian, 2019) and many others. Positive results are obtained in confirmed antidiabetic plants like Acacia spps. (Xie et al., 2007); Allium spps. (Malla et al., 2021); *Momordica charantia* L. (Narra et al., 2018; Thiruvengadam et al., 2012) and others when transformed with agrobacterium or biolistic facilitated delivery systems. With this backing of successful transformation protocols, the delivery of TALENs can be pursued. Transformed cells are then obtained using suitable selectable markers. Since these cells are totipotent, plants with desired characteristics can be developed (Demorest et al., 2016; Yong Zhang et al., 2013).

15.4 TALENS ASSISTED GENOME EDITING MECHANISMS

Once a pair of TALENs are designed, they bind at two adjacent sites on opposite strands. *FokI* is then dimerized and cleaves the target site. The

DNA repair system is activated due to this double-strand break. Repair can be carried out by either non-homologous end joining (NHEJ) or homology directed recombination (HDR) (K. Chen and Gao, 2013).

NHEJ pathway requires slight to no homology to join the broken ends (Pruett-Miller, 2014). An error prone pathway which may lead to frame-shift mutation due to small insertion or deletion of nucleotides (indels) at or near the DSBs (Bi and Yang, 2017; Chen and Gao, 2013). NHEJ is the most frequent cellular response to break repair. Since the pathway brings about minimal change at the repaired site, they are preferred to execute targeted mutagenesis for gene expression or function (Puchta, 2005). Alternatively, HDR can repair DSBs with the help of a homologous DNA template. This will result in gene modification by sequence insertion. Therefore, through these two systems, TALENs are used to edit target genes by creating mutation, insertion, replacement, and chromosome rearrangements (Chen and Gao, 2013).

15.4.1 MUTATIONS

To introduce mutation in a gene, TALENs must be designed for its open reading frame. Specifically, they should be targeted to the exons in the first 1/3 to 1/2 of the reading frame. The indels introduced by the NHEJ pathway can cause frame-shift mutation, resulting in non-functional protein, i.e., knockout gene function. This method is used to target many plant species, which includes:

- Targeting *Arabidopsis* gene *ADH1* in Voytas's laboratory;
- Disrupting the effector-binding element (EBE) in the promoter region of S gene (the sucrose-efflux transporter gene) of rice, without changing its expression, imparting resistance to the cells against bacteria blight disease (Chen and Gao, 2013).

15.4.2 GENE INSERTION AND REPLACEMENT

Traditionally gene insertion methods were based on the integration of foreign genes into the host cell at random. TALENs along with the HDR pathway, integrates genes at a specific site therefore avoids random insertion. Using this approach one can select a gene which has a strong promoter and can insert a gene of interest at that site. This will result in an enhanced level of gene expressions. Furthermore, multiple gene insertion can be carried out at the same target site.

TALENs-HDR pathway can also be employed for target gene replacement. In this case, along with TALENs, homologous sequences of the donor DNA templates flanking the target site are transferred to the cell. Once DSB is induced by TALENs, HDR immediately repairs the break using the donor DNA, resulting in gene replacement. Gene insertion and replacement is carried out in *SurA* and *SurB* genes in tobacco (Chen and Gao, 2013).

15.4.3 CHROMOSOMAL REARRANGEMENT

Large deletions, inversions, and translocations can be obtained by introducing two pairs of TALENs simultaneously in the cell. When these two pairs of TALENs are targeted at two loci on the same chromosome, deletion, and inversion is observed. However, when they are directed to two different chromosomes then it results in translocation. Since these modifications require the introduction of two DSBs at once, less chromosomal rearrangements are obtained to date (Figure 15.4) (Chen and Gao, 2013).

15.5 ANTICIPATED USE OF TALENS TO MANIPULATE ANTIDIABETIC ACTIVITY

Exploitation of genome editing tools for medicinal plants have been hindered due to limited genetic information. To execute sequence specific breaks and avoid any off target effect, it is safer to know the entire genome for precise editing (Alok et al., 2020). Using CRISPR-Cas9 targeted mutagenesis has been executed in *Salvia miltiorrhiza, Papaver somniferum L., Taraxacum kok-saghyz and Solanum pimpinellifoilium* considered as medicinal plants. We enlist possible mechanisms using TALENs.

15.5.1 KNOCKOUTS

One of the main hurdles in engineering plants with ethnomedicinal records is the lack of understanding of the pathways involved in the complex secondary metabolic scheme which is characteristically species/genus/family specific throughout the plant kingdom. TALENS are effective systems to create knockout mutants. Such mutants are a key to understanding and elucidating biochemical pathways as well as to manipulate a desired trait that may lead to the overproduction of required secondary metabolites.

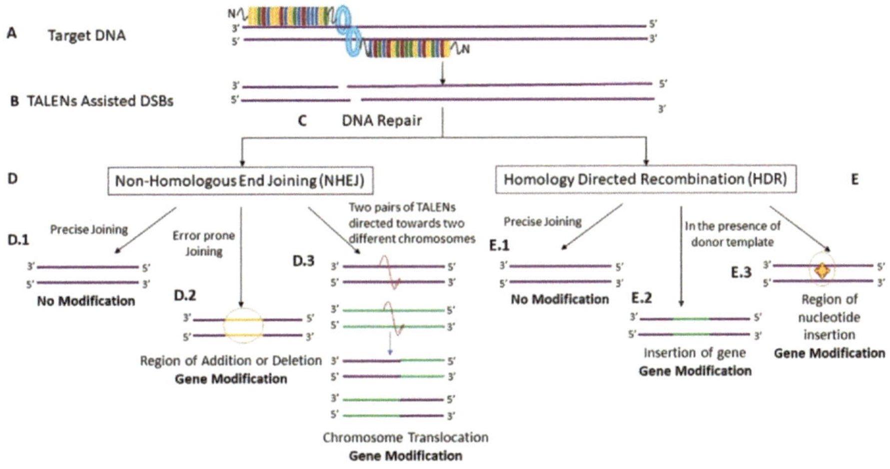

FIGURE 15.4 TALENs assisted genome editing mechanisms. (A) Recognition and binding of pair of TALENs to target DNA. (B) TALENs assisted double-strand breaks leading to staggered cuts. (C) Once DNA is cleaved, the cell's inherent DNA repair system tries to repair the damage via non-homologous end joining (NHEJ) and homology directed recombination (HDR). (D) For NHEJ, there are three possible outcomes. (D.1) In case of precise joining of the sequence, there will be no modification. (D.2) Since NHEJ is an error prone repair system, it may lead to random addition or deletion of nucleotides. This may cause gene modification resulting in gene knockout. (D.3) When two pairs of TALENs are directed towards two different chromosomes, there is a mutual exchange of chromosome fragments. This is called the translocation of chromosome and gene modification is observed. (E) For HDR repair, a donor gene with intended modification, is inserted along with TALENs in a vector for homologous recombination. HDR can also produce one of the three outcomes. (E.1) No gene modification in case of precise joining. (E.2) A novel gene can be inserted at the cleaved site which is provided by the vector. (E.3) Specific nucleotide insertion can be achieved when repair is accomplished with a similar donor gene with modified nucleotides.

15.5.1.1 ELUCIDATION OF TARGET GENES

Since the pathway for the synthesis of a particular compound of interest may be a part of a hugely complex network of secondary metabolism, identifying enzymes that directly affect a particular metabolite concentration is the key to genetic manipulation. RNAi has been used very often to knockdown certain genes to understand the properties and elucidate the synthesis of natural products. However, incomplete suppression and lack of absolute selectability of the target, make RNAi knockdown mutants extremely tedious to screen.

Programmable genome-editing tools like TALEN can be used with evidently high specificity for mutagenesis sites. Such mutants enable qualitative interrogation of characters and give leads to design metabolic engineering experiments. This approach to understand gene function is categorized under the reverse genetics approach. Similar approach to understand gene functions through knockouts has been attempted in model organisms like *Arabidopsis* and rice (Kitaoka et al., 2015; Miao et al., 2013; Thorneycroft et al., 2001)

Studies on secondary metabolism by knocking out genes have been implemented in a diverse range of microorganisms. Site-specific saturation mutagenesis, in *S. cerevisiae* led to the identification of three activity-enhancing mutations, namely V13D, S148I, and V301E on mevalonate kinase which enhanced the lycopene production (Chen et al., 2018). Molecular targeted mutagenesis has also been used to obtain a 11 fold increase in carotene content in *S. cerevisiae* through knockouts at mevalonate kinase (Jakočiūnas et al., 2018). These studies indicate the possible site for manipulations for overexpression of a desired phytochemical

15.5.1.2 TRIGGERING DESIRED ACTIVITY

Overcoming inadequate enzymatic activity by targeting the regulatory mechanism can be an obvious and effective method of phytochemical overproduction. Mevalonate pathway associated negative regulators genes have been a frequent site of targeted knockouts in *S. cerevisiae*, leading to overproduction of carotenoids (Bröker et al., 2018). Enhanced isoprene production by genetic manipulation of negative regulator linked genes has been reported in *Rhodobacter sphaeroides*. By knocking out the gene associated with glucose-6-phosphate dehydrogenase, a higher beta-carotene titer was obtained (Qiang et al., 2019)

Knockout mutants for undesired toxic metabolites can also be created using TALENs providing a desirable cultivar. Steroidal glycoalkaloids (SGAs) in potatoes like solanine and α-chaconine have been aimed in such studies. Potato cultivars free of SGA are desired traits for plant breeders. A TALEN expression vector system was constructed targeting the SSR2, which encodes a key enzyme for SGA biosynthesis. The resultant SSR-knockout potato mutants were negative for the common precursors of SGA without any change in the phenotypic characteristic (Yasumoto et al., 2019).

Reducing sugars accumulation increases in potato tubers under cold environment conditions, a necessary step charted to reduce sprouting and extending post-harvest shelf life. Use of RNA interference (RNAi) technology to control the expression of vacuolar invertase gene (VInv), encoding a protein that breaks down storage sugar sucrose to glucose and fructose was developed. Incomplete gene silencing is one of the major drawbacks of this technology along with plants labeled as transgenic. TALENs technology when used created some plant lines with all allele VInv knockout mutants resulting in undetectable levels of reducing sugars. Furthermore, some of the modified plant lines showed no presence of TALEN DNA insertion in the potato genomes (Clasen et al., 2016).

Natural antinutrients like phytic acid present in cereal grains chelate the essential micronutrients such as calcium, iron, and potassium, making them unavailable and lowering the nutritive value of the cereals. To lower the concentration of phytic acid, it is necessary to understand its biosynthetic pathway and determine the genes responsible for the same. TALENs have been engineered against candidate genes, which could lower the phytic acid and provide low phytate wheat (Pandey et al., 2021).

These targeted mutations generally have been found to be heritable and follow Mendelian traits in most of the experimental plants like rice (Zhang et al., 2016), tomatoes (Čermák et al., 2015) and *Arabidopsis thaliana* (Christian et al., 2013). Studies on truncated TALENs especially with its C-terminus domain demonstrate improved efficiency of mutational rates in tobacco and *Arabidopsis* plants (Christian et al., 2013; Zhang et al., 2016). Nevertheless, to ensure the presence of TALEN-induced mutations, a large number of transgenic lines must be produced (Zhang et al., 2016) and most of the time, the mutant lines are chimeric or mosaic in nature (Yasumoto et al., 2019; Zhang et al., 2016).

15.5.2 ENHANCED PRODUCTION OF SECONDARY METABOLITES

Applications of genome editing tools like TALENs include gene expression regulation beside gene knockouts and knockins. Regulatory regions of the target gene interact with transcriptional repressors or activators. Fusing them with DNA-binding domains of TALENs can bring in regulation of the target gene (Yi Zhang et al., 2019). Precise modification in the tomato genomes was executed by engineered TALENs for their pigment anthocyanin increased production. DSBs formed in the initial region of the ANT1 gene resulted in NHEJ-induced mutations. ANT1 codes for a Myb transcription factor, which increases anthocyanin accumulation developing intensely purple plant tissue (Čermák et al., 2015). MYB Transcription factor family is involved in controlling various cellular processes, one being the plant cell response to biotic and abiotic stress (Liu et al., 2015). In plants, MYB proteins like R2R3-MYBs control Phenylpropanoid metabolism (Liu et al., 2015; Ma and Constabel, 2019). Phenylpropanoid is central to the synthesis of secondary metabolites like flavonoids, phenolic acids, and many others (Liu et al., 2015). Many antidiabetic plants active phytochemical belong to flavonoids (Mohan and Nandhakumar, 2014; Mutha et al., 2021). Flavonoids like morin, naringenin, luteolin, daidzein, chrysin, and others show antihyperglycemic effect (Al-Ishaq et al., 2019) and hence MYB Transcription factor genes play an important role in defining the antidiabetic property to plants. Similarly phenolic acid-like ellagic acid which is an α-glucosidase inhibitor (Ali Asgar, 2013), gallic acid a β-cell regenerator, gluconeogenesis lowering agent (Latha and Daisy, 2011) can be the target for their enhanced production. Upregulating the biosynthesis pathways of such secondary metabolites using TALENs can be sought to enhance the antidiabetic property of the plant. Developing plants as bioreactors and optimizing the extraction process for enhanced production and isolation of such active phytochemicals could be economically viable options.

15.5.3 INTRODUCTION OF GENES IN MICROBIAL SYSTEMS

Microbial systems are one of the robust, commercially profitable, and easier systems to manipulate due to their less complex genome, detailed information, and technology available to modify and control the gene expression and ease of growth and product purification. Hence, it makes it easier to express constructs of eukaryotic nature into simpler prokaryotic models like Escherichia coli or unicellular eukaryotes like yeast and use them as production factories.

Lycopene belongs to the Isoprenoid class and is a useful phytochemical which has a huge commercial value for its antioxidant property. Literature cites many studies on metabolic engineering methods to enhance the lycopene production in E. coli and Yeast with success (Yu Zhang et al., 2021). Similarly, flavonoids like naringenin have been produced in *E. coli* by overexpressing the three enzymes; phenylalanine ammonia-lyase, 4-coumarate: CoA ligase, and chalcone synthase (CHS) and exogenous addition of 4-coumaric acid to induced cultures (Watts et al., 2004). Metabolic engineering of phenylpropanoid pathways have been accomplished in *E. coli, S. cerevisiae and Streptomyces spps.* for production of varied flavonoids (Sheng et al., 2020). Such studies suggest the feasibility of expressing phytochemicals in such model organisms.

The shortcoming of such manipulation is the low titer obtained in such arrangements (Shah et al., 2019). Persuasion of TALENs technology intertwined with the metabolic pathway engineering can prime enhanced production of such secondary metabolites.

15.6 ADVANTAGES AND LIMITATIONS OF TALENS OVER OTHER GENE EDITING SYSTEMS

For a very long time, selective breeding was the main tool for domestication of agriculture as well as for medicinal plants. This led to the selection of a better variety of plants for desirable outcomes. Later on, using hybridization technology, a major revolution took place in plant breeding. Even though these were effective methods, they were not gene-specific. In the early 1970s, the discovery of Ti plasmid opened endless opportunities for gene editing in plants as well. This fronted employment of restriction enzymes for gene editing increased target specificity but limited the site for modification.

Compared to these technologies, TALENs technology is more target site-specific in cutting and insertion of genes. It shows high engineering feasibility and is highly efficient (Khan, 2019). Also, TALENs are more mutagenic and less toxic in eukaryotic systems as compared to that of their counterpart genome editing tool ZFN (Christian et al., 2013; Jankele and Svoboda, 2014).

Use of TALENs technology in medicinal plants can also be beneficial seen from experiments on knockout mutants, wherein TALENs technology is more efficient when compared with RNAi technology. RNAi can successfully knock-down gene expression, but might result in incomplete suppression. It can also affect expression of other genes which might interfere with interpretation of the studies (Kitaoka et al., 2015). Since TALENs can be used to specifically alter a few nucleotides, the resulting plant line is similar to the original plant, or it can be one of the natural variants which may be already established in nature. This minimizes those risk factors which might occur due to the introduction of GM plants in nature.

Even though TALENs are used for gene manipulation, there is still scope for further research. The major limitations of TALENs are as follows:

- Cloning of TALENs is challenging, due to the presence of tandem repeat sequences (Hommelsheim et al., 2014).
- Compared to CRISPR, TALENs designing is complex; therefore, it may pose technical difficulties in generation also it is costly technology (Khan, 2019).
- Since TALENs are larger in size, delivery, and expression is tricky (Juillerat et al., 2014). Experimental setups have used truncated TALENs, which showed better stability, folding, and activity (Christian et al., 2013; Yong Zhang et al., 2013) nullifying its limitation.
- Even though the specificity of TALENs is high, efficiency is low, compared to other genome editing tools (Khan, 2019).

Ethical concerns associated with genome editing have been dealt in this book elsewhere.

15.7 FUTURE PROSPECTS

The 2020 Nobel prize in Chemistry was awarded to the discovery of CRISPR-Cas9 which suggests the revolution, the genome editing tools will bring in the field of plant genomics. In the backdrop of CRISPR-Cas9 its predecessors like ZFN and TALENs are also simultaneously and popularly being explored to manipulate genomes. As per a recent study in view of site-specific search and editing the heterochromatin regions in mammalian cells, TALENs performed better than CRISPR-Cas9 systems when compared (Jain et al., 2021). In addition to nuclease activity, TALE proteins have been employed along with other effector domains for diverse purposes like gene activation, gene repression and epigenetic modifications. Medicinal properties of plants are essentially dependent on a complex network of secondary metabolism. To resolve this complexity and enhance the desirable compound production, multifaceted genome editing tools like TALE give a promising road map. With the addition to the currently limited genome data of medicinal plants, the significance of these tools are also set to gain prominence.

Traditional medicinal records prescribe several plants to manage the symptoms of diabetes. These records are being revisited in the light of the prevalence of the global nature of this lifestyle disease. The knowledge of ethnomedicine is being combined with modern genome editing tools to enhance their efficacy. Customizable tools like TALENs are the imminent way ahead.

KEYWORDS

- **antidiabetic phytochemicals**
- **antidiabetic plants**
- **antihyperglycemic**
- **enhanced activity**
- **genome editing**
- **hypoglycemic**
- **plant engineering**
- **TALENs**
- **α-amylase inhibitors**
- **α-glucosidase inhibitors**

REFERENCES

Aggarwal, N., & Shishu, (2011). A review of Recent Investigations on Medicinal Herbs Possessing Antidiabetic Properties. *J. Nutr. Disord. Ther., 01*(01), 1–10.

Ali Asgar, M., (2013). Antidiabetic potential of phenolic compounds: A review. *Int. J. Food Prop., 16*(1), 91–103.

Al-Ishaq, R. K., Abotaleb, M., Kubatka, P., Kajo, K., & Büsselberg, D., (2019). Flavonoids and their antidiabetic effects: Cellular mechanisms and effects to improve blood sugar levels. *Biomolecules, 9*(9).

Alok, A., Jain, P., Kumar, J., Yajnik, K., & Bhalothia, P., (2020). *Genome Engineering in Medicinally Important Plants Using CRISPR / Cas9 Tool*. Elsevier Inc.

Altpeter, F., Springer, N. M., Bartley, L. E., Blechl, A. E., Brutnell, T. P., Citovsky, V., Conrad, L. J., et al., (2016). Advancing crop transformation in the era of genome editing. *Plant Cell, 28*(7), 1510–1520.

Ansari, P., Hannon-Fletcher, M. P., Flatt, P., & Abdel-Wahab, Y. H. A., (2021). Effects of 22 traditional antidiabetic medicinal plants on DPP-IV enzyme activity and glucose homeostasis in high-fat-fed obese diabetic rats. *Biosci. Rep 29, 41*(1), 1–15.

Bacman, S. R., Williams, S. L., Pinto, M., & Moraes, C. T., (2014). The use of mitochondria-targeted endonucleases to manipulate MtDNA. *Methods Enzymol., 547*, 373–397.

Behzad, M., Saber, A., Somayeh, S., Mohsen, A., & Beyranvand F., (2018). The most useful medicinal herbs to treat diabetes. *Biomed. Res. Ther., 5*(8), 2538–2551.

Beurdeley, M., Bietz, F., Li, J., Thomas, S., Stoddard, T., Juillerat, A., Zhang, F., et al., (2013). Compact designer TALENs for efficient genome engineering. *Nat. Commun., 4*. https://doi.org/10.1038/ncomms2782.

Bi, H., & Yang, B., (2017). Gene editing with TALEN and CRISPR/Cas in rice. In: *Progress in Molecular Biology and Translational Science* (Vol. 149, pp. 81–98). Elsevier Inc.

Bitinaite, J., Wah, D. A., Aggarwal, A. K., & Schildkraut, I., (1998). Foki dimerization is required for DNA cleavage. *Proc. Natl. Acad. Sci. U. S. A., 95*(18), 10570–10575.

Boch, J., Scholze, H., Schornack, S., Landgraf, A., Hahn, S., Kay, S., Lahaye, T., et al., (2009). Breaking the code of DNA binding specificity of TAL-type III effectors. *Science, 326*(5959), 1509–1512.

Bröker, J. N., Müller, B., Van, D. N., Prüfer, D., & Schulze, G. C., (2018). Upregulating the mevalonate pathway and repressing sterol synthesis in *Saccharomyces cerevisiae* enhances the production of triterpenes. *Appl. Microbiol. Biotechnol., 102*(16), 6923–6934.

Čermák, T., Baltes, N. J., Čegan, R., Zhang, Y., & Voytas, D. F., (2015). High-frequency, precise modification of the tomato genome. *Genome Biol., 16*(1).

Chen, C. Y. C., (2011). TCM Database@Taiwan: The world's largest traditional Chinese medicine database for drug screening in silico. *PLoS one*, *6*(1), 1–5.

Chen, H., Liu, C., Li, M., Zhang, H., Xian, M., & Liu, H., (2018). Directed evolution of mevalonate kinase in: *Escherichia coli* by random mutagenesis for improved lycopene. *RSC Adv., 8*(15021), 15021–15028.

Chen, K., & Gao, C., (2013). TALENs: Customizable molecular DNA scissors for genome engineering of plants. *J. Genet. Genomics, 40*(6), 271–279.

Christian, M. L., Demorest, Z. L., Starker, C. G., Osborn, M. J., Nyquist, M. D., Zhang, Y., Carlson, D. F., et al., (2012). Targeting G with TAL effectors: A comparison of activities of TALENs constructed with NN and NK repeat variable Di-residues. *PLoS One*, *7*(9), 1–9.

Christian, M., Qi, Y., Zhang, Y., & Voytas, D. F., (2013). Targeted mutagenesis of *Arabidopsis thaliana* using engineered TAL effector nucleases. *G3 Genes Genomes Genet., 3*(9), 1697–1705.

Clasen, B. M., Stoddard, T. J., Luo, S., Demorest, Z. L., Li, J., Cedrone, F., Tibebu, R., et al., (2016). Improving cold storage and processing traits in potato through targeted gene knockout. *Plant Biotechnol. J., 14*(1), 169–176.

Cong, L., Ran, F. A., Cox, D., Lin, S., Barretto, R., Hsu, P. D., Wu, X., et al., (2013). Multiplex Genome engineering using CRISPR/Cas systems. *Science, 339*(6121), 819–823. New York, N.Y.

De Lange, O., Binder, A., & Lahaye, T., (2014). From dead leaf, to new life: Tal effectors as tools for synthetic biology. *Plant J., 78*(5), 753–771.

Demorest, Z. L., Coffman, A., Baltes, N. J., Stoddard, T. J., Clasen, B. M., Luo, S., Retterath, A., et al., (2016). direct stacking of sequence-specific nuclease-induced mutations to produce high oleic and low linolenic soybean oil. *BMC Plant Biol., 16*, 1–8.

Dey, A., (2021). CRISPR/Cas genome editing to optimize pharmacologically active plant natural products. *J. Hazard. Mater., 164*(105359).

Engler, C., Gruetzner, R., Kandzia, R., & Marillonnet, S., (2009). Golden gate shuffling: A one-pot DNA shuffling method based on type IIs restriction enzymes. *PLoS One*, *4*(5).

Gogolok, S., Garcia-Diaz, C., & Pollard, S. M., (2016). STAR: A simple TAL effector assembly reaction using isothermal assembly. *Sci. Rep., 6*, 1–9.

Grover, J. K., Yadav, S., & Vats, V., (2002). Medicinal plants of India with antidiabetic potential. *J. Ethnopharmacol., 81*(1), 81–100.

Hommelsheim, C. M., Frantzeskakis, L., Huang, M., & Ülker, B., (2014). PCR amplification of repetitive DNA: A limitation to genome editing technologies and many other applications. *Sci. Rep., 4*, 1–13.
Hussein, R. A., & El-Anssary, A. A., (2019). Plants secondary metabolites: The key drivers of the pharmacological actions of medicinal plants. In: *Herbal Medicine* (pp. 11–30). IntechOpen.
Jain, S., Shukla, S., Yang, C., Zhang, M., Fatma, Z., Lingamaneni, M., Abesteh, S., et al., (2021). TALEN outperforms Cas9 in editing heterochromatin target sites. *Nat. Commun., 12*(606), 1–10.
Jakočiūnas, T., Pedersen, L. E., Lis, A. V., Jensen, M. K., & Keasling, J. D., (2018). CasPER, a method for directed evolution in genomic contexts using mutagenesis and CRISPR/Cas9. *Metab. Eng., 48*, 288–296.
Jankele, R., & Svoboda, P., (2014). TAL effectors: Tools for DNA targeting. *Brief. Funct. Genomics, 13*(5), 409–419.
Joung, J. K., & Sander, J. D., (2013). TALENs: A widely applicable technology for targeted genome editing. *Nat. Rev. Mol. Cell Biol., 14*(1), 49–55.
Juillerat, A., Beurdeley, M., Valton, J., Thomas, S., Dubois, G., Zaslavskiy, M., Mikolajczak, J., et al., (2014). Exploring the transcription activator-like effectors scaffold versatility to expand the toolbox of designer nucleases. *BMC Mol. Biol., 15*(13), 1–10.
Kayarohanam, S., (2015). Current trends of plants having antidiabetic activity: A review. *J. Bioanal. Biomed., 07*(02).
Kelliher, T., Starr, D., Richbourg, L., Chintamanani, S., Delzer, B., Nuccio, M. L., Green, J., et al., (2017). Matrilineal, a sperm-specific phospholipase, triggers maize haploid induction. *Nature, 542*(7639), 105–109.
Khan, S. H., (2019). Genome-editing technologies: Concept, pros, and cons of various genome-editing techniques and bioethical concerns for clinical application. *Mol. Ther. – Nucleic Acids, 16*, 326–334.
Kitaoka, N., Lu, X., Yang, B., & Peters, R. J., (2015). The application of synthetic biology to elucidation of plant mono-, sesqui-, and diterpenoid metabolism. *Mol. Plant, 8*(1), 6–16.
Kumar, A., Kumar, R., Sharma, M., Kumar, U., Prasad, G. M. N. V., & Singh, K. P., (2018). Uttarakhand medicinal plants database (UMPDB): A platform for exploring genomic, chemical, and traditional knowledge. *Data, 3*(7), 1–10.
Lankatillake, C., Huynh, T., & Dias, D. A., (2019). Understanding glycaemic control and current approaches for screening antidiabetic natural products from evidence-based medicinal plants. *Plant Methods, 15*(1), 1–35.
Latha, R. C. R., & Daisy, P., (2011). Insulin-secretagogue, antihyperlipidemic and other protective effects of gallic acid isolated from *Terminalia bellerica* roxb. in streptozotocin-induced diabetic rats. *Chem. Biol. Interact., 189*(1, 2), 112–118.
Liang, Z., Zhang, K., Chen, K., & Gao, C., (2014). Targeted mutagenesis in Zea mays using TALENs and the CRISPR/Cas system. *J. Genet. Genomics.*
Liu, J., Osbourn, A., & Ma, P., (2015). MYB transcription factors as regulators of phenylpropanoid metabolism in plants. *Mol. Plant, 8*(5), 689–708.
Luo, S., Li, J., Stoddard, T. J., Baltes, N. J., Demorest, Z. L., Clasen, B. M., Coffman, A., et al., (2015). Non-transgenic plant genome editing using purified sequence-specific nucleases. *Mol. Plant, 8*(9), 1425–1427.
Ma, D., & Constabel, C. P., (2019). MYB repressors as regulators of phenylpropanoid metabolism in plants. *Trends Plant Sci., 24*(3), 275–289.

Ma, R., Yu, Z., Cai, Q., Li, H., Dong, Y., Oksman-Caldentey, K. M., & Rischer, H., (2020). Agrobacterium-mediated genetic transformation of the medicinal plant *Veratrum dahuricum*. *Plants, 191*(2), 1–12.

Mak, A. N. S., Bradley, P., Bogdanove, A. J., & Stoddard, B. L., (2013). TAL effectors: Function, structure, engineering and applications. *Curr. Opin. Struct. Biol., 23*(1), 93–99.

Malla, A., Shanmugaraj, B., Srinivasan, B., & Sharma, A., (2021). Metabolic engineering of isoflavonoid biosynthesis by expressing glycine max isoflavone synthase in *Allium cepa* L. for genistein production. *Plants, 10*(52), 1–10.

Miao, J., Guo, D., Zhang, J., Huang, Q., Qin, G., Zhang, X., Wan, J., Gu, H., & Qu, L. J., (2013). Targeted mutagenesis in rice using CRISPR-Cas system. *Cell Res., 23*(10), 1233–1236.

Mohan, S., & Nandhakumar, L., (2014). Role of various flavonoids: Hypotheses on novel approach to treat diabetes. *J. Med. Hypotheses Ideas, 8*(1), 1 6.

Mohanraj, K., Karthikeyan, B. S., Vivek-Ananth, R. P., Chand, R. P. B., Aparna, S. R., Mangalapandi, P., & Samal, A., (2018). IMPPAT: A curated database of Indian medicinal plants, phytochemistry and therapeutics. *Sci. Rep., 8*(1), 1–17.

Mumtaz, A., Ashfaq, U. A., Ul Qamar, M. T., Anwar, F., Gulzar, F., Ali, M. A., Saari, N., & Pervez, M. T., (2017). MPD3: A useful medicinal plants database for drug designing. *Nat. Prod. Res., 31*(11), 1228–1236.

Mutha, R. E., Tatiya, A. U., & Surana, S. J., (2021). Flavonoids as natural phenolic compounds and their role in therapeutics: An overview. *Future J. Pharm. Sci., 7*(25), 1–13.

Narra, M., Ellendula, R., Kota, S., Kalva, B., Velivela, Y., & Abbagani, S., (2018). efficient genetic transformation of *Momordica charantia* L. by microprojectile bombardment. *3 Biotech, 8*(2), 1–8.

Khan, M. A., (2016) Introduction and Importance of Medicinal Plants and Herbs. National Health Portal https://www.nhp.gov.in/introduction-and-importance-of-medicinal-plants-and-herbs_mtl (accessed on 18 December 2021).

Niazian, M., (2019). Application of genetics and biotechnology for improving medicinal plants. *Planta, 249*(4), 953–973.

Nicolia, A., Proux-Wéra, E., Åhman, I., Onkokesung, N., Andersson, M., Andreasson, E., & Zhu, L. H., (2015). Targeted gene mutation in tetraploid potato through transient TALEN expression in protoplasts. *J. Biotechnol., 204*, 17–24.

Ozougwu, J. C., & Eyo, J. E., (2011). Evaluation of the activity of *Zingiber officinale* (Ginger) aqueous extracts on alloxan-induced diabetic rats. *Pharmacologyonline, 1*, 258–269.

Pagare, S., Bhatia, M., Tripathi, N., Pagare, S., & Bansal, Y. K., (2015). Secondary metabolites of plants and their role: Overview. *Curr. Trends Biotechnol. Pharm., 9*(3), 293–304.

Pandey, A. K., Aggarwal, S., Meena, V., & Kumar, A., (2021). Phytic acid reduction in cereal grains by genome engineering. *Genome Eng. Crop Improv.*, 146–156.

Patel, D. K., Prasad, S. K., Kumar, R., & Hemalatha, S., (2012). An overview on antidiabetic medicinal plants having insulin mimetic property. *Asian Pac. J. Trop. Biomed., 2*(4), 320–330.

Prasad, R. B., & Groop, L., (2015). Genetics of type 2 diabetes—pitfalls and possibilities. *Genes, 6*(1), 87–123.

Pruett-Miller, S. M., (2014). Genome editing in somatic cells using zinc-finger nucleases and transcription activator-like effector nucleases. In: *Principles of Cloning* (2nd edn., pp. 369–378). Elsevier Inc.

Puchta, H., (2005). The repair of double-strand breaks in plants: Mechanisms and consequences for genome evolution. *J. Exp. Bot., 56*(409), 1–14.

Qiang, S., Su, A. P., Li, Y., Chen, Z., Hu, C. Y., & Meng, Y. H., (2019). Elevated β-carotene synthesis by the engineered *Rhodobacter sphaeroides* with enhanced CrtY expression. *J. Agric. Food Chem., 67*(34), 9560–9568.

Ramachandra, R. S., & Ravishankar, G. A., (2002). Plant cell cultures: Chemical factories of secondary metabolites. *Biotechnol. Adv., 20*(2), 101–153.

Reitsma, W. D., & Terpstra, J., (1981). *WHO Expert Committee on Diabetes Mellitus*, 646.

Richard, K., Gavin, J. R., Alberti, K. G. M. M., Davidson, M. B., DeFronzo, R. A., Drash, A., Gabbe, S. G., et al., (2000). Report of the expert committee on the diagnosis and classification of diabetes mellitus. *Diabetes Care, 23*(1 Suppl.), 1183–1197.

Shah, F. L. A., Ramzi, A. B., Baharum, S. N., Noor, N. M., Goh, H. H., Leow, T. C., Oslan, S. N., & Sabri, S., (2019). Recent advancement of engineering microbial hosts for the biotechnological production of flavonoids. *Mol. Biol. Rep., 46*(6), 6647–6659.

Shanak, S., Saad, B., Zaid, H., & Carvalho, J. C. T., (2019). Metabolic and epigenetic action mechanisms of antidiabetic medicinal plants. *Evid. Based Complement. Alternat. Med., 2019.*

Sheng, H., Sun, X., Yan, Y., Yuan, Q., Wang, J., & Shen, X., (2020). Metabolic engineering of microorganisms for the production of flavonoids. *Front. Bioeng. Biotechnol., 8*, 1–15.

Shrivastav, M., De Haro, L. P., & Nickoloff, J. A., (2008). Regulation of DNA double-strand break repair pathway choice. *Cell Res., 18*(1), 134–147.

Sidhu, M. C., & Sharma. T., (2013). Medicinal plants from twelve families having antidiabetic activity: A review. *Am. J. PharmTech. Res., 3*(5), 36–52.

Stoddard, B. L., (2005). Homing endonuclease structure and function. *Q. Rev. Biophys., 38*(1), 49–95.

Takasu, Y., Sajwan, S., Daimon, T., Osanai-Futahashi, M., & Uchino, K., (2013). Efficient TALEN construction for bombyx mori gene targeting. *PLoS One*, *8*(9), 73458.

Thirumurugan, D., Cholarajan, A., Raja, S. S. S., & Vijayakumar, R., (2018). An introductory chapter: Secondary metabolites. In: *Secondary Metabolites – Sources and Applications* (pp. 3–22). IntechOpen.

Thiruvengadam, M., Praveen, N., & Chung, I. M., (2012). An efficient agrobacterium tumefaciens-mediated genetic transformation of bitter melon (*Momordica charantia* L.). *Aust. J. Crop Sci., 6*(6), 1094–1100.

Thorneycroft, D., Sherson, S. M., & Smith, S. M., (2001). Using gene knockouts to investigate plant metabolism. *J. Exp. Bot., 52*(361), 1593–1601.

Tran, N., Pham, B., & Le, L., (2020). Bioactive compounds in antidiabetic plants: From herbal medicine to modern drug discovery. *Biology, 9*(9), 1–31.

Tung, C. W., Lin, Y. C., Chang, H. S., Wang, C. C., Chen, I. S., Jheng, J. L., & Li, J. H., (2014). TIPdb-3D: The three-dimensional structure database of phytochemicals from Taiwan indigenous plants. *Database, 2014*, 1–5.

Wachtel-Galor, S., & Iris, F. F. B, (2011). Herbal medicine. In: *Herbal Medicine: Biomolecular and Clinical Aspects*. CRC Press/Taylor & Francis.

Watts, K. T., Lee, P. C., & Schmidt-Dannert, C., (2004). Exploring recombinant flavonoid biosynthesis in metabolically engineered *Escherichia coli*. *ChemBioChem, 5*(4), 500–507.

Weber, E., Engler, C., Gruetzner, R., Werner, S., & Marillonnet, S., (2011). A modular cloning system for standardized assembly of multigene constructs. *PLoS One*, *6*(2).

White, F. F., Potnis, N., Jones, J. B., & Koebnik, R., (2009). The type III effectors of *Xanthomonas*. *Mol. Plant Pathol., 10*(6), 749–766.

Wink, M., (2015). Modes of action of herbal medicines and plant secondary metabolites. *Medicines, 2*(3), 251–286.

Wood, A. J., Lo, T. W., Zeitler, B., Pickle, C. S., Ralston, E. J., Lee, A. H., Amora, R., et al., (2011). Targeted genome editing across species using ZFNs and TALENs. *Science*, 307.

Xie, X., Yang, M., & He, X., (2007). Somatic embryogenesis and genetic engineering of *Acacia* species. *Transgenic Plant J., 1*(1), 244–249.

Yasumoto, S., Umemoto, N., Lee, H. J., Nakayasu, M., Sawai, S., Sakuma, T., Yamamoto, T., et al., (2019). Efficient genome engineering using platinum TALEN potato. *Plant Biotechnol., 36*(3), 167–173.

Zhang, H., Gou, F., Zhang, J., Liu, W., Li, Q., Mao, Y., Botella, J. R., & Zhu, J. K., (2016). TALEN-mediated targeted mutagenesis produces a large variety of heritable mutations in rice. *Plant Biotechnol. J., 14*(1), 186–194.

Zhang, Y., Chiu, T. Y., Zhang, J. T., Wang, S. J., Wang, S. W., Liu, L. Y., Ping, Z., et al., (2021). Systematical engineering of synthetic yeast for enhanced production of lycopene. *Bioengineering, 8*(1), 14.

Zhang, Y., Massel, K., Godwin, I. D., & Gao, C., (2019). Applications and potential of genome editing in crop improvement. *Genome Biol., 20*(1), 1–11.

Zhang, Y., Zhang, F., Li, X., Baller, J. A., Qi, Y., Starker, C. G., Bogdanove, A. J., & Voytas, D. F., (2013). Transcription activator-like effector nucleases enable efficient plant genome engineering. *Plant Physiol., 161*(1), 20–27.

Index

D

E

H